Superlattices and Microstructures of Dielectric Materials

介电体超晶格

朱永元 王振林 陈延峰 陆延青 祝世宁

南京大学出版社

——谨以此书献给我们尊敬的老师闵乃本先生
庆祝闵先生八十华诞！

闵乃本,男,汉族,1935年8月生,江苏如皋人,日本东北大学理学博士,南京大学教授,中国科学院院士,第三世界科学院院士。曾任南京大学固体微结构国家重点实验室主任、学术委员会主任,江苏省自然科学基金会主任,教育部科学技术委员会副主任,教育部"材料科学与工程教学指导委员会"主任,中国晶体学会理事长,全国政协第九、十、十一届常委,江苏省政协第八、九届副主席,九三学社第十、十一届中央副主席,江苏省九三学社第四、五届主委。于1998年获"何梁何利科学与技术进步奖",1999年获第三世界科学院基础科学奖——物理奖,2000年获美国科学信息研究所(ISI)经典引文奖,2006年获国家自然科学一等奖,1982年、2005年、2007年,三次获得国家自然科学二等奖。1995年获"全国优秀教师"称号及奖章,2001年获"全国模范教师"称号及奖章,2009年被评为"新中国成立以来江苏省十大杰出科技人物"。

Foreword

The properties of bulk materials are understood and controlled by the chemical elements and bonds in the material. The development in ab initio solid state science allows the study of the electronic, optical, thermal and mechanical properties starting from the atom-or molecule-level, and has helped expand the range of materials accessible to us.

In last few decades, advanced nanofabrication and computational techniques has offered a new route to design material properties at will. Through controlled interactions with electromagnetic (EM) waves, it became possible to produce fascinating physical properties unavailable in naturally occurring or chemically synthesized materials. Photonic crystals and metamaterials are good examples of employing structural design to manipulate electromagnetic wave propagation in an unprecedented way.

Over last thirty years, Professor Naiben Min of Nanjing University, has developed the novel concept of "dielectric superlattices". Analogous to the superlattice for electrons in semiconductors, (for instance, an array of coupled quantum wells that have controllable electronic characteristics), dielectric superlattices are composed of a periodic of dielectric media processing unprecedented optical properties. For example, electro-optic and acoustic-optic modulation of optical superlattices can be used for compact solid-state pulsed lasers. Since optical superlattices have the advantage of high nonlinear gain and phase shift over broad wavelengths, they can act as mode locking devices, generating high repetition rate semiconductor laser pulses. In quantum optics, optical superlattices can exhibit higher photon single photon yield, and entangled photon pair production compared with conventional nonlinear crystals. Min's scientific contributions have greatly influenced the science of functional optical materials, and has shaped modern Optical Physics in China.

This book is based on representative publications on dielectric superlattices by Prof. Min's research team in Nanjing University. His original work has had a profound impact in science and technology. The team was honored by the First-Class State Natural Science Award (the highest award in scientific research in China) in 2006, for their significant achievements on the fundamental and applied nonlinear optics, laser physics, and technology. Prof. Min has extended the optical superlattice concept from periodic structures to quasi-periodical, multi-periodic, and even completely random structures, in 1D, 2D, and 3D.

The phase matching techniques developed by Min were extended from the original single quasi-phase matching; to multiple quasi-phase matching; local quasi-phase matching; and, eventually the nonlinear "Huygens principle", a generic rule for phase matching in complex nonlinear optical nanostructures. In 30 years of effort, three kinds of dielectric superlattices have been systematically developed for different functionalities, including, optical superlattices, acoustic superlattices, and ionic-type phononic crystals. It is noteworthy that soon after I proposed the idea of photonic crystals in 1987, Professor Min's team had proposed and demonstrated phononic crystals for acoustic waves. The acoustic forbidden band is similar to photonic bands. In addition, ionic-type phononic crystals with superlattices of piezoelectric materials can efficiently couple acoustic waves with electromagnetic wave. Such dielectric superlattices have now been used in microwave electromagnetic devices and electronic devices.

As we enter the new age of "designer's materials", this book provides a reference source for scientists or engineers practicing in the field of optical materials and device applications. I believe the readers in both academic and industrial sectors will find this book helpful and beneficial to their understanding of the exciting research in material science and photonics.

Eli Yablonovitch

Berkeley, August 2016

序

　　介电体超晶格是一种在光电子学、声电子学、量子信息学领域有重大应用前景的新型功能材料,是我们南京大学介电体超晶格研究团队经过近30年的努力开拓出来的研究领域。我们团队在围绕介电体超晶格材料体系开展的基础研究和应用探索过程中,还为畴工程学的建立做出了奠基性的贡献。本书从我们团队发表的数百篇论文中精选出90篇代表论文,根据研究内容整理、编辑成九个章节。第一章是绪论,第二至第八章分别介绍了介电体超晶格研究的七个分支领域。为了方便读者,编者在每章前都增加了中英文序言作为导读。第九章为"总结与展望"。全书的编辑力图能清晰地描绘出该领域的发展轨迹、学术系统和科学内涵,期待本书出版后能成为为在该领域和相关领域从事基础研究和应用研究科技人员提供借鉴的有益读物。

　　光波或超声波在介电晶体中传播,其波长甚大于晶格常数,或者说,其波矢甚小于晶格倒格矢,光波或声波/超声波在介电晶体中的行为完全等价于在连续介质中的行为,晶格的周期性对其没有任何影响。如果在介电晶体中引入可与光波或声波/超声波波长比拟的超周期-超晶格,亦即引入可与光波或声波/超声波的波矢比拟的倒格矢,则情况迥然不同。铁电晶体是一种介电晶体。对铁电晶体的铁电畴进行人为的调制,调制周期与光波或声波/超声波波长可比拟,形成超周期,我们称之为介电体超晶格。在介电体超晶格中,除了非线性光学系数外,压电系数和电光系数等也相应地被调制。依据研究对象的不同,介电体超晶格也被称为光学超晶格、声学超晶格和离子型声子晶体等。

　　我们对介电体超晶格的研究按两条思路展开。其一是由于超晶格的倒格矢可以参与光波(光子)、声波(声子)的波矢量守恒(动量守恒),在无法满足动量守恒的所有物理过程中,如存在折射率色散的光参量过程中,倒格矢的引入必然产生新颖的物理效应。其二是光波(光子)、声波(声子)在超晶格的周期结构中传播,可类比于电子在晶格周期势场中运动,这种周期结构必然导致不同类型能带,于是可通过能带的设计与裁剪,实现对作为信息载体的光波或光子、声波或声子的调控。

　　介电体超晶格研究可追溯到1986年,经过19年的努力,该项研究于2006年获国家自然科学一等奖。获奖后我们曾写过一篇总结性论文:"介电体超晶格的研究"(物理 37 (2008) 1-10)。在这之前,我也曾应英文刊物"先进材料"之约写过一篇综述文章,"Superlattices and Microstructures of Dielectric Materials" (Adv. Mater. 11 (1999) 1079-1090)。这两篇文章分别用中、英文发表,虽然年代不同,都在一定程度上系统地介绍了介电体超晶格的研究内容与学术体系。本书将这两篇论文选作第一章,代"绪论"。

　　介电晶体中折射率色散造成了发生在其中的光参量过程的位相失配,即动量守恒

（波矢量守恒）不能满足。而介电体超晶格中超晶格结构所提供的倒格矢可以补偿折射率色散所引起的位相失配，使参量光通过相长干涉（相干叠加）得到有效增强，这被称之为准相位匹配，或准波矢量守恒。一维周期超晶格具有一组倒格矢，其初基倒格矢可以满足一个单一光参量过程的准位相匹配。一维准周期超晶格具有多组倒格矢，可高效地同时实现多个相互独立的或相互耦合的光参量过程的准位相匹配，于是多个参量光能同时有效输出，这被称之为多重准位相匹配。我们发展了多重准位相匹配理论及其应用，拓展了非线性光学的研究领域。我们还研究了二维光学超晶格，实现了共线、非共线准位相匹配，获得了弹性散射、非弹性散射（拉曼散射）以及Čerenkoverenkov辐射的非线性增强。这部分工作放在第二章介绍，并标以题目"准相位匹配概念的拓展和非线性光学新效应"。

在二维正方对称性介电常数（或折射率）周期调制的超晶格中，利用两组倒格矢，能激发四个Floquet-Bloch波，据此我们发展了四（多）波动力学理论，并在此基础上进一步考虑介电常数的Kerr非线性，发现了新的光学双稳机制-折射率调制机制，让多束光同时进入双稳态。这部分工作以及后续发展的用折射率调制的光子晶体工作放在第三章"克尔非线性光学超晶格与光子晶体"中介绍。

超声波可以在正负铁电畴交替排列的声学超晶格中被激发，由于相邻电畴压电常数符号相反，畴界处可看为δ声源。声学超晶格中适当选取相邻δ声源的间距即电畴单元的宽度，可使所有δ声源发出的超声波相长干涉。用声学超晶格研制成的声学器件不仅输出强度高、插入损耗低，还填补了超声工程中体波声学器件从几百兆周到几千兆周的空白频段。这部分工作在第四章"声学超晶格与声子晶体"中介绍。

对于像铌酸锂、钽酸锂等仅具有180度畴界的铁电晶体，如果以一对自发极化相反的铁电畴作为超晶格的基本构造单元，可以造成在两种畴的构型：畴界平行于自发极化矢量，畴界上不存在束缚电荷；畴界垂直于自发极化矢量，则畴界上存在等效束缚电荷，且相邻畴界的等效束缚电荷符号相反。对于第二种情况，形式上可以将超晶格看为一维正、负离子链，电磁波通过超晶格必然引起畴界（束缚电荷）振动。这就是说，电磁波能激发超晶格集体振动，畴界（束缚电荷）的振动将激发电磁波，于是超晶格振动和电磁波的耦合导致了极化激元的产生。这种情况类似于1951年黄昆提出的离子晶体中电磁波与晶格振动耦合所激发的极化激元。我们将这类超晶格称之为离子型声子晶体。不过在离子晶体中的极化激元波长是在红外波段，而这里发生在微波波段。研究表明，超晶格中的光、声耦合实际上源于压电效应。即使对于畴界处不存在束缚电荷的超晶格，电磁波在其中传播，亦可能存在纵向超晶格声子与光子耦合形成的声子极化激元，这就扩展了传统的极化激元的概念。而且，由于我们可以通过对压电或压磁超晶格的结构进行人工设计，使得超晶格在特定波段的介电常数或磁导率为负，这与新近提出的电磁超构材料出现了有趣的一致，因而可看作一种压电或压磁超构材料。这部分工作放在第五章"离子型声子晶体与超构材料"中介绍。

自 1986 年到 2006 年,我们关于超晶格的研究,其基本构造单元的材料仅限于介电体,其研究的思路主要是沿着准波矢量守恒所产生的相干叠加增强效应展开。2006 年前后,我们一方面将超晶格构造单元的材料由介电体扩展到金属和半导体,这就包括了表面等离激元和超构材料(metamaterials),使介电体超晶格内涵有所拓展。另一方面,研究的思路由通过准波矢量守恒产生相干叠加的增强效应,扩展到通过能带结构研究来发现新效应和新应用。如研究了离子型声子晶体中极化激元的能带结构,发现了两种类型的极化激元,即寻常极化激元和异常极化激元。极化激元的带隙禁止寻常极化激元通过,而异常极化激元则可以部分地通过极化激元带隙。又如,对以金属棱柱体和空气棱柱体为构造单元的声学超晶格(声超构材料,声子晶体),发展了声子能带的计算和设计方法,实现了第一和第二能带的负折射效应,提出并验证了"能带交叠"导致的双负折射效应。关于光、声和极化激元能带研究的相关工作分别在第三、四、五章中有所介绍。

当代信息技术中,用于信息处理与运算的信息载体是电子的电荷属性(电荷流量密度,电流密度),用于信息远程传输的是光的波动属性(频率、振幅、位相、偏振态)。信息处理与运算和信息远程传输使用了不同的信息载体,这成了当代信息技术发展的瓶颈。另一方面,集成电路愈做愈小,终究要达到它的物理极限。这表明用电子的电荷属性作为信息载体的时代,即摩尔时代的终结。后摩尔时代信息处理、运算以及传输的信息载体是什么?这是新一轮信息革命所面临的基本问题。电子的属性有电荷、自旋、质量、能量、动量,除了电荷,电子自旋也可作为的信息载体。总的说来,当前人们开始关注粒子(电子、光子),准粒子(声子、极化激元、等离激元……)的量子态。量子信息是以量子态作为信息载体的,量子信息在增大信息传输容量、提高信息处理和运算速度、确保信息安全等方面将突破经典信息的瓶颈。另一方面,纠缠的光子作为信息载体的研究引人注目,这是由于产生纠缠光子的全固态光子芯片技术和远程传输纠缠光子的光纤技术日趋成熟。我们在光的量子信息领域的工作放在第六章"准位相匹配量子光学和光子芯片"中介绍。

基础科学中的物质科学(凝聚态物理学与化学)的研究路径是,化合物的合成和材料的制备→组分、结构的测定→化学性能和物理效应的揭示→其间内在联系和客观规律的阐明。工程学则相反,从应用目的出发,瞄准某种性能和效应,利用组分、结构与性能、效应间的内在联系与规律,设计并制备出具有所需结构和性能的材料或器件。当前,分子设计与组装以及微结构设计与制备就具有工程学研究的特征。畴工程学是微结构设计与制备的分支领域。畴工程学始于铁电畴工程学,稍后扩展到铁磁畴和铁弹畴等。但由于精确控制铁磁畴和铁弹畴组态和稳定性问题一直没有得到解决,迄今,只有铁电畴工程学发展得较为完善。通过铁电畴工程学,可以实现具有设定功能的光电子声电子材料、器件和量子信息器件等。铁电畴工程学也为研制全固态集成的铌酸锂光量子芯片开拓了全新的技术途径。我们在第七章"介电体超晶格与畴工程学"中着重介绍铁电畴工程学。

光学超晶格作为激光频率转换晶体可用于研制新型全固态激光器件。我们将准位相匹配、多重准位相匹配理论与全固态激光技术相结合，拓展了全固态激光器件的研究领域。通过准位相匹配，光学超晶格可以在晶体的整个透明波段实现参量光的高效输出。另外，对于具有 3m 点群的非线性光学晶体如铌酸锂、钽酸锂等，它们最大的非线性光学常数 d_{33} 是不能位相匹配的，但可以实现准相位匹配。与普遍采用的 d_{31} 位相匹配相比，利用晶体 d_{33} 的准位相匹配转换效率能增加数十倍。利用准相位匹配和多重准位相匹配，我们研制出同时输出多个波长的激光器及红绿蓝三基色激光器，研制出高功率、宽调谐光学超晶格中红外激光系统，以及芯片上的多波长激光器阵列。这部分工作将在第八章"光学超晶格的应用研究"中介绍。

第九章则是对全书的简略总结和对介电体超晶格研究的展望。

本书的编辑与出版是祝世宁和朱永元博士策划的，编者中还有王振林、陈延峰、陆延青博士。南京大学出版社的领导给予大力支持，王南雁、吴汀等编辑提出了许多有益的建议与帮助。该书内容是团队成员集体努力的结晶，反映了我学术研究工作的一个重要方面，该书的整理过程也让我有机会回顾了我的科研生涯以及那令人难忘的挑战与机遇，因此我十分感谢促成本书完成的我的学生和同事。我真诚地感谢我的团队成员及合作者，虽然在此序言中我没有一一列出他们的名字，但读者可以通过阅读本书所选文章了解他们对这一学术体系所做的贡献，可以说没有他们出色的研究工作就没有此书。我要借此机会向我的同事和朋友表示敬意，感谢他们长期以来对我们的工作给予的热情鼓励和积极支持。

闵乃本

2015 年 8 月 9 日，于南京碧树园

Preface

Dielectric superlattice has been extensively explored by the research team from Nanjing University for more than 30 years. With a series of breakthrough, it has been developed into a variety of novel functional materials with significantly potential applications in photo-electronics, acousto-electronics and quantum informatics. Meanwhile, in order to designing and preparing the dielectric superlattice, a new research field, domain engineering, has been developed. This book is a collection of original papers from more than hundreds of works that we have published in peer reviewed journals. These papers are classified into the first 8 chapters according to their contents, with a summary and outlook provided in Chapter 9. With this organization, we try to provide a solid and detailed description on the intrinsic characteristics and the scientific framework of dielectric superlattice. We expect this book to be helpful to various audiences, including scientists, engineers and graduates in physics, material science, optics and optical engineering, and other related fields.

Electromagnetic waves and ultrasonic waves can be excited and propagate in a dielectric crystal. The wavelengths of these waves are much larger than the lattice constants of a dielectric crystal, i.e., their wave vectors are much smaller than the reciprocal lattice vectors of a crystal. Thus their propagation in a dielectric crystal is completely equivalent to in a continuous medium, not being affected by the lattice periodicity. Among the dielectric crystals, ferroelectric crystals show excellent nonlinear optical, electro-optical, piezoelectric properties. By artificial modulation of ferroelectric domains, that is, introducing a periodicity comparable to the wavelength of the light or ultrasonic waves into the dielectric crystals to create an artificial reciprocal lattice vector to compensate the dispersion of wave vectors, a dielectric superlattice (sometimes we also call it optical superlattice or acoustic superlattice or ionic-type phononic crystal, depending on which physical process is involved) can be fabricated. Thus the scenario of wave propagation in a dielectric superlattice would be totally different from that in a homogeneous dielectric crystal.

Our study on dielectric superlattices can be divided into two categories: one is known as quasi-phase matching engineering, the other is the artificial energy band gap engineering. Both of them can bring some novel physical effects those are not able to be produced in homogeneous dielectric crystals. We begin Chapter 1 in which two invited review papers are included. The first one with the title of "介电体超晶格的研究" was published in a Chinese journal ("物理"(Physics), 37, 1-10 (2008)). The other one titled "Superlattices and Microstructures of Dielectric Materials" appeared in an international journal (Adv. Mater. 11, 1079-1089 (1999)).

In a dielectric crystal, the refractive index dispersion usually causes a phase mismatch and therefore breaks the momentum conservation (the wave vector conservation) in an optical parametric process, which results in a destructive interference parametric waves, therefore, low conversion efficiency. The reciprocal lattice vectors in dielectric superlattices can provide a phase compensation for this mismatch process, converting the destructive interference into a constructive interference (coherent superposition). This is called quasi-phase matching (QPM) or quasi-wave vector conservation. An one-dimensional (1D) periodic superlattice has a set of reciprocal lattice vectors in which the largest primary reciprocal lattice vector can efficiently implement a QPM optical parametric process. An 1D quasi-periodic superlattice has two or more sets of reciprocal lattice vectors, some of them can efficiently support several independent or mutually coupled QPM optical parametric processes. This process, involving the intervention of multiple reciprocal lattice vectors, is called multiple quasi-phase-matching (MQPM). MQPM may also take place in a two-dimensional (2D) ferroelectric optical superlattice. For example, we designed a 2D superlattice with a hexagonal symmetry, in which three sets of reciprocal lattice vectors simultaneously participate in the different nonlinear processes in some geometries. Using such a 2D superlattice, we respectively demonstrated three kinds of novel nonlinear effects: non-collinear QPM enhanced elastic scattering, QPM enhanced in elastic scattering (Raman Scattering) and QPM enhanced Čerenkov radiations. These works above are involved in Chapter 2 with title "Nonlinear Optical Phenomena in Optical Superlattice and Some Concepts Extended".

In Chapter 3, we describe another type of 2D superlattice with the periodic modulation of dielectric constant (or refractive index). For example, a 2D superlattice with a square symmetry offers two sets of reciprocal lattice vectors, leading to the excitation of four Floquet-Bloch waves. The phenomenon was explained by the multi-wave dynamics theory we developed. On this basis, with further consideration of nonlinear Kerr effect, we proposed a new kind of optically bistable mechanism modulated by index. It leads to a unique multi-beam optical bistability. This part of the work is introduced in Chapter 3, with the title of "Kerr-type Nonlinear Optical Superlattices and Photonic Crystals".

The dielectric superlattice that can excite ultrasonic waves by piezoelectric effect is called as an acoustic superlattice. In an acoustical superlattice, there is an abrupt change for the sign of the piezoelectric constants of crystal at the boundary of two adjacent domains, therefore, each domain's boundary is equivalent to a δ sound source. By appropriate selection of the interval between the adjacent domains, one can effectively engineer all δ sound sources in-phase to generate constructive interference and coherent superposition for ultrasonic exciting and propagating inside. The emergence of acoustic superlattices not only promises novel ultrasound devices with high output power and low insertion losses, but also unprecedentedly covers the ultrasonic frequency spectrum from several hundred MHz to a few GHz. This part of the work is described in Chapter 4 with the title "Acoustic Superlattices and Sonic Crystals".

In dielectric superlattices with 180° antiphase domain boundary, only there are two different kinds of domain configurations: one with the domain boundaries parallel to the spontaneous polarization, therefore, no charge bounded at the boundaries; the other with the domain boundaries perpendicular to the spontaneous polarization, with bounded charges of opposite signs accumulated alternately at the domain boundaries across the superlattice. In this case, the 1D superlattice effectively serves as an ionic chain with positive and negative ions. While electromagnetic waves propagating through the superlattice induce the vibrations of the charged domain boundaries, i. e., the electromagnetic waves excite the superlattice vibrations. The vibrations of the domain boundaries (bounded charges) also simultaneously excites and radiates electromagnetic waves. Such superlattice vibrations and electromagnetic waves consequently couple with each other, leading to the emergence of polaritons, which shares similar fundamentals with the coupling of electromagnetic waves and lattice vibrations excited by polarization in an ionic crystal proposed by Huang Kun in 1951. Therefore, we named it ionic-type phononic crystal. In conventional ionic crystals, the polaritons are excited in the far-infrared regime. However, they occur in the microwave regime in the ionic-type phononic crystals we defined. Further studies demonstrated that, for the superlattices without the bound charges at the domain boundaries, a transverse polarization still can be induced by a longitudinal wave due to piezoelectric effect. This type of polariton does not exist in a real ionic crystal. Furthermore, even negative permittivity or permeability could be achieved in some frequency regimes by the couple of piezoelectric and piezomagnetic effects in a few composite superlattices. These phenomena just coincide with the properties of metamaterials in some extent. This part of the work is presented in Chapter 5 with the title "Ionic-Phononic Crystals and Metamaterials".

From 1986 to 2006, the study on superlattices was limited to dielectric materials. The fundamental principle depends mainly on the enhanced coherent superposition effects originated from the conservation of quasi-wave vectoror quasi-phase-matching. Since 2006, two explorations on the study of superlattice have appeared: the first one is that the superlattice structure has been extend to be used to metals and semiconductors, which means that the study has penetrated into the fields of surface plasmonics and metamaterials; the second one is that the artificial band gap engineering is used to various superlattices. These two explorations have brought the study of superlattice many intriguing effects and applications. For the research on the polariton band structure of ionic-type phononic crystals, we found two types of polaritons: ordinary polaritons and extraordinary polaritons. The band gap prohibits ordinary polaritons, but partially permits to extraordinary polaritons. For another example, in an acoustic superlattice (phonon crystal), whose structural unit is composed of metallic cylinders in an air matrix or air cylinders in a solid-state matrix, we developed the corresponding calculation and design methodologies for effective manipulation of sound/ultrasound. In experiment, we observed the negative refraction effects in the first and second energy bands and further proposed and validated a negative birefraction effect caused by "band overlapping". Related works on the band structures of light, sound and polariton are introduced in Chapters 3, 4 and 5, respectively.

In the modern information technology, the operation and processing of information utilize the charges of electrons (charge flow density, current density), while the information remote transfer relies on light or photon, relying on their wave properties, such as frequency, amplitude, phase and polarization etc. This creates a significant barrier due to the fact that different information carriers are utilized for information network. On the other hand, currently the integrated circuits are eventually approaching its physical limits. This suggests the end of the Moore era in which the electron charges serve as the information carriers. We face an open question in fundamentals: what will become the information carriers for the next round of information revolution? The intrinsic properties of electrons at the quantum level include charge, spin, mass, energy and momentum, in which electron spin is being frequently investigated as a potential information carrier. Overall, the quantum states of particles (electron, photon) and quasi-particles (polariton, phonon, plasmon, and so on) have drawn significant efforts in the exploration of quantum information. By exploiting quantum states as information carriers, quantum information technology can break through the bottleneck of classical information technology in several aspects: increasing information transmission capacity, increasing the speed of the information processing and computing, and enhancing the information security and so on. It is compelling that the entangled photons serve as information carriers instead of classic light. Benefiting from the well-developed solid-state photon chip and optical fiber technologies, we may realize the generation and the remote transfer of entangled photons, respectively. The photonics-based quantum information technology can in further eliminate the fundamental flaw of using two different carriers in information processing and remote transfer in current information technology. Our work about quantum information based on single or entangled photon is presented in Chapter 6 with the title "Review Article: Quasi-phase-matching Engineering of Entangled Photons".

The research strategy in basic sciences, such as in condensed matter physics, material science, and chemistry, is divided into different layers, that is, the sample preparation or compound synthesis, the identification of constituent and structure, the characterization of physical and chemical properties and the clarification of the intrinsic links between them and the objective laws. On the contrary, the engineering is originated from the application goals, aiming at the performance and corresponding effects, thus designing and preparing the useful materials and devices with the desired structures and properties. During this process, one must obey the intrinsic links and objective laws among constituents, structures, properties, and performances. At this moment, molecular design and assembly as well as micro-structural design and preparation bear such essential characteristics of engineering. As a branch of micro-structural engineering, domain engineering started from the study on ferroelectric domains and extended later to the ferromagnetic and ferroelastic domains. Due to the difficulty for accurately controlling the configuration and stability of the ferromagnetic and ferroelastic domains, so far, only the ferroelectric domain engineering is sophisticatedly developed. By ferroelectric domain engineering and other modern state-of-the-art technologies, one can integrate various functions for the manipulation of photon on a $LiNbO_3$ chip, which promises a new venue towards solid-state integrated optical quantum chips. We emphatically describe domain engineering of ferroelectrics in Chapter 7 with the title "Domain Engineering for Dielectric Superlattice".

Chapter 8 deals with the synergy of QPM, MQPM and solid-state laser technology to tremendously explore and advance the research of solid-state laser and other photo-electronic devices. Using the MQPM scheme, we have developed novel solid-state lasers with simultaneous output of the three-primary colors (RGB) and multi-wavelength output on-demand. Based on QPM, we achieved highly efficiently optically parametric outputs, which may almost cover the entire transparent region of the implemented nonlinear optical crystals. Moreover, based on conventional technologies, for the nonlinear optical crystals with 3m point group, such as $LiNbO_3$ crystals, their largest nonlinear optical coefficient d_{33} cannot be used to realize conventional birefringence phase matching, but can be utilized by QPM. This will increase the output power several tens of times more than that of the widely used phase matching of d_{31}. By means of the optical superlattice and QPM, we have developed high-power widely-tunable mid-infrared laser systems. We have also efficiently integrated multi-functional entangled photonic sources on the LiNbO$_3$ chips by utilizing domain engineering. This part of works is highlighted in Chapter 8 with the title "Engineered Quasi-phase-matching for Laser Techniques".

Chapter 9 is a summary and outlook for the whole book. It briefly looks back on the history about the research of dielectric superlattice, and predicts possible applications and intriguing development directions in future. Of course, these predictions are based on the authors' own opinionsand do not represent a formal recommendation.

Dr. Shi-Ning Zhu and Dr. Yong-Yuan Zhu conceived the editing and publishing of this book. The editorial team also includes Dr. Zhen-Lin Wang, Dr. Yan-Feng Chen, and Dr. Yan-Qing Lu. I very much appreciate their efforts as well as the contributions from many others in editing and publishing of this book, such as Dr. Nan-Yan Wang and Dr. Ting Wu, the executive editors of this book. This event gives me a chance to recall my scientific career and my own life, as the old saying goes: it is a special time in which opportunity and challenge coexist. I wish to sincerely thank my students and coauthors for their excellent works presented in this book, and my colleagues and friends for their invaluable advices and assistance as well.

Nai-Ben Ming

August 9, 2015, in Nanjing Bishu Park

目 录

第一章 介电体超晶格的研究 ··· 003
Chapter 1 Superlattices and Microstructures of Dielectric Materials ··· 016

第二章 准相位匹配概念的拓展和非线性光学新效应 ··· 039
Chapter 2 Nonlinear Optical Phenomena in Optical Superlattice and Some Concepts Extended ··· 042

    2.1 Harmoic Generations in an Optical Fibonacci Superlattice ··· 046

    2.2 Second-harmonic Generation in a Fibonacci Optical Superlattice and the Dispersive Effect of the Refractive Index ··· 054

    2.3 Quasi-Phase-Matched Third-Harmonic Generation in a Quasi-Periodic Optical Superlattice ··· 061

    2.4 Experimental Realization of Second Harmonic Generation in a Fibonacci Optical Superlattice of $LiTaO_3$ ··· 069

    2.5 Crucial Effects of Coupling Coefficients on Quasi-Phase-Matched Harmonic Generation in an Optical Superlattice ··· 076

    2.6 Wave-Front Engineering by Huygens-Fresnel Principle for Nonlinear Optical Interactions in Domain Engineered Structures ··· 082

    2.7 Conical Second Harmonic Generation in a Two-Dimensional $\chi^{(2)}$ Photonic Crystal: A Hexagonally Poled $LiTaO_3$ Crystal ··· 089

    2.8 Experimental Studies of Enhanced Raman Scattering from a Hexagonally Poled $LiTaO_3$ Crystal ··· 096

    2.9 Nonlinear Čerenkov Radiation in Nonlinear Photonic Crystal Waveguides ··· 105

    2.10 Nonlinear Volume Holography for Wave-Front Engineering ··· 113

    2.11 Nonlinear Talbot Effect ··· 121

    2.12 Diffraction Interference Induced Superfocusing in Nonlinear Talbot Effect ··· 128

    2.13 Cavity Phase Matching via an Optical Parametric Oscillator Consisting of a Dielectric Nonlinear Crystal Sheet ··· 139

第三章　克尔非线性光学超晶格与光子晶体 ········· 149
Chapter 3　Kerr-type Nonlinear Optical Superlattices and Photonic Crystals ········· 152
   3.1　Light Transmission in Two-dimensional Optical Superlattices ········· 157
   3.2　Optical Bistability in a Two-Dimensional Nonlinear Superlattice ········· 170
   3.3　Experimental Observations of Bistability and Instability in a Two-Dimensional Nonlinear Optical Superlattice ········· 177
   3.4　Gap Shift and Bistability in Two-dimensional Nonlinear Optical Superlattices ········· 184
   3.5　Optical Bistability in Two-dimensional Nonlinear Optical Superlattice with Two Incident Waves ········· 191
   3.6　Three-dimensional Self-assembly of Metal Nanoparticles: Possible Photonic Crystal with a Complete Gap Below the Plasma Frequency ········· 196
   3.7　Parity-time Electromagnetic Diodes in a Two-dimensional Nonreciprocal Photonic Crystal ········· 203
   3.8　Nonreciprocal Light Propagation in a Silicon Photonic Circuit ········· 216
   3.9　Experimental Demonstration of a Unidirectional Reflectionless Parity-time Metamaterial at Optical Frequencies ········· 225
   3.10　Plasmonic Airy Beam Generated by In-Plane Diffraction ········· 237
   3.11　Collimated Plasmon Beam: Nondiffracting versus Linearly Focused ········· 245
   3.12　The Anomalous Infrared Transmission of Gold Films on Two-Dimensional Colloidal Crystals ········· 254
   3.13　Localized and Delocalized Surface-plasmon-mediated Light Tunneling Through Monolayer Hexagonal-close-packed Metallic Nanoshells ········· 265
   3.14　Experimental Observation of Sharp Cavity Plasmon Resonances in Dielectric-metal Core-shell Resonators ········· 280
   3.15　Magnetic Field Enhancement at Optical Frequencies Through Diffraction Coupling of Magnetic Plasmon Resonances in Metamaterials ········· 290

第四章　声学超晶格和声子晶体 ········· 301
Chapter 4　Acoustic Superlattices and Sonic Crystals ········· 304
   4.1　Acoustic Superlattice of $LiNbO_3$ Crystals and Its Applications to Bulk-wave Transducers for Ultrasonic Generation and Detection up to 800 MHz ········· 309
   4.2　High-frequency Resonance in Acoustic Superlattice of $LiNbO_3$ Crystals ········· 314
   4.3　Ultrasonic Spectrum in Fibonacci Acoustic Superlattices ········· 319
   4.4　Ultrasonic Excitation and Propagation in an Acoustic Superlattice ········· 327

4.5　High-frequency Resonance in Acoustic Superlattice of Periodically Poled LiTaO$_3$ ·········· 347

4.6　Bulk Acoustic Wave Delay Line in Acoustic Superlattice ············· 353

4.7　Negative Refraction of Acoustic Waves in Two-dimensional Sonic Crystals ·········· 359

4.8　Acoustic Backward-Wave Negative Refractions in the Second Band of a Sonic Crystal ············· 367

4.9　Negative Birefraction of Acoustic Waves in a Sonic Crystal ·············· 375

4.10　Extraordinary Acoustic Transmission through a 1D Grating with Very Narrow Apertures ·············· 385

4.11　Acoustic Surface Evanescent Wave and its Dominant Contribution to Extraordinary Acoustic Transmission and Collimation of Sound ········· 393

4.12　Tunable Unidirectional Sound Propagation through a Sonic-Crystal-Based Acoustic Diode ············· 401

4.13　Acoustic Asymmetric Transmission Based on Time-dependent Dynamical Scattering ············· 409

4.14　Acoustic Cloaking by a Near-zero-index Phononic Crystal ·············· 423

4.15　Acoustic Phase-reconstruction near the Dirac Point of a Triangular Phononic Crystal ············· 432

4.16　Topologically Protected One-way Edge Mode in Networks of Acoustic Resonators with Circulating Air Flow ············· 441

第一章　介电体超晶格的研究
Chapter 1　Superlattices and Microstructures of Dielectric Materials

第一章　介电体超晶格的研究*

闵乃本　朱永元　祝世宁　陆亚林　陆延青　陈延峰　王振林　王慧田　何京良

（南京大学固体微结构物理国家重点实验室　南京　210093）

介电体超晶格是一种新型的有序微结构材料.它具有通常均质材料所不具有的独特优异性能,展现出重要的应用前景.文章介绍了南京大学研究组关于介电体超晶格研究所取得的进展,如将多个光参量过程集成于一块介电体超晶格之中获得了多波长激光同时输出,研制成超晶格全固态三基色激光器原型,在介电体超晶格中将拉曼散射强度增强到 $10^4 \sim 10^5$ 倍,用超晶格研制的器件填补了体波超声器件从几百 MHz 到几千 MHz 的空白频段,发现了微波与超晶格振动的强烈耦合以及极化激元(polariton)的激发与传播等.

1. 引言

半导体晶体中存在电子能带,通过能带设计与裁剪,实现了电子调控,奠定了当代信息技术的基础[1].然而,光子能带不存在于均匀晶体中.

20 世纪 70 年代,将周期微结构引入半导体晶体,构成半导体超晶格[1].在这一成就的启发下,我们于 20 世纪 80 年代初,将微结构引入介电晶体[2],构成介电体超晶格.超晶格的周期可和光波、超声波的波长比拟.光波、超声波在介电体超晶格中传播,就类似于电子在晶格周期势场中运动.于是,光子能带、声子能带及其他准粒子能带就出现在介电体超晶格中[3].

介电体中引入有序微结构,可以实现不同物理常数的有序调制.介电常数（或折射率）周期调制的介电体超晶格,称为光子晶体,具有光子能带[4—6].弹性常数周期调制的称为声子晶体,具有声子能带[7].压电常数周期调制的称为离子型声子晶体（ionic-type phononic crystals）,具有极化激元能带[8—10].非线性光学常数被调制的称为准位相匹配材料（quasi-phase-matching materials）,在激光变频方面有着广泛的应用[11—13].

表 1　介电体超晶格与粒子或准粒子能带

半导体晶格	介电体超晶格		
电子	光子	声子	极化激元
晶格周期势场	介电周期结构	弹性周期结构	压电周期结构
电子能带	光子能带	声子能带	极化激元能带
信息技术	光电子技术	声电子技术	微波技术

表 1 列举出半导体晶格和介电体超晶格所具有的不同粒子（如电子和光子）、准粒子（如声子和极化激元）能带,周期结构的类型以及可能的应用领域.我们期待介电体超晶格能像

---

* 物理,2008,37(1):1

半导体超晶格一样,通过不同能带的设计、裁剪,实现对粒子、准粒子的调控,为光电子技术、声电子技术以及微波技术作出贡献.

1984年发现了准晶[14].1986年我们将准周期引入介电体超晶格.周期超晶格只有一组倒格矢,只能高效完成一个光参量过程.准周期超晶格具有多组倒格矢,能同时高效完成多个光参量过程.我们于1990年建立了多重准位相匹配理论[15].1995年发展了能够制备准周期介电体超晶格的室温电场极化技术[16].1997年完成了实验验证,将两个光参量过程高效地集成于一块准周期超晶格之中,实现了两种波长激光同时输出[17,18].2003年实验上实现了超晶格中三基色的产生[19—21].2005年研制成超晶格三基色和白光全固态激光器原型[22].此外,在光散射的准位相匹配增强效应[23,24]、高频超声激发[25]、微波与超晶格振动耦合[8]等方面也有所发现.

从真空电子管到半导体场效应晶体管,实质上是将电子发射、电子调控、电子收集诸过程,全固态地集成于半导体微结构中.这一全固态集成,使实现电子调控所需空间缩小到原来的 7.2×10^{-13},这为集成电路四十多年来能按摩尔定律发展提供了可能.我们关注于光电子学中的类似的全固态集成,如上所述,将多个光参量过程集成于一块介电体超晶格之中,这只是第一步.将不同的物理过程(如变频、会聚、偏转、开关等)集成于一块有序微结构材料之中,是我们努力的目标.

2. 介电体超晶格的组成、制备与表征

介电体超晶格的组成:以有序铁电畴为例,简要说明介电体超晶格的组成.属3m点群的铌酸锂($LiNbO_3$,LN)、钽酸锂($LiTaO_3$,LT)铁电晶体中,铁电畴自发极化矢量,或是平行于 z 轴(正畴),或是反平行于 z 轴(负畴).取一对正、负畴作为构造单元.如果取一种构造单元重复排列,这就构成了周期超晶格,如图1(a)所示.如果取两种构造单元按Fibonacci序列排列,这就构成了二单元Fibonacci准周期超晶格[26—31],如图1(b)所示.通常用倒格矢来描述晶格或超晶格.对周期超晶格,其倒格矢可表示为

$$G_m = m \frac{2\pi}{\Lambda} g, (m=1,2,3\cdots) \quad (1)$$

式中 Λ 为超晶格结构参量,即超晶格周期,g 是 G_m 的单位矢量,垂直于畴界,见图1(a).当 $m=1$,$G_1 = \frac{2\pi}{\Lambda} g$ 为超晶格的初基倒格矢.在光参量过程中,使用初基倒格矢,其转换效率最高.对二单元Fibonacci准周期超晶格,其倒格矢可表示为

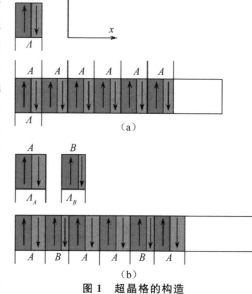

图1 超晶格的构造

(a) 一种构造单元(A)无限重复就构成周期超晶格;(b) 两种构造单元(A 和 B)按fibonacci序列就构成准周期超晶格.[11—13]

$$G_{mn} = \frac{2\pi(m+n\tau)}{\tau\Lambda_A + \Lambda_B}\boldsymbol{g}, (m,n = 1,2,3,\cdots) \tag{2}$$

其中 $\tau = \frac{1+\sqrt{5}}{2}$ 为黄金分割数，Λ_A 和 Λ_B 是超晶格结构参量，$\boldsymbol{g}$ 为 $\boldsymbol{G}_{mn}$ 的单位矢量，垂直于畴界，见图1(b).可以看出，设计准周期超晶格，有更多的结构参量可调，更多的倒格矢可选用.因而在一块准周期超晶格中，可以同时高效地实现多个光参量过程[15,17,18].从理论和实验方面研究了二单元Fibonacci准周期超晶格的超声激发谱[26]、激光倍频谱[27—29]和电光透射谱[30].发现了超声谱结构的自相似性，这是实空间结构自相似性在倒空间中的反映.而激光倍频谱和电光透射谱，由于折射率的色散效应，谱结构的自相似性遭到破坏[27—30].最近还研究了极化激元的色散谱和介电谱，都得到了理论与实验相符的结果[31].对二构造单元按intergrowth序列[32]和按Thue-Morse序列[33]，三构造单元按准周期序列构成的超晶格[34—36]，以及双周期[37]、非周期[38,39]超晶格都进行了系统的研究.在此基础上建立了设计介电体超晶格的专家系统，已能按需优化设计介电体超晶格.

通过构造单元的有序排列，实现了铁电畴自发极化矢量的调制.现在简要说明，如何通过铁电畴自发极化矢量的调制实现物性常数的调制.从图1(a)可以看出，构造单元是由一对正畴、负畴组成，其自发极化矢量相反.其中负畴的形成可想象为，将原为正畴的第二片及描述物性张量坐标系的 z 轴，绕 x 轴相对于第一片旋转180°，故第二片畴中奇数阶张量变号，而偶数阶张量将保持不变.因而在这类超晶格中，对应的三阶张量如非线性光学常数、电光常数、压电常数不再保持不变，而空间坐标的函数，属二阶张量的介电常数和属四阶张量的弹性常数在整个超晶格中保持不变[11].

介电体超晶格的制备：如何实现铁电畴自发极化矢量的有序调制，我们发展了两种技术，就是生长层技术和室温电场极化技术.前者用于制备周期性超晶格，后者用于制备周期、非周期以及二维超晶格.

晶体生长过程中，如果出现周期性温度起伏，就出现周期性生长层，就如树木生长中出现的年轮.周期性生长层是晶体(固溶体)中溶质浓度沿生长方向的周期性分布[40].通常溶质以离子状态存在于晶体中，周期性溶质分布就等价于周期性空间电荷分布，于是在晶体中产生了周期性内电场.在降温过程中，当温度通过居里点时，即从顺电相向铁电相转变时，该内电场将引起金属氧化物晶体(如LN，LT)中在顺电相处于对称中心的金属离子(如Nb，Ta)沿 z 轴的正向或反向的择优位移，这就决定了这类晶体中铁电畴自发极化取向.于是在晶体生长过程中，人为地引入周期性温度起伏，通过周期性生长层的产生，实现了铁电畴自发极化矢量的周期调制，并发展成制备周期超晶格的生长层技术[41,42].

为了制备准周期超晶格，借用半导体平面工艺.取一片 z 切割的单畴LN或LT晶片，在其上、下晶面沉积金属膜作为电极.下晶面的金属膜为平面电极，上晶面的金属膜通过半导体平面工艺刻蚀成具有周期、准周期或其他有序图案的电极，两电极与高压脉冲发生器联结，如图2(a)所示(图中 $\boldsymbol{P}_s$ 为自发极化矢量).在高压脉冲电场作用下，单畴晶片中的反向畴首先成核于上电极界面处，逐步向下延伸.适当选择脉冲强度、脉冲宽度和脉冲重复频率，可使反向畴贯穿整个样品，获得与图案电极相同的有序畴结构.这被称为室温电场极化技术[16].图2(b)、(c)和(d)是用该技术制备的周期、准周期和二维周期超晶格，图2(e)是具有特定图案的有序畴结构,该技术能用于批量生产,图 2(f)表明晶片上同时制备的多个相同和不同的超晶格.

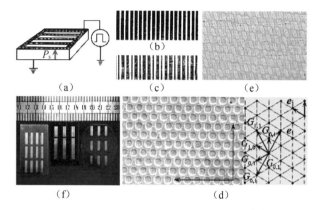

图 2 制备超晶格的室温电场极化技术[16]
(a)室温电场极化装置;(b)周期超晶格;(c)准周期超晶格;(d)二维周期超晶格;(e)二维有序畴结构;(f)超晶体的批量制备.

介电体超晶格的表征:为了实现对介电体超晶格的无损检测,我们发展了两种表征技术,即介电常数与介电损耗的微波近场显微成像技术[43],以及环境扫描二次电子成像技术[44].

通常,可方便地测得体块材料的介电常数与介电损耗,我们更关心局域表征.我们发展了介电常数与介电损耗的微波近场显微成像技术,其分辨率达亚微米量级.我们知道,介电极化来源于电子极化和离子(位移)极化.在光频波段只有电子极化作出贡献,而在微波波段两者都有贡献.因此在微波波段,近场显微像能提供更多的信息.通常复介电常数可表示为

$$\varepsilon = \varepsilon' + i\varepsilon'', \tag{3}$$

其中实部 ε' 为介电常数,虚部 ε'' 为介电损耗.系统的共振频率为 f_0,有 $\Delta f_0/f_0 = g\Delta\varepsilon'$,其中常数 $g = \sim 7\times 10^{-5}$(给定实验系统的测量值),故共振频率像反映了介电常数的变化,因而能反映晶体中的掺杂浓度和内应力,参阅图 3(a)的左图.系统的品质因子为 $(1/Q)$,有 $\Delta(1/Q) = g'\Delta\varepsilon''$,其中 $g' = \sim 7.1\times 10^{-2}$,品质因子像反映了介电损耗(微波吸收)的空间变化,与微波作用下的离子位移和畴界振动有关,显然,畴界处介电损耗最大,参阅图 3(a)的右图.图 3(b)对比了 LN 超晶格的同一区域的两种像,上图为介电常数像,下图为介电损耗像.可以看出,在该区域的右上部,介电常数像(上图)与介电损耗像(下图)中的周期结构遭到破坏,在介电损耗像(下图)中出现绿色区(其介电损耗最大),这对应于岛屿畴分布区.岛屿畴区内畴界密度较大,因而介电损耗亦大.岛屿畴区的出现是晶体生长过程中界面失稳所致.微波近场显微术不仅能表征超晶格中的周期畴结构,还能提供超晶格中的溶质分布、内应力分布、双折射的均匀性以及微波吸收分布的有关信息[43].

介电材料通常都是绝缘体,用电镜观察时,入射电子束在绝缘体表面形成电荷积累,使电子像逐渐模糊.通常必须在表面蒸镀导电膜,以防止电荷积累.近年来,环境扫描电镜的发展,有可能直接观察绝缘样品而不需要表面导电膜.这是由于环境扫描电镜腔内的气压较通

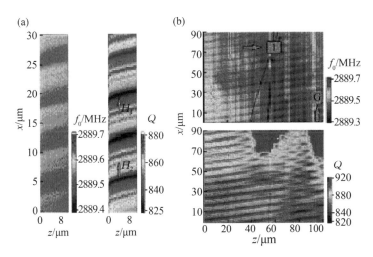

图 3　介电常数与介电损耗的微波近场显微成像
（a）左为介电常数像，右为介电损耗像；（b）同一区域的两种像，其中上部为介电常数像，下部为介电损耗像.[43]

常扫描电镜内高一万倍，利用入射电子束与中性分子碰撞产生的自由离子，在样品表面邻近形成的通道，来消除表面电荷积累.另一方面，二次电子成像技术的发展，能使不作任何处理（如浸蚀）的表面直接成像.我们在环境扫描电镜中，在未作处理的LT超晶格表面，实现了周期铁电畴的二次电子成像，而且图像稳定，与传统方法获得的像相符[44].

3. 介电体超晶格的光学效应

介电体以介电极化为特征，介电极化来源于电子极化和离子（位移）极化.在光频波段离子的位移跟不上光波的变化，因此只有电子极化作出贡献.这里考虑二阶非线性极化.当共线的频率为 ω_1 和 ω_2 的光作用于非线性介质时，引起的二阶非线性极化强度含有 $2\omega_1, 2\omega_2$，$\omega_1 \pm \omega_2$ 等不同频率分量，相应的极化分量将辐射上述频率分量的光波.但要获得这些频率分量的有效输出，其相应的光参量过程必须满足能量守恒和动量守恒.以和频为例，为了使能量有效地从频率为 ω_1 和 ω_2 的基波光转换到频率为 ω_3 的和频光，必须满足的能量守恒为 $\omega_3 = \omega_1 + \omega_2$ 和动量守恒为 $\boldsymbol{k}_3 = \boldsymbol{k}_1 + \boldsymbol{k}_2$，或 $\boldsymbol{k}_3 - (\boldsymbol{k}_1 + \boldsymbol{k}_2) = 0$. 其中 $k_i = \dfrac{n(\omega_i)}{c}\omega_i \boldsymbol{\kappa}, (i=1,2,3)$，$\boldsymbol{k}_i$ 是频率为 ω_i 的光的波矢，$\boldsymbol{\kappa}$ 是 $\boldsymbol{k}_i$ 的单位矢量，三波矢是共线的.这里动量守恒用波矢量守恒表达，通常称为位相匹配.若 $\omega_1 = \omega_2 = \omega$，即 $\boldsymbol{k}_{\omega_1} = \boldsymbol{k}_{\omega_2} = \boldsymbol{k}_{\omega}$，由能量守恒，有 $\omega_3 = 2\omega$，由波矢量守恒，即 $\boldsymbol{k}_{2\omega} - 2\boldsymbol{k}_{\omega} = 0$，有 $n(2\omega) = n(\omega)$，此为倍频情况.可以看出，位相匹配要求基波光与倍频光的折射率相等，亦即要求在介质中基波光与倍频光的相速度相等.然而由于折射率色散效应，$n(2\omega) \neq n(\omega)$，必然产生位相失配.利用晶体的双折射效应，在特定的方向上可以实现基波光和倍频光的相速度相等，从而得到高效输出.这就是光学工程中通常使用的位相匹配技术.

3.1 准位相匹配

1962年,Bloembergen提出的准位相匹配理论[45]指出,在任何光参量过程中,由于折射率色散所引起的波矢量失配(位相失配),可用超晶格提供的倒格矢来补偿.例如,倍频过程,其折射率色散引起波矢量失配为 $\Delta k = k_{2\omega} - 2k_\omega = \frac{2\omega}{c}[n(2\omega) - n(\omega)]\boldsymbol{\kappa}$,设计周期超晶格,使其初基倒格矢 $\boldsymbol{G}_1 = \Delta \boldsymbol{k} = \frac{2\omega}{c}[n(2\omega) - n(\omega)]\boldsymbol{\kappa}$,故有

$$k_{2\omega} - 2k_\omega - G_1 = 0. \tag{4}$$

(4)式被满足,即准位相匹配的条件被满足.此时,周期超晶格的所有构造单元产生的倍频光都满足相长干涉,其振幅相互叠加,而其强度比例于叠加振幅的平方.若周期超晶格的构造单元数(周期数)为 N,构造单元产生的倍频光强为 I_0,则超晶格的倍频输出为 $I_{2\omega} = I_0 N^2$.我们曾用实验定量地验证了上述结果的正确性[46,47].对给定的光参量过程,设计超晶格,使其倒格矢与失配矢量的矢量和为零,这样就能使准位相匹配条件被满足.这不仅对诸波矢共线的光参量过程是如此,对诸波矢非共线的光参量过程也是如此[23].

3.2 多重准位相匹配

我们仍考虑二阶非线性极化.如上所述,当两共线的频率为 ω_1 和 ω_2 的光作用于非线性介质,引起的二阶非线性极化强度含有 $2\omega_1, 2\omega_2, \omega_1 \pm \omega_2$ 频率分量.如考虑一次级联效应,还存在 $3\omega_1, 3\omega_2, 2\omega_2 \pm \omega_1, 2\omega_1 \pm \omega_2$ 的频率分量.相应的极化分量将辐射上述频率的光波.这些频率分量是通过级联光参量过程产生的.要获得上述频率分量的有效输出,级联的诸参量过程的动量守恒必须同时满足,这对传统的位相匹配技术是不可能的,即使对上述准位相匹配技术也是不可能的.我们于1990年将周期超晶格拓展为准周期超晶格,建立了多重准位相匹配理论[15].结果表明,准周期超晶格可提供多个倒格矢,分别参与相互级联的,或相互独立的多个光参量过程,能同时实现多个光参量过程的准位相匹配,获得多波长激光同时输出.如果 $\omega_1 = \omega_2 = \omega$,上述一次级联效应产生的六种频率分量退化为一种,即三倍频 3ω.三倍频是通过抽运光 ω 的倍频产生 2ω,接着再与基波光 ω 和频产生 3ω.这就是说,三倍频是通过倍频与和频的级联过程产生的,参阅图4(a).要获得高效三倍频输出,必须同时满足倍频、和频的准位相匹配条件.设计准周期超晶格,使之具有两个特定的倒格矢 G_{11} 和 G_{23},分别等于倍频与和频过程的失配矢量,于是有

$$k_{2\omega} - 2k_\omega - G_{11} = 0, \tag{5}$$
$$k_{3\omega} - (k_\omega + k_{2\omega}) - G_{23} = 0. \tag{6}$$

(5)式和(6)式分别是倍频与和频过程的准位相匹配条件.在准周期超晶格中,(5)式、(6)式同时被满足,级联过程得以实现,可直接得到高效三倍频输出.

我们于1997年实现了上述级联过程[17],直接得到了三倍频,转换效率为 23%,同时还获得了倍频输出,转换效率为 20%,参阅图4(b).在此基础上,通过基波($\lambda = 1342$ nm)的倍频得到红光,三倍频得到蓝光,实现了红-蓝双波长激光输出[48].进一步研制成红-蓝双波长全固态激光器原型,其红光输出为1 W、蓝光输出为140 mW.全固态激光技术的发展,在激光晶体 $Nd:YVO_4$ 中实现了波长1342 nm 和1064 nm 的同时振荡.通过它们的倍频、和频、三

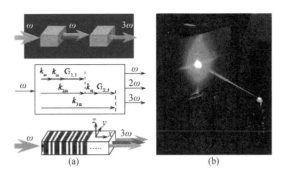

图 4 多重准位相匹配理论及实验[17,18]
（a）多重准位相匹配示意图；（b）三倍频实验.

倍频等就能得到多波长同时输出.我们通过 1342 nm 基波光的倍频得到红光、通过 1342 nm 和 1064 nm 的和频得到黄光、通过 1064 nm 的倍频得到绿光,实现了红-黄-绿三波长的同时输出[49].我们还实现了红-黄-绿-蓝四波长的同时输出[50].

3.3 三基色与白光激光器

在上述工作基础上,我们采用了两种方案,实现了红、绿、蓝三基色同时输出,并研制成全固态准白光激光器原型.第一方案[19—21]是,用半导体抽运的掺 Nd^{3+} 激光晶体双波长激光器作为基波光源,通过对其 1342 nm 输出的倍频得到红光,对其 1064 nm 输出的倍频得到绿光,对其 1342 nm 输出的三倍频得到蓝光,见图 5(a).通过三基色的能量分配得到了准白光,其功率大于 1 W,并进一步研制成全固态三基色准白光激光器原型,参阅图 5(b)和(c).第二方案[22]是,以 $\lambda_p=532$ nm 的绿光为抽运光,通过 OPO 产生 $\lambda_s=633$ nm 的信号光（红光）和 $\lambda_i=3342$ nm 的闲置光,再由抽运光与闲置光和频产生 $\lambda_{p+i}=459$ nm 的蓝光,其转换效率大于 30%.该方案的理论转换效率可达 100%.有望获得高功率白光.第二方案涉及频率下转换,可能赋予白光全新的纠缠态的特性.

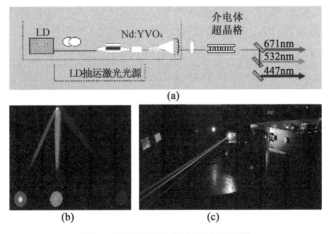

图 5 超晶格三基色全固态激光器
（a）三基色激光器示意图；（b）实验观测结果；（c）三基色全固态激光器实物图.

上面讨论的是一次级联效应和频率上转换.在激光与非线性介质交互作用时,高次级联效应同样存在,通过上述理论与技术,可以得到高阶谐波的有效输出,如四倍频等.频率下转换同样要求能量守恒和动量守恒,多重准位相匹配理论与技术同样有效,这就为研制量子通信的光源提供了新途径.

3.4 二维周期超晶格

我们研究了两种二维周期超晶格.其一为铁电畴反转构成,具有二维六方对称性,在其中实现了非共线准位相匹配,获得了弹性散射和非弹性散射(拉曼散射)的准位相匹配的增强[23,24].其二是折射率二维周期调制的,具有二维正方对称性,在其中发现新的光学双稳机制,并实现了多束光同时进入双稳态[51-56].铁电畴反转构成的二维超晶格,其点阵参数为 $a = 9.05~\mu m$,初基倒格矢为垂直于密排方向的 G_{01} 的六个等效矢量,其大小为 $4\pi/\sqrt{3}a$,参阅图 2(d).若入射波矢 k_ω 沿 LT 晶体的 x 轴,当入射波长为 1064 nm 时,在屏幕上出现的是其倍频光的圆环,环的中心是 k_ω 在屏幕上的投影点,见图 6(a).当入射波长为 933 nm,在屏幕上为倍频光的具有镜面对称性的双环,如图 6(c)所示.若入射波矢 k_ω 沿 LT 晶体的 y 轴,随着入射光频率的变化,在投影屏上出现的倍频图样如图 6(d)、(e)、(f)所示.详细分析表明,这是来自基波光的弹性散射导致的非共线准位相匹配倍频增强效应.在非线性介质中,入射光与散射光的和频 $\omega + \omega'$ 是普遍存在的.在通常情况下,该参量过程无法满足波矢量守恒,其效应过于微弱.弹性散射满足能量守恒,故入射光与散射光频率相等,$\omega = \omega'$,弹性散射波矢与入射光波矢大小相等,有 $|k_\omega| = |k_{\omega'}|$.但入射光波矢的方向是给定的,而弹性散射波矢沿所有可能的方向.其中只有那些满足准位相匹配条件的散射波矢,才能与入射光和频得到增强.其准位相匹配条件是,诸非共线矢量 $k_\omega, k_{\omega'}, k_{\omega+\omega'}$ 与倒格矢 G_{mn} 的矢量和为零,即

$$k_{\omega+\omega'} - (k_\omega + k_{\omega'}) - G_{mn} = 0. \quad (7)$$

如果入射波矢平行于 LT 晶体的 x 轴,入射波

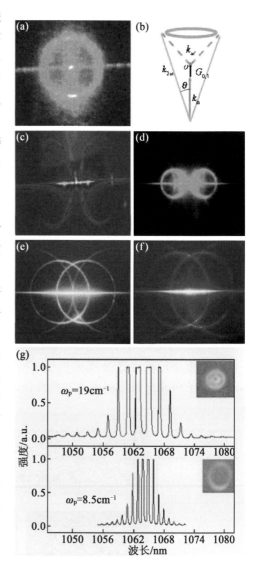

图 6 LT 晶体二维铁电畴超晶格中的弹性散射和拉曼散射的准位相匹配增强效应
(a)-(f)是弹性散射的增强效应,其中(a)-(c)是入射光平行于 LT 晶体的 x 轴,(d)-(f)是入射光平行于 LT 晶体的 y 轴,(g)为拉曼散射增强效应.[23,24]

长为 1064 nm,初基倒格矢 G_{01} 参与准位相匹配,其准位相匹配条件为 $k_{2\omega}-(k_\omega+k_{\omega'})-G_{01}=0$,满足上述条件的诸波矢 $k_{2\omega}$ 的集合,构成了半顶角为 θ 的圆锥面,见图 6(b).因而在屏幕上的投影为一圆,见图 6(a).如果入射波矢平行于 LT 晶体的 x 轴,入射光的波长为 933 nm,一对镜面对称的非初基倒格矢 G_{11} 参与准位相匹配,在屏幕上的投影是一对具有镜面对称的双环,参阅图 6(c).值得注意的是,初基倒格矢 G_{10} 和非初基倒格矢 G_{11} 及其相应的等效倒格矢,都镜面对称地分布在 y 轴两侧,而沿 y 轴不存在初基倒格矢,见图 2(d).如果入射波矢平行于 LT 晶体的 y 轴,一对初基倒格矢 G_{10} 参与准位相匹配,或一对初基倒格矢 G_{10} 和一对非初基倒格矢 G_{11} 同时参与准位相匹配,其倍频光在屏幕上的投影如图 6(d),(e),(f)所示[23].关于非弹性散射,我们仔细研究了拉曼散射,证实了同样存在准位相匹配增强效应,并通过实验将散射强度提高了 $10^4\sim10^5$ 倍[24].这就为设计新型拉曼激光器和发展高分辨拉曼谱学提供了新途径.

研究的第二类是折射率被调制的二维周期超晶格,具有二维正方对称性.设计超晶格时调节其点阵面间距,对给定的入射光,使相互正交的两组点阵面都满足 Bragg 条件.因此二维超晶格内有 4 个 Floquet-Bloch 波被激发.于是我们发展了四波动力学理论[5].进一步考虑了 Kerr 型介电非线性,发现了一种新的光学双稳机制,称之为折射率调制机制,观测到四束出射光同时达到光学双稳态[51,52],并对非 Bragg 条件下普适的多波动力学行为进行了系统的研究[53—56].

4. 介电体超晶格的声学效应

前面已经说明,在以一对正、负畴为构造单元所构成的超晶格中,相邻铁电畴的压电常数变号,即畴界处压电常数不连续.我们知道,压电应力是压电常数与电场强度的乘积.在恒稳电场中,畴界两侧的正、负畴内的压电应力不等(符号相反),故畴界处产生内应力.在交变电场中,畴界处产生交变内应力.交变内应力则产生交变应变,并以弹性波的形式在超晶格中传播,故超晶格中的畴界都可看为 δ 声源.在考虑超声激发与传播时,超晶格本身可以看为周期排列的一列 δ 声源,这类超晶格我们称为声学超晶格(acoustic superlattice).设计声学超晶格时,适当选择相邻 δ 声源的间距,使所有 δ 声源产生的超声波相长干涉,即相干叠加[25,57,58].满足相长干涉的诸 δ 声源,产生的超声波的振幅相互叠加,而其强度比例于总振幅的平方.若构造单元中 δ 声源所激发的超声强度为 I_0,超晶格周期数为 N,该超晶格产生的超声波强度为 $I_n=I_0N^2$.通常超声工程中超声换能器都是用均质晶片制成的,可以看出,用声学超晶格研制的换能器明显地提高了超声强度和转换效率.通过求解声学超晶格的弹性波波动方程,可得声学超晶格的电抗,从而得到谐振器、换能器的谐振频率[25,59,60]为

$$f_m=mv/\Lambda\quad(m=1,2,3\cdots),\tag{8}$$

其中 v 是固体中声速,Λ 是超晶格周期,参阅图 1(a).值得注意的是,谐振频率只决定于超晶格周期 Λ,而与晶片的总厚度无关.如我们所知,如果用通常均质材料来制备工作频率为几百兆周到几千兆周的体波超声器件,其晶片总厚度要减薄到几十微米到几微米,这对当前超声工程中的工艺系统是有困难的.然而制备周期为数微米的声学超晶格是很方便的.另一方面,通过求解声学超晶格的弹性波波动方程,所得换能器的电抗是超晶格周期数 N 和电极面积 A 的函数[25,60].用通常均质材料制备换能器,电抗只是电极面积 A 的函数.我们知道,在

高频下静态电容是电抗的主要部分,因而换能器的插入损耗很高.但若使用超晶格,适当地选用 N 和 A,可使电抗的实部等于甚至大于虚部.这样,换能器的插入损耗在 50 Ω 的测量系统中可接近于零[60].我们利用声学超晶格的上述优点,研制成或提供了超声换能器[60,61]、谐振器[59,62,63]、声光偏转器[64]、集成声开关[65]的原型器件和设计原理,填补了当前超声工程中体波声学器件从几百兆周到几千兆周的空白频段.

5. 超晶格振动与微波耦合以及极化激元激发

在 LN、LT 这类晶体中,铁电畴自发极化矢量只能平行或反平行于 z 轴,因而超晶格中存在两类典型畴界:一类是平行于 z 轴的畴界,如图 1 所示,在畴界上不存在束缚电荷;另一类是垂直 z 轴的畴界,如图 7(a)所示,畴界上存在束缚电荷,且其面密度最大.我们先讨论第二种情况,如图 7(a)所示,畴界上存有束缚电荷,且相邻畴界束缚电荷的符号相反.这样,畴界既可看为电荷中心,又可看为质量中心,于是,近似地将这类超晶格看为一维正、负离子链,如图 7(a)所示.若 y 偏振的电磁波沿 z 轴通过超晶格,将引起畴界(束缚电荷)沿 y 方向振动,这就是说,电磁波激发了超晶格振动.反之,畴界(束缚电荷)沿 y 方向振动,将激发沿 z 轴传播沿 y 偏振的电磁波.这表明超晶格振动与电磁波耦合,并导致极化激元的产生.这十分类似于 1951 年黄昆先生提出的离子晶体中电磁波与晶格振动的格波互相耦合形成的极化激元[66],因而我们将这类超晶格称为离子型声子晶体.我们研究了离子型声子晶体中相邻畴界的运动规律,发现它满足的运动方程与反映离子晶体中正负离子相对运动的黄昆方程在形式上完全一样[8].所得的色散曲线和介电函数曲线完全类似,参阅图 7(b)(图中 ω_0 为介电函数的零点频率,ω_L 为超晶格中纵声波的谐振频率,V_c 为尚未与超晶格振动耦合的微波波速),还得到了完全类似的联系纵、横光学波频率的 Lyddane-Sachs-Teller 关系式[8].只不过在离子晶体中上述长波光学性质是发生在红外波段,这里发生在微波波段.同时,我们在实验中发现了由于极化激元激发而引起的强烈的微波吸收(见图 7(c)),也就是说发现了微波吸收的新机制.

进一步研究发现,畴界上是否存在束缚电荷,并非产生超晶格振动与微波耦合的根本原因,其根本原因在于压电效应,在于超晶格中压电系数的周期跃变[9,10].我们考虑如图 1(a)所示的超晶格,可以看到,超晶格中畴界上不存在束缚电荷.在前面讨论超晶格的声学效应时,已经说明,超晶格中畴界处压电系数变号,因而可将畴界看为 δ 声源,故在微波的交变电场中必然产生超晶格振动.另一方面,在超晶格振动时,由于正、负畴中应变的符号相反,通过压电效应产生的感生极化矢量在正负畴中方向相同,在同一瞬间超晶格中所有畴的感生极化矢量方向相同[9,67],于是超晶格振动引起感生极化矢量振动,从而引起微波激发.这就是说,由于压电效应产生了微波与超晶格振动的耦合.基于超晶格的压电效应可以得出,不仅畴界的横振动可与微波耦合,畴界的纵振动也可与微波耦合导致极化激元激发.这是由于畴界的纵振动通过压电效应产生横向感生极化矢量振动的原故[9,67],在实验上测量了纵振动与微波耦合产生的极化激元带隙的中心频率与带宽,其结果与理论估计相符[67].我们还发现了两种不同偏振态的微波耦合而产生两类极化激元的新效应.参阅图 1(a),若 y 偏振的微波在超晶格中沿 x 方向播,通过压电效应,激发畴界沿 x 方向振动,从而产生沿 x 方向传播的纵超声波.沿 x 方向传播的纵超声波通过压电效应不仅产生 y 偏振沿 x 方向传播的微波,还

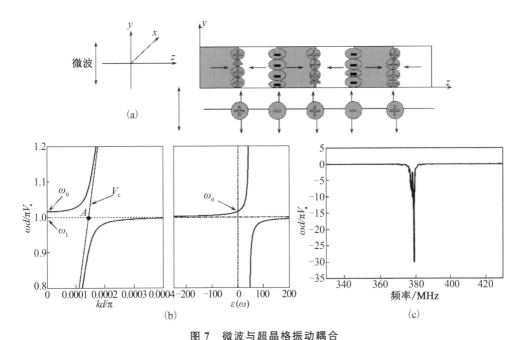

图7 微波与超晶格振动耦合
(a) 离子型声子晶体示意图；(b) 极化激元的色散与介电异常；(c) 极化激元激发引起的微波吸收。[8—10]

同时产生 z 偏振沿 x 方向传播的微波．y 偏振和 z 偏振的电磁波的耦合，引起了介电椭球主轴的旋转．在这种情况下，出现了两类极化激元，一种不能通过带隙，称寻常极化激元(ordinary polariton)，另一种能通过带隙，称异常极化激元(extraordinary polariton)[10]．这一有趣的发现，对进一步通过极化激元能带设计与裁剪，实现极化激元调控，以及研制新型的微波器件，可能具有十分重要的意义．

6. 结束语

二十多年的理论与实验研究表明，介电体超晶格已经成为一种具有重要应用前景的人工有序微结构材料．在我们对铁电畴超晶格研究的基础上，准位相匹配材料已经成为国际热门领域，以及在光电子学、声电子学、材料科学的交叉领域中催生了称之为"畴工程学"的新生学科[68—70]．实践表明，已能将多个光参量过程、平行光束的汇聚过程(透镜功能)和偏转过程(棱镜功能)集成于一块有序畴结构材料之中．相信随着工作的持续深入，在粒子(如光子)、准粒子(如声子、极化激元、表面等离激元等)能带设计、裁剪方面，在实现上述粒子、准粒子调控方面，在级联物理过程全固态集成方面，介电体超晶格必将对未来光电子技术、声电子技术、微波技术的发展作出贡献．

References and Notes

[1] 黄昆.物理,1993,22:257[Huang K. Wuli(Physics), 1993,22:257(in Chinese)]

[2] 冯端.物理,1989,18:1[Feng D. Wuli(Physics),1989,18:1 (in Chinese)]

[3] Ming N B. Facets, 2003, 2 (2):6

[4] Yablonovitch E, Gmitter T J. Phys. Rev. Letter., 1989, 63:1950

[5] Feng J, Ming N B. Phys. Rev. A, 1989, 40:7047

[6] Wang Z L, Zhu Y Y, Yang Z J et al. Phys. Rev. B, 1996, 53:6984

[7] Economou E N, Sigalas M. J. Acous. Soc. Am., 1994, 95:1734

[8] Lu Y Q, Zhu Y Y, Chen Y F et al. Science, 1999, 284:1822

[9] Zhu Y Y, Zhang X J, Lu Y Q et al. Phys. Rev. Lett., 2003, 90:053903

[10] Huang C P, Zhu Y Y. Phys. Rev. Lett., 2005, 94:117401

[11] 闵乃本.自然科学进展,1994,4:543[Ming N B. Progress in Natural Science, 1994, 4:543(in Chinese)]

[12] Zhu Y Y, Ming N B. Opt. Quant. Electron., 1999, 31:1093

[13] Ming N B. Adv. Mater., 1999, 11:1079

[14] Shechtman D, Blech I, Gratias D et al. Phys. Rev. Lett., 1984, 53:1951

[15] Feng J, Zhu Y Y, Ming N B. Phys. Rev. B, 1990, 41:5578

[16] Zhu S N, Zhu Y Y, Zhang Z Y et al. J. Appl. Phys., 1995, 77:5481

[17] Zhu S N, Zhu Y Y, Ming N B. Science, 1997, 278:843

[18] Zhu Y Y, Xiao R F, Fu J S et al. Appl. Phys. Lett., 1998, 73:432

[19] He J L, Liao J, Iiu H et al. Appl. Phys. L. ett., 2003, 83:228

[20] Liao J, He J L, Liu H et al. Appl. Phys. Lett., 2003, 82:3159

[21] Li H X, Fan Y X, Xu P et al. J. Appl Phys., 2004, 96:7756

[22] Gao Z D, Zhu S N, Tu S Y et al. Appl. Phys. Letter., 2006, 89:181101

[23] Xu P, Ji S H, Zhu S N et al. Phys. Rev. Lett., 2004, 93:133904

[24] Xu P, Zhu S N, Yu X Q et al. Phys. Rev. B, 2005, 72:064307

[25] Zhu Y Y, Ming N B. J. Appl. Phys. 1992, 72:904

[26] Zhu Y Y, Ming N B, Jiang W H. Phys. Rev. B, 1989, 40:8536

[27] Zhu Y Y, Ming N B. Phys. Rev. B, 1990, 42:3676

[28] Zhu S N, Zhu Y Y, Qin Y Q et al. Phys. Rev. Lett., 1997, 78:2752

[29] Zhu Y Y, Xiao R F, Fu J S. Optics Letters, 1997, 22:1382

[30] Zhu Y Y, Ming N B. J. Phys:Condens Matter, 1992, 4:8073

[31] Zhang X J, Lu Y Q, Zhu Y Y et al. Appl. Phys. Lett., 2004, 85:3531

[32] Liu X J, Wang Z L, Jiang X S et al. J. Phys. D:Appl. Phys., 1998, 31:2502

[33] Liu X J, Wang Z L, Wu J et al. Chin. Phys. Lett., 1998, 15:426

[34] Liu X J, Wang Z L, Wu J et al. Phys. Rev. B, 1998, 58:12782

[35] Chen Y B, Zhu Y Y, Qin Y Q et al. J. Phys.: Condens Matter, 2000, 12:529

[36] Chen Y B, Zhang C, Zhu Y Y et al. Appl. Phys. Lett., 2001, 78:577

[37] Liu Z W, Du Y, Liao J et al. J. Opt. Soc. Am. B, 2002, 19:1676

[38] Liu H, Zhu Y Y, Zhu S N et al. Appl. Phys. Lett., 2001, 79:728

[39] Liu H, Zhu S N, Zhu Y Y et al. Appl. Phys. Lett., 2002, 81:3326

[40] 闵乃本.晶体生长的物理基础.上海:上海科技出版社,1982.第四章[Ming N B. The Fundamentals of Crystal Growth Physics, Shanghai: Shanghai Scientific and Technical Publishers, 1982. Chapter 4(in Chinese)]

[41] Ming N B, Hong J F, Feng D. J. Mater. Sci., 1982, 17:1663

[42] Lu Y L, Lu Y Q, Chen X F et al. Appl. Phys. Lett., 1996, 68:2642

[43] Lu Y L, Wei T, Duewer F et al. Science, 276, 1997:2004

[44] Zhu S N, Cao W W. Phys. Rev. Lett., 1997, 79:2558

[45] Armstrong J A, Bloembergen N, Dncuing J et al. Phys. Rev., 1962, 127:1918
[46] Feng D, Ming B B, Hong J F et al. Appl. Phys. Lett., 1980, 37:607
[47] 薛英华,闵乃本,朱劲松等.物理学报,1983,32:1515 [Xue Y H, Ming N B, Zhn J S et al. Acta Physica Sinica, 1983,32:1515 (in Chinese)]
[48] Luo G Z, Zhu S N, He J L et al. Appl. Phys. Lett., 2001, 78:3006
[49] He J L, Liao J, Liu H et al. Appl. Phys. Lett., 2003, 83:228
[50] Liao J, He J L, Liu H et al. Appl. Phys. B, 2004, 78:265
[51] Xu B, Ming N B. Phys. Rev. Lett., 1993,71:1003
[52] Xu B, Ming N B. Phys. Rev. Lett., 1993,71:3959
[53] Wang Z L, Wu J, Liu X J et al. Phys. Rev. B, 1997,56:9185
[54] Chen X F, Lu Y L, Wang Z L. Appl. Phys. Lett., 1995, 67:3538
[55] Wang Z L, Zhu Y Y, Xu N et al. J. Appl. Phys., 1996, 80:25
[56] Wang Z L, Yang Z J, Zhu Y Y et al. Optics Commun., 1996, 123:649
[57] Zhu Y Y, Zhu S N, Qin Y Q et al. J. Appl. Phys., 1996, 79:2221
[58] Zhu Y Y, Zhn S N, Ming N B. J. Phys. D:Appl. Phys. 1996, 29:185
[59] Zhu Y Y, Ming N B, Jiang W H et al. Appl. Phys. Lett., 1988, 53:2278
[60] Zhu Y Y, Ming N B, Jiang W H et al. Appl. Phys. Lett., 1988, 53:1381
[61] Cheng S D, Zhu Y Y, Lu Y L et al. Appl. Phys. Lett., 1995, 66:291
[62] Chen Y F, Zhu Y Y, Zhu S N et al. Appl. Phys. Lett., 1997, 70:592
[63] Xu H P, Jiang G Z, Mao L et al. J. Appl. Phys., 1992, 71:2450
[64] Zhu Y Y, Cheng S D, Ming N B. Ferroelectrics, 1995, 173:207
[65] Zhang X J, Zhu Y Y, Chen Y F et al. J. Phys. D:Appl. Phys., 2002, 35:1414
[66] Huang K. Proc. R. Soc. London A, 1951, 208:352
[67] Zhang X J, Zhu R Q, Zhao J et al. Phys. Rev. B, 2004, 69:085118
[68] Rosenman G, Skliar A, Arie A. Ferroelectrics Review, 1999, 1:263
[69] Fousek J, Litvin D B, Cross L E. J. Phys:Condens Mattert, 2001, 13:133
[70] Kutuzov A V G, Kutuzov V A G, Kalimullin. Physics-Uspekhi, 2000, 42:647
[71] 该项目获得国家攀登计划(批准号:85-02-02,95-预-04-03)、国家重点基础研究发展计划(批准号:G19980614-03,2004CB619003)、国家高技术研究发展计划(批准号:863-715-28-03-01,863-715-001-0200,2002AA31090:02-04)、国家自然科学基金(批准号:10534020)资助项目

Chapter 1   Superlattices and Microstructures of Dielectric Materials[*]

Nai-Ben Ming

National Laboratory of Solid States Microstructures,
Nanjing University, Nanjing Jiangsu 210093(P.R.China)

1. Introduction

For the past 40 years, semiconductor technology, in which electrons are the information carriers, has played an important role in almost every aspect of our daily lives. Scientists are now turning to light instead of electrons as the information carrier. Light has several advantages over electrons. It can travel in a dielectric material at a much greater speed than an electron in a metallic wire. Light can also carry a larger amount of information per second. The bandwidth of a dielectric material is about 10^9 times larger than that of a metal. Furthermore, light or photons do not interact as strongly as electrons, which helps to reduce energy losses.

In dielectric crystals, including piezoelectric, pyroelectric, and ferroelectric crystals, the most important physical processes are the propagation and excitation of classical waves (light and ultrasonic waves). The behavior of classical waves in a homogeneous dielectric crystal is the same as that in a continuous medium, because the wavevector of a classical wave is much smaller than the reciprocal vectors of the crystal lattice. However, if some microstructures are introduced into a dielectric crystal, forming a superlattice, and if the reciprocal vectors of the superlattice are comparable with the wavevectors of classical waves, the situation will be quite different. The propagation of classical waves in a superlattice (classical system) is similar to the motion of electrons in the periodic potential of a crystal lattice (quantum system). Thus, some concepts in solid-state electronics – for example, the reciprocal space, the Brillouin zone, and the dispersion relation – may be used in classical wave processes. This is the case for photonic crystals.[1-3] On the other hand, the interactions between wavevectors of classical waves and reciprocal vectors of a

---

[*]   Adv. Mater., 1999, 11(13):1079

superlattice may generate some new physical effects. It is these interactions that have led to the new methods of laser frequency generation with quasi-phase-matching (QPM) schemes,[4—7] optical bistability with new mechanisms,[8,9] and ultrasonic generations with frequency up to several gigahertz[10] in dielectric superlattices.

2. Constructions of Dielectric Superlattices

Dielectric superlattices can be realized by the modulation of microstructures, such as domain structures, phase structures, compositions, crystallographic orientations, and heteroepitaxial structures. The modern technology of crystal growth,[11—22] the assembly of small dielectric spheres,[23—26] atomic-layer-controlled epitaxy and heteroepitaxy, including selective epitaxy and fractional epitaxy by molecular beam epitaxy (MBE), chemical beam epitaxy (CBE), and metal-organic chemical vapor deposition (MOCVD),[27,28] or the use of the acousto-optic effect,[29] electro-optic effect,[30] or photo-refractive effect[9] make the construction of 1D (one-dimensional), 2D, or 3D dielectric superlattices possible. In the following, we will describe the construction of dielectric superlattices by modulation of ferroelectric domains and by the use of photorefractive effects.

2.1 Microstructures and Reciprocal Vectors of Dielectric Superlattices

Hereafter we shall take $LiNbO_3$ (LN) and $LiTaO_3$ (LT) crystals as examples to discuss the construction of a dielectric superlattice by modulation of ferroelectric domains or by use of photorefractive effects. In general, the 1D periodic superlattice consists of one kind of fundamental block A arranged along the x-axis or z-axis, such as $AAAA...$, as shown in Figure 1(a). The 1D quasi-periodic superlattice consists of two or more kinds of fundamental blocks A and B, arranged along the x-axis or z-axis according to the production rule $S_j = S_j - 1 : S_j - 2$, for $j > 3$ with $S_1 = A$ and $S_2 = AB$, where : stands for concatenation. The sequence of blocks $ABABAA...$ produces a 1D quasi-periodic superlattice, as shown in Figure 1(b).[6,7,30—33] All the fundamental blocks are composed of one positive domain and one negative domain. In such structures, the spontaneous polarizations of successive domain lamellae are opposite, as are the signs of nonlinear optic coefficients, electro-optic coefficients, and piezoelectric coefficients.[34—37] As a result, these microstructures form a 1D superlattice for nonlinear optic (NLO) effects, electro-optic effects, and piezoelectric effects.

In general, the superlattice is described by a reciprocal vector. For the periodic structure shown in Figure 1(a), the magnitude of the reciprocal vector is given by Equation (1) and the direction is perpendicular to the domain lamellae, L_A is the period of the superlattice, i.e., the thickness of the fundamental block A, $2\pi/L_A$ is a primitive reciprocal vector, and m is an integer. It is worth noting that L_A is an adjustable structural parameter during the design of the periodic superlattice.

$$G_m = m2\pi/L_A \quad (1)$$

For the quasi periodic structure as illustrated in Figure 1b, L_A and L_B represent the thickness of block A and B, respectively, where $L_A = L_{A1} + L_{A2}$ and $L_B = L_{B1} + L_{B2}$. Let $L_{A1} = L_{B1} = L$, $L_{A2} = L(1+\eta)$, $L_{B2} = L(1-\tau\eta)$, where L and η are adjustable structure parameters during the design of the quasi-periodic superlattice and $\tau = (1+\sqrt{5})/2$ is the golden ratio. The sequence of blocks, as shown in Figure 1(b), produces a quasiperiodicity with $L_A/L_B = \tau$. Even if $L_A/L_B \neq \tau$, quasi-periodic properties of this structure are still preserved. Unlike the periodic one, which has reciprocal vectors derived from an integer times a primitive vector, see Equation (1), a quasi-periodic superlattice with an infinite number of blocks provides reciprocal vectors governed by two integers, which have the form given in Equation (2),[6,7,30-33] where $D = \tau L_A + L_B$ is the "average structure parameter".

$$G_{m,n} = (m + n\tau)2\pi/D \quad (2)$$

It is clear that the reciprocal vectors of a quasi-periodic superlattice can be adjusted by L, η and selected by m, n. Because of this property, some coupled optical parametric processes or multi-wavelength processes occur in the quasiperiodic superlattice with efficient conversions.

For a 2D periodic superlattice, as shown in Figure 1c, the reciprocal vectors

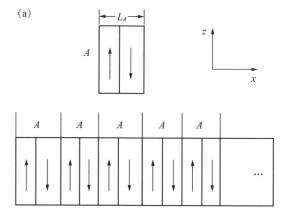

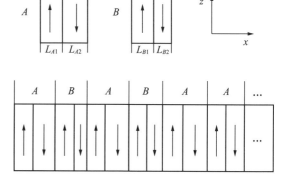

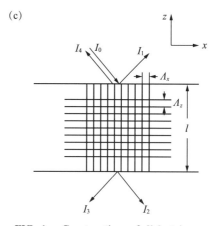

FIG. 1. Constructions of dielectric superlattices: (a) 1D periodic superlattice composed of one kind of fundamental block A. (b) 1D quasi-periodic superlattice composed of two kinds of fundamental blocks A and B. (c) 2D periodic superlattice constructed by the photorefractive effect.

along x and z are given by Equation 3, where L_x and L_z are the periods of the superlattice along x and z, respectively.

$$G_m^x = m2\pi/L_x \tag{3a}$$

$$G_m^z = m2\pi/L_z \tag{3b}$$

2.2. Dielectric Superlattices Prepared by Modulation of Ferroelectric Domains

As a typical example of 1D periodic dielectric superlattices, bulk ferroelectric crystals with periodic laminar domains have been successfully prepared by the growth striation technique and electric poling method. Examples include LN,[11-15] LT,[19] and $Ba_2NaNb_5O_{15}$ (BNN)[22] grown from the melt, $(NH_2CH_2COOH)_3 \cdot H_2SO_4$ (TGS)[20] grown from aqueous solution, and LN, LT,[38,39] and $Sr_{0.8}Ba_{0.4}Nb_2O_6$ (SBN)[43,44] prepared by electric poling.

The growth of LN, LT, and BNN crystals with periodic laminar ferroelectric domains (PLFEDs) was accomplished by the Czochralski growth method.[11-17,19,21,22] The key point for the growth process is to build up pronounced growth striations, i.e., a periodic solute concentration fluctuation in the crystal along the direction of growth. The melt should be doped with solute, such as yttrium, indium, or chromium, at a concentration of about 0.1–0.5 wt.%. PLFEDs may be automatically induced by the growth striations when the growing crystal cools through the Curie temperature.[13-17] The growth striations may be induced by the following two methods. The first method is eccentric rotation, i.e., intentionally displacing the rotation axis off the symmetric axis of the temperature field, so that temperature fluctuation at the solid–liquid interface may be induced during each cycle of crystal rotation.[13,40,41] It is well known that there is a linear dependence of the growth speed on the temperature fluctuation at a solid–liquid interface, as well as the dependence of the solute concentration in the crystal on growth speed.[18] Figures 2(a) and (b) show the temperature fluctuations at the solid–liquid interface actually measured when growth striations formed. Figures 2(c) and d show the correspondence between growth striations and domain lamellae. The close correspondence between the temperature fluctuations, growth striations, and domain lamellae is clearly demonstrated. The period of the PLFEDs may be adjusted by varying the pulling speeds and the rotation frequencies. Figure 3 shows a typical example of the PLFEDs in LN crystals with different periods.

The alternative method of achieving growth striations is to apply a modulated electric current during Czochralski growth. In our experiment we use a rectangular current pulse with alternate signs, periods of modulation in the range from several seconds to several tens of seconds, and current density at the interface of about $10-20 mA/cm^2$.[15] The

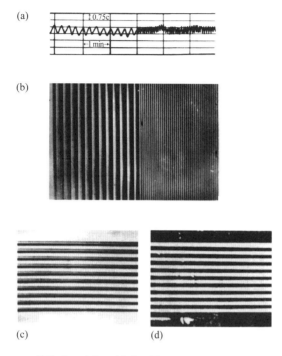

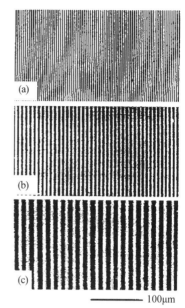

FIG. 2. (a), (b) The correspondence between temperature fluctuations and growth striations. (c), (d) The correspondence between growth striations and periodic laminar ferroelectric domains.

FIG. 3. Optical micrographs of dielectric superlattices of LN with period (a) 2.7 μm, (b) 5.2 μm, and (c) 15 μm. These samples were prepared by the growth striation technique and revealed by etching.

response of the domain structure is quite regular, so using this method we can fabricate PLFEDs with periods in the range of several micrometers to several tens of micrometers. The mechanism for the build-up of the solute concentration fluctuation can also be traced to the temperature fluctuation at the solid-liquid interface arising from the Peltier effect of electric current across the solid-liquid interface.[15,18] In order to reveal the formation mechanism of the PLFEDs, we have used energy dispersive X-ray analysis in a scanning microscope to measure the solute fluctuation in growth striations[13,14] and a scanning-tip microwave near-field microscope to image the profile of solute concentration distribution in cooperative research recently.[42] The experimental results show that the direction of spontaneous polarization of the ferroelectric domain is determined by the gradient of solute concentration in the growth striations.[13] This fact may be explained as follows: the non-uniform distribution of solute, which in a crystal is generally ionized, is equivalent to a non-uniform space-charge distribution in the crystal and a non-uniform field is produced in it. Although the field is comparatively small, it can induce the metallic ions within the lattice to displace preferentially at a temperature close to the Curie point and thus a crystal with PLFEDs is formed.[13] Further studies pointed out that the local internal electric field induced by the solute concentration gradient consists of two parts, one is the space-charge field and the other is the elastic equivalent field. The Gibbs free energy is minimized only when the spontaneous polarization has the same direction as the internal field.[16]

An electric poling technique has also been used to prepare LN,[39] LT,[38] and BSN[43,44] crystals with periodically or quasi-periodically inverted ferroelectric domain lamellae. Let us take a c-cut LT or LN single-domain crystal wafer (0.5 mm thick) as an example. Both c faces of the wafer are polished, upon which 0.2 μm thick Al films are deposited as electrodes. On the $-c$ face of the wafer there is a plane electrode, and on the $+c$ face there is a periodic, quasi-periodic, or patterned electrode fabricated by a standard photolithographic and wet-etching technique, as shown in Figure 4(a). The wafer is connected to a high-voltage pulse generator. The pulsed-field treatment is carried out at room temperature. By applying a field strength of 22 kV/mm with 1:3 duty cycle for the LT wafer and a field strength of 26.5 kV/mm with 1:2 duty cycle for the LN wafer, we successfully fabricated ~0.5 mm thick LT and LN wafers with periodic[38,39] and quasi-periodic[6,7] domain structures. The periodic domain structure of an LT wafer with a period of 23.7 μm and total length of 7 mm is shown in Figure 4(b). The quasi-periodic domain structure produced in the LT crystal wafer is shown in Figure 4(c). In the quasi-periodic superlattice, ferroelectric domain lamellae are arranged along the x-axis of the LT crystal and the domain boundaries parallel to the y-z plane, block A and block B consist nominally of 11 μm, 13 μm and 11 μm, 6.5 μm, respectively. The wafer has 13 generations, i.e., S_{13}, 377A and B blocks, with a total length of ~8 mm and a thickness of ~0.5 mm.[7]

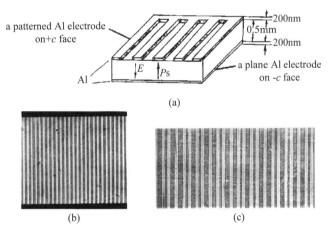

FIG. 4. (a) Schematic diagram of periodical or quasi-periodical poling for ferroelectric crystals. (b) Optical micrograph of a periodic dielectric superlattice of LT with a period of 23.7 μm. (c) Optical micrograph of a quasi-periodic dielectric superlattice of LT with $L_{A1}=L_{B1}=11$ μm, $L_{A2}=13$ μm, $L_{B2}=6.5$ μm, $L_A=24$ μm, $L_B=17.5$ μm. The samples in (b) and (c) were prepared by electric poling and revealed by etching.

2.3. Dielectric Superlattices Prepared Using Photorefractive Effects

Using volume holographic recording, followed by thermal fixation, 1D, 2D, or 3D modulations of refractive index in a photorefractive material can be easily realized and thus 1D, 2D, or 3D dielectric superlattices can be constructed.[9,45,46] The photorefractive material used in our experiment is an oxidized Fe-doped LN (0.1 wt.% Fe) single-domain crystal. A beam of a blue line (488 nm, TEM_{00} mode) of an argon-ion laser was split into three nearly equal-intensity beams, S_1, S_2, and S_3, which were recombined and interfered to cause spatial variations of the refractive index as shown in Figure 5. Thus a 2D square periodic grating was recorded in the crystal. The intensities of the beams $S_i (i=1, 2, 3)$ were set to be about 2.5 W/cm². The diameter of the

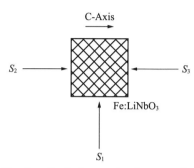

Fig. 5. Optical geometry for recording a 2D refractive index grating in an Fe:LN single-domain crystal for the construction of a 2D orthogonal periodic superlattice.

beams, stipulating the size of the grating, was 0.2 cm. Despite the difficulty in determining precisely the grating strength in this holographic writing process, after an exposure time of about 30–40 s the grating's diffraction efficiency reached a value in the range 10%–15%. This oxidized Fe:$LiNbO_3$ sample showed high resistance to relaxation and the grating remained stable in the sample after the writing beams were moved.[9] If a beam of an argon laser is split into four equal-intensity ones and S_4 is parallel to the normal of the face of Figure 3(d), a cubic dielectric superlattice can be formed.

3. Nonlinear Optic Effects in Dielectric Superlattices

As we know, there are two methods to obtain frequency conversion with high efficiency. One is phase matching (PM), which is commonly used, and the other is quasi phase matching (QPM), first proposed by Bloembergen and co-workers.[4] QPM can be realized only in a dielectric superlattice, because the reciprocal vectors provided by the superlattice can participate in the wave process, that is, the mismatch of wavevectors in the optical parametric process due to the dispersive effect of the refractive index can be compensated by the reciprocal vectors of the dielectric superlattice.

3.1 Second Harmonic Generation in Dielectric Superlattices

The QPM concept has been developed in quasi-periodic dielectric superlattices.[32,33] For second harmonic generation (SHG), the QPM conditions are as in Equation 4, where

$k(2\omega)$ and $k(\omega)$ are the wavevectors of the second harmonic and fundamental wave, respectively. The expressions for G_m and $G_{m,n}$ are shown in Equations 1 and 2.

$$k(2\omega) - 2k(\omega) = G_m \quad \text{(periodic superlattice)} \tag{4a}$$

$$k(2\omega) - 2k(\omega) = G_{m,n} \quad \text{(quasi-periodic superlattice)} \tag{4b}$$

In a periodic dielectric superlattice, there is one set of reciprocal vectors, which comprises one primitive reciprocal vector and the others that are derived from an integer times the primitive one, see Equation 1. The primitive vector may be used to attain first-order QPM and the others may be used to attain higher-order QPM.[50,51,54] In general, first-order QPM will provide the greatest conversion efficiency and the maximum attainable power. In a quasi-periodic superlattice, there are more sets of reciprocal vectors, which may be used for multi-wavelength SHG and a coupled optical parametric process.[6,7]

Early in 1980, QPM SHG was realized in a LN periodic superlattice.[5] The intensity of SHG of a LN periodic superlattice fulfilling QPM for d_{33} has been experimentally compared with that of a single-domain LN crystal fulfilling PM for d_{31}. An order-of-magnitude enhancement predicted theoretically has been realized.[47] Periodic superlattices of LN and LT prepared by Czochralski growth and electric poling have been used for SHG by QPM to obtain green, blue, and violet light.[47-54] Typical experimental results are shown in Table 1. These experiments were performed using a picosecond automatic tunable optical parametric oscillator (OPO) as a source of the fundamental. This laser has a pulse rate of 1 Hz and a pulse duration of 30 ps. The linewidth of the output light is less than 1 nm except at the OPO's degeneracy point of 106.4 nm. It is worth noting that green and violet light can be obtained using the same sample (sample number 5 in Table 1) when the first-order and third-order QPM conditions are satisfied, respectively. First-order QPM blue light generation in a LN superlattice has been performed also by direct frequency doubling of light from a continuous wave 978 nm InGaAs diode laser. 1.27 mW output power of blue light has been obtained with an incidence power of 500 mW and conversion efficiency of 0.25%.[49] For direct frequency doubling of 810 nm continuous wave (cw) output of a GaAlAs diode laser with third-order QPM, 0.35 mW blue light has also been obtained with an incidence power of 250 mW.[54]

TABLE 1. Experimental results of QPM SHG in periodic superlattices of LN.

Sample no.	Thickness [mm]	Period length [μm]	Number of periods	Type of QPM	$\lambda(2\omega)$ [nm]	η(SHG) [%]
1	0.62	2.8	220	1st order	407.5	3.0
2	0.78	3.4	230	1st order	430.0	4.2
3	1.56	5.2	300	1st order	490.0	24.0
4	2.20	6.4	310	1st order	513.0	17.0

续 表

Sample no.	Thickness [mm]	Period length [μm]	Number of periods	Type of QPM	λ(2ω) [nm]	η(SHG) [%]
5	0.98	6.8	144	1st order	531.0	11.4
5	0.98	6.8	144	3rd order	385.0	1.6
6	1.50	8.3	180	1st order	565.0	19.8

Recently, a quasi-periodic superlattice of LT has been used to realize multi-wavelength SHG.[7] In the superlattice, block A consists of 11 and 13 μm, while block B consists of 11 μm and 6.5 μm. This superlattice has 13 generations (S_{13}), 377 A and B blocks, with a total length of ~8 mm and a thickness of ~5 mm, as shown in Figure 4(c). The fundamental light source is a picosecond OPO with a repetition of 1 Hz and a duration of 23 ps. The fundamental wave is polarized along the z-axis of the sample. It is weakly focused and coupled into the polished end face of the sample and propagates along the x-axis of the sample. The radius of the waist inside the sample is $\omega_0 \approx 0.1$ mm. The confocal parameter for the system is $Z_0 \approx 6$ cm. Because Z_0 is much greater then the length of the sample, the plane-wave theory of SHG can be applied to the Gaussian beam. The SHG spectrum of the quasi-periodic superlattice of LT is measured in the range from 0.9 to 1.4 μm and from 1.55 to 1.7 μm, as shown in Figure 6(b). Figure 6(a) is the theoretical result. When the fundamental wavelength was tuned to 0.9726, 1.0846, 1.2834, 1.3650, and 1.5699 μm, we obtained QPM second harmonic blue, green, red, and infrared output with conversion efficiencies up to ~5%-20%, see Table 2. It is worth noting that the peak indexed (2,2) does not appear in either the calculated or the measured spectrum (Fig. 6) because the extinction condition is satisfied for this peak.[7,32] It can be seen from this example that we can adjust the structure parameters L and η select the indices m and n to design a set of special reciprocal vectors, thus obtaining a set of special frequencies that are doubled.

TABLE 2. Experimental results of multi-wavelength SHG in a quasi-periodic superlattice of LT.

Reciprocal vector $G_{m,n}$ (m,n)	Fundamental wavelength[a]		Harmonic wavelength Measured [μm]	Input energy [μJ]	Output energy [μJ]	FWHM [nm]	Efficiency [%]
	Calcd. [μm]	Measured [μm]					
(3,4)	0.9720	0.9726	0.4863	40	3	≈0.3	≈7.5
(2,3)	1.0820	1.0846	0.5423	40	7	≈0.4	≈17.5
(1,2)	1.2830	1.2834	0.6417	33	3	≈0.85	≈9.1
(2,1)	1.3640	1.3650	0.6825	30	2	≈1.1	≈6.7
(1,1)	1.5687	1.5699	0.7845	54	11	≈2.5	≈20.4

[a] Fundamental wave source: ps-OPO with 1 Hz repetition rate and 22 ps pulse duration.

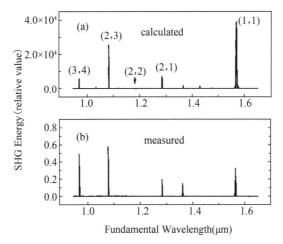

FIG. 6. The SHG spectra in a quasi-periodic superlattice of LT: (a) calculated, (b) measured.

3.2 Coupled Parametric Process in Quasi-Periodic Dielectric Superlattices

A coupled parametric process, that is, third harmonic generation (THG), has been realized in a quasi-periodic superlattice of LT.[6] In THG, two parametric processes – the SHG process and the sum-frequency generation process, which mixes the fundamental frequency with the second harmonic – are coupled in one material. In this case, the coupled QPM conditions are as in Equation 5; both of these two QPM conditions must be satisfied simultaneously.

$$k(2\omega) - 2k(\omega) = G_{m,n} \tag{5a}$$

$$k(3\omega) - k(2\omega) - k(\omega) = G_{m',n'} \tag{5b}$$

The THG process can be analyzed by solving the coupled nonlinear equations that describe the interaction of the three fields E_ω, $E_{2\omega}$, and $E_{3\omega}$ in the quasi-periodic superlattice. We can obtain an analytical result for the boundary condition $E_{2\omega}(0) = 0$ and $E_{3\omega}(0) = 0$. In the small signal approximation and QPM condition, the second-and third-harmonic intensities are given by Equation 6[6,33] with the effective nonlinear coefficient $d_{m,n} = d_{1,1}$ and $d_{m',n'} = d_{2,3}$.[7,32,33] Here, n_ω, $n_{2\omega}$, and $n_{3\omega}$ represent the refractive indices of the fundamental, second, and third harmonics, respectively; c is the speed of light in vacuum; λ is the fundamental wavelength; ε_0 is the dielectric constant of vacuum and L is the total length of the sample. From Equation 6 it is evident that $I_{3\omega}$ would depend more strongly on the fundamental intensity I_ω and the length of sample L than $I_{2\omega}$.

$$I_{2\omega} = \frac{8\pi^2 d_{m,n}^2 L^2}{n_\omega^2 n_{2\omega} c \varepsilon_0 \lambda^2} I_\omega^2 \tag{6a}$$

$$I_{3\omega} = \frac{144\pi^4 d_{m,n}^2 d_{m',n'}^2 L^4}{n_\omega^3 n_{2\omega}^2 n_{3\omega} c^2 \varepsilon_0^2 \lambda^4} I_\omega^3 \tag{6b}$$

The quasi-periodic superlattice was fabricated by quasiperiodically poling a z-cut LiTaO$_3$ wafer at room temperature.[6] The structure parameters were selected to be $L_{A1} =$

$L_{B1} = L_{c2} = \tau^2 L_{c3} = 10.7$ μm and $\eta = 0.23$, so that blocks A and B were 23.9 μm and 19.9 μm thick, respectively. The total length of the sample was ~8 mm. In this design, the reciprocal vector $G_{1,1}$ was used for QPM SHG, whereas $G_{2,3}$ was used for QPM sum-frequency generation. Here, the two processes no longer proceed alone but couple each other. This coupling led to a continuous energy transfer from fundamental to second-to third-harmonic fields; thus a direct third-harmonic was generated from the quasi-periodic superlattice with high efficiency.

THG was tested with a tunable OPO pumped with an yttrium aluminum garnet (YAG) laser with a pulse width of 8 ns and a repetition rate of 10 Hz. The linewidth of the fundamental wave was about 1.5 nm at 1.6 μm. In order to use the largest nonlinear optical coefficient d_{33} of $LiTaO_3$, the fundamental wave was z-polarized and propagated along the x-axis of the sample. It was weakly focused and coupled into the polished end of the sample. For the system, the confocal parameter was $Z_0 = 2$ cm, and the radius of the waist spot inside the sample was $\omega_0 = 30$ μm. When the fundamental wavelength was tuned to 1.570 μm, green light at 0.523 μm wavelength was generated from the sample. 6 mW green light was obtained for a fundamental power of 26 mW (with an average power density ~1 kW/cm^2). The conversion efficiency was close to 23%. Together with THG, SHG was observed in the same range of fundamental wavelength. The output of SHG and THG were stable, and no optical damage was observed at this fundamental intensity level. The relationship of the average power of second- and third-harmonic fields to the average power of the fundamental field was studied for the quasiperiodic superlattice. Figure 7(a) shows the dependence of second- and third-harmonic average power measured at the output end of the sample on the average fundamental power. The theoretical results according to Equation 6, as shown in Figure 7(b), are in good agreement with the experimental ones, Figure 6a. Aside from the THG, some other higher-order harmonics can be also generated using the same QPM scheme. For example,

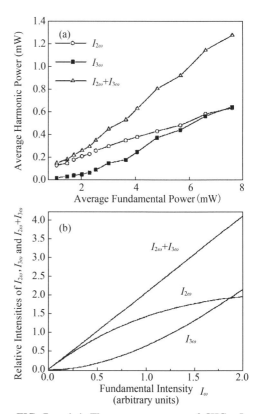

FIG. 7. (a) The average powers of SHG, $I_{2\omega}$, and THG, $I_{3\omega}$, versus the average power of the fundamental wave, I_ω, for a quasi-periodic superlattice of LT. The light source is a nanosecond OPO with a repetition rate of 10 Hz. (b) The calculated relative intensities of $I_{2\omega}$, $I_{3\omega}$, and $I_{2\omega} + I_{3\omega}$ versus fundamental intensity I_ω.

the fourth harmonic can be generated by two coupled, cascaded SHG processes. And more interesting effects such as tunable multi-wavelength optical parametric oscillation can be expected, where a number of QPM conditions must be satisfied simultaneously.

3.3　Optic Bistability in a Two-Dimensional Periodic Dielectric Superlattice

Here we only consider a 2D orthogonal dielectric superlattice similar to the sinusoidal cross grating shown in Figure 1c. One method for the construction of this kind of 2D periodic superlattice is the following: Couple two ultrasonic transducers to two side faces of a crystal that are perpendicular to each other; according to acousto-optic effects, its refractive index will be modulated periodically along the two orthogonal directions. To adjust the periodicity of the 2D superlattice, i.e., to adjust the wavelength of the two acoustic waves propagating in the crystal, Bragg conditions must be satisfied in two orthogonal directions at the same time. In this case, four Floquet-Bloch waves will be excited in the 2D superlattice. This is a four-wave problem, which cannot be explained by either the coupled-mode theory[55] or the two-wave dynamical theory.[56] A four-wave dynamical theory that can deal with interaction and propagation of light in both 1D and 2D superlattices has been developed by the author.[29] According to this theory, several interesting conclusions can be drawn. Let us discuss one of them.

If M_x and M_z are modulation strengths of refractive index and $M = M_z/M_x$ is the modulation strength ratio, the intensity of each exciting wave can be controlled by modulating the index strength M_x, M_z or strength ratio M. As mentioned just above, an acoustic wave is used to produce the periodically modulated index in the crystal; the acousto-optic interaction provides a new way to probe the acoustical field in a crystal and manipulate the optical radiation. Such effects in 2D superlattices can be used to construct acoustooptic devices in areas such as light modulation, beam deflection, and signal processing. Let us look at a simple example of using a 2D superlattice to design an acousto-optical switch with modulation strength ratio M. Generally, four beams of light emerge from the superlattice. However, the incident beam will be diffracted into the direction of one of the other beams when M takes some particular values. If we choose $M=0$, i.e., the acoustic field is applied only along the x-axis, the intensities of exiting waves are $I_3 \sim I_0$, $I_1 = I_2 = I_4 = 0$, where I_0 is the intensity of the incident beam, see Figure 1c. If now another acoustic field is applied in the z-direction, and a suitable intensity of acoustic field is chosen to make the strength ratio $M=0.745$, then I_2 reaches its maximum, i.e., $I_2 \sim I_0$, $I_1 = I_3 = I_4 = 0$. When the strength ratio M is changed from 0.745 to 1, the intensities are $I_4 \sim I_0$ and $I_1 = I_2 = I_3 = 0$; that is, the incident beam will reflect back in the direction opposite to the incident direction and other exiting waves are very weak. When the strength ratio M is changed from 1 to ∞, e.g., $M_x = 0$, in this case, the acoustic field is applied only along the z-axis, thus, $I_1 \sim I_0$ and $I_4 = I_3 = I_2 = 0$. It is clear that with this acousto-optical switch we could have the incident beam diffracted into one of the four propagating directions.

A nonlinear system with positive feedback will exhibit bistable behavior under suitable conditions. That is to say, two necessary elements are expected to coexist in a dielectric system that possesses optical bistability. One is the nonlinear response element, requiring that the transmission light field responds in nonlinear form to parameters that are related to the transmission process; the other is the positive feedback element, requiring that the values of these parameters change with the transmission field. In common dispersive bistability systems, it is the phase mismatch between the propagating wavevector and the cavity-matching condition or the Bragg condition that brings the incident wave from a forbidden transmission state to an allowed state, or from a low-transmission state to a high-transmission state in the allowed band. This kind of bistability is thus attributed to the phase-mismatch mechanism. Our theoretical and experimental work[8,9] has revealed that a novel bistable mechanism, i.e., the index-modulation mechanism, which is characteristically different from the phase-mismatch mechanism, exists in 2D superlattices containing Kerr-form dielectric nonlinearity. Compared with the phase-mismatch mechanism, the relevant parameter in the index-modulation mechanism is not the wavevector's mismatch, but the index-modulation strength of the 2D superlattice. It is the value of the index-modulation strength that will be perturbed by the interference of the transmission and diffraction light fields in the medium, forming a positive feedback element to establish bistable behavior in the relationships between the incident and diffracted light in the 2D superlattice. Figure 8, one of our results, shows the intensities of four diffracted waves, I_1, I_2, I_3, and I_4, as functions of the incident intensity I_0. Curves A, B, and C are three different stable convergent self-consistent solutions obtained by numerical computation. The shapes of the diffracted-against-incident intensity curves are determined by the index-modulation strengths. Discontinuous jumps of intensity always appear in the bistable or multistable regions and the jumps can be either from higher to lower values or from lower to higher values. Figure 9 illustrates results recorded in experiments that demonstrate bistable behavior with discontinuous jumps of the diffracted intensities. The intensity jumps occur simultaneously for different diffracted waves.[9] In fact, the transmission diffraction field and the perturbed index-modulation diffraction field are interacting with each other. The observed bistable region indicates that this kind of interaction can reach a stable state, of which there can be more than one; if this interaction does not reach a stable state, instability is exhibited.[9] It is worth noting that the bistable states of the four beams of diffracted light can be reached simultaneously in an orthogonal 2D superlattice, and the bistable states of more beams can be obtained in a hexagonal 2D superlattice or a 3D superlattice, which may be favorable for the design of miniaturized and compact bistable devices.[57—63]

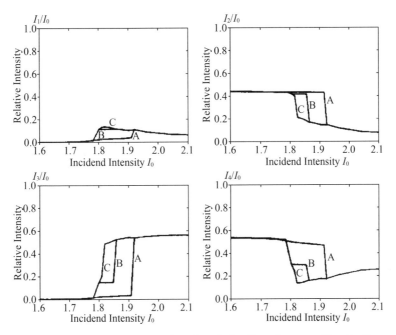

FIG. 8. With dielectric nonlinearity of the Kerr form considered, the intensities of four diffracted waves I_1, I_2, I_3, and I_4 as a function of the incident intensity I_0. Curves A, B, and C are three different stable convergent self-consistent solutions obtained by numerical computation.

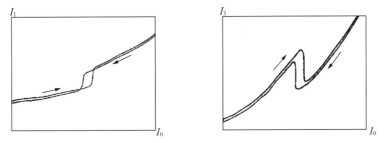

FIG. 9. With the incident wave satisfying the Bragg condition of four-wave diffraction, and its intensity I_0 being cycled, the diffracted intensities I_1 and I_3 were recorded. Intensity jumps occur simultaneously in the diffracted waves.

4. Piezoelectric Effects of Dielectric Superlattices

In an acousto-electric field, a dielectric superlattice consisting of periodic or quasi-periodic laminar ferroelectric domains can be viewed as a structure composed of a series of δ-function-like sound sources arranged along the normal of the domain lamellae, as shown in Figure 10. These sources are generated by the discontinuity of the piezoelectric stress, which is the product of a piezoelectric coefficient and electric field. As mentioned above, all the odd-rank tensors in this kind of superlattice will change signs from one domain to the next, and the piezoelectric tensor is a third-rank tensor. The sign of the piezoelectric coefficient in the superlattice changes alternately.[34—36] In this case, the piezoelectric coefficient is discontinuous at the ferroelectric domain boundaries. Under the action of an alternating external electric field, the discontinuity

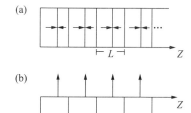

FIG. 10. Schematic diagram of an LN dielectric superlattice with periodic laminar ferroelectric domains (a), and the corresponding δ-function-like sound sources (b).

in piezoelectric stresses at the ferroelectric domain boundaries at any time must be balanced by a strain. This strain will propagate as a wave in the superlattice. So each domain boundary may be regarded as a δ-function-like sound source. Assuming that the superlattice is arranged along the z-axis of an LN or LT crystal perpendicular to the lamellae and that the (0001) planes are the electrode faces that are parallel to the lamellae, as shown in Figure 10a, a longitudinal planar wave propagating along the z-axis will be excited by the action of an alternating external electric field.

4.1 Resonator Made from LN Superlattices

For a periodic superlattice, by using the Green's function method to solve the elastic wave equations, the electric impedance of a resonator can be derived,[10] and then the resonance frequency can be obtained as in Equation 7, where v is the velocity of the longitudinal wave propagating along the z-axis; a and b are the thickness of positive and negative domains, respectively, and $a+b=L$ is the period of the superlattice.

$$f_n = nv/L = nv/(a+b), \quad n=1, 2, 3,\ldots \tag{7}$$

It is clear that the resonance frequency is determined only by the period of the superlattice, not by the total thickness of the wafer. As we know, the thickness of a resonator working at several hundred megahertz with ordinary materials, such as a single-domain LN crystal, is too thin to be fabricated by ordinary processing techniques. However, it is easy to prepare a superlattice with each lamella several micrometers thick by the growth striation technique[11—22] or electric poling.[38,39,64] Therefore, it is possible to fabricate acoustic devices operating at frequencies of hundreds of megahertz to several gigahertz by using superlattices. A set of resonators with working frequencies in the range from 500 MHz to 1000 MHz has been made from a LN superlattice.[10,64,66] The measured and calculated resonance frequencies are listed in Table 3. The experimental values are very close to the theoretical ones.

TABLE 3. Relationship between the resonance frequency (f)
and period ($a+b$) of the periodic superlattice of LN.

Resonator no.	Period of superlattice $a+b$ [μm]	Resonance frequency f [MHz]		Error [%]
		Calculated	Measured	
1	7.4	989	975	1.4
2	8.2	892	882	1.1
3	10.0	732	710	3.0
4	11.0	665	686	3.0
5	13.2	554	552	0.4
6	13.3	550	552	0.5

4.2 Resonators Made from BNN Superlattices

The BNN crystal possesses not only electro-optic and nonlinear optical properties but also good acoustic properties. In particular, the z-cut plate of this material has a thickness longitudinal mode coupling factor of 0.57, which is about three times larger than that of LN. Hence, this material is useful for ultrasonic device applications. As we know, microtwinning often appears in as-grown BNN crystals and the detwinning procedure is complex. After studying the influence of microtwins on the performance of these materials, that is, by transformation of the component of the piezoelectric tensor from the untwinned coordinate system to the twinned coordinate system, we concluded that the component of the piezoelectric tensor h_{33} remains constant from one microtwin lamella to the next. Namely, microtwins have no influence on the acoustic resonance properties of a superlattice arranged along the z-axis, and a longitudinal planar wave is excited and propagates along the z-axis in this case. A set of resonators with working frequencies in the range from 200 MHz to 400 MHz have been made from BNN superlattices. Both experimental and theoretical results show that resonators of BNN superlattices have low acoustic loss.[22]

4.3 Transducers Made of LN Superlattices

After studying the propagation and excitation of elastic waves in superlattices, the electrical impedance of a transducer made of a superlattice has been obtained which is a function of the number of laminar domains N and the area of the electrode face A.[10,67] For transducers made of ordinary material, such as a single-domain LN wafer, the static capacitance is the main part of the impedance at the resonance frequency under high-frequency operation. As a result, the insertion loss of the transducer is very high. In our case, the real part of the impedance can be made equal to or even larger than the imaginary part by choosing N and A suitably. The transducer will thus have an insertion loss near 0 dB in a 50 Ω measurement system. If the normal of the laminar domains does not coincide with the z-axis but lies in x-y-plane, and the electrodes are still made parallel to the lamellae, one quasi-longitudinal wave (QLW) and one quasi-shear wave (QSW) will be excited by the transducer. A typical experimental result is shown in the inset in Figure 11. Two resonance frequencies are observed, one for QLW and the other for QSW, with frequencies of 555 MHz and 279 MHz, respectively. Theoretical values are 553 MHz for QLW and 273 MHz for QSW. Figure 11 also gives the insertion loss

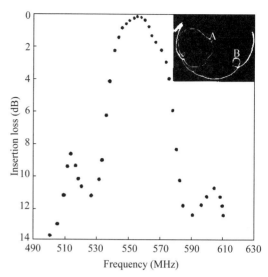

FIG. 11. Insertion loss vs. frequency for an ultrasonic transducer made of an LN periodic dielectric superlattice. Its Smith chart is shown in the inset: A) QLW, B) QSW.

versus frequency for this transducer. An insertion loss of near 0 dB at 555 MHz is achieved, which is in good agreement with the theory. The 3 dB bandwidth of this transducer is 5.8%. The results obtained up to now[64—75] suggest that the dielectric superlattice can be used to generate and detect ultrasonic waves of very high frequencies. Hence, the dielectric superlattice will have potential applications in acoustic devices.

4.4 Ultrasonic Excitation in Quasi-Periodic Dielectric Superlattices

Ultrasonic excitation and propagation in quasi-periodic superlattices have been studied both experimentally and theoretically.[75] The ultrasonic spectrum in quasi-periodic superlattices was obtained from theoretical analysis[75] as Equation 8, where $X_{m,n} = \pi\tau(m\tau - n)/(1+\tau^2)$, $D = \tau L_A + L_B$, and k is the wavevector.

$$H(k) \propto \sin(kL/2) \sum (\sin X_{m,n}/X_{m,n}) \delta\{k - 2\pi(m + n\tau)/D\} \quad (8)$$

Figure 12(a) is the spectrum calculated numerically and Figure 12(b) shows the experimental results. It is clear that there exists a one-to-one correspondence between the resonant peaks in Figures 12(a) and (b). After studying the spectrum in detail, we found that resonant peaks densely fill reciprocal space in a self-similar manner, which is a reflection of the self-similar structure of quasi-periodic superlattices in reciprocal space.

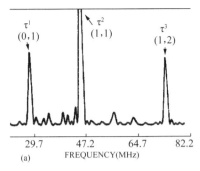 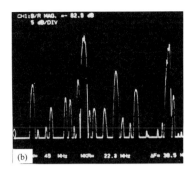

FIG. 12. Ultrasonic spectrum for an LN dielectric quasi-periodic superlattice: (a) calculated, (b) measured.

5. Conclusions

Over the past 20 years, we have taken the dielectric superlattice as an example in a study of the relationship between microstructures and physical properties of dielectrics both experimentally and theoretically. It has been demonstrated that microstructured materials of this kind produce significant effects on optic and acoustic wave processes and can be used as a new type of device for a variety of applications in the opto-electric and acousto-electric fields. We believe that it is possible to use modern experimental techniques to control the microstructures and thus to develop new types of synthetic materials with pre-designed microstructures.

References and Notes

[1] E. Yablonovitch, *Phys. Rev. Lett.* **1987**, *58*, 2095.
[2] S. John, *Phys. Rev. Lett.* **1987**, *58*, 2486.
[3] J. D. Joannopoulos, P. R. Villeneuve, S. Fan, *Nature* **1997**, *386*, 143.
[4] J. A. Armstrong, N. Bloembergen, J. Ducuing et al., *Phys. Rev.* **1962**, *127*, 1918.
[5] D. Feng, N.-B. Ming, J.-F. Hong et al., *Appl. Phys. Lett.* **1980**, *37*, 607.
[6] S.-N. Zhu, Y.-Y. Zhu, N.-B. Ming, *Science* **1997**, *278*, 843.
[7] S.-N. Zhu, Y.-Y. Zhu, Y.-Q. Qin, H.-F. Wang, C.-Z. Ge, N.-B. Ming, *Phys. Rev. Lett.* **1997**, *78*, 2752.
[8] B. Xu, N.-B. Ming, *Phys. Rev. Lett.* **1993**, *71*, 1003.
[9] B. Xu, N.-B. Ming, *Phys. Rev. Lett.* **1993**, *71*, 3959.
[10] Y.-Y. Zhu, N.-B. Ming, *J. Appl. Phys.* **1992**, *72*, 904.
[11] Y. L. Lu, Y. Q. Lu, C. C. Xue, N.-B. Ming, *Appl. Phys. Lett.* **1996**, *68*, 1467.
[12] S. D. Chen, Y. Y. Zhu, Y. L. Lu, N.-B. Ming, *Appl. Phys. Lett.* **1995**, *66*, 291.
[13] N.-B. Ming, J.-F. Hong, D. Feng, *J. Mater. Sci.* **1982**, *17*, 1663.
[14] Y.-L. Lu, Y.-Q. Lu, X.-F. Chen, G.-P. Luo, C.-C. Xue, N.-B. Ming, *Appl. Phys. Lett.* **1996**, *68*, 2642.
[15] J.-F. Hong, Y.-S. Yang, *Acta Opt. Sin.* (in Chinese) **1984**, *4*, 821.
[16] J. Chen, Q. Zhou, J.-F. Hong, W.-S. Wang, N.-B. Ming, D. Feng, *J. Appl. Phys.* **1989**, *66*, 336.

[17] Y.-Y. Zhu, J.-F. Hong, N.-B. Ming, *Ferroelectrics* **1993**, *142*, 31.
[18] N.-B. Ming, *Physical Fundamentals of Crystal Growth*, Shanghai Scientific & Technical Publisher, Shanghai **1982**, pp. 45, 265, 415 (in Chinese).
[19] D. Feng, W.-S. Wang, Q. Zhou, et al., *Chin. Phys. Lett.* **1986**, *3*, 181.
[20] W.-S. Wang, M. Qi, *J. Cryst. Growth* **1986**, *79*, 758.
[21] Y. L. Lu, Y. Q. Lu, X. F. Cheng, C. C. Xue, N.-B. Ming, *Appl. Phys. Lett.* **1996**, *68*, 2781.
[22] H. P. Xu, G. Z. Jiang, L. Mao, Y. Y. Zhu, N.-B. Ming, *J. Appl. Phys.* **1992**, *71*, 2480.
[23] X.-Y. Lei, P. Wen, C.-H. Zhou, N.-B. Ming, *Phys. Rev. E* **1996**, *54*, 5298.
[24] X.-Y. Lei, Q.-H. Wei, P. Wen, C.-H. Zhou, N.-B. Ming, *Phys. Rev. E* **1995**, *52*, 5161.
[25] Q.-H. Wei, C.-H. Zhou, N.-B. Ming, *Phys. Rev. E* **1995**, *52*, 1877; Q.-H. Wei, C.-H. Zhou, N.-B. Ming, *Phys. Rev. E* **1995**, *51*, 1586.
[26] Q.-H. Wei, X.-H. Liu, C.-H. Zhou, N.-B. Ming, *Phys. Rev. E* **1993**, *48*, 2786.
[27] J. Nishizawa, *J. Cryst. Growth* **1991**, *115*, 12.
[28] S. Ando, S. S. Chang, T. Fukui, *J. Cryst. Growth* **1991**, *115*, 69.
[29] J. Feng, N.-B. Ming, *Phys. Rev. A* **1989**, *40*, 7047.
[30] Y.-Y. Zhu, N.-B. Ming, *J. Phys.: Condens. Mater.* **1992**, *4*, 8078.
[31] Y.-Y. Zhu, N.-B. Ming, W.-H. Jiang, *Phys. Rev. B* **1989**, *40*, 8536.
[32] Y.-Y. Zhu, N.-B. Ming, *Phys. Rev. B* **1990**, *42*, 3676.
[33] J. Feng, Y.-Y. Zhu, N.-B. Ming, *Phys. Rev. B* **1990**, *41*, 5578.
[34] N.-B. Ming, *Prog. Nat. Sci.* **1994**, *4*, 554.
[35] N.-B. Ming, in *Proc. of 8th IEEE Int. Symp. on Applications of Ferroelectrics*, Greenville, SC, Aug. 30 – Sept. 2, **1992**, pp. 35 – 38.
[36] N.-B. Ming, in *Proc. of the 1st Pacific Rim Int. Conf. on Advanced Materials and Processing* (Eds: C. Shi, H. Li, A. Scott), The Minerals, Metals and Materials Society, Warrendale, PA **1992**, pp. 29 – 36.
[37] N.-B. Ming, Y. Y. Zhu, D. Feng, *Ferroelectrics* **1990**, *106*, 99.
[38] S. N. Zhu, Y. Y. Zhu, X. Y. Zhang, H. Shu, H. W. Wang, H. F. Hong, C. Z. Ge, N. B. Ming, *J. Appl. Phys.* **1995**, *77*, 5461.
[39] S. N. Zhu, Y. Y. Zhu, H. F. Wang, Z. Y. Zhang, N. B. Ming et al., *J. Phys. D: Appl. Phys.* **1996**, *29*, 76.
[40] N.-B. Ming, J.-F. Hong, Z.-M. Sun et al., *Acta Phys. Sin.* (in Chinese) **1981**, *30*, 1672.
[41] Z.-M. Sun, N.-B. Ming, D. Feng, *J. Inorg. Mater.* (in Chinese) **1986**, *1*, 207.
[42] Y. L. Lu, T. Wei, F. Duewer, Y. Q. Lu, N. B. Ming, P. G. Schultz, X. D. Xiang, *Science* **1997**, *276*, 2004.
[43] Y.-Y. Zhu, J. S. Fu, R.-F. Xiao, G. K. L. Wong, *Appl. Phys. Lett.* **1997**, *70*, 1793.
[44] Y.-Y. Zhu, R.-F. Xiao, J. S. Fu, G. K. L. Wong, N.-B. Ming, *Opt. Lett.* **1997**, *22*, 1382.
[45] Z. L. Wang, J. Wu, Z. J. Yang, Y. Y. Zhu, N.-B. Ming, *Solid State Commun.* **1996**, *98*, 1057.
[46] Z.-L. Wang, J. Wu, Z.-J. Yang, Y.-Y. Zhu, N.-B. Ming, *Chin. Phys. Lett.* **1996**, *13*, 440.
[47] Y. H. Xue, N. B. Ming, J. S. Zhu, D. Feng, *Chin. Phys.* **1984**, *4*, 554.
[48] Y. Q. Lu, Y. L. Lu, C. C. Xue, N. B. Ming et al., *Appl. Phys. Lett.* **1996**, *69*, 3155.
[49] Y. L. Lu, Y. Q. Lu, C. C. Xue, N. B. Ming et al., *Appl. Phys. Lett.* **1996**, *69*, 1160.
[50] S. N. Zhu, Y. Y. Zhu, Z. J. Zhang, N. B. Ming et al., *Appl. Phys. Lett.* **1995**, *67*, 320.
[51] Y. L. Lu, L. Mao, N. B. Ming, *Appl. Phys. Lett.* **1994**, *64*, 3092.

[52] Y. L. Lu, L. Mao, N. B. Ming, *Appl. Phys. Lett.* **1991**, *59*, 516.
[53] S. N. Zhu, Y. Y. Zhu, N. B. Ming et al., *J. Phys. D: Appl. Phys.* **1995**, *28*, 2389.
[54] Y. L. Lu, L. Mao, N. B. Ming, *Opt. Lett.* **1994**, *19*, 1037.
[55] A. Yariv, *IEEE J. Quantum Electron.* **1973**, *QE-9*, 919.
[56] P. St. J. Russell, *Phys. Rev. Lett.* **1986**, *56*, 596.
[57] B. Xu, N.-B. Ming, *Phys. Rev. A* **1994**, *50*, 5197.
[58] Z. L. Wang, J. Wu, X. J. Liu, Y. Y. Zhu, N.-B. Ming, *Phys. Rev. B* **1997**, *56*, 9185.
[59] Z.-L. Wang, Y.-Y. Zhu, Z.-J. Yang, N.-B. Ming, *Phys.Rev. B* **1996**, *53*, 6984.
[60] X.-F. Chen, Y.-L. Lu, Z.-L. Wang, N.-B. Ming, *Appl.Phys.Lett.* **1995**, *67*, 3538.
[61] Z. L. Wang, Y. Y. Zhu, N. Xu, N.-B. Ming, *J. Appl. Phys.* **1996**, *80*, 25.
[62] Z.-L. Wang, Z.-J. Yang, Y.-Y. Zhu, N.-B. Ming, *Opt. Commun.* **1996**, *123*, 649.
[63] Z.-L. Wang, Z. J. Yang, N.-B. Ming, *Chin. Phys. Lett.* **1996**, *13*, 109.
[64] Y.-F. Chen, S.-N. Zhu, Y. Y. Zhu, N. B. Ming, *Appl. Phys. Lett.* **1997**, *70*, 592.
[65] S.-D. Cheng, Y.-Y. Zhu, Y.-L. Lu, N.-B. Ming, *Appl. Phys. Lett.* **1995**, *66*, 291.
[66] Y. Y. Zhu, N. B. Ming, W. H. Jiang, Y. A. Shui, *Appl. Phys. Lett.* **1988**, *53*, 2278.
[67] Y. Y. Zhu, N. B. Ming, W. H. Jiang, Y. A. Shui, *Appl. Phys. Lett.* **1988**, *53*, 1381.
[68] Y.-Y. Zhu, S.-N. Zhu, N.-B. Ming, *J. Appl. Phys.* **1996**, *79*, 2221.
[69] H.-P. Xu, G. Z. Jiang, L. Mao, Y.-Y. Zhu, N.-B. Ming, *J. Appl. Phys.* **1992**, *71*, 2480.
[70] Y.-Y. Zhu, S. N. Zhu, N. B. Ming, *J. Phys. D: Appl. Phys.* **1996**, *29*, 185.
[71] Y. Y. Zhu, Y. F. Chen, S. N. Zhu, Y. Q. Qin, N. B. Ming, *Mater. Lett.* **1996**, *28*, 503.
[72] S.-D. Cheng, Y.-Y. Zhu, N.-B. *Ming*, *Ferroelectrics* **1995**, *173*, 153.
[73] Y.-Y. Zhu, S.-D. Cheng, N.-B. Ming, *Ferroelectrics* **1995**, *173*, 207.
[74] Y.-Y. Zhu, N.-B. Ming, *Ferroelectrics* **1993**, *142*, 231.
[75] Y.-Y. Zhu, N.-B. Ming, W. H. Jiang, *Phys. Rev. B* **1989**, *40*, 8536.
[76] This work was supported by the National Natural Science Foundation of China and the National Climb-Up Project of China for basic science.

第二章 准相位匹配概念的拓展和非线性光学新效应
Chapter 2 Nonlinear Optical Phenomena in Optical Superlattice and Some Concepts Extended

2.1 Harmoic Generations in an Optical Fibonacci Superlattice

2.2 Second-harmonic Generation in a Fibonacci Optical Superlattice and the Dispersive Effect of the Refractive Index

2.3 Quasi-Phase-Matched Third-Harmonic Generation in a Quasi-Periodic Optical Superlattice

2.4 Experimental Realization of Second Harmonic Generation in a Fibonacci Optical Superlattice of $LiTaO_3$

2.5 Crucial Effects of Coupling Coefficients on Quasi-Phase-Matched Harmonic Generation in an Optical Superlattice

2.6 Wave-Front Engineering by Huygens-Fresnel Principle for Nonlinear Optical Interactions in Domain Engineered Structures

2.7 Conical Second Harmonic Generation in a Two-Dimensional $\chi^{(2)}$ Photonic Crystal: A Hexagonally Poled $LiTaO_3$ Crystal

2.8 Experimental Studies of Enhanced Raman Scattering from a Hexagonally Poled $LiTaO_3$ Crystal

2.9 Nonlinear Čerenkov Radiation in Nonlinear Photonic Crystal Waveguides

2.10 Nonlinear Volume Holography for Wave-Front Engineering

2.11 Nonlinear Talbot Effect

2.12 Diffraction Interference Induced Superfocusing in Nonlinear Talbot Effect

2.13 Cavity Phase Matching via an Optical Parametric Oscillator Consisting of a Dielectric Nonlinear Crystal Sheet

第二章 准相位匹配概念的拓展和非线性光学新效应

朱永元

激光频率转换自二十世纪六十年代激光发现以来一直是非线性光学领域的重要研究内容。激光频率转换需要满足两个条件：一是能量守恒，二是动量守恒或称之为相位匹配；包括双折射相位匹配、准相位匹配（QPM）和腔相位匹配（CPM）。由于制备技术的进展，准相位匹配（QPM）是近年来的研究热点。利用光学超晶格，我们在准相位匹配（QPM）频率转换方面作了系统的研究以及对一些概念进行了拓展，并对腔相位匹配进行了实验探索：

1. 从周期光学超晶格到准周期，从单一的准相位匹配到多重准相位匹配：(2.1～2.5)

1962年布洛姆伯根等提出了利用周期结构（我们称之为周期光学超晶格）来实现准位相匹配的概念。一般来说，周期结构只能完成单一的准位相匹配。受准晶发现的启发，我们将周期光学超晶格拓展至准周期超晶格，提出了多重准相位匹配的概念。理论研究了多重准相位匹配情况下的耦合参量过程：倍频过程和和频过程的耦合。发现了在多重准相位匹配条件下各参量波之间能量的动力学演化取决于各耦合系数之间的比值（与超晶格的结构参数有关），因而可以通过超晶格结构参数的设计，控制各参量波之间能量转换的方向与速率。实验研究了准周期超晶格中的二次倍频谱和高效直接三倍频的产生。

2. 从线性惠更斯原理到非线性，从传统的准位相匹配到局域准位相匹配：(2.6)

通常的准相位匹配处理的是单纯的频率转换。我们将线性惠更斯原理拓展至非线性：即基频光在光学超晶格中传播时，将其波前上的每一点既看作基频光的次波源、也看作谐波的波源。利用非线性惠更斯原理设计的光学超晶格可以对谐波的波前进行调控，从而同时完成多个功能：如将倍频、偏转与聚焦集于一身。上述方案得到了实验验证。该方法或可用作光集成，但与传统的方法又有所区别。传统方法是多区域功能集成，如欲将倍频、偏转和聚焦三功能集成在一起，则需将光学超晶格的微结构分成三个不同的区域：第一个区域利用非线性光学效应将基频光转换成倍频光，第二个区域利用电光效应将倍频光进行偏转，第三个区域利用电光效应将倍频光聚焦。易见，该器件在工作时，除了要有基频光入射外，还需在第二和第三区域外加一定大小的直流电场，因此使用上不是很方便。此外，某一微结构区域的破损会导致聚焦功能的丧失。利用非线性惠更斯原理，通过对光学超晶格微结构的特殊设计对倍频波的波前进行调控，超晶格的任何区域都能够同时实现倍频、偏转和聚焦三个功能，从而克服了传统方法的不足。在整个过程中只利用了非线性光学效应，无需利用电光效应。其物理机制可以用局域准相位匹配概念来解释。

3. 从一维光学超晶格到二维，从共线准相位匹配到非共线准相位匹配：(2.7～2.8)

在1D光学超晶格中通常的准相位匹配指的是共线的准相位匹配。在2D光学超晶格中非共线的准相位匹配占了主要地位即基波波矢、倒格矢、倍频波矢不共线但组成封闭矢量多边形。

在二维六角结构的光学超晶格中,我们发现了一种新的二次谐波效应——由光散射导致的锥形二次谐波。当满足非共线准位相匹配条件时,锥形二次谐波可被大大增强。随着基波频率的变化,锥的大小和颜色会发生相应的变化。光的弹性散射来源于密度起伏、结构缺陷等空间不均匀性。

同时还发现非弹性散射可以通过级联准位相匹配获得增强,产生高强度的拉曼谱及高阶的斯托克斯和反斯托克斯峰。研究表明该拉曼效应起源于入射光与介质中的声子极化激元之间的非弹性散射,产生高强度的、间隙为声子极化激元频率的多峰斯托克斯和反斯托克斯谱,高阶峰产生于多声子过程导致的拉曼频移。

4. 从块体光学超晶格到波导,从完全准相位匹配到部分准相位匹配:(2.9)

传统的完全准相位匹配模式(即基波波矢、倒格矢、倍频波矢可以组成封闭矢量多边形)可以等价为纵向与横向两个准相位匹配过程,原则上它们可以分别独立匹配(部分准相位匹配):若仅为纵向准相位匹配则为非线性切伦科夫辐射;若仅为横向准相位匹配,则通常为非线性拉曼-奈斯(Raman-Nath)衍射。在线性光学中,当有界面存在时,界面两侧的波矢量沿界面的切向分量要连续。在波导光学超晶格的非线性切伦科夫辐射中表现为部分准相位匹配:基波导模波矢、倒格矢和倍频波矢沿界面的切向分量组成封闭矢量多边形(基波在非线性波导中传播,倍频在衬底中传播。)

我们在二维六角结构的光学超晶格光波导中实现了光的切伦科夫倍频,还新发现了和频切伦科夫效应。这一发现证实了光学超晶格可以调控介质中非线性极化波的相速(相当于控制带电粒子的运动速度),从而改变非线性切伦科夫辐射在空间的分布和传播方向。特别是当有两种红外光入射时,能同时产生多组、多色的切伦科夫辐射。实验上观察到了犹如圣诞树状的美丽光斑。

5. 超越 QPM,从线性全息到非线性全息:(2.10)

传统的 QPM 通常要求基波和谐波都是平面波。光学超晶格的结构参数由 QPM 条件决定,由此得到的是周期或准周期光学超晶格。如果基波或者谐波不是平面波,譬如基波是平面波,倍频波是球面波,则光学超晶格的具体结构可以用非线性惠更斯原理来设计,其结果可以用局域准位相匹配来理解。而非线性全息可用于更一般情况下的超晶格设计,即基波和谐波可以是任意波形。

全息术,不同于摄影,是一种记录和再现光场的特殊方法。在线性全息中,利用干涉原理,将物光以干涉条纹的形式记录在一定的介质底片上(全息板),从而得到全部的强度和相位信息。全息板通常为卤化银等感光材料,通过曝光,干涉条纹转化为透过率的变化。或在光折变材料中,干涉条纹转化为折射率的变化。将线性全息拓展至非线性全息,其原理是通过非线性极化波和非线性物光的干涉,干涉条纹可以以二阶非线性系数的调制存储在非线性材料——非线性全息板(即光学超晶格)中。非线性全息可以是 Fourier 全息,也可以是 Fresnel 全息。我们提出了非线性波前调制技术——非线性菲涅尔体积全息技术。实验上演示了倍频 Airy 光束的产生,研究了该光束的无衍射、自加速和自愈性能。

6. 从垂直到平行,从线性泰保(Talbot)效应到非线性泰保效应:(2.11~2.12)

用于制备光学超晶格的主要是铁电晶体 LN,LT 和 KTP,其铁电畴的极化方向是沿 c 轴的。在准位相匹配和非线性全息中,基波的传播方向垂直于极化方向;而在非线性泰保效应中,基波则平行于极化方向传播。

Talbot 效应,又叫自成像效应,是指当激光照射到具有周期结构的物体时在物体后面有限距离处出现物体自身的像。我们将微结构功能材料引入 Talbot 效应,在周期光学超晶格中第一次实验观测到二次谐波 Talbot 效应。与线性光学 Talbot 效应相比,非线性 Talbot 效应反映的是微结构晶体中非线性系数的周期性调制,同时对畴结构成像的分辨率有很大提高。对非线性 Talbot 效应的进一步研究发现了一种新的实现突破衍射极限超聚焦的方法。其基本原理是利用干涉衍射效应实现光的空间相位调控,进而在远场得到超聚焦。

7. 从光学超晶格到微腔晶片,从准位相匹配到腔相位匹配:(2.13)

1962 年布洛姆伯根等提出了腔相位匹配(CPM)的原理:采用相干长度量级的光学微腔,通过腔内反射实现高效率的非线性频率变换。该位相匹配方式自提出以来,由于工艺技术的限制,一直缺乏实验验证。2011 年,我们对 CPM 参量过程进行了实验验证,采用 II 类匹配的非线性晶体制成片状微腔振荡器,在单纵模激光泵浦条件下获得了窄线宽、单纵模、高峰值功率、高转化效率的近简并参量光输出。

Chapter 2 Nonlinear Optical Phenomena in Optical Superlattice and Some Concepts Extended

Yongyuan Zhu

Since the invention of Laser in 1960, frequency conversion has been being important in nonlinear optics. It requires fulfilling two conservation laws: energy conservation and momentum conservation or phase matching. The latter includes birefringence phase matching, quasi-phase matching (QPM) and cavity phase matching (CPM). Due to the development of modern technology, QPM becomes a hot topic recently. We have studied the frequency conversion with QPM method both theoretically and experimentally with some concepts extended. And also we have performed the CPM for optical parametric oscillator experimentally.

1. From periodic to quasiperiodic, from single QPM process to multiple QPM processes:(2.1~2.5)

In 1962 Bloembergen et al. proposed the QPM for efficient harmonic generation where a periodic nonlinear dielectric (we call it periodic OSL) is needed. Usually for 1D periodic OSL only one process such as SHG can be quasi-phase matched (QPMed). Inspired by the discovery of quasicrystal, we extended the 1D periodic OSL to 1D quasiperiodic OSL where multiple processes can be QPMed simultaneously. Compared with the periodic structure, a 1D quasiperiodic structure has a low space group symmetry. But its symmetry is higher than that of an aperiodic structure. Its reciprocal vectors are governed by two integers rather than by one integer as in the case of the periodic one. By using this kind of OSL, some coupled optical parametric processes may be realized with efficient conversion. Theoretically we studied the multiple QPM for coupled SHG and SFG processes. It is found that dynamical evolution exists among the fundamental, second harmonic and third harmonic which depends on the structural parameters of the structure. Experimentally we observed the second harmonic spectrum and efficient direct third harmonic generation.

2. From linear Huygens principle to nonlinear Huygens principle, from conventional QPM to local QPM:(2.6)

Conventional QPM deals solely with frequency conversion. We extended the Huygens principle from linear optics to nonlinear optics. That is, each point on the primary wavefront acts as a source of secondary wavelets of the fundamental as well as a source of, for example, the SH wave. With this method multiple optical functions (such as frequency conversion, beam splitting and focusing) can be achieved simultaneously. As an example, here the HFP is used to design the OSL structure, in which SH wave-front can be controlled and focused into several points (we call it optical functional integration). The experimental results agree well with the theoretical ones. The concept of local QPM is put forward to explain the observed phenomenon.

3. From 1D OSL to 2D OSL, From collinear QPM to noncollinear QPM: (2.7~2.8)

In 1D OSL QPM is usually collinear. Whereas in 2D OSL noncollinear QPM becomes dominating. Here noncollinear QPM means that the vectors of the fundamental, second harmonic and the reciprocal form a closed vector polygon.

In a hexagonally poled 2D OSL, a new type of conical second-harmonic generation was discovered. It reveals the presence of another type of nonlinear interaction-an elastic-scattering involved optical parametric generation in a nonlinear medium. Such a nonlinear interaction can be enlarged in OSL by a noncollinear QPM process. As the wavelength of the fundamental changes, the size and the color of the conical change accordingly. The results disclose the structure information and symmetry of the OSL.

Further studies on the hexagonally poled 2D OSL show that phonon-polariton Raman scattering (including the anti-Stokes and Stokes Raman signal) can be significantly enhanced by cascading a couple of QPM processes.

4. From bulk OSL to waveguide OSL, from complete QPM to partial QPM: (2.9)

In linear optics, when there is an interface between medium 1 and 2, the tangential components of the wave vectors (one in medium 1 and the other in medium 2) should be equal. In the case of nonlinear Čerenkov Radiation in a nonlinear optical waveguide, the conventional QPM (that is, the vectors of the fundamental, second harmonic and the reciprocal are not collinear but form a closed vector polygon) should be extended to partial QPM. The wave vectors of fundamental guided modes and the tangential component of the wave vector of the second harmonic together with the reciprocal vector form a closed polygon. Here the fundamental propagates inside the nonlinear waveguide whereas the second harmonic propagates in the substrate.

We studied nonlinear Čerenkov radiation generated from a hexagonally poled nonlinear OSL waveguide. Nonlinear polarization driven by an incident light field may emit coherently harmonic waves at new frequencies along the direction of Čerenkov angles determined by the partial QPM. Multiple radiation spots with different azimuth angles are simultaneously exhibited from such a hexagonally poled OSL waveguide. As two fundamental beams were collinearly coupled into the waveguide, SHG and SFG occur simultaneously. A beautifully colorful pattern like a "Christmas tree" was observed. Čerenkov radiation associated with partial QPM leads to these novel nonlinear phenomena.

5. Beyond QPM, From linear hologram to nonlinear hologram: (2.10)

QPM requires that the fundamental and the generated harmonic waves be plane waves.

With QPM conditions the structure of OSL can be determined which is either periodic or quasiperiodic. Nonlinear Huygens principle can be used to design complicated OSL structures to perform multiple functions simultaneously. However, in some cases, such as nonlinear generation of special beams, the nonlinear Huygens principle is found difficult to work with practically. In such cases nonlinear holography should be used for OSL structure designing.

Holography, being different from photography, is a technique that can store and reconstruct both the amplitude and phase of an object wave. In linear optics, the holography is based on the interference fringes of the reference beam and the object beam recorded on a photographic film through corrugated surface gratings. Or in a photorefractive crystal the fringes can be imprinted as refractive index variation. The concept of holography can be extended from linear optics to nonlinear optics. In nonlinear optics, this can be done by transferring the fringes into the modulation of the nonlinear coefficient in a ferroelectric crystal (That is just the OSL). Here the reference beam is the nonlinear polarization wave driven by the fundamental and the object beam is the required second harmonic. Thus the nonlinear hologram can be realized.

Two kinds of nonlinear holography have been studied: Fourier holography and Fresnel holography. With Fresnel holography the OSL structure for nonlinear generation of special beams can be designed. The generated nonlinear wave can be considered as a holographic image caused by the incident fundamental wave. As an example, we experimentally realized a second-harmonic Airy beam, and the results are found to agree well with numerical simulations.

6. From perpendicular to parallel, from linear Talbot effect to nonlinear Talbot effect: (2.11 – 2.12)

The crystals used for OSL are ferroelectric such as LN, LT and KTP with their domain polarization along the z axis. For QPM and nonlinear hologram, the fundamental wave propagates perpendicular to the domain polarization; whereas for nonlinear Talbot effect, parallel to.

In linear optics, the Talbot effect is a near-field diffraction phenomenon in which self-imaging of a grating or other periodic structure replicates at certain imaging planes. We demonstrated the nonlinear Talbot effect, i.e., the formation of second-harmonic self-imaging instead of the fundamental one. We observed second-harmonic Talbot self-imaging from 1D and 2D periodic OSLs. Both integer and fractional nonlinear Talbot effects were investigated. Further study on nonlinear Talbot effect shows that superfocusing can be realized due to diffraction interference.

7. From OSL to nonlinear crystal sheet, From QPM to CPM (2.13)

In 1962 Armstrong et al. proposed the idea of CPM in nonlinear optics. The principle of this so-called CPM can be interpreted in comparison with the well-known QPM method. To compensate the phase mismatch QPM requires engineering an OSL structure; while in the case of CPM, the resonance recirculation in cavity can ensure that the traveling light and reflected light are exactly in phase in every circling. This is equivalent to extend the effective nonlinear interaction path, and therefore, increase the frequency conversion efficiency. However this idea has not been experimentally verified since then. We experimentally demonstrated CPM using a sheet optical parametric oscillator which is made of a nonlinear crystal sheet. Although the thickness of the sheet is less than one coherence length, such a sheet oscillator exhibits high slope efficiency and peak power output, as well as near-transform-limited spectral and near-diffraction-limited spatial features.

Harmonic Generations in an Optical Fibonacci Superlattice[*]

Jing Feng and Yong-yuan Zhu

Laboratory of Solid State Microstructures, Nanjing University, Nanjing 210008, Jiangsu, People's Republic of China

Nai-ben Ming

Laboratory of Solid State Microstructures, Nanjing University, Nanjing 210008, Jiangsu, People's Republic of China and Center of Condensed Matter Physics and Radiation Physics, China Center of Advanced Science and Technology(World Laboratory), P.O. Box 8730, Beijing 100 080, People's Republic of China

An optical Fibonacci superlattice has been proposed to produce the second-harmonic generation and the third-harmonic generation, which is the sum frequency of the second-harmonic and the fundamental frequency in the same material. Because of the quasiperiodicity of the optical Fibonacci superlattice, the phase mismatches of the optical parametric processes caused by the frequency dispersion of the refractive index can be compensated with the reciprocal vectors which the optical Fibonacci superlattice provides. A theory which analyzes the second-harmonic generation and the third-harmonic generation processes in the material and the calculations applied to the optical Fibonacci superlattice made from a single $LiNbO_3$ crystal is presented in detail. The calculations show that the efficiencies of the second-harmonic generation and the third-harmonic generation are comparable to, or even larger than, those obtained with commonly used phase-matching methods.

1. Introduction

In conventional methods, only when the phase-matching condition is satisfied does an optical parametric interaction proceed efficiently. Normally, phase matching may be realized in nonlinear optical crystals with birefringence. This method can only be applied to some of the nonlinear processes in uniaxial or biaxial crystals. Another method, quasi-phase-matching,[1,2] can be applied to both the nonbirefringent crystals and some birefringent crystals with optical coefficients that are phase unmatchable.[3-7] The key to quasi-phase-matching is to construct a one-dimensional periodic structure with the phase sign of the nonlinear polarization shifted from one plate to the consecutive plate by π radians. This one-dimensional periodic structure can provide a series of reciprocal vectors, each of which is an integer times a primitive vector. It is the reciprocal vectors which make the optical parametric processes in the material phase matched.

---

[*] Phys.Rev.B, 1990, 41(9):5578

Compared with the periodic structure, a one-dimensional quasiperiodic structure has a low spacegroup symmetry. But its symmetry is higher than that of an aperiodic structure.[8—10] Its reciprocal vectors are governed by two integers rather than by one integer as in the case of the periodic one. By using this kind of material, some coupled optical parametric processes may be realized with efficient conversion.

Applying our theory to, as an example, a single $LiNbO_3$ crystal with quasiperiodic laminar ferroelectric-domain structures or simply termed an optical Fibonacci superlattice (OFS), where the nonlinear coefficient d_{33} is to be used, we find the enhancement of the second-harmonic generation (SHG) in the OFS is larger than that of an aperiodic structure, but less than that of the periodic one. We also find that the third-harmonic generation (THG), which is coupled with the SHG, can be obtained because the reciprocal vectors of such a material have more chosen values than those of the periodic one and its intensity will be large enough to be used practically and efficiently if the parameters of the building blocks are properly designed.

2. Theoretical Analysis

Hereafter we shall take $LiNbO_3$ crystals with laminar ferroelectric-domain structures as an example. In such a material, the directions of polarization vectors in successive domains are opposite, as are the signs of nonlinear optical coefficients. This structure forms a one-dimensional superlattice for the nonlinear optical effect. On this basis, an optical Fibonacci superlattice can be constructed. It consists of two fundamental blocks of A and B arranged according to the production rule $S_j = S_{j-1} | S_{j-2}$, for $j \geqslant 3$ with $S_1 = A$ and $S_2 = AB$, where | stands for concatenation. Both blocks are composed of one positive and one negative ferroelectric domain as shown in Fig. 1(a), where l_A^+ and l_B^+ represent the thicknesses of the positive domains in blocks A and B, and l_A^- and l_B^- represent the thicknesses of the negative ones. Let

$$l_A^+ = l_B^+ = l,$$
$$l_A^- = l(1+\delta), \quad (1)$$
$$l_B^- = l(1-t\delta),$$

l, δ, and t, are adjustable structure parameters. The sequence of the blocks, $ABAABABA\ldots$, produces an OFS, see Fig.1(b).

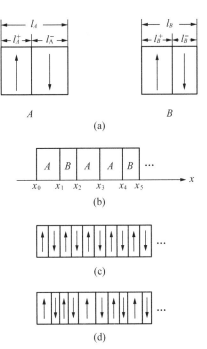

FIG. 1. Optical superlattice of $LiNbO_3$ crystals (the arrows indicate the directions of the spontaneous polarization). (a) The two blocks of an OFS, each composed of one positive and one negative ferroelectric domain. (b) Schematic diagram of an OFS. (c) Schematic diagram of a POS. (d) Schematic diagram of an APOS.

With the OFS, a series of reciprocal vectors can be provided to compensate the phase mismatches of the optical parametric processes in the material. Unlike the periodic optical superlattice (POS) [Fig. 1 (c)] which has reciprocal vectors derived from an integer times a primitive vector, an OFS with an infinite number of blocks provides reciprocal vectors governed by two integers, which has the form:

$$k_{m,n} = 2\pi(m + n\tau)/D, \quad (2)$$

where $D = \tau l_A + l_B$, with the golden ratio $\tau \equiv (1+\sqrt{5})/2$; l_A, l_B are block thicknesses as shown in Fig. 1(a). The reciprocal vectors of the OFS can be adjusted by l, δ, and t. Because of this property, some coupled optical parametric processes will be likely to occur in the OFS with efficient conversion.

Consider a case in which a single laser beam with $\omega_1 = \omega$ is incident from the left onto the surface of an OFS and through the nonlinear optical effect, the SHG and the THG exist simultaneously in the OFS. In order to make use of the largest nonlinear coefficient d_{33}, which cannot be used in an ordinary phase-matching regime, let the interfaces of each domain be parallel to the y-z plane, the optical propagation direction be along the x axis, and the directions of electric fields be along the z axis (see Fig. 2). Here, three optical fields must be taken into account, one for $\omega_1 = \omega$, one for $\omega_2 = 2\omega$, and one for $\omega_3 = 3\omega$. The three optical fields, described in terms of their electric field components, are given by

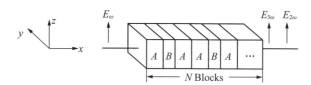

FIG. 2. The polarization orientation of electric fields with respect to the superlattice.

$$E_i(x,t) = E_i(x)\exp[i(\omega_i t - k_i x)], i = 1,2,3 \quad (3)$$

which satisfy the wave equation:

$$\nabla^2 E = \frac{1}{c^2}\frac{\partial^2}{\partial t^2}(\varepsilon E + 4\pi p_{NL}) \quad (4)$$

The presence of these electric fields can give rise to nonlinear polarizations at frequencies ω_2 and ω_3, etc., which are

$$P_{2\omega}(x,t) = 2d(x)E_1^2(x)\exp[i(2\omega_1 t - 2k_1 x)],$$
$$P_{3\omega}(x,t) = 4d(x)E_1(x)E_2(x) \times \exp\{i[(\omega_1 + \omega_2)t - (k_1 + k_2)x]\}, \quad (5)$$

where

$$d(x) = \begin{cases} d_{33} & \text{if } x \text{ is in the positive domains,} \\ -d_{33} & \text{if } x \text{ is in the negative domains.} \end{cases}$$

Before going into a detailed analysis, we must make some assumptions. We assume that the variation of the field amplitudes with x is small enough so that $k_i dE_i/dx \gg d^2 E_i/dx^2$ and that the amount of power lost from the input beam (ω_1) is negligible, i.e., $dE_1(x)/dx = 0$. We also assume that $E_1 \gg E_2, E_3$; this is the so-called small-signal approximation.

Under these conditions, using Eqs. (3) – (5) and carrying out the indicated differentiation, we can get[11]

$$dE_1(x)/dx = 0, \tag{6a}$$

$$dE_2(x)/dx = -i\frac{8\pi\omega_2^2}{k^{2\omega}c^2}d(x)E_1^2(x) \times \exp[i(k^{(2\omega)} - 2k^{(\omega)})x], \tag{6b}$$

$$dE_3(x)/dx = -i\frac{16\pi\omega_3^2}{k^{3\omega}c^2}d(x)E_1(x)E_2(x) \times \exp[i(k^{(3\omega)} - k^{(2\omega)} - k^{(\omega)})x]. \tag{6c}$$

In Eqs. (6), only the largest terms have been kept.

By integrating Eqs. (6), the electric fields after passing through the OFS can be represented as

$$E_1(x_N) = E_1, \tag{7a}$$

$$E_2(x_N) = \sum_{i=1}^{N} E_2^i \exp(i\Delta k x_{i-1}), \tag{7b}$$

$$E_3(x_N) = -i\frac{144\pi\omega^2}{k^{3\omega}c^2}E_1\int_0^{x_N} d(x)E_2(x') \\ \times \exp(i\Delta k'x')dx', \tag{7c}$$

where

$$\Delta k = k^{(2\omega)} - 2k^{(\omega)},$$
$$\Delta k' = k^{(3\omega)} - k^{(2\omega)} - k^{(\omega)},$$
$$0.25K_1 = (32\pi\omega^2 d_{33}/k^{(2\omega)}c^2\Delta k)E_1^2,$$
$$E_2^i = 0.25K_1[1 - 2\exp(i\Delta kl) + \exp(i\Delta kL)].$$

When $L = l_A = l_A^+ + l_A^-$, $E_2^i = E_2^A$, which suits block A, and when $L = l_B = l_B^+ + l_B^-$, $E_2^i = E_2^B$, which suits block B. In deriving Eq. (7), the boundary conditions have been used, which are $E_1(0) = E_1$, $E_2(0) = 0$, $E_3(0) = 0$.

These equations constitute the basis of our numerical calculations and discussions of this paper. We will discuss them in detail in the following section.

3. Numerical Calculations and Discussions

We have performed numerical computations for both SHG and THG with the pump beam at a wavelength 1.318 μm of a neodymium-doped yttrium aluminum garnet (Nd: YAG) laser. For LiNbO$_3$ crystals, under room temperature, the refractive indices, according to Hobden and Warner's equation,[12] are

$n_o = 2.2215$, $n_e = 2.1436$ at $\lambda = 1.318$ μm,
$n_o = 2.2839$, $n_e = 2.1953$ at $\lambda = 0.659$ μm,
$n_o = 2.3913$, $n_e = 2.2882$ at $\lambda = 0.439$ μm.

3.1 Second-Harmonic Generation

SHG is the result of two intense pump beams mixing. Under the condition of the small-signal approximation, the second-harmonic intensity depends completely on the structures of the superlattice.

Figure 3 shows the relationship between the second-harmonic intensity and the block nubmer with $l=l_c^{(2\omega)}=\pi/\Delta k$, $t=\tau$, and δ taking various values. Note that when $\delta=0$, the enhancement of the second-harmonic intensity is proportional to the square of the block number as curve a of Fig. 3 indicates. It is just the result of a periodic one. For when $\delta=0$, the OFS turns back to a periodic optical superlattice (POS) [see Eq. (1)]. Curves b and c represent the enhancement of the second-hamonic intensity with $\delta=0.15$ and 0.30. The curve of $\delta=0.15$ grows more slowly than the square dependence curve a, but more rapidly than the curve of $\delta=0.30$.

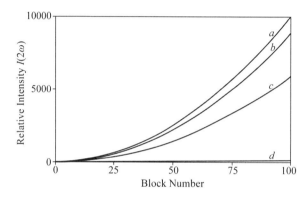

FIG. 3. The dependence of the second-harmonic intensity on the block number with $l=6.37$ μm in different cases, a, $\delta=0$, i.e., in a POS; b, $\delta=0.15$, i.e., in an OFS; c, $\delta=0.30$, i.e., in an OFS; d, in an APOS.

If the thicknesses of the domains do not have regularities like those of OFS and POS, but are radomly distributed around the coherent length as illustrated in Fig. 1(d), i.e., an aperiodic optical superlattice (APOS), the second-harmonic intensity will be linearly dependent on the number of blocks[7] (see Fig. 3, curve d).

Compared with those of POS and APOS, the enhancement of the second-harmonic intensity of the OFS with an arbitrary value of δ can be represented as

$$I_{2\omega}(N) \propto N^{\alpha}, \tag{8}$$

with $1<\alpha<2$.

We know that the symmetry of the POS is the highest of the three, and the next highest is the OFS. The symmetry of the APOS is the lowest. From our discussion, the enhancement of the second-harmonic intensity is clearly related to the symmetry of the superlattice in which the parametric process takes place.

3.2 Third-Harmonic Generation

The process of THG discussed here is a coupled parametric process; that is, two parametric processes, the SHG process and the frequency up-conversion process (FUP), which mixes the fundamental frequency with the second harmonic, are coupled in this material.

Taking $l = l_c^{(2\omega)} = \pi/\Delta k$ and $l = 5.98$ μm, we have calculated the dependence of the third-harmonic intensity on the block number, which is shown in Fig. 4. The two curves differ from each other in nature completely; one fluctuates drastically while the other increases steadily with the block number. The explanation is as follows.

As discussed above, when $l = l_c^{(2\omega)} = 6.37$ μm and $\delta = 0$, the second harmonic is quasi-phase-matched. However, when δ is small and $l = l_c^{(2\omega)}$, the second-harmonic intensity still increases with the block number and reaches a degree to be used practically, as Fig. 3 shows. But then the third harmonic is not quasi-phase-matched. In some parts of the superlattice, the third harmonic is constructive, and in other parts of the superlattice it is destructive. So its intensity fluctuates drastically as the block number varies. Curve a of Fig. 4 shows this feature clearly. But from Eq. (7) we can see that the THG depends not only on the structure parameters but also on the second-harmonic intensity. So as the block number increases, the third-harmonic intensity undulates more severely while the second-harmonic intensity increases steadily.

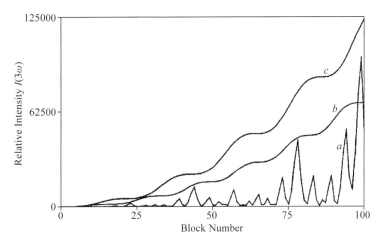

FIG. 4. The dependence of the third-harmonic intensity on the block number under different conditions. a, $l = 6.37$ μm, $\delta = -0.10$, $t = 1.62$; b, $l = 5.98$ μm, $\delta = -0.02$, $t = 1.78$; c, $l = 6.08$ μm, $\delta = 0.01$, $t = 1.90$.

When $l = 5.98$ μm, we find that
$$\Delta k' l = 3\pi. \tag{9}$$

Therefore, $l = 3l_c^{(3\omega)}$; here $l_c^{(3\omega)}$ represents the coherent length for the THG in a single FUP. The reason why l should assume a value three times $l_c^{(3\omega)}$ is obvious. Because, if $l = l_c^{(3\omega)}$ the THG in a single FUP is quasi-phase-matched, but the SHG is severely phase mismatched. The result is that it is impossible to get efficient THG because of its relation to the SHG. We know the effect of $l = 3l_c^{(3\omega)}$ is the same as the effect of $l = l_c^{(3\omega)}$ for THG. And when $l = 3l_c^{(3\omega)}$, the mismatch of the SHG becomes smaller. Thus the third-harmonic intensity increases with the block number (Fig. 4 curve b).

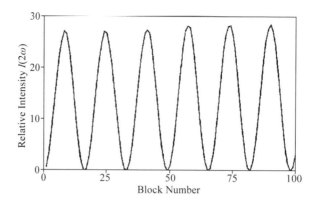

FIG. 5. The dependence of the second-harmonic intensity on the block number with $l=5.98$ μm, $\delta=-0.02$, and $t=1.78$.

Curve b of Fig. 4 has a steplike shape. In this case, the second harmonic is phase mismatched. Figure 5 reveals this feature. The second-harmonic intensity fluctuates almost sinusoidally. There is a one-to-one correspondence between Fig. 5 and curve b of Fig. 4. Whenever the second-harmonic intensity decreases, a platform appears on the third-harmonic intensity. This strongly indicates the dependence of THG on SHG.

By adjusting the parameters properly, the optimum condition has been found, which is $l=6.08$ μm, $\delta=0.01$, and $t=1.90$ with the third-harmonic intensity $I_{3\omega}\approx 122000\, K_2^2$; here $K_2=576\pi\omega^2 d_{33} K_1 E_1/(k^{(3\omega)} c^2 \Delta k')$ (Fig. 4 curve c).

To obtain an appreciation for the enhancement of the third harmonic available in our case, consider a commonly used two-step process.[13] The second harmonic is generated in the first LiNbO$_3$ crystal of $100 l_c^{(2\omega)}$ length using the nonlinear coefficient d_{31} with phase matching; then it mixes with the fundamental frequency in the second LiNbO$_3$ crystal of $100 l_c^{(3\omega)}$ length using the same nonlinear coefficient; here $l_c^{(3\omega)}=\pi/\Delta k'$, and the relative output intensity of THG is $I_{3\omega}\approx 11\,000 K_2^2$. Compared with it, the enhancement of THG in our case is increased by an order of magnitude, which is favorable to the practical applications.

4. Conclusion

We have presented a detailed theoretical analysis of SHG and THG in the OFS, and particularly considered an OFS made from a single LiNbO$_3$ crystal with quasi-periodic laminar ferroelectric-domain structures as an example.

The analysis used here can be carried over to the OFS of other materials. The deal with OFS's which consist of different materials with different refractive indices along the optical propagation direction, the reflection by the interfaces must be considered.

The OFS discussed here is a new solution to the phase mismatch of the optical parametric processes. With this material, not only might SHG and THG have applicable enhancement, but also other parametric processes might proceed with large enhancement.

References and Notes

[1] J. A. Armstrong, N. Bloembergen, J. Ducuing, and P. S. Pershan, Phys. Rev. **127**, 1918 (1962).

[2] N. Bloembergen and A. J. Sievers, Appl. Phys. Lett. **17**, 483 (1970).

[3] D. Feng, N. B. Ming, J. F. Hong, Y. S. Yang, J. S. Zhu, Z. Yang, and Y. N. Wang, Appl. Phys. Lett. **37**, 607 (1980).

[4] Y. H. Xue, N. B. Ming, J. S. Zhu, and D. Feng, Wuli Xuebao (Acta Phys. Sin.) **32**, 1515 (1983) [Chin. Phys. **4**, 554 (1984)].

[5] D. E. Thompson, J. D. McMullen, and D. B. Anderson, Appl. Phys. Lett. **29**, 113 (1973).

[6] M. Okada, K. Takizawa, and S. Ieiri, NKH (Nippon Hoso Kyokai) Tech. J. **29**, 24 (1977).

[7] C. F. Dewey and L. O. Hocker, Appl. Phys. Lett. **26**, 442 (1975).

[8] D. Levine and P. J. Steinhardt, Phys. Rev. Lett. **53**, 2477 (1984).

[9] R. K. Zia and W. J. Dallas, J. Phys. A **18**, L341 (1985).

[10] Veit. Elser, Phys. Rev. B **32**, 4892 (1985).

[11] F. Zernike and J. M. Midwinter, *Applied Nonlinear Optics* (Wiley, New York, 1973).

[12] M. V. Hobden and J. Warner, Phys. Lett. **22**, 243 (1966).

[13] R. Piston, Laser Focus **14**, 66 (1978).

[14] This work was supported by the Chinese National Natural Science Foundation.

Second-harmonic Generation in a Fibonacci Optical Superlattice and the Dispersive Effect of the Refractive Index*

Yong-yuan Zhu

Laboratory of Solid State Microstructures, Nanjing University, Nanjing 210008, People's Republic of China

Nai-ben Ming

Laboratory of Solid State Microstructures, Nanjing University, Nanjing 210008, People's Republic of China and Center for Condensed Matter Physics and Radiation Physics, Chinese Center of Advanced Science and Technology (World Laboratory), P.O. Box 8730, Beijing 100080, People's Republic of China

A Fibonacci optical superlattice is analyzed which is made from a single crystal with quasiperiodic laminar ferroelectric domain structures. The second-harmonic generation in this system is studied. Because of the dispersive effect of the refractive index, the second-harmonic spectrum does not reflect the symmetry of the quasiperiodic structure and thus does not exhibit self-similarity. The existence of the extinction phenomenon constitutes one major difference between our system and heterostructure systems. A general extinction rule is also obtained.

One of the most striking events in condensed-matter physics in recent years has been the discovery and development of quasiperiodic crystals which show many unused physical properties. Many subsequent researchers have focused on its linear phenomena[1—3] and its third-order nonlinearity.[4,5] In these works the physical parameters such as dielectric coefficients and elastic coefficients were taken to be nondispersive. Little has been done on the second-order nonlinear-optical phenomena because of lack of proper materials.

The Fibonacei optical superlattice (FOS) made from a single $LiNbO_3$ crystal with quasiperiodic laminar ferroelectric domain structures provides a useful tool for the study of second-order nonlinear-optical phenomena. Previously, we investigated second-harmonic generation in a periodic optical superlattice,[6,7] which is made from a single $LiNbO_3$ crystal with periodic laminar ferroelectric domain structures, and verified the theory of quasiphase-matching proposed by Bloembergen et al.[8,9] In this paper we report our theoretical results of the second-harmonic generation in a FOS. Considering dispersive effects of the refractive index, we find that the spectrum of the second-harmonic intensity in a FOS doe not possess self-similarity. We also find that if the structure parameter of the FOS is properly selected, a phenomenon somewhat similar to the extinction phenomenon in solid-state physics will occur.

Traditionally, the Fibonacci superlattice is constructed as follows. First, define two

---

* Phys. Rev. B, 1990,42(6):3676

building blocks A and B, each composed of two layers of different constituent materials. Then arrange them according to the concatenation rule $S_j = S_{j-1} | S_{j-2}$ for $j \geq 3$, with $S_1 = A$ and $S_2 = AB$. The Fibonacci superlattice thus formed is a heterostructure. In our case each block consists of one positive ferroelectric domain and one negative ferroelectric domain. The two domains are interrelated by a dyad axis in the x direction. Since the nonlinear-optical coefficients form a third-rank tensor, it is easy to prove that they will change their signs from positive domains to negative domains.[10] The superlattice thus formed is not a heterostructure, but still a single crystal. This kind of structure can be fabricated by a special crystal-growth technique developed by us.[11-13] Here we term this structure a Fibonacci optical superlattice, or FOS, which is shown in Fig. 1.

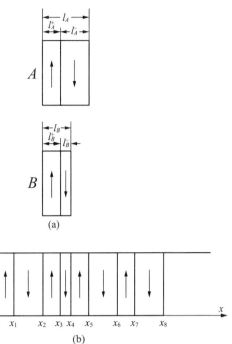

FIG. 1. FOS made of a single LiNbO$_3$ crystal (the arrows indicate the directions of the spontaneous polarization). (a) The two building blocks of a FOS, each composed of one positive and one negative, ferroelectric domain. (b) Schematic diagram of a FOS.

In Fig. 1 we can see that $l_A = l_A^+ + l_A^-$ and $l_B = l_B^+ + l_B^-$. In our treatment we have set $l_A^+ = l_B^+ = l$, $l_A^- = l(1+\eta)$, and $l_B^- = l(1-\tau\eta)$ with the golden ratio $\tau \equiv (1+\sqrt{5})/2$. Here, l and η are two adjustable parameters. Through the variation of the value of η, the structure can be either a periodic superlattice or a quasiperiodic one. For example, if $\eta = 0$, then the structure is a periodic one; otherwise it is a quasiperiodic one. In our system, l plays an important role, and so we call it a structure parameter.

As will be seen below, the phase-matching regime cannot be used to study the effect of the FOS on the nonlinear-optical phenomena, and so the quasi-phase-matching regime[8,9] should be used. In order to use the largest nonlinear-optical coefficient d_{33} of LiNbO$_3$ crystals, we assume that the domain boundaries are parallel to the y-z plane [Fig. (1)] and that the polarizations of the electric fields are along the z axis, with their propagating directions along the x axis.

In what follows we will restrict ourselves to the case of second-harmonic generation (SHG) with a single laser beam incident onto the surface of the FOS. According to Refs. 10 and 14, these two electric fields, E_1 and E_2, satisfy the wave equation under the small-signal approximation:

$$\frac{dE_2(x)}{dx} = -i\frac{32\pi\omega^2}{k^{2\omega}c^2}d(x)E_1^2\exp[i(k^{2\omega}-2k^{\omega})x], \tag{1}$$

with $d(x)=d_{33}$ in positive domains and $d(x)=-d_{33}$ in negative domains, where ω and k^{ω} are the angular frequency and wave number of the fundamental beam, respectively, $k^{2\omega}$ is the wave number of the second harmonic, and c is the speed of light in vacuum.

By integrating Eq. (1), the second-harmonic electric field after passing through N blocks of the FOS can be represented as

$$E_2(N) = -\frac{64\pi\omega^2}{k^{2\omega}c^2\Delta k}d_{33}E_1^2\left\{\sum_{j=0}e^{i\Delta k x_{2j+1}} + \exp(i\pi)\sum_{j=0}e^{i\Delta k x_{2j}}\right\}, \tag{2}$$

where $\Delta k=k^{2\omega}-2k^{\omega}$. $\{x_n\}$, $n=0,1,2,\ldots$, are the positions of the ferroelectric domain boundaries [Fig. (1)].

In Eq. (2) the terms inside the curly braces comprise the structure factor, which is divided into two parts, with one part lagging behind the other by a phase $\exp[i(\Delta kl+\pi)]$.

For an infinite array with $l_A/l_B=\tau$, i.e., $\eta\approx 0.34$, by use of the direct[15] or the projection method,[16] Eq. (2) can be written in the form

$$E_2(\Delta k) \propto -\frac{128\pi\omega^2}{k^{2\omega}c^2\Delta k}d_{33}E_1^2\exp\left(i\frac{1}{2}\Delta kl\right)\sin\left(\frac{1}{2}\Delta kl\right)\sum\frac{\sin X_{m,n}}{X_{m,n}}$$
$$\cdot \exp(iX_{m,n})\delta(\Delta k - 2\pi(m+n\tau)/D). \tag{3}$$

Here, $X_{m,n}=\pi\tau(m\tau-n)/(1+\tau^2)$, $D=\tau l_A+l_B$.

The appearance of Δk is due to the energy coupling between the fundamental beam and the second harmonic through the nonlinear-optical effect. Obviously the FOS cannot be used in the study of phase-matched SHG ($\Delta k=0$). It can be only used in the study of quasiphase-matchable SHG.

The peaks of the second-harmonic intensity can be obtained from the δ function in Eq. (3), which is

$$\Delta k_{m,n} = 2\pi(m+n\tau)/D,$$

or
$$(\Delta kD)_{m,n} = 2\pi(m+n\tau). \tag{4}$$

The factor $(\sin X_{m,n})/X_{m,n}$ in Eq. (3) is important. It determines which peaks are stronger. Here we know that the smaller the $X_{m,n}$, the larger the value of $(\sin X_{m,n})/X_{m,n}$. This means that n/m must be close to τ. It is well known that the best rational approximants to τ occur when n and m are successive Fibonacci numbers, F_k.[17] Therefore, the sequence of the most intense peaks corresponds to $(m,n)=(F_{k-1},F_k)$, where $(F_0,F_1)=(0,1)$. Note that the second-harmonic intensity peaks are indexed by two integers, even though the structure is one dimensional. The appearance of more indices than the dimensionality is typical of incommensurate crystals and quasicrystals.

We will discuss some interesting phenomena in both real space and reciprocal space. All calculational results are valid only under room temperature and, without loss of generality, only the results with $N=100$ have been presented.

In real space we study the dependence of the second-harmonic intensity on the structure parameter l of the FOS. In this case the wavelength λ_0 is kept unchanged, as are the refractive indices n_{10} and n_{20}. Here, n_{10} and n_{20} are the refractive indices for the fundamental beam and the second harmonic, respectively. Thus the dispersion of the refractive indices has no effect on the second-harmonic spectrum. Equation (4) can be rewritten as

$$l_{m,n} = \frac{(m+n\tau)\lambda_0}{4(n_{20}-n_{10})(1+\tau)} \tag{5}$$

Here, λ_0 represents the wavelength of the fundamental frequency in vacuum.

For those intense peaks, Eq. (5) becomes

$$l(s,p) = \frac{\lambda_0}{4(n_{20}-n_{10})(1+\tau)} s\tau^p. \tag{6}$$

Here, s and p are integers.

Obviously, here the relation $l(s, p+1) = l(s,p) + l(s, p-1)$ holds, and thus the spectrum of the second harmonic exhibits self-similarity. Figure 2 shows the relation between the second-harmonic intensity and the structure parameter l with the pump beam at wavelength $\lambda_0 = 1.318$ μm and $n_{10} = 2.1453$ and $n_{20} = 2.1970$ for LiNbO$_3$ crystals under the condition $l_A/l_B = \tau$. The result conforms to the discussion above. The intense peaks take the form of $\lambda_0 s\tau^p/[4(n_{20}-n_{10})(1+\tau)]$. Under general conditions, i.e., $l_A \neq \tau l_B$, calculations have shown that the peak positions of the second-harmonic intensity remain unchanged, except for their strengths. In our case the change of the value of η does not affect the value of D. This can be seen easily from the relation $D = \tau l_A + l_B = 2l(1+\tau)$. This indicates that D is a characteristic parameter of the FOS. The result is consistent with that of Merlin et al.[4] They found that for all $l_A \neq l_B$ the Fourier spectrum of the structure factor of a Fibonacci superlattice consists of δ-function peaks at $k = 2\pi(m+n\tau)/D$, with $D = \tau l_A + l_B$. We may deduce from these results that the FOS possesses certain space

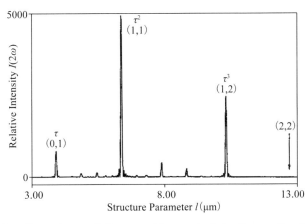

FIG. 2. Dependence of the second-harmonic intensity on the structure parameter l in real space. Note that $l(s, p+1) = l(s,p) + l(s, p-1)$.

symmetry. The symmetry is determined by the order of arrangement of the blocks, not by the block thicknesses.

In reciprocal space we study the dependence of the second-harmonic intensity on the wavelength λ. In this case, the structure parameter l is kept constant. Equation (4) can be rewritten as

$$(1/\lambda)_{m,n} = \frac{m + n\tau}{4[n_2(\lambda) - n_1(\lambda)](1+\tau)l}. \tag{7}$$

Here, $n_1(\lambda)$ and $n_2(\lambda)$ are functions of λ.[18] Equation (7) indicates that here the dispersive effect of the refractive index on the second-harmonic spectrum must be taken into account; moreover, for these intense peaks Eq. (7) becomes

$$(1/\lambda)_{s,p} = \frac{s\tau^p}{4[n_2(\lambda) - n_1(\lambda)](1+\tau)l}. \tag{8}$$

The relation $(1/\lambda)_{s,p+1} = (1/\lambda)_{s,p} + (1/\lambda)_{s,p-1}$ no longer holds because of the dispersion of the refractive index, whereas in linear phenomena and the third-order nonlinear-optical phenomena this relation is valid.

Figure 3 shows the relation between the second-harmonic intensity and the wavelength with $l = l_c = \pi/\Delta k_0$ and $l_A = \tau l_B$. Here, $\Delta k_0 = 4\pi(n_{20} - n_{10})/\lambda_0$. As expected, the intense peaks occur at $(m,n) = (F_{k-1}, F_k)$, but their positions have shifted markedly. For example, in Fig. 3 we can see three intense peaks occurring at $\lambda_{s,p}$, as indicated by τ^p. They are $\lambda_{1,2} = 1.318$ μm, $\lambda_{1,3} = 1.115$ μm, and $\lambda_{1,4} = 0.960$ μm. Obviously, $(1/\lambda)_{1,4} \neq (1/\lambda)_{1,3} + (1/\lambda)_{1,2}$, and thus the spectrum of the second-harmonic intensity in reciprocal space does not exhibit self-similarity.

Here, another interesting phenomenon should be noted, one somewhat similar to the extinction phenomenon in solid-state physics. In both Figs. 2 and 3 the mode (2,2) does not appear. As discussed above, after the fundamental light passing through the entire superlattice, the resultant second-harmonic light can be viewed as being composed of two parts with a phase difference of $\exp[i(\Delta kl + \pi)]$. When the two parts of second-harmonic light satisfy the condition $(\Delta kl + \pi) = (2j+1)\pi$ or

$$l = 2j(\pi/\Delta k), \tag{9}$$

they interfere destructively. Here, $\pi/\Delta k$ is the coherence length for SHG.[10] That is to say, when the structure parameter l equals an even number times the coherence length, the corresponding SHG will disappear. This can be also deduced from Eq. (3) easily. In Eq. (3), for mode (2,2), we can obtain $\Delta kl = 2\pi$ from the δ function, but then the factor $\sin(\Delta kl/2) = 0$. Thus the second-harmonic intensity is zero. By substituting for Δkl from Eq. (9) into the δ function in Eq. (3), we can obtain the general extinction rule, which is

$$(m,n) = (2j, 2j). \tag{10}$$

Namely, all peaks with indices $(m,n) = (2j, 2j)$ are absent in the spectrum of the second-harmonic intensity. This property may not be restricted to the nonlinear-optical effect

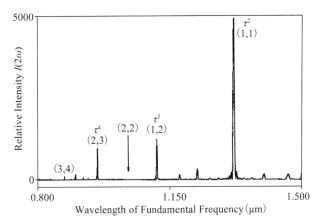

FIG. 3. Dependence of the second-harmonic intensity on the fundamental wavelength in reciprocal space. Note that $(1/\lambda)_{s,p+1} \neq (1/\lambda)_{s,p} + (1/\lambda)_{s,p-1}$.

exclusively. In our previous study[3] on ultrasonic excitation in a Fibonacci superlattice, which is the same as the one discussed here (in this case, the linear effect is discussed), the extinction phenomenon also exists, provided the structure parameter is selected properly. Nevertheless, in a conventional heterostructure superlattice this phenomenon does not exist. This constitutes one major difference between the FOS system and conventional heterostructure superlattices.

In conclusion, we have presented a FOS made of a single $LiNbO_3$ crystal and theoretically second-harmonic generation in the FOS. In our system we find that because of the dispersive effect of the refractive index, the spectrum of the second-harmonic intensity in reciprocal space does not reflect the symmetry of the quasiperiodic structure, and thus does not exhibit self-similarity. Also found is an extinction phenomenon which constitutes one major difference between our system and conventional heterostructure systems. A general extinction rule has also been obtained.

References and Notes

[1] F. Nori and J. P. Rodriguez, Phys. Rev. B **34**, 2207 (1986).

[2] S. Tamura and J. P. Wolfe, Phys. Rev. B **36**, 3491 (1987).

[3] Y. Y. Zhu, N. B. Ming, and W. H. Jiang, Phys. Rev. B **40**, 8536 (1989).

[4] R. Merlin, K. Bajema, R. Clarke, F. Y. Juang, and P. K. Bhattacharya, Phys. Rev. Lett. **55**, 1768 (1985).

[5] X. K. Zhang, H. Xia, G. X. Cheng, A. Hu, and D. Feng, Phys. Lett. A **136**, 312 (1989).

[6] D. Feng, N. B. Ming, J. H. Hong, Y. S. Yang, J. S. Zhu, Z. Yang, and Y. N. Wang, Appl. Phys. Lett. **37**, 607 (1980).

[7] Y. H. Xue, N. B. Ming, J. S. Zhu, and D. Feng, Acta Phys. Sin. 32, 1515 (1983) [Chin. Phys. **4**, 554 (1984)].

[8] J. A. Armstrong, N. Bloembergen, J. Ducuing, and P. S. Pershan, Phys. Rev. **127**, 1918 (1962).

[9] N. Bloembergen and A. J. Sievers, Appl. Phys. Lett. **17**, 483 (1970).

[10] A. Yariv and P. Yeh, *Optical Waves in Crystals* (Wiley, New York, 1984).

[11] N. B. Ming, J. F. Hong, Z. M. Sung, and Y. S. Yang, Acta Opt. Sin. **4**, 821 (1981).

[12] N. B. Ming, J. F. Hong, and D. Feng, J. Mater. Sci. **27**, 1663 (1982).

[13] J. F. Hong and Y. S. Yang, Opt. Sin. **4**, 821 (1986).

[14] F. Zernike and J. M. Midwinter, *Applied Nonlinear Optics* (Wiley, New York, 1973).

[15] D. Levine and P. J. Steinhardt, Phys. Rev. B **34**, 596 (1986).

[16] R. K. P. Zia and W. J. Dallas, J. Phys. A **18**, L341 (1985).

[17] V. Hoggatt, *Fibonacci and Lucas Numbers* (Houghton-Mifflin, Boston, 1969).

[18] Smith, Opt. Commun. **17**, 332 (1976).

[19] This work was supported by the National Natural Science Foundation of China.

Quasi-Phase-Matched Third-Harmonic Generation in a Quasi-Periodic Optical Superlattice*

Shi-ning Zhu, Yong-yuan Zhu, Nai-ben Ming

National Laboratory of Solid State Microstructures, Nanjing University, Nanjing 210093, China, and Center for Advanced Studies in Science and Technology of Microstructures, Nanjing 210093, China

Quasi-periodic structure can be introduced into nonlinear optical materials such as $LiTaO_3$ crystals. Such structures were used for quasi–phase-matching second-harmonic generation. These materials are now shown to be able to couple second-harmonic generation and sum-frequency generation through quasi–phase-matching. The approach led to a direct third-harmonic generation with high efficiency through a coupled parametric process. The result verifies that high-order harmonics may be generated in a quadric nonlinear medium by a number of quasi–phase-matching processes, and therefore, exhibits a possible important application of quasi-periodic structure materials in nonlinear optics.

In dielectric crystals, the most important physical processes are the propagation and excitation of classical waves (optical and ultrasonic waves). The behavior of classical waves in a homogeneous dielectric crystal is the same as that in a continuous medium, because the wave vector of a classical wave is much smaller than the reciprocal vectors of crystal lattice. However, if some microstructure is introduced into a dielectric crystal, forming a superlattice, and if the reciprocal vectors of the superlattice are comparable with the classical wave vectors, the situation is quite different. The propagation of classical waves in a superlattice (classical system) is similar to the electron motion in a periodic potential of crystal lattice (quantum system). Thus, some ideas in solid-state electronics—for example, the reciprocal space, Brillouin zone, dispersion relation, and the like—may be used in classical wave processes. Such is the case for photonic band-gap materials[1]. With classical systems, eigenvalues and eigenfunctions were measured directly[2]. These are difficult if not impossible to obtain in quantum systems. On the other hand, the interactions between wave vectors of classical waves and reciprocal vectors of the superlattice may generate some new physical effects. In nonlinear optical fields, the interactions have led to new laser frequency generations in quasi–phase-matching (QPM) schemes from a number of optical superlattice crystals such as $LiNbO_3$, $LiTaO_3$, and $KTiOPO_4$[3,4].

The above concepts may be equally applied to the quasiperiodic structure. Despite the

---

* Science, 1997, 278(31):843

large amount of research on the quasiperiodic structure since its discovery in 1984[5], whether this kind of structure can be of any practical use remains undetermined. It was proposed that the quasi–phase-matching theory can be extended from periodic structures to quasiperiodic structures[6], which may find applications in nonlinear optics through the QPM method. With the development of the electric poling technique, ferroelectric crystals such as $LiTaO_3$, $LiNbO_3$, $KTiOPO_4$, and the like with quasiperiodically domain-inverted structure (hereafter we call it quasiperiodic optical superlattice, or QPOS) can be fabricated. We previously reported the experimental results of multiwavelength second-harmonic generation (SHG) in a Fibonacci QPOS $LiTaO_3$[6]. Because more reciprocal vectors can be provided by a QPOS, not only the quasi–phase-matched (QPM) multiwavelength SHG but also some coupled parametric processes, such as the third-harmonic generation (THG) and fourth-harmonic generation, can be realized with high efficiency. Taking THG as an example, we present our results using the second-order nonlinear optical processes in a QPOS $LiTaO_3$ crystal.

THG has a wide application as a means to extend coherent light sources to short wavelengths. The creation of the third harmonic directly from a third-order nonlinear process is of little practical importance because of the intrinsic low third-order optical nonlinearity. Conventionally, an efficient THG was achieved by a two-step process. Two nonlinear optical crystals are needed: the first one for SHG and the second one for sum-frequency generation[7]. In this regard, QPOS has some advantages over the conventional method. Here, only one crystal is needed and the harmonic generation can be realized with high efficiency by using the largest nonlinear optical coefficient over the entire transparency range of the material.

A QPOS may be thought to contain two or more linear independent periods, whose ratios are given by irrational numbers[8]. Its reciprocal vectors are indexed by two or more integers, which is different from a periodic structure's reciprocal G_n indexed by one integer. For a Fibonacci QPOS, the QPM conditions for THG in a colinear interaction are

$$\Delta k_1 = k_2 - 2k_1 - G_{m,n} = 0 \tag{1}$$

for SHG and

$$\Delta k_2 = k_3 - k_2 - k_1 - G_{m',n'} = 0 \tag{2}$$

for sum-frequency generation process, respectively, where k_1, k_2, and k_3 are the wave vectors of the fundamental, second-, and third-harmonic fields, respectively; $G_{m,n}$ and $G_{m',n'}$ are predesigned two different reciprocal vectors of the superlattice.

The QPOS used here consists of two fundamental blocks, A and B, arranged according to the Fibonacci sequence: ABAABABAABAAB …[9]. For $LiTaO_3$, each block (A or B) contains a pair of antiparallel 180° domains. The widths of A and B are l_A and l_B, respectively, where $l_A = l_{A1} + l_{A2}$ and $l_B = l_{B1} + l_{B2}$. We assumed that $l_{A1} = l_{B1} = l$ for the width of the positive domain, and $l_{A2} = l(1+\eta)$, and $l_{B2} = l(1-\tau\eta)$ for the width of the negative domain. Here, l and η are two adjustable structure parameters and $\tau=(1+\sqrt{5})/2$ is the golden ratio [Fig. 1(a)]. The reciprocal vector $G_{m,n}=2\pi D^{-1}(m+n\tau)$, where $D=\tau l_A + l_B$ is an "average structure parameter". Theoretically, $l=sl_{c2}=vl_{c3}$, where l_{c2} and l_{c3} are, respectively, the coherence length of SHG and the sum-frequency generation in a homogeneous crystal and where $s=1,3,5$, and so on, and $v=\tau, \tau^2, \tau^3$, and so on. When l is fixed, the variation of η can be used to adjust the magnitude of the effective nonlinear coefficient to obtain the most efficient THG[6].

The QPOS was fabricated by quasi-periodically poling a z-cut LiTaO$_3$ wafer at room temperature[10]. The structure parameters were selected to be: $l=l_{c2}=\tau^2 l_{c3}=10.7$ μm and $\eta=0.23$, so that block A and B were 24 μm and 17.5 μm in width, respectively. The quasi-periodic structure can be confirmed by observing the etched y surface of the sample[Fig. 1(c)]. It consisted of 13 generations and had a total length of ~8 mm. The sample thickness was 0.5 mm. In this design, the reciprocal vector $G_{1,1}$ was used for QPM SHG, whereas $G_{2,3}$ was used for QPM sum-frequency generation. Here, the two processes no longer proceed alone but couple each other. This coupling led to a continuous energy transfer from fundamental to second- to third-harmonic fields, thus, a direct third-harmonic was generated from the QPOS with high efficiency [Fig. 1(d)].

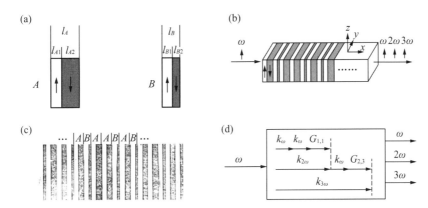

FIG. 1. QPOS made from a single LiTaO$_3$ crystal. The arrows indicate the directions of spontaneous polarization. (a) Two building blocks, A and B, each composed of one positive and one negative ferroelectric domain. (b) Schematic diagram shows a QPOS composed of two blocks, A and B, arranged in Fibonacci sequence and the polarization orientation of electric fields in the THG process with respect to the superlattice. (c) Optical micrograph shows a QPOS of a single LiTaO$_3$ crystal revealed by etching. (d) Schematic diagram of the process of THG in a QPOS material. The QPOS has two specially designed reciprocal vectors: $G_{1,1}$ is used to compensate the mismatch of wave vectors in the SHG process, and $G_{2,3}$ is used to compensate the mismatch of wave vectors in the SFG process. Two QPM conditions, $\Delta k_1=0$ and $\Delta k_2=0$, are simultaneously satisfied in the coupled parametric process, which leads to a THG with high efficiency.

THG was tested with a tunable optical parametric oscillator pumped with an yttrium-aluminum-garnet laser (Nd-YAG, NY81-10, Continuum, Santa Clara, California) with a pulsewidth of 8 ns and a repetition rate of 10 Hz. The linewidth of the fundamental wave was about 1.5 nm at 1.6 μm. In order to use the largest nonlinear optical coefficient d_{33} of LiTaO$_3$, the fundamental wave was z-polarized and propagated along the x axis of the sample (Fig. 1B). It was weakly focused and coupled into the polished end of the sample. For the system, the confocal parameter $Z_0 = 2$ cm, and the radius of the waist spot inside the sample was $\overline{\omega}_0 = 30$ μm. When the fundamental wavelength was tuned to 1.570 μm, green light at 0.523 μm wavelength was generated from the sample (Fig. 2). The third-harmonic peak has a shift of 1.5 nm to the theoretical value calculated from the dispersion relation of bulk LiTaO$_3$ at room temperature (25℃). A green light of 6 mW was obtained for an average fundamental power of 26 mW (with an average power density $\sim$1 kW/cm^2). The conversion efficiency was close to 23%. Together with the THG, an SHG was observed in the same range of fundamental wavelength. The output of SHG and THG were stable, and no optical damage was observed at this fundamental intensity level. The second-and third-harmonic tuning curves, calculated [according to (6)] and measured, are shown in Fig. 3(a) and (b), respectively. The measured peak shifts and widening from calculated values may arise from some uncertainties in the Sellmeier equation used to estimate the dependence of the refractive index on wavelength. A focused fundamental beam with wide linewidth was used, and some imperfections in the domain pattern could also widen the bandwidth.

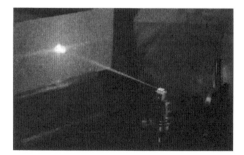

FIG.2. A third-harmonic beam of green light was generated when an infrared light from an optical parametric oscillator passed through a 8-mm-long QPOS LiTaO$_3$ crystal.

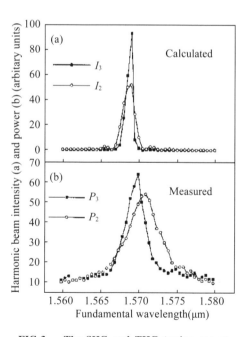

FIG.3. The SHG and THG tuning curves for the QPOS sample, (a) calculated and (b) measured by using an ns-optical parametric oscillator.

Figure 4(a) shows the dependence of second – and third-harmonic average power on the fundamental average power measured at the output end of the sample. I_3 grows slower than I_2 when I_1 is low [Fig. 4(a)]. With the increase of I_1, the increase rate of I_3 grows faster, whereas that of I_2 becomes slower, because more and more I_2 participates in the sum-frequency process. Finally, I_3 approaches and even exceeds I_2. These results imply a coupling effect of three waves when the phases are matched. The situation continues until conversion efficiency exceeds a certain level under which a nondepletion approximation will be no longer valid.

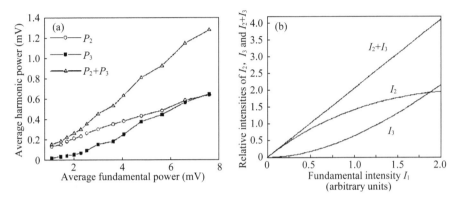

FIG. 4. (a) The average powers of second – and third-harmonic fields versus the average power of the fundamental field for the QPOS sample. The light source is an ns-optical parametric oscillator with a repetition of 10 Hz. (b) The calculated relative intensities of I_2, I_3, and $I_2 + I_3$ versus fundamental intensity I_1.

The THG process can be analyzed by solving the coupled nonlinear equations that describe the interaction of the three fields E_ω, $E_{2\omega}$, and $E_{3\omega}$ in QPOS. We can get an analytical result for boundary conditions $E_{2\omega}(0) = 0$ and $E_{3\omega}(0) = 0$. Under small signal approximation and QPM conditions[11], the second – and third-harmonic intensities are

$$I_2 = \frac{8\pi^2 d_{m,n}^2 L^2}{n_1^2 n_2 c \varepsilon_0 \lambda^2} I_1^2 \tag{3}$$

and

$$I_3 = \frac{144\pi^4 d_{m,n}^2 d_{m',n'}^2 L^4}{n_1^3 n_2^2 n_3 c^2 \varepsilon_0^2 \lambda^4} I_1^3 \tag{4}$$

with the effective nonlinear coefficient $d_{m,n} = d_{1,1}$ and $d_{m',n'} = d_{2,3}$[6]. Here, n_1, n_2, and n_3 represent the refraction indices of the fundamental, second, and third harmonics, respectively; c is the speed of light in a vacuum; λ is the fundamental wavelength; and ε_0 is the dielectric constant of vacuum. From Eqs. 3 and 4 it is evident that I_3 would depend more strongly on the fundamental intensity I_1 and the length of sample L than I_2. The theoretical results according to Eqs. 3 and 4 [Fig. 4(b)] are in excellent agreement with the experimental ones [Fig. 4(a)].

The result obtained here can be compared with that from a third-harmonic generator constructed by two periodic superlattices, each with a length $L/2$. The second harmonic is generated in the first superlattice with period Λ_1, using the first-order QPM; it then mixes with the fundamental wave in the second superlattice with period Λ_2, using the third-order QPM (if a first-order QPM is used in sum-frequency process, the grating period is too small to fabricate a bulk sample with the thickness of 0.5 mm by poling). Compared with the two-step process, the conversion efficiency in a QPOS is increased by a factor of

$$\frac{d_{1,1}^2 d_{2,3}^2 L^4}{d_1^2 d_3^2 \left(\frac{L}{2}\right)^2 \left(\frac{L}{2}\right)^2} \cong 8 \tag{5}$$

Here, $d_{1,1} \approx 0.86 d_1$ and $d_{2,3} \approx 0.88 d_3$, where d_1 and d_3 are the effective nonlinear coefficients of SHG and sum-frequency generation of two periodic superlattices, respectively[6]. The reason for this is that in the traditional two-step scheme, second-harmonic and sum-frequency processes are not coupled and achieved in two separate steps, respectively. Therefore, its effective interaction length is only one-half of a QPOS with the same total length where two processes are coupled.

The Fibonacci QPOS is only a subclass of quasi-periodic structures, and LiTaO$_3$ is a ferroelectric. In fact, the quasi-periodic structure can be extended outside the Fibonacci sequence. Various quasi-periodic sequences may be generated by some inflation rules, for example, $A \rightarrow A^m B$, and $B \rightarrow A$ for a class of sequence with two blocks A and B, where $A^m B$ means a sequence of m basic building blocks of type A followed by one block of type B. Some more complex sequences, such as Thue-Morse sequences and fractal sequences, can also be chosen to construct the structures to obtain the required phase-matching wavelength[9]. Furthermore, the quasi-periodic structure can also be introduced into nonlinear optical materials that are not ferroelectric. For example, by using some special epitaxial growth method[12] or diffusion-bonding technique[13], even LiB$_3$O$_5$ and β-BaB$_2$O$_4$ crystals[14], which are widely used for ultraviolet (UV) harmonic generations, with a polar axis changing its direction quasi-periodically might be realized. For a superlattice with a particular material and quasi-periodic sequence type, there is only a finite number of choices for THG with two QPM conditions satisfied simultaneously. However, for a required fundamental wavelength, use of different quasi-periodic structures may be considered, because superlattices with different quasi-periodic modulations can provide different reciprocal vectors for phase matching. In this way, efficient THG might be realized in the UV region[15].

Generally speaking, a third harmonic may always generate at any required wavelength as long as structure parameter l satisfies the QPM condition of the sum-frequency process, $\Delta k_2 = 0$, even if $\Delta k_1 \neq 0$[6]. Because the efficiency of the THG depends on the second-harmonic intensity, the smaller the mismatch of the second harmonic, the higher the conversion efficiency of the third harmonic. As long as the chosen structure parameter l of the superlattice makes $\Delta k_2 = 0$, and second-harmonic generation is not severely phase-mismatched or Δk_1 is small, a third harmonic can always be efficiently generated. So, for a given material and quasi-periodic type, the structure parametric l is no longer some fixed value, but may be adjusted in certain regions adjacent to $\Delta k_1 = 0$ in order to get the required frequency of third harmonic. This characteristic is beneficial to the design of a practical third-harmonic device.

Aside from the THG, some other higher-order harmonics can also be generated using the same QPM scheme. For example, the fourth harmonic can be generated by two coupled, cascaded SHG processes. And more interesting effects such as tunable multiwavelength optical parametric oscillation can be expected, where a number of QPM conditions must be satisfied simultaneously. Exploring physical process in various microstructures may offer an opportunity to develop some novel materials with potential applications. We provide an example of possible technical applications of quasi-periodic structure materials in nonlinear optics.

References and Notes

[1] J. D. Joannopoulos, P. R. Villeneuve, S. Fan, *Nature* **386**, 143 (1997).

[2] S. He and J. D. Maynard, *Phys. Rev. Lett.* **62**, 1888 (1989).

[3] J. A. Armstrong *et al.*, *Phys. Rev.* **127**, 1918 (1962); S. Somkeh and A. Yariv, *Opt. Commun.* **6**, 301 (1972).

[4] D. Feng *et al.*, *Appl. Phys. Lett.* **37**, 607 (1980); R. L. Byer, *Nonlinear Opt.* **7**, 234 (1994); V. Pruneri, J. Webjorn, J. Russell, D. C. Hanna, *Appl. Phys. Lett.* **65**, 2126 (1995); S. N. Zhu *et al.*, *ibid.* **67**, 320 (1995); H. Ito, C. Takyu, H. Inaba, *Electron. Lett.* **27**, 1221 (1991); M. C. Gupta, W. Kozlovesky, A. C. G. Nutt, *Appl. Phys. Lett.* **64**, 3210 (1994); J. D. Bierlein *et al.*, *ibid.* **56**, 1725 (1990).

[5] P. J. Steinhardt and S. Ostlund, *The Physics of Quasi-crystals* (World Scientific, Singapore, 1997); C. Janot, *Quasicrystals* (Clarendon Press, Oxford, UK, 1992).

[6] S.N.Zhu *et al.*, *Phys. Rev. Lett.* **78**, 2752 (1997); Y. Y. Zhu and N. B. Ming, *Phys. Rev. B* **42**, 3676 (1990); J. Feng, Y. Y. Zhu, N. B. Ming, *ibid.* **41**, 5578 (1990).

[7] Y. R. Shen, *The Principles of Nonlinear Optics* (Wiley, New York, 1984).

[8] For standard Fibonacci sequence, the ratio is $\tau = (1+\sqrt{5})/2$. The ratios other than τ can be derived from other 1D quasi-periodic subclasses. All possible subclasses have been classified [D. Levine and P. J. Steinhardt, *Phys. Rev. B* **34**, 596 (1986)].

[9] The blocks A and B can also be arranged according to other quasi-periodic subclasses and can be composed of one or more layers of different materials [R. Merlin *et al.*, *Phys. Rev. Lett.* **55**, 1768 (1985); A. Behrooz *et al.*, *ibid.* **57**, 368 (1986)].

[10] S. N. Zhu *et al.*, *J. Appl. Phys.* **77**, 5481 (1995).

[11] Some essential simplifications of the theoretical treatment were considered, including the approximation of slowly varying envelope for fundamental, second-, and third-harmonic fields and ignoring depletion of the fundamental and harmonic field.

[12] By using a special deposition method, B. Hadimioglu *et al.* [*Appl. Phys. Lett.* **50**, 1642 (1987)] fabricated multiple-layer ZnO films with the crystallographic orientation changing for alternating

layers.

[13] L. A. Gordon *et al.*, *Electron. Lett.* **29**, 1942 (1993).

[14] H. W. Mao, F. C. Fu, B. C. Wu, C. T. Chen, *Appl. Phys. Lett.* **61**, 1148 (1992).

[15] It was reported that $LiTaO_3$ has short-wavelength transparency from 280 nm [K. Mizuuchi, K. Yamamoto, T. Taniuchi, *ibid.* **58**, 2732 (1991)]. Thus, by a suitable choice of a QPOS, harmonic generation in the UV region is possible.

[16] We thank D. Feng for stimulating discussions. Supported by a grant for the Key Research Project in Climbing Program from the National Science and Technology Commission of China.

[17] S.N.Zhu, Y.Y.Zhu, N.B.Ming, these authors contributed equally to this work.

Experimental Realization of Second Harmonic Generation in a Fibonacci Optical Superlattice of LiTaO$_3$*

Shi-ning Zhu, Yong-yuan Zhu, Yi-qiang Qin, Hai-feng Wang, Chuan-zhen Ge, and Nai-ben Ming

National Laboratory of Solid State Microstructures, Nanjing University, Nanjing 210093, China and Center for Advanced Studies in Science and Technology of Microstructures, Nanjing 210093, China

We have designed and fabricated a novel nonlinear optical superlattice of LiTaO$_3$ in which two antiparallel 180° domains building blocks A and B were arranged as a Fibonacci sequence. We measured the quasi-phase-matched second-harmonic spectrum of the superlattice. The second-harmonic blue, green, red, and infrared light generation, with energy conversion efficiencies of ~5%—20%, was demonstrated experimentally, which efficiencies are comparable with those of a periodic superlattice. Destruction of self-similarity and extinction phenomenon have also been observed in the spectrum. The experiment results are in good agreement with theory.

An important development in condensed-matter physics is the discovery of quasicrystalline structure[1]. Much effort has been devoted to the studies of structure and physical properties of quasicrystal[2,3]. A quasiperiodic superlattice is an analog of one-dimensional quasicrystal. The first quasiperiodic semiconductor superlattice was fabricated by Merlin et al. by molecular-beam epitaxy in 1985[4]. Since then metallic and dielectric Fibonacci superlattices have been produced by various techniques[5—7]. These superlattices have shown many unusual physical properties depending on their composition and layer thickness.

In dielectric crystals, the most important physical processes are the excitation and the propagation of classical waves, including optical waves and acoustic waves. Ultrasonic excitation and propagation in the quasiperiodic acoustic superlattices have been studied both theoretically and experimentally[8]. More recently, the localization of optical waves in a quasiperiodic optical superlattice (QPOS) of SiO$_2$ and TiO$_2$ has been reported[9]. For second-order nonlinear optical effects of the QPOS, some preliminary theoretical work has been carried out[10]. It has been discovered that the second harmonic spectrum of a QPOS is different from that of a periodic optical superlattics (POS) due to its lower space-group symmetry. According to the theory of quasiphase-matching (QPM) proposed by Armstrong et al.[11], the phase matching condition in the second harmonic process of a QPOS can be written into

$$\Delta k = k_{2\omega} - 2k_\omega - G_{m,n} = 0, \qquad (1)$$

---

* Phys.Rev.Lett., 1997, 78(14):2752

where $k_{2\omega}$, k_ω are the wave vectors of the second harmonic and fundamental waves, respectively, $G_{m,n}$ is the reciprocal vector (called the "grating wave vector" in nonlinear optics) which depends on the structure parameter of a QPOS. In a Fibonacci system, two incommensurate periods with ratio τ are superimposed. The indexing of $G_{m,n}$ requires two integers m, n, which is different from the POS's reciprocal vector G_n indexed with only one integer. Therefore, a QPOS can provide more reciprocal vectors to the QPM optical parametric process, which results in the second harmonic spectrum of a QPOS showing more plentiful spectrum structure than that of a POS. This characteristic of the QPOS may be used in multiwavelength laser frequency conversion application. However, up to now, this has not been experimentally proved due to the lack of proper material.

In this paper, we report for the first time the second harmonic generation experiment on a Fibonacci optical superlattice of $LiTaO_3$ (LT). The superlattice was fabricated by the external field poling technique at room temperature. We measured the QPM second-harmonic spectrum of the QPOS and calculated its main effective second-order nonlinear optical coefficients $d_{m,n}$. Two different extinction rules were found. We confirmed that the second-harmonic spectrum of the QPOS does not reflect the symmetry of the quasiperiodic structure due to the dispersive effect of the refractive index and, consequently, self-similarity destructs in this spectrum.

The QPOS with Fibonacci sequence is constructed as follows. We first define two fundamental blocks A and B, which are arranged according to the production rule $S_j = S_{j-1} | S_{j-2}$ with $j \geq 3$, $S_1 = A$, and $S_2 = AB$. Both block A and block B are composed of one positive and one negative ferroelectric domain, so that neighboring domains are interrelated by a dyad axis in the x direction. As illustrated in Figs. 1(a) and 1(b), l_A and l_B represent the thickness of block A and block B, respectively, where $l_A = l_{A1} + l_{A2}$, $l_B = l_{B1} + l_{B2}$. Let $l_{A1} = l_{B1} = l$, $l_{A2} = l(1+\eta)$, $l_{B2} = l(1 - \tau\eta)$, where l, η are adjustable structure parameters, $\tau = (1+\sqrt{5})/2$ is the golden ratio. The sequence of the blocks, $ABAABABA...$, produces a QPOS with $l_A/l_B = \tau$; see Fig. 1(b). Even if $l_A/l_B \neq \tau$, the quasiperiodic properties of the superlattics are still preserved. Since the

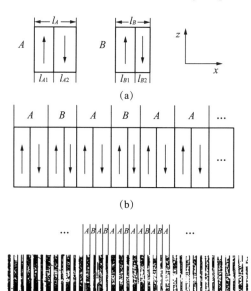

FIG. 1. Quasiperiodic optical superlattic (QPOS) made from a LT single crystal. (a) Two building blocks: A and B, each composed of one positive and one negative ferroelectric domain. (b) Schematic diagram of a QPOS with Fibonacci sequence. (c) The optical micrograph of a QPOS of LT single crystal revealed by etching.

second-order nonlinear-optical coefficients form a third-rank tensor, they will change their signs from positive domains to negative domains, so the nonlinear coefficients in the superlattice are modulated with quasiperiodic sign reversal. In order to utilize the largest nonlinear optical coefficient d_{33} ($=26$ pm/V for LT), the ferroelectric domain lamellae are arranged along the x axis of the LT crystal and the domain boundaries parallel to the y-z plane, the z-polarized fundamental wave propagates along the x axis of crystal. For an infinite array, the modulated nonlinear coefficient $d(x)$ can be written by use of Fourier transform approach[10,12] as

$$d(x) = \sum_{m,n} d_{m,n} e^{iG_{m,n}x}, \tag{2}$$

where the reciprocal vector $G_{m,n} = 2\pi D^{-1}(m+n\tau)$, $D = \tau l_A + l_B$ is the "average structure parameter" of the superlattice. The corresponding Fourier coefficients $d_{m,n}$ can be defined as the effective nonlinear coefficients of the QPOS.

The sample was fabricated by poling a z-cut LT single-domain wafer at room temperature[13]. Figure 1(c) is the optical micrograph of the cross section of the poled sample revealed by etching; the observed surface was perpendicular to the y axis and the lamellae were perpendicular to the x axis. The figure shows that a volume quasiperiodic domain grating has been produced in the sample. In the QPOS block A and block B consist nominally of 11 μm, 13 μm and 11 μm, 6.5 μm, respectively. The sample has 13 generations (S_{13}), 377 A and B blocks, with a total length of ~ 8 mm and a thickness of ~ 0.5 mm.

We used a tunable optical parametric oscillator (OPO) as the fundamental light source. Its pulse duration was 23 ps and the repetition rate was 1 Hz. The fundamental wave was polarized along the z axis of the sample. It was weakly focused and coupled into the polished end face of the sample and propagated along the x axis of the sample. The radius of the waist inside the sample was $\omega_0 \approx 0.1$ mm. The confocal parameter for the system was $Z_0 \approx 6$ cm. Because Z_0 is much greater than the length of the sample, the plane-wave theory of second-harmonic generation (SHG) can be applied to the Gaussian beam.

The SHG spectrum of the QPOS of LT was measured in the range from 0.9 μm to 1.4 μm and from 1.55 μm to 1.7 μm [Fig. 2(a)], respectively. When the fundamental wavelength was tuned to 0.9726 μm, 1.0846 μm, 1.2834 μm, 1.3650 μm, and 1.5699 μm, we obtained QPM second harmonic blue, green, red, and infrared light output with conversion efficiencies up to $\sim 5\%-20\%$ (Table 1). According to Eq. (1), the position of second harmonic peaks may be marked with fundamental wavelength as

$$\left[\frac{1}{\lambda}\right]_{m,n} = \frac{G_{m,n}}{4\pi[n_2(\lambda) - n_1(\lambda)]}, \tag{3}$$

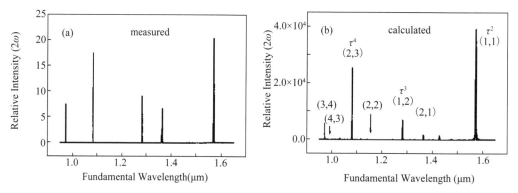

FIG.2. The SHG spectra measured and calculated in a QPOS of LT. Note that (i) $(1/\lambda)_{1,4} \neq (1/\lambda)_{1,2} + (1/\lambda)_{1,3}$ and (ii) the (2,2) peak and (4,3) peak are absent in the spectra.

TABLE 1. SHG experiment results of quasiperiodic poling LiTaO$_3$. The fundamental source is a ps-OPO with the repetition of 1 Hz and the duration of 23 ps.

Reciprocal vectors $G_{m,n}$	Fundamental wavelength (μm)		Harmonic wavelength(μm) Measured	Input energy (μJ)	Output energy (μJ)	FWHM (nm)	Efficiency %
	Calculated	Measured					
(3,4)	0.9720	0.9726	0.4863	40	3	~0.3	~7.5
(2,3)	1.0820	1.0846	0.5423	40	7	~0.4	~17.5
(1,2)	1.2830	1.2834	0.6417	33	3	~0.85	~9.1
(2,1)	1.3640	1.3650	0.6825	30	2	~1.1	~6.7
(1,1)	1.5687	1.5699	0.7845	54	11	~2.5	~20.4

where $n_2(\lambda)$, $n_1(\lambda)$ are the refractive indexes of fundamental and harmonic of LT crystal, respectively. The participating of the multireciprocal vector leads to the multipeak structure of the spectrum. If m, n are successive Fibonacci numbers, or $(m,n) = (F_{k-1}, F_k)$, the reciprocal vector can be rewritten as $G_{m,n} = G_{s,p} = 2\pi D^{-1} s \tau^p$, where s, p are integers, and $G_{s,p} = G_{s,p-1} + G_{s,p-2}$, thus the $G_{s,p}$ is self-similar. However, because of the dispersion effect, although $G_{s,p}$ presents the self-similarity in reciprocal space, the relation $(1/\lambda)_{s,p+1} = (1/\lambda)_{s,p} + (1/\lambda)_{s,p-1}$ will no longer hold. By careful analysis to the measured spectrum, we do find $(1/\lambda)_{1,4} \neq (1/\lambda)_{1,2} + (1/\lambda)_{1,3}$. This differs from the X-ray diffraction and Raman spectra of quasiperiodic superlattice[4] in which the spectrum structures exhibit self-similarity. Moreover, Eq. (3) can be rewritten as

$$\left[\frac{1}{\lambda}\right]_{s,p} = \frac{s\tau^2}{4[n_2(\lambda) - n_1(\lambda)](1+\tau)l}. \qquad (4)$$

The equation shows that the position of peak (or wavelengths of phase matching) depends on the structure parameter l, and does not depend on the thickness of blocks A and B and their ratio. Figure 2(b) shows the result of numerical calculation for the QPOS of LT with $l = 11$ μm and $l_A/l_B = 1.37$. Indeed from Figs. 2(a) and 2(b) we find a close correspondence between the calculated and measured results on the positions and intensities of peaks. For $l_A/l_B = \tau$, calculation has also shown the positions of corresponding peaks remain unchanged, except for their strengths.

The intensities of peaks in Fig. 2 are related to the effective nonlinear coefficients $d_{m,n}$. Fourier transferring Eq. (2), we can get

$$d_{m,n} = d_{33} \frac{\sin\left(\frac{1}{2}G_{m,n}l\right)}{\frac{1}{2}G_{m,n}l} * \frac{\sin X_{m,n}}{X_{m,n}}, \quad (5)$$

here $X_{m,n} = \pi D^{-1} \tau^2 (ml_A - nl_B)$. For a plane-wave interaction, ignoring depletion of the fundamental field, the second harmonic intensity can be written as[14]

$$I_{2\omega} = \frac{8\pi^2 d_{m,n}^2 L^2 I_\omega^2}{n_1^2 n_2 c \varepsilon_0 \lambda^2} \sin c^2 \left(\frac{1}{2}\Delta k L\right), \quad (6)$$

where LI_ω is the length-intensity product associated with a particular device geometry; n_1, n_2 are the refractive indices of the fundamental and harmonic, respectively; λ is the fundamental wavelength; c is the speed of light; and ε_0 is the dielectric constant of vacuum. When the phase matching condition is satisfied ($\Delta k = 0$), the sinc factor in Eq. (6) is unity. The intensities of peaks are proportional to the square of $d_{m,n}$. In Eq. (5), $d_{m,n}$ contains the two factors $\frac{\sin(1/2 G_{m,n}l)}{1/2 G_{m,n}l}$ and $\frac{\sin X_{m,n}}{X_{m,n}}$. For $\frac{\sin(1/2 G_{m,n}l)}{1/2 G_{m,n}l}$, the smaller the indexes m and n, the larger its value. While the value of $\frac{\sin X_{m,n}}{X_{m,n}}$ depends strongly on the indices m, n and the ratio l_A/l_B, $\frac{\sin X_{m,n}}{X_{m,n}}$ is larger when the ratio n/m is closer to ratio l_A/l_B. We calculated the magnitude of the main $d_{m,n}$ for $l_A/l_B = 1.37$ and $l_A/l_B = \tau$ according to Eq. (5); the results are shown in Table 2. If $l_A/l_B = \tau$, it is well known that the best rational approximations to τ occur when n and m are successive Fibonacci numbers F_k. The larger $d_{m,n}$ corresponds to $(m, n) = (F_{k-1}, F_k)$. However, this does not mean that the $d_{m,n}$ has the largest value when $l_A/l_B = \tau$. The magnitude of $d_{m,n}$ changes with the ratio of l_A/l_B. The calculated values versus the ratio l_A/l_B from 1 to 2 for some $d_{m,n}$ are shown in Fig. 3. The curves in Fig. 3 show that $d_{m,n}$ with low indices, such as $d_{1,1}$, $d_{1,2}$,... exhibit a monotonic dependence on the ratio l_A/l_B in this range, whereas those with high indices, for example $d_{2,3}$, $d_{3,4}$,... oscillate in the same range. From Table II and Fig. 3 we can find the two cases for which $d_{m,n} = 0$. One is $(m, n) = (2j, 2j)$, in which j is an integer, which corresponds to that the structure parameter l equals an even number times the coherence length of SHG. The other is when the ratio l_A/l_B equals a specified value, e. g., $l_A/l_B = 1.36$ for $G_{4,3}$, which lead to $X_{4,3} = 2\pi$, and $\sin X_{4,3} = 0$. Thus all peaks, with indices (m, n) in accord with the two conditions above will disappear in the spectrum if even the condition of phase matching $\Delta k = 0$ is satisfied. In Fig. 2 the peaks indexed $(2, 2)$ and $(4, 3)$ do not appear in both calculated and measured spectra because these two peaks satisfy extinction conditions: $d_{2,2} = 0$ and $d_{4,3} \approx 0$ for $l_A/l_B = 1.37$, respectively. Since $d_{m,n}$

is a function of ratio l_A/l_B, we may significantly increase some $d_{m,n}$ by optimizing the structure design.

TABLE 2. The effective nonlinear coefficients $d_{m,n}$ of quasiperiodic poling $LiTaO_3$.

m,n	s,p	$|d_{m,n}/d_{33}|$	
		$l_A/l_B=1.37$	$l_A/l_B=\tau$
0,1	1,1	0.156	0.286
1,1	1,2	0.546	0.447
1,2	1,3	0.138	0.195
2,3	1,4	0.184	0.191
3,5	1,5	0.018	0.052
3,4		0.098	0.048
4,3		0.001	0.020
2,2		0	0
4,4		0	0

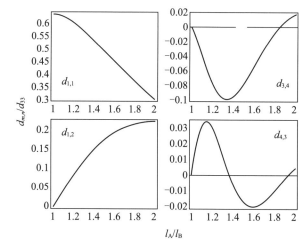

FIG. 3. The dependence of the effective nonlinear coefficients $d_{m,n}$ on the ratio of l_A and l_B.

In Table 1, we present the values of full width at half maximum (FWHM) of the SHG signal for various phase-matching wavelengths. They are close to the values predicated by theory, which shows that the effective interaction is over the entire sample length in the SHG process. This verifies that the sample was poled uniformly.

We can compare the conversion efficiency $\eta_{m,n}$ of a QPOS with the η_n of a POS through comparing $d_{m,n}$ with d_n. In main $d_{m,n}$, $d_{1,1}$ is maximum, for $l_A/l_B=1.37$, $d_{1,1}\approx 0.55d_{33}$. In contrast with the largest effective nonlinear coefficient of a POS in the first-order QPM, $d_1=0.64d_{33}$. For the equal length-strength parameter LI_ω, the conversion efficiency $\eta_{1,1}\approx 0.75\eta_1$, which is 75% of a POS in a first-order QPM process. The rest is comparable with the third-order QPM's $d_3\approx 0.2d_{33}$ (see Table 2). This can also be seen from Fig. 3. When the ratio $l_A/l_B \to 1$, $d_{1,1} \to d_1 = 2d_{33}/\pi$, the rest $\to 0$, thus the system becomes a periodic superlattice, and the multipeak structure in the spectrum will disappear.

In summary, a one-dimensional QPOS of LT consisting of positive and negative ferroelectric domain with Fibonacci sequence has been fabricated using the pulse field poling technique at room temperature. The second harmonic spectrum of the superlattice has been studied theoretically and experimentally. The spectrum does not exhibit self-similarity because of the dispersion of the refractive index of LT. This is in contrast to the X-ray diffraction and Raman spectra of quasiperiodic superlattices in which spectrum structure reflects the symmetry of the quasiperiodic structure. The extinction phenomenon has also been verified experimentally. Frequency conversion efficiencies as high as 5%–20% for SHG at some fundamental wavelengths were measured using a ps-OPO laser, which efficiencies are comparable with that of a POS. Our results show the QPOS may be applied to some multiwavelength SHG devices.

References and Notes

[1] D. Shechtman, I. Blech, D. Gratias, and J. W. Cahn, Phys. Rev. Lett. **53**, 1951 (1984).
[2] *The Physics of Quasicrystals*, edited by P. J. Steinhardt and S. Ostlund (World Scientific, Singapore, 1987).
[3] C. Janot, *Quasicrystals* (Clarendon Press, Oxford, 1992).
[4] R. Merlin, K. Bajema, and R. Clarke, Phys. Rev. Lett. **55**, 1768 (1985).
[5] A. Behrooz et al., Phys. Rev. Lett. **57**, 368 (1986).
[6] W. Gellermann, M. Kohmoto, B. Sutherland, and P. C. Taylor, Phys. Rev. Lett. **72**, 633 (1994).
[7] R. W. Peng, A. Hu, and S. S. Jiang, Appl. Phys. Lett. **59**, 2512 (1991).
[8] Y. Y. Zhu, N. B. Ming, and W. H. Jiang, Phys. Rev. B **40**, 8536 (1989).
[9] W. Gellermann, M. Kohmoto, B. Sutherland, and P. C. Taylor, Phys. Rev. Lett. **72**, 633 (1994).
[10] Y. Y. Zhu and N. B. Ming, Phys. Rev. B **42**, 3676 (1990).
[11] J. A. Armstrong et al., Phys. Rev. **127**, 1918 (1962).
[12] D. Levine and P. J. Steinhardt, Phys. Rev. B **34**, 596 (1986).
[13] S. N. Zhu et al., J. Appl. Phys. **77**, 5481 (1995).
[14] A. Yariv and P. Yeh, *Optical Wave in Crystal* (Wiley, New York, 1984), pp. 504–551.
[15] This work was supported by a grant for the Key Research Project in Climbing Program from the National Science and Technology Commission of China. The authors acknowledge assistance in the experiments by P. N. Wang, Q. Guo, and Z. N. Wang of the Anhui of the Institute of Fine Mechanics and Optics.

Crucial Effects of Coupling Coefficients on Quasi-Phase-Matched Harmonic Generation in an Optical Superlattice[*]

Chao Zhang, Yong-yuan Zhu, Su-xia Yang, Yi-qiang Qin, Shi-ning Zhu,
Yan-bin Chen, Hui Liu, and Nai-ben Ming

*National Laboratory of Solid State Microstructures, Nanjing University, Nanjing 210093, China,
and Center for Advanced Studies in Science and Technology of Microstructures, Nanjing 210093, China*

Coupling of optical parametric processes in an optical superlattice through quadratic nonlinearity was analyzed theoretically. Solving the coupled equations, we found that efficient quasi-phase-matched third-harmonic (TH) generation depends not only on the magnitude of the coupling coefficients but also on their ratio. Theoretically, all the fundamental energy can be transferred to the TH at a particular ratio. In other cases, there exists an optimum condition that corresponds to a maximum TH conversion efficiency. The result is of practical importance for the design of TH devices.

Studies of nonlinear optical effects in optical superlattices (OSL's) has opened up a new field for frequency conversion. Much research has been conducted on subjects such as second-harmonic generation (SHG),[1-3] optical parametric oscillation,[4-6] engineerable compression of ultrashort pulses,[7] amplitude squeezing,[8] wavelength-division multiplexing, and cascaded nonlinearity.[9] Recently there has been interest in high-harmonic generation. Quasi-phase-matched third-harmonic generation (THG) has been demonstrated through cubic nonlinearity [$\chi^{(3)}$] by use of a simple silica structure of six modulation periods.[10] Continuous-wave frequency tripling by simultaneous three-wave mixing has been realized in a periodically poled $LiNbO_3$ crystal, and fourth-harmonic generation has been predicted.[11] In a $LiNbO_3$ waveguide, ultraviolet 355 nm THG has been observed.[12] When quasi-phase matching (QPM) is extended from a periodic to a quasi-periodic structure, interesting phenomena, such as multiwavelength SHG and direct THG, can be produced with high conversion efficiency.[13-17] The theory previously proposed is based on the assumption that the fundamental wave is undepleted during the conversion process. In view of energy conservation, depletion of the fundamental wave must be taken into consideration, especially when the conversion efficiency is high. Then some questions may be raised: What happens when the conversion efficiency becomes

---

[*] Opt.Lett., 1997, 25(7):436

high? What are the optimum material requirements for efficient THG? The answers to these questions are of fundamental interest in physics and are helpful in device design for microelectronics.

As we know, the key to QPM is to construct a structure that provides reciprocal vectors to compensate for the mismatch of wave vectors owing to the dispersive effect of the refractive index.[18] It is the reciprocal vectors that make the optical parametric processes in the material phase matched. From this point of view, the structure can be periodic, quasi-periodic, or of any other type only if reciprocal vectors can be provided. The most commonly used materials at present are those with ferroelectric domain-inverted structures.

We treated the coupled parametric processes for THG, both numerically and analytically, taking into consideration the depletion of the fundamental wave. A critical value of the ratio of the coupling coefficients was obtained at which the conversion efficiency of the THG can reach its maximum. Above the critical value, all the energy will finally be transferred to the second harmonic (SH), whereas below that value the conversion efficiency will oscillate periodically. These interesting phenomena arise from the coupling effect among the three optical fields.

To use the largest nonlinear optical coefficient of the proposed structure (a domain-inverted structure), a single fundamental wave propagates along the x axis with its polarization along the optical axis. Through the nonlinear optical effect, different optical fields (fundamental, SHG, and THG) can be coupled in the following way[19]

$$\frac{dA_1}{dx} = -i\kappa_1 A_1^* A_2 \exp(-i\Delta k_1 x) - i\kappa_2 A_2^* A_3 \exp(-i\Delta k_2 x),$$

$$\frac{dA_2}{dx} = -1/2 i\kappa_1 A_1^2 \exp(i\Delta k_1 x) - i\kappa_2 A_1^* A_3 \exp(-i\Delta k_2 x),$$

$$\frac{dA_3}{dx} = -i\kappa_2 A_1 A_2 \exp(i\Delta k_2 x), \quad (1)$$

with

$$A_i = \sqrt{n_i/\omega_i} E_i, \quad i = 1, 2, 3,$$

$$\Delta k_1 = k^{2\omega} - 2k^{\omega} - G_{a'}$$

$$\Delta k_2 = k^{3\omega} - k^{2\omega} - k^{\omega} - G_{a''},$$

$$\kappa_1 = \frac{f_{a'} d_{\text{eff}}}{c} \sqrt{\frac{\omega_1^2 \omega_2}{n_1^2 n_2}}, \kappa_2 = \frac{f_{a''} d_{\text{eff}}}{c} \sqrt{\frac{\omega_1 \omega_2 \omega_3}{n_1 n_2 n_3}}.$$

E_i, ω_i, and n_i are the electric field [$i = 1, 2, 3$ refer to fundamental, SH, and third harmonic (TH), respectively], the angular frequency, and the refractive index, respectively; c is the speed of light in vacuum, d_{eff} is the effective nonlinear optical coefficient of the proposed structure, f_a and G_a are the Fourier coefficient and the reciprocal vector of the structure, respectively, and $*$ denotes complex conjugation.

In Eqs. (1) there are five distinct parametric process: two sum-frequency processes and three difference-frequency processes. Under QPM conditions ($\Delta k_1=0$ and $\Delta k_2=0$), the coupling effect of the three waves will be greatly enhanced.

Numerical calculations of Eqs. (1) under QPM conditions were made with the following boundary conditions: $A_1(0)=A_{10}$, $A_2(0)=0$, and $A_3(0)=0$. In the calculations we found that efficient THG depends not only on the magnitude of the coupling coefficients κ_1 and κ_2 but also on their ratio. There is a critical ratio, $t=\kappa_1 : \kappa_2=0.8858$. Different ratios correspond to different conditions for harmonic generation. Figure 1(a) shows the case for $t<0.8858$. The intensities of the three fields oscillate with a period $\kappa_2 A_{10} L$. The TH is much stronger than the SH, indicating that here sum-frequency generation (SFG) is more efficient than SHG because of the large value of κ_2. It can be seen from Fig. 1(a) that, when the conversion efficiency of the TH attains its maximum, the SH is depleted completely, with the intensity of the fundamental still high. After that, the difference-frequency generation process dominates, resulting in a decrease of the TH and an increase of the fundamental and the SH. For $t>0.8858$, the result is quite different [see Fig. 1(b)]. Owing to the large ratio, the SH has an obvious enhancement. Correspondingly, the TH can also obtain much energy from the SH and the fundamental through the SFG process. When the TH reaches its maximum, the fundamental has been exhausted completely, and a dip appears in the SH curve. The closer the ratio $\kappa_1 : \kappa_2$ is to 0.8858, the deeper the dip. However, so long as the ratio is not equal to 0.8858 and $\kappa_2 A_{10} L$ is large enough, all the energy can be converted to the SH. The above discussion implies that for $t<0.8858$ the SFG process dominates, whereas for $t>0.8858$ the SHG process dominates. In both cases, there is an optimum condition at which maximum TH conversion efficiency can be obtained. When $t=0.8858$ the energy can be transferred completely to the TH [see Fig. 1(c)]. Deviation from this value will result in a decrease of THG.

We found that under QPM conditions, if A_{10} is real (hereafter $A_{10}=1$), it remains real all the time, with A_2 being pure imaginary and A_3 being real. Also, during the process when THG approaches its maximum, A_1 and A_3 take opposite signs. In this case, some analytical results can be obtained from Eqs. (1). With $y_1=A_1$, $y_2=(1/i)A_2$, and $y_3=-A_3$ (therefore, y_i are all real), the energy conservation law can be obtained:

$$y_1^2+2y_2^2+3y_3^2=A_{10}^2=1. \qquad (2)$$

Furthermore, with $u=y_3/y_1$ and $t=\kappa_1/\kappa_2$ (the ratio of the coupling coefficients), we can get

$$\ln|y_1| = \begin{cases} -\dfrac{1}{2}\ln|(1+tu+u^2)| + \dfrac{t/2}{\left[1-\left(\dfrac{t}{2}\right)^2\right]^{1/2}} \cdot \\ \left[-\arctan\dfrac{u+t/2}{\left[1-\left(\dfrac{t}{2}\right)^2\right]^{1/2}} + \arctan\dfrac{t/2}{\left[1-\left(\dfrac{t}{2}\right)^2\right]^{1/2}}\right] & t<2 \quad (3a) \\ -\dfrac{1}{2}\ln|(1+tu+u^2)| + \dfrac{t/2}{u+t/2} - 1, & t=2. \quad (3b) \\ -\dfrac{1}{2}\ln|(1+tu+u^2)| + \dfrac{1}{2}\dfrac{t/2}{\left[\left(\dfrac{t}{2}\right)^2-1\right]^{1/2}} \cdot \\ \left\{\ln\left|\dfrac{u+\dfrac{t}{2}+\left[\left(\dfrac{t}{2}\right)^2-1\right]^{1/2}}{u+\dfrac{t}{2}-\left[\left(\dfrac{t}{2}\right)^2-1\right]^{1/2}}\right| - 2\ln\left(\dfrac{t}{2}\right) + \left[\left(\dfrac{t}{2}\right)^2-1\right]^{1/2}\right\} & t>2 \quad (3c) \end{cases}$$

To obtain the best TH conversion efficiency [which corresponds to Fig. 1(c)], here we need y_1, $y_2 \to 0$ and $u \to \infty$. In this case the optimum value of the ratio can be obtained (hereafter denoted τ) from Eq. (3a):

$$\dfrac{\tau}{2} = \cos\left\{\dfrac{\left[1-\left(\dfrac{\tau}{2}\right)^2\right]^{1/2}}{\tau}\ln 3\right\}. \tag{4}$$

Its solution is $\tau \approx 0.8858$. It may be taken as a critical ratio for THG.

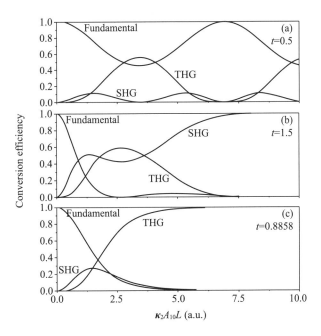

FIG. 1. Dependence of the optical intensities of the parameter $\kappa_2 A_{10} L$ on the ratio of the two coupling coefficients: (a) 0.5, (b) 1.5, (c) 0.8858.

For $t \geqslant \tau$ [which corresponds to Figs. 1(b) and 1(c)], the maximum TH conversion efficiency can be determined to be

$$\eta_{max}(t) = \begin{cases} 3\exp\left\{-\dfrac{t}{\left[1-\left(\dfrac{t}{2}\right)^2\right]^{1/2}}\arccos(t/2)\right\} & \tau \leqslant t < 2 \\ 3\exp(-2), & t = 2. \\ 3\left\{\dfrac{t}{2}+\left[\left(\dfrac{t}{2}\right)^2-1\right]^{1/2}\right\}^{-\dfrac{t}{\left[\left(\dfrac{t}{2}\right)^2-1\right]^{1/2}}} & t > 2 \end{cases} \quad (5)$$

For $t < \tau$ [which corresponds to Fig. 1(a)], $\eta_{max}(t)$ cannot be written in analytical form but can be numerically calculated from Eq. (3a).

The dependence of the maximum THG conversion efficiency on the ratio t is shown in Fig. 2. The figure is divided into two regions by $\tau = 0.8858$. Below τ, η_{max} decreases with decreasing t; above τ, η_{max} decreases with increasing t. Only when $t = \tau$, does η_{max} reach its peak value. Physically, this implies that only when the SHG process is balanced by the SFG process can 100% THG conversion efficiency be realized theoretically.

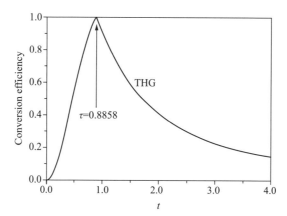

FIG. 2. Dependence of the conversion efficiency η_{max} on the ratio of the coupling coefficients.

The above results are of practical importance for the design of a THG device. Equations (1) show that the coupling coefficients are related to the material parameters as well as to the microstructure of the OSL. For efficient TH, the value of ratio t should be as close to that of τ as possible, which can be accomplished in the design of the structure parameters of the OSL. For the general cases $t \neq \tau$, one cannot ensure an increase in THG conversion efficiency by simply increasing the length of the sample or raising the intensity of the fundamental, even though the QPM conditions of both the SHG and the SFG processes are satisfied simultaneously. These results are totally different from a single quasi-phase-matched parametric process, such as SHG in a periodic OSL.

At present, OSL's with different microstructures[7] and different materials[17] are used for nonlinear optical frequency conversion. By an appropriate choice, it is possible to approach the optimum condition in which the TH can be generated with high conversion efficiency.

References and Notes

[1] G. D. Miller, R. G. Batchko, W. M. Tulloch, D. R. Weise, M. M. Fejer, and R. L. Byer, Opt. Lett. **22**, 1834 (1997).

[2] A. Arie, G. Rosenman, A. Korenfeld, A. Skliar, M. Oron, M. Katz, and D. Eger, Opt. Lett. **23**, 28 (1998).

[3] V. Pruneri, S. D. Betterworth, and D. C. Hanna, Opt. Lett. **21**, 390 (1996).

[4] L. E. Myers, R. C. Eckardt, M. M. Fejer, and R. L. Byer, Opt. Lett. **21**, 591 (1996).

[5] K. C. Burr, C. L. Tang, M. A. Arbore, and M. M. Fejer, Opt. Lett. **22**, 1458 (1997).

[6] T. Kartaloglu, K. G. Koprulu, O. Aytur, M. Sundheimer, and W. P. Risk, Opt. Lett. **23**, 61 (1998).

[7] M. A. Arbore, A. Galvanauskas, D. Harter, M. H. Chou, and M. M. Fejer, Opt. Lett. **22**, 1341 (1997).

[8] D. K. Serkland, P. Kumar, M. A. Arbore, and M. M. Fejer, Opt. Lett. **22**, 1497 (1997).

[9] G. D. Landry and T. A. Maldonado, Opt. Lett. **22**, 1400 (1997).

[10] D. L. Williams, D. P. West, and T. A. King, Opt. Commun. **148**, 208 (1998).

[11] O. Pfister, J. S. Wells, L. Hollberg, L. Zink, D. A. Van Baak, M. D. Levenson, and W. R. Bosenberg, Opt. Lett. **22**, 1211 (1997).

[12] K. Kintaka, M. Fujimura, T. Suhara, and H. Nishihara, Electron. Lett. **33**, 1459 (1997).

[13] J. Feng, Y. Y. Zhu, and N. B. Ming, Phys. Rev. B **41**, 5578 (1990).

[14] Y. Y. Zhu and N. B. Ming, Phys. Rev. B **42**, 3676 (1990).

[15] S. N. Zhu, Y. Y. Zhu, Y. Q. Qin, H. F. Wang, C. Z. Ge, and N. B. Ming, Phys. Rev. Lett. **78**, 2752 (1997).

[16] S. N. Zhu, Y. Y. Zhu, and N. B. Ming, Science **278**, 843 (1997).

[17] Y. Y. Zhu, R. F. Xiao, J. S. Fu, and G. K. L. Wong, Appl. Phys. Lett. **73**, 432 (1998).

[18] J. A. Armstrong, N. Bloembergen, J. Ducuing, and P. S. Pershan, Phys. Rev. **127**, 1918 (1962).

[19] Y. Y. Zhu and N. B. Ming, Opt. Quantum Electron. **31**, 1093 (1999).

[20] This study was supported by a grant from the State Key Program for Basic Research of China, by the National Advanced Materials Committee of China, and by the National Natural Science Foundation of China (grant 69938010).

Wave-Front Engineering by Huygens-Fresnel Principle for Nonlinear Optical Interactions in Domain Engineered Structures*

Yi-qiang Qin, Chao Zhang, Yong-yuan Zhu, Xiao-peng Hu, and Gang Zhao

National Laboratory of Solid State Microstructures, Nanjing University, Nanjing 210093, China

Wave-front engineering for nonlinear optical interactions was discussed. Using Huygens-Fresnel principle we developed a general theory and technique for domain engineering with conventional quasi-phase-matching (QPM) structures being the special cases. We put forward the concept of local QPM, which suggests that the QPM is fulfilled only locally not globally. Experiments agreed well with the theoretical prediction. The proposed scheme integrates three optical functions: generating, focusing, and beam splitting of second-harmonic wave, thus making the device more compact.

The ferroelectric domain structure has been widely used for a variety of applications in both linear and nonlinear optics due to its multifunctional properties. In linear optics, it can be used for wave-front controlling by electro-optic effect with lens-or prism-like domain morphology[1,2]. In nonlinear optics, the study of quasi-phase-matching (QPM) has become a hot topic recently[3-9]. QPM can be realized in a ferroelectric crystal by an artificial modulation of its second-order nonlinearity. The structure may be one-dimensional (1D) or two-dimensional (2D), periodic or quasiperiodic[10-17]. Recently, a homocentrically poled LiNbO$_3$ and annular symmetry nonlinear frequency converters have been used to increase the angle acceptance and widen phase mismatch tolerance of second-harmonic generation (SHG)[18,19]. Parametric interactions in transversely patterned QPM gratings (with the periodicity still holding in the longitudinal direction)[20,21] and even disordered domain structures[22] have attracted considerable attention. In Ref. [21], the classical Fraunhofer diffraction and the focusing of SH beam have been demonstrated. Until now, basically, most of parametric interactions realized in different domain engineered structures are totally treated in the framework of conventional QPM configuration. For QPM to be realized, the wave vectors (including the reciprocal vectors) should all be well defined. However, if the wave vectors are not well defined, would it be possible to develop a method with which nonlinear optical parametric interaction can be realized efficiently?

It is well known that Huygens-Fresnel principle (HFP) plays an important role in

---

* Phys. Rev. Lett., 2008, 100(6):063902

classical optics. Normally, this process is performed in linear optics regime and in real space. In our present Letter, we extend the HFP to nonlinear optical parametric processes induced by second-order nonlinearity. That is, each point on the primary wave-front acts as a source of secondary wavelets of the fundamental as well as a source of, for example, the SH wave. As an example of proposed method, here the HFP is used to design the domain structure, in which SH output can be controlled and focused into several points. The system acts as a complex lens for SH output; thus, the proposed scheme may be also called wave-front engineering. Both theoretical and experimental results are reported. The concept of local QPM is put forward to explain the observed phenomenon. The proposed scheme can overcome the difficulties confronted either in structures without well-defined reciprocal vector or in optical interacting waves without well-defined wave vectors such as for Gaussian beams.

In order to elucidate the above idea, we consider, as an example, the case of SHG in a domain engineered $LiTaO_3$ crystal. The input fundamental wave is a plane wave propagating along the x direction in the structure, and the generated SH is focused into n points on the focal plane, all z polarized. Usually, the domain pattern can be taken to be uniform in depth (the z direction); the system then simplifies to a 2D system, denoted as xy-plane. Within the plane, the light propagation is isotropic. The multiple reflections, leading to photonic band-gap effects, are not present in this system due to the linear dielectric constant being constant in the whole structure. In such a 2D structure, the problem can be considered as scalar[23], which simplifies the notation. Under the slowly varying envelope approximation[24], the evolution of the second-harmonic amplitude can be written as a function of the pump field and the second-order coefficient $\chi^{(2)}$[13]:

$$\boldsymbol{k}^{2\omega} \cdot \nabla [E^{2\omega}(\boldsymbol{r})] = -2i(\omega^2/c^2)[E^{\omega}(\boldsymbol{r})]^2 \chi^{(2)}(\boldsymbol{r}) \exp\{-i[2\boldsymbol{k}^{\omega}(\boldsymbol{r}) - \boldsymbol{k}^{2\omega}(\boldsymbol{r})] \cdot \boldsymbol{r}\}. \quad (1)$$

Here, $\boldsymbol{r} \equiv (x, y)$ is the 2D spatial coordinate. Generally, $E^{2\omega}$, E^{ω}, $\boldsymbol{k}^{2\omega}$, and $\boldsymbol{k}^{\omega}$ are all spatial dependent. The nonlinear harmonic components can be described physically by HFP. Because of nonlinearity in the atomic response, each atom develops an oscillating dipole moment which contains a component at frequency 2ω. Here, a small part of crystal containing an enormous number of atomic dipoles can be regarded as a point source which emits SH wave through stimulated dipoles oscillation. The initial phases of these radiations are determined by the phases of the incident fields, and modulated by the 2D domain engineered structure. In the case of focused SHG, the wave vectors $\boldsymbol{k}^{2\omega}(\boldsymbol{r})$ are spatial-dependent in the crystal. The SH wave focused at point (X_i, Y_i) propagated from the source at (x, y) with sample size $dxdy$ is given:

$$dA_i^{2\omega} = -i \frac{1}{\sqrt{R_i(x,y)}} K f(x,y) (A^{\omega})^2 \exp[-i\Delta\varphi_i(x,y)] dxdy. \quad (2)$$

Here, K is the coupling coefficient, and $f(x, y)$ is a 2D domain structural function. The distance between a point source and focused SHG point is $R_i(x, y) = \sqrt{(X_i-x)^2 + (Y_i-y)^2}$. The factor $1/\sqrt{R_i}$ represents the decay law for amplitude of cylindrical wave and $A_j = \sqrt{n_j/\omega_j}\, E_j$ ($j = \omega, 2\omega$), respectively. It is noted that $\Delta\varphi_i(x,y) = [2\,\boldsymbol{k}^\omega(\boldsymbol{r}) - \boldsymbol{k}^{2\omega}(\boldsymbol{r})] \cdot \boldsymbol{r}$, the phase mismatch between the fundamental and harmonic waves, depends on the positions of both the point sources and the focused SHG points. For determination of structural function $f(x,y)$, we derived the following correlation function:

$$F(x,y) = \sum_{i=1,n} \frac{C_i}{\sqrt{R_i}} \exp\{-\mathrm{i}[2\,\boldsymbol{k}^\omega \cdot \boldsymbol{r}(x,y) - \boldsymbol{k}^{2\omega} \cdot \boldsymbol{r}(x,y)]\}, \quad (3)$$

where C_i are the adjustable parameters related to the intensity distribution among focused points; n is the number of multifocused SHG points. Here, the "correlation" means that the change of the coordinates of any one focused point will change the value of $F(x,y)$, even change its sign, and thus change the whole domain structure as can be seen below.

For the input being a plane wave, its phase function reduces to $k^\omega x$ and

$$\Delta\varphi_i(x,y) = 2k^\omega x + k^{2\omega} R_i - k^{2\omega}(X_i \cos\theta_{1i} + Y_i \cos\theta_{2i}) \quad (i=1,\ldots,n). \quad (4)$$

Here, k^ω is the well defined wave vector of the fundamental; $\cos\theta_{1i}$ and $\cos\theta_{2i}$ are the directional cosines of $k^{2\omega}(\boldsymbol{r})$. Here, the correlation function plays the same role as the continuous-valued function[25]. Setting the solutions of $F(x,y) = 0$ to be the locations of the domain wall, the structural function $f(x,y)$ is determined:

$$f(x,y) = \begin{cases} 1, & \mathrm{Re}[F(x,y)] \geq 0 \\ -1, & \mathrm{Re}[F(x,y)] < 0 \end{cases}. \quad (5)$$

By calculating the integral over the whole system, the focused SHG at spatial points can be obtained.

Formulas (1)–(5) constitute the basis of a general theory for domain engineering with conventional QPM structures being the special case. As an example, wave-front engineering for the multifocused SHG by HFP is discussed.

Based on the above theory, the 2D domain structures for the focused SHG are calculated. As a demonstration, Fig. 1 shows the ferroelectric domain structures with the focused points designed to be on the exit surface of the crystal. Obviously, the domain modulation is quite different from the conventional 1D and 2D structures. Figure 1(a) shows the domain structure for the single focused SHG. Near the focused point, the domain boundary curves strongly, whereas far away from the focused point, the domain morphology looks like the conventional 1D periodic structure [as shown in the inset of Fig. 1(a)] which approaches the case investigated in Ref.[21].

Figure 1(b) shows the domain structures for the dual focused SHG. The conventional stripelike domains break into small bricks. The domain structures for the decal focused SHG becomes strange as shown in Fig. 1(c). Because of these domain structures, the SH wave-front generated from a fundamental plane wave is no longer a plane wave-front; rather, it becomes very complicated depending on the number of focused points, resulting in that the conventional QPM is no longer valid.

In order to verify the above scheme, we performed the experiments on focused SHG.

The sample was fabricated by poling a 0.5 mm thick z-cut $LiTaO_3$ single domain wafer at room temperature. The domain structure was designed such that the single focused SHG or dual focused SHG for the fundamental 1319 nm z-polarized can be realized with the focused points located 10 cm away from the exit surface. For dual focused SHG, the two focused points are 2 mm apart. In general $f(x,y)$ can have a number of domains whose sizes are shorter than the critical size (about 2 μm in our experiment). Because of the electric-poling limits of domain size, in the design, we merge those domains smaller than 2 μm with the adjacent domains. According to our calculation, such technical limits will influence the SH conversion efficiency[10], but keep almost the same focusing features.

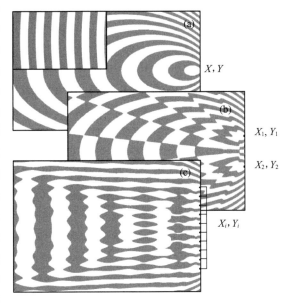

FIG. 1. The schematic diagram of ferroelectric domain structures for (a) single focused SHG (The inset is the domain morphology far from the focal point), (b) dual focused SHG, and (c) decal focused SHG. The focused points are indicated by (X_i, Y_i).

SHG of fabricated samples with 10 mm length and 3 mm width was tested using a mode-locked Nd:YAG laser system with a pulse width of 150 ns and a repetition rate of 4 kHz. The fundamental wave was coupled into the polished end face and propagated along the x axis of the sample. Three samples were tested: periodic sample, samples with single focused SHG, and dual focused SHG, whose corresponding domain structures are exhibited in the bottom insets of Fig. 2.

Using a weakly focused fundamental beam by a 15 cm focal-length lens, the waist inside the sample is ~300 μm where the intensity drops to $1/e^2$ of the maximum value and the confocal parameter for the system is Z_0 ~10 cm much larger than the sample length 10 mm. Thus the fundamental beam can be considered as a plane wave. Figure 2 shows the SHG output CCD images for three different samples. Figure 2(a) exhibits the SHG image in periodic sample, corresponding to the conventional 1D QPM scheme. The beam waist of red SHG is a little bit smaller (~80%) than the waist of the input fundamental beam. Figures 2(b) and 2(c) show the experimental results of the single and dual focused SHG (The calculated results are shown in the top insets). Figure 2(b) shows the single focal SHG output with its beam waist considerable smaller. Figure 2(c) corresponds to the situation of dual focused SHG. By using bricklike domains with neither translational nor annular symmetry, wave-fronts of SHG are controlled to be convergent towards the two focal points. In both cases, the measured SH minimum waists are about 120 μm, agreeing well with the simulation result.

The SH powers are measured using a Si-calibrated field master detector. In the measurement with the fundamental power 260 mW, the average SH outputs are 110 mW and 63 mW in 1D periodic and bricklike structures, respectively. Here, 42% SH conversion efficiency is achieved in periodic structure, 24% in bricklike structures. Experimental data also indicate almost the same SHG efficiency is obtained in 1D periodic structure and single focused SHG structure.

For deep understanding of the physical nature of wave-front engineering by HFP, the Fourier spectra of the 2D domain structure for dual focused SHG are studied theoretically and experimentally. A He-Ne laser beam (632.8 nm) is used to scan the z surface of the sample horizontally and vertically. The diffraction pattern is projected onto a screen and recorded by a CCD camera. The Fourier transformation is performed numerically. Figure 3 is the measured (a) and calculated (b) diffraction patterns obtained along the vertical direction. From the results, some distinct features can be revealed. The most noticeable diffraction spots lay mainly in the second and forth quadrant for the upper patterns and in

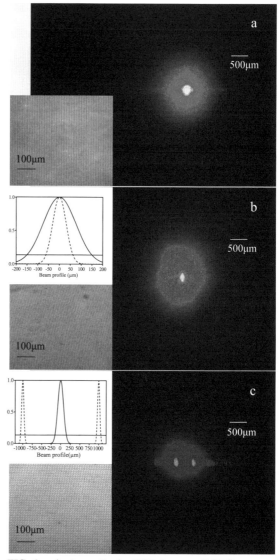

FIG. 2 (color online). CCD images of SHG generated from (a) a periodic domain structure, (b) a domain structure for single focused SHG, and (c) a domain structure for dual focused SHG, respectively. The bottom insets are their corresponding optical microscopic images of domain structures revealed by etching. The top insets in (b) and (c) are the calculated beam profiles of input fundamental wave (solid lines) and output SH wave (dashed lines) at focusing plane.

the first and third quadrant for the bottom patterns. They show mirror symmetry with each other, which reflects the symmetry of domain pattern in real space. The symmetries of these patterns are totally different from the symmetry of the middle ones. This indicates the reciprocal vectors are spatial dependent. Actually, in the experiments, when moving the laser beam vertically from the upper part to the bottom part, the diffraction pattern has been observed to change gradually. The diffraction pattern changes slightly along the horizontal direction due to the fact that the focused points are designed rather far from the exit face of the sample. From the diffraction pattern exhibited above, it seems possible to define the so-called local QPM condition although the QPM condition could not be fulfilled globally. That is, the phase mismatch can be compensated locally with reciprocal vectors provided by local structure. In Fig. 3(b), we schematically show the reciprocal vectors used for local QPM, where G_1 is used for the upper focal point and G_2 for the lower one.

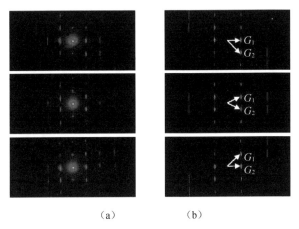

(a)　　　　(b)

FIG. 3 (color online). Fourier spectra of domain structure shown in the inset of Fig. 2(c) for dual focused SHG obtained along the vertical direction: (a) the experimental and (b) the calculated results. The schematic diagrams of local QPM are indicated.

It is interesting to compare the conventional QPM structure to current domain structure. When the single focused point of SHG is designed to be located infinitely, 2D modulated structure function $f(x,y)$ degenerate into that of conventional 1D QPM structure with the period of $2\pi/(k^{2\omega}-2k^{\omega})$. The structure can also degenerate into the conventional 1D structure when the focused points on the focal plane approach infinity. Moreover, when multifocused points of SHG are located infinitely at two perpendicular directions, the structure becomes the 2D periodic structure.

The advantage of the proposed scheme lies in its capability to perform several functions with a special designed domain structure. As one of the perspectives for application, the example presented above combines three functions: SHG, focusing, and beam splitting, thus making the device more compact. The method can also be used for the tight focused Gaussian beam with which the efficient frequency conversion does not occur at phase-matching condition[24,26] and extended to other parametric processes such as down conversion. Studies on these processes by using local QPM are of potential interest in photonic applications. Thus, HFP may play an important role in nonlinear optical field with engineered domain structures.

References and Notes

[1] Y. Chiu, D. D. Stancil, and T. E. Schlesinger, Appl. Phys. Lett. **69**, 3134 (1996).

[2] M. Yamada, M. Saitoh, and H. Ooki, Appl. Phys. Lett. **69**, 3659 (1996).

[3] G. D. Miller R. G. Batchko, W. M. Tulloch, M. M. Fejer, and R. L. Byer, Opt. Lett. **22**, 1834 (1997).

[4] M. A. Arbore A. Galvanauskas, D. Harter, M. H.Chou, and M. M. Fejer, Opt. Lett. **22**, 1341 (1997).

[5] S. N. Zhu et al., Phys. Rev. Lett. **78**, 2752 (1997); L. Noirie, P. Vidakovic, and J. A. Levenson, J. Opt. Soc. Am. B **14**, 1 (1997); P. Unsbo, *ibid*. **12**, 43 (1995); G. I. Stegeman M. Bahae, E. W. Van Stryland, and G. Assanto, Opt. Lett. **18**, 13 (1993).

[6] M. Tiihonen, V. Pasiskevicius, and F. Laurell, Opt. Express **12**, 5526 (2004).

[7] L. Torner, C. R. Menyuk, and G. I. Stegeman, Opt. Lett. **19**, 1615 (1994).

[8] A. Stabinis, G. Valiulis, and E. A. Ibragimov, Opt. Commun. **86**, 301 (1991).

[9] G. Assanto et al., Appl. Phys. Lett. **67**, 2120 (1995).

[10] M. M. Fejer et al., IEEE J. Quantum Electron. **28**, 2631 (1992).

[11] Y. Y. Zhu and N. B. Ming, Quantum Electron. **31**, 1093 (1999).

[12] K. Fradkin-Kashi, A. Arie, and G. Rosenman, Phys. Rev. Lett. **88**, 023903 (2001).

[13] V. Berger, Phys. Rev. Lett. **81**, 4136 (1998).

[14] N. G. Broderick et al., Phys. Rev. Lett. **84**, 4345 (2000).

[15] L. H. Peng et al., Appl. Phys. Lett. **84**, 3250 (2004).

[16] B. Q. Ma et al., Appl. Phys. Lett. **87**, 251103 (2005).

[17] R. Lifshitz, A. Arie, and A. Bahabad, Phys. Rev. Lett. **95**, 133901 (2005).

[18] T. Wang et al., Opt. Commun. **252**, 397 (2005).

[19] D. Kasimov et al., Opt. Express, **14**, 9371 (2006).

[20] G. Imeshev, M. Proctor, and M. M. Fejer, Opt. Lett. **23**, 673 (1998).

[21] J. R. Kurz et al., IEEE J. Sel. Top. Quantum Electron. **8**, 660 (2002).

[22] M. Baudrier-Raybaut et al., Nature (London) **432**, 374 (2004).

[23] P. R. Villeneuve and M. Piche, Phys. Rev. B **46**, 4969 (1992).

[24] R. W. Boyd, *Nonlinear Optics* (Academic Press, New York, 2003).

[25] T. Kartaloglu, Z. G. Figen, and O. Aytur, J. Opt. Soc. Am. B **20**, 343 (2003).

[26] G. D. Xu et al., J. Opt. Soc. Am. B **20**, 360 (2003).

[27] This work was supported by the State Key Program for Basic Research of China (Grants No. 2004CB619003 and No. 2006CB921804) and the National Natural Science Foundation of China (Grants Nos. 10523001, 10504013, and 10674065).

Conical Second Harmonic Generation in a Two-Dimensional $\chi^{(2)}$ Photonic Crystal: A Hexagonally Poled LiTaO$_3$ Crystal [*]

P. Xu, S. H. Ji, S. N. Zhu, X. Q. Yu, J. Sun, H. T. Wang, J. L. He, Y. Y. Zhu, and N. B. Ming

National Laboratory of Solid State Microstructures, Nanjing University, Nanjing, 210093, China

A new type of conical second-harmonic generation was discovered in a 2D $\chi^{(2)}$ photonic crystal—a hexagonally poled LiTaO$_3$ crystal. It reveals the presence of another type of nonlinear interaction—a scattering involved optical parametric generation in a nonlinear medium. Such a nonlinear interaction can be significantly enlarged in a modulated $\chi^{(2)}$ structure by a quasi-phase-matching process. The conical beam records the spatial distribution of the scattering signal and discloses the structure information and symmetry of the 2D $\chi^{(2)}$ photonic crystal.

Recently several novel classes of noncollinear parametric interactions were discovered, such as the ring-shaped second-harmonic generation (SHG) excited by a Bessel beam in a periodically poled KTP crystal[1], the hollow beam generated by the frequency difference in a periodically poled LiTaO$_3$ crystal[2], the ring-shaped SH generated from a SBN crystal having an antiparallel domain when a fundamental wave is propagated in its crystallographic c axis[3], and so on. On the other hand, ring-shaped harmonic patterns were even produced from either a single domain crystal[4] or an isotropic material[5]. In all the cases above, conical patterns indicated that either the single fundamental beam was particularly prepared or an additional beam was required in addition to the incident fundamental beam.

The nonlinear crystal with a modulated nonlinear susceptibility tensor $\chi^{(2)}$ was called a quasi-phase-matching (QPM) material. The concept was first referred to by Armstrong[6] and Franken and Ward[7] independently. In 1998, Berger extended the QPM study from one dimension (1D) to 2D[8], and proposed the concept of an $\chi^{(2)}$ photonic crystal in order to contrast and compare it with a regular photonic crystal having a periodic linear susceptibility. Since then, the concept of a $\chi^{(2)}$ photonic crystal has been gradually accepted in the nonlinear optics community. A number of theoretical and experimental results show that the $\chi^{(2)}$ photonic crystal, in particular, the 2D one, provides a valuable platform to study light-matter interaction in a highly nonlinear regime[9–12].

In this Letter, we report a novel nonlinear optical phenomenon—a conical second-

---

[*] Phys. Rev. Lett., 2004, 93(13):133904

harmonic (SH) beam, generated from a 2D $\chi^{(2)}$ photonic crystal—a hexagonally poled LiTaO$_3$(HPLT) crystal. The conical SH beam emerged when the HPLT was illuminated with a z-polarized fundamental beam with no special requirements. The conical beams were visualized as rings when projected onto a screen behind the crystal, and were presented in axial symmetry or mirror symmetry when the fundamental beam propagated along a symmetrical axis of the hexagonal structure. This reveals the presence of another type of nonlinear interaction—a scattering involved optical parametric generation. The interaction is greatly enhanced by a QPM process in the 2D $\chi^{(2)}$ photonic crystal; thus, the infrared scattering signal (ω) is converted to a visible band (2ω) through mixing with the incident wave (ω). Further study confirms that the conical beam records the spatial distribution of the scattering signal and reveals the structure information of the 2D $\chi^{(2)}$ photonic crystal.

In the experiments the 2D $\chi^{(2)}$ photonic crystal is a hexagonally poled LiTaO$_3$ crystal fabricated by a field poling technique[13]. Figure 1(a) shows its domain structure slightly etched in acids. The nearly circularly inverted domains (with $-\chi^{(2)}$) distribute regularly in a $+\chi^{(2)}$ background with $a = 9.05$ μm and a reversal factor of $\sim 40\%$. No domain merging was found across the sample dimensions of 1.55 cm (x) × 1.55 cm (y). Figure 1(b) is the corresponding reciprocal space, in which each reciprocal vector has six equivalent ones[9].

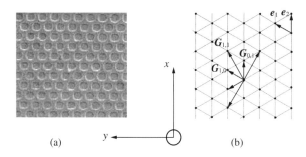

FIG. 1. (a) HPLT crystal and its domain structure. (b) Reciprocal lattice and reciprocal vectors of the HPLT crystal. In our experiments the fundamental beam propagated in the $+x$ and the $+y$ directions, respectively. Here are two sets of the mentioned reciprocal vectors clarified by the magnitude, $4\pi/\sqrt{3}a$ and $4\pi/a$, respectively.

The phase-matching condition for SHG in a 2D $\chi^{(2)}$ photonic crystal is generally written into

$$\boldsymbol{k}_2 - 2\boldsymbol{k}_1 - \boldsymbol{G}_{m,n} = 0, \qquad (1)$$

where $\boldsymbol{G}_{m,n} = \dfrac{4\pi}{\sqrt{3}a}(\sqrt{m^2+n^2+m \cdot n})$ is the reciprocal vector of the 2D hexagonal lattice with the lattice parameter of a, and the subscripts m and n are integers, representing the order of the reciprocals. In Eq. (1), k_1 represents the incident beam providing two fundamental waves, and three vectors, k_1, k_2, and $G_{m,n}$, may be collinear or noncollinear to one another.

We performed a detailed experimental study of SHG for this HPLT in a single-pass scheme. The fundamental wave source is a tunable optical parametric oscillator (PL9010, Continuum, Santa Clara, CA). Its output, having a nearly Gaussian profile, can be tuned from ultraviolet (200 nm) to infrared (1700 nm) with the pulse width of ~5 ns, the linewidth ≤0.075 cm^{-1}, and the repetition rate of 10 Hz. The z-polarized fundamental beam was weakly focused with a beam waist of 0.1 mm inside the crystal. The confocal parameter $Z_0 \approx 6$ cm for the system. It is much longer than the length of the sample of 1.55 cm; therefore, the divergence of k_1 inside the crystal is negligible.

At the beginning, the fundamental k_1 propagated along the x direction of the HPLT crystal. By changing the wavelength of k_1, two kinds of axial SH beams k_2 were found as expected according to Eq.(1). The first one was a single axial beam collinear with k_1, and the involved reciprocal $G_{m,n}$ was parallel to k_1. The other one was a pair of axial beams symmetrical from k_1 in the x-y plane[9]. In this case, a pair of symmetrical reciprocals from k_1 participated in the noncollinear QPM process. In the visible band, the number of such an axis beam at different wavelengths adds up to be about 20. It is caused by the participation of different reciprocals. All results are in good agreement with the theoretical estimations by Eq. (1). For simplicity, two of them are studied in detail. One occurred at SH $\lambda_2 = 532$ nm in a single axial beam by $G_{0,1}$ and the other at $\lambda_2 = 466.5$ nm in a pair of axial beams by $G_{1,1}$ and its mirror image.

In a subsequent experiment, new effects were exhibited as k_1 was detuned from the above matching points toward a longer wavelength. We found that the axial SH beams disappeared, yet, instead, a new type of beam, a conical SH beam, emerged from this HPLT. These beams exhibited circular rings when projected onto a screen as shown in Fig. 2. The inset of Fig. 2(a) is a single circular ring at $\lambda_2 = 533$ nm. It developed from the single axial SH beam at 532 nm. Its center overlapped with the projection point of k_1. Its radius increased with the increase of the fundamental wavelength λ_1.

Another kind of conical SH beams, k_2, emerged from the crystal in pairs as shown in the inset of Fig. 2(b). Two rings have the identically symmetrical angles from k_1. Similarly, these two rings decreased their radiuses synchronously with the decreasing of λ_1. During this process, a bright spot appeared on the inner side of each ring at a specified $\lambda_1 = 933$ nm. They were the SH of λ_1, fulfilling Eq. (1) in noncollinear scheme. As the rings reduced in radius with the further decrease of λ_1, these two bright spots on the rings disappeared due to their phase-matched condition to be destroyed. The rings shrunk to their centers as $\lambda_1 = 930$ nm, forming another two weaker bright spots at $\lambda_2 = 465$ nm. Both the ring and the spot disappeared as λ_1 was tuned down to shorter than 930 nm. α_1 and α_2 are defined in Fig. 3(b) as the angles of k_2 to k_1 on the x-y plane and the dependence on λ_2 is shown in Fig. 2(b). The conical SH beams and latter two weak bright spots mentioned above cannot be explained using Eq. (1). Obviously, they occur through a different mechanism.

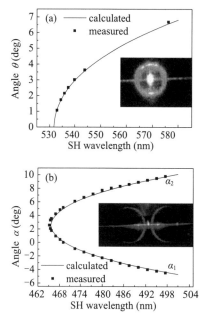

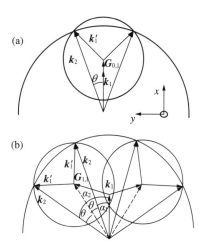

FIG. 2. (color online)(a) The angle θ as a function of λ_2 for a single ring. (b) The angle α as a function of λ_2 for a pair of rings (see text). The insets of (a) and (b) are the axially symmetrical ring and the mirror-symmetrical ring, respectively. The bright spot at the ring center in (a) is the projection of k_1.

FIG. 3. Phase-matching geometries for (a) the axially symmetrical ring and (b) the mirror-symmetrical rings.

A comprehensive interpretation for the conical SH beams can be given by introducing an additional fundamental wave k_1'. Taking k_1' to be noncollinear to k_1, Eq. (1) can be more generally written as

$$k_2 - k_1 - k_1' - G_{m,n} = 0. \quad (2)$$

The equation degenerates into Eq. (1) when $k_1' = k_1$. In order to obtain a conical SH beam, k_1' has to present a continuously and symmetrically spatial distribution from k_1. In our experiment k_1 was loosely focused and no second fundamental beam was directed onto the crystal; therefore, k_1' can be provided from k_1 by scattering. k_1' has a small scattering angle from k_1. The scattering may arise mainly from domain walls and other imperfections on the surface and inside the crystal and does not change the wavelength of light[14]. The scattering effect has been observed by nonlinear interactions in some homogeneous crystals[15]. Now, however, it is clarified in a 2D $\chi^{(2)}$ photonic crystal in a QPM scheme, which leads to a new class of SH pattern—a reciprocal-dependent conical SH beam.

Geometries for the generation of two kinds of conical beams at normal incidence onto the HPLT are shown in Figs. 3(a) and 3(b), respectively. In Fig. 3(a), $\boldsymbol{k}_1 \parallel \boldsymbol{G}_{0,1}$, and $\boldsymbol{k}_1'$ exhibits a forward axis-symmetrical distribution around $\boldsymbol{k}_1$. The generated SH $\boldsymbol{k}_2$ obeying Eq. (2) forms a conical beam. The angle θ between $\boldsymbol{k}_1$ and $\boldsymbol{k}_2$ is given by

$$\cos(\theta) = \frac{|\boldsymbol{k}_2|^2 + |\boldsymbol{k}_1 + \boldsymbol{G}_{m,n}|^2 - |\boldsymbol{k}_1|^2}{2|\boldsymbol{k}_2| \cdot |\boldsymbol{k}_1 + \boldsymbol{G}_{m,n}|}, \tag{3}$$

where $\boldsymbol{G}_{m,n}$ is $\boldsymbol{G}_{0,1}$ for the present case. Because $\cos\theta \leqslant 1$, the equation implies that the conical beam just occurs as λ_1 is longer than a critical wavelength at which $\theta = 0$. The angle θ increases with λ_1. The curve calculated based on Eq. (3) fits the measured values well as shown in Fig. 2(a). Figure 3(b) shows another phase-matching geometry that leads to the generation of the pair of rings on the screen shown as the inset of Fig. 2(b), where $\boldsymbol{G}_{m,n}$ is $\boldsymbol{G}_{1,1}$ or its mirror image. These two rings are symmetrical from $\boldsymbol{k}_1$, as well as from the x axis. Equation (3) is still valid for this phase-matching geometry. However, here θ is the angle between $\boldsymbol{k}_2$ and $\boldsymbol{k}_1 + \boldsymbol{G}_{1,1}$ instead of $\boldsymbol{k}_1$[see Fig. 3(b)]. Obviously, the projection spot of $\boldsymbol{k}_1 + \boldsymbol{G}_{1,1}$ on the screen is the center of rings, and it is invisible generally. The center changes its position with λ_1 because $\boldsymbol{k}_1 + \boldsymbol{G}_{1,1}$ is a function of λ_1. The pair of rings reduce in radius with the shortening of λ_1, and, lastly, the conical beams $\boldsymbol{k}_2$ shrink into two axial beams with $\theta = 0$ when the $\boldsymbol{k}_2$, $\boldsymbol{k}_1'$, and $\boldsymbol{k}_1 + \boldsymbol{G}_{1,1}$ three vectors are collinear. At that moment two symmetrical bright spots appear on the screen.

It should be mentioned that $LiTaO_3$ is a birefringence crystal. The ring on the screen should be elliptical and not circular. However, the birefringence of the crystal is so small that the calculated result shows the deviation from a circle amounts to about $10^{-4°}$, and, therefore, is negligible in the experiment.

In addition to the two cases of rings studied above, there are many other rings observed in the visible band. They differ little but correspond to different reciprocals. As the HPLT was rotated by $90°$ around the z axis and $\boldsymbol{k}_1$ propagated in the y direction, two pairs of conical SH beams with different azimuths occurred simultaneously. They are, respectively, associated with two sets of reciprocals, $\boldsymbol{G}_{1,0}$ and its mirror image, and $\boldsymbol{G}_{1,1}$ and its mirror image, the former for two inner rings and the latter for two outer rings. Two pairs of rings and their evolution with λ_1 is shown in Figs. 4(a)–4(d).

We measured the intensity dependence of an axially symmetric ring [Fig.2(a)] on θ. The intensity integral along the ring decreased considerably with the increasing of θ as shown in Fig.5, which implies the SHG efficiency is strongly dependent on the distribution of scattering light. Strong scattering occurs when $\theta < 4°$ in the sample. For the pair of rings, the intensity of the each part of the ring is different, depending on the azimuth relative to $\boldsymbol{k}_1$[see Fig. 3(b)]. Solving the nonlinear wave equation, the SH intensity can be expressed as $I(L, 2\omega) \propto I(\omega)I'(\omega)L^2$, where L is the interaction volume and $I(\omega)$ the intensity of $\boldsymbol{k}_1$. Scattering intensity $I'(\omega) = \gamma I(\omega)$, where γ is the scattering coefficient. Thus harmonic intensity $I(2\omega)$ is directly proportional to $I^2(\omega)$. The inset of Fig. 5 shows

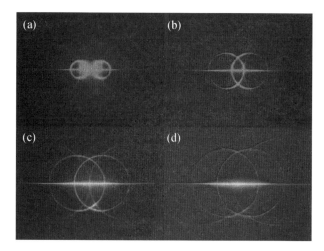

FIG. 4. (color online) The projection of the mirror-symmetrically conical SH beams on the screen ($k_1 \parallel y$ axis): (a) $\lambda_1 = 1.12$ μm; (b) $\lambda_1 = 1.13$ μm; (c) $\lambda_1 = 1.15$ μm; (d) $\lambda_1 = 1.17$ μm.

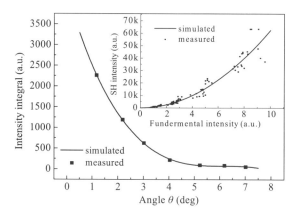

FIG. 5. The intensity integral of the axially symmetrical ring as a function of θ. The inset shows the dependence of $I(2\omega)$ on $I(\omega)$ at $\theta = 0.5°$.

the measured total intensity of ring $I(2\omega)$ dependence on $I(\omega)$ at $\theta = 0.5°$. The fit curve in the figure confirms the square law dependence of $I(2\omega)$ on $I_1(\omega)$. We obtained a maximum average power of ~3.55 mW at the input power of 25.7 mW with an efficiency of ~14%.

Recent parametric scattering patterns, such as the ring and the line, were found in photorefractive crystals, such as Y, Fe codoped $LiNbO_3$[16], Cr doped SBN[17], etc. $LiTaO_3$ is one of the photorefractive crystals. We performed an elevated temperature measurement up to 170℃. No obvious change was found in the ring intensity except that the ring reduced in radius with temperature. It was reported that the photorefractive effect in $LiTaO_3$ could be greatly depressed over 80℃, so the conical SH beam could not originate from this effect.

The conical SH beam generated in a 2D $\chi^{(2)}$ photonic crystal contains useful information about the quality and microstructure of crystal and, therefore, can be viewed as a sensitive, contact-free probe for the characterization of a nonlinear photonic crystal and for the study of light scattering in a crystal. For example, one may obtain the angular distribution of scattering light inside the crystal from the results as shown in Fig. 5. We also found the conical beams generated from 1D $\chi^{(2)}$ crystals like PPLN and PPLT. For a 1D quasiperiodical structure, the pattern on the screen is a set of concentric rings. They exhibit a quasiperiodical ratio at radius. Therefore, the patterns can disclose the structure information of the $\chi^{(2)}$ crystal.

It is worth noting that the scheme in this Letter provides new possibilities for engineering nonlinear interactions. It also demonstrates a new approach for generating a conical beam by frequency conversion. However, the brightness of the conical beam can be significantly enhanced by intentionally inducing the imperfections on the surface and inside the crystal or by using another laser with proper divergence, for some particular purposes.

References and Notes

[1] A. Piskarskas et al., Opt. Lett. **24**, 1053 (1999).
[2] G. Giusfredi et al., Phys. Rev. Lett. **87**, 113901 (2001).
[3] A. R. Tunyagi et al., Phys. Rev. Lett. **90**, 243901 (2003).
[4] J. A. Giordmaine, Phys. Rev. Lett. **8**, 19 (1962).
[5] K. D. Moll et al., Phys. Rev. Lett. **88**, 153901 (2002).
[6] J. Armstrong et al., Phys. Rev. **127**, 1918 (1962).
[7] P. A. Franken and J. F. Ward, Rev. Mod. Phys. **35**, 23 (1963).
[8] V. Berger, Phys. Rev. Lett. **81**, 4136 (1998).
[9] N. G. R. Broderick et al., Phys. Rev. Lett. **84**, 4345 (2000).
[10] N. G. R. Broderick et al., J. Opt. Soc. Am. B **19**, 2263 (2002).
[11] L. H. Peng et al., Appl. Phys. Lett. **83**, 3447 (2003).
[12] S. N. Zhu, Y.Y. Zhu, and N. B. Ming, Science **278**, 843 (1997).
[13] S. N. Zhu et al., J. Appl. Phys. **77**, 5481 (1995).
[14] B. Bittner et al., J. Phys. Condens. Matter **14**, 9013 (2002).
[15] K.U. Kasemir and K. Betzler, Appl. Phys. B **68**, 763 (1999).
[16] M. Goulkov et al., Phys. Rev. Lett. **86**, 4021 (2001).
[17] M. Goulkov et al., Phys. Rev. Lett. **91**, 243903 (2003).
[18] This work was supported by grants for the National Advanced Materials Committee of China and for the State Key Program for Basic Research of China, as well as by the National Natural Science Foundation of China under Grant No. 90201008 and Natural Science Foundation of Jiangsu under Grant No. BK2002202.

Experimental Studies of Enhanced Raman Scattering from a Hexagonally Poled LiTaO₃ Crystal[*]

P. Xu, S. N. Zhu, X. Q. Yu, S. H. Ji, Z. D. Gao, G. Zhao, Y. Y. Zhu, and N. B. Ming

National Laboratory of Solid State Microstructures, Nanjing University, Nanjing, 210093, China

The study of phonon-polariton Raman scattering continues to be a challenge, in part because of the difficulty of measuring relatively low intensity of scattered signals. This paper reports the experimental results on phonon-polariton Raman scattering in a hexagonally poled LiTaO₃ crystal, showing the anti-Stokes and Stokes Raman signal intensities are significantly enhanced by cascading a couple of quasi-phase-matching processes where the coherent polariton fields are driven and the enhanced scattering signals are further amplified.

Quasi-phase-matching (QPM) makes nonlinear light wave mixing active in a more general stage through phase compensation with the modulation of secondorder nonlinear susceptibility $\chi^{(2)}$ of the medium,[1—3] which usually leads to generating laser at a new frequency efficiently by sum frequency,[4] difference frequency,[5] or parametric amplification.[6] In addition, the QPM approach has promising applications, by modulating the phase and amplitude of an optical field,[7] compression of a tunable-chirp pulse,[8] generation of an optical soliton,[9] realization of optical bistability,[10] optical chaos,[11] etc. Other progresses, such as use of QPM material as a coherent THz radiation source[12] and bright entanglement pair source,[13] have emerged.

In addition, the QPM approach can also enhance the scattering signal intensity in a nonlinear medium. In a recent work,[14] we reported the generation of a conical second-harmonic beam in a hexagonally poled LiTaO₃ (HPLT) crystal by a quasi-phase-matched frequency adding of incident light and elastic scattering light. The information relating to elastic scattering in the HPLT as well as its symmetry was revealed from the nonlinear diffractive pattern. QPM, for the first time, showed its elegant function in the study of light scattering.

Raman scattering, as a sort of inelastic scattering, is always accompanied by frequency change. Raman scattering can be mediated by many different types of elementary excitations in a medium, such as TO phonons, phonon-polaritons, or plasmons. Compared to elastic scattering, Raman scattering is much weaker, so it brings more challenge in detection. One immediate question can be proposed, that is, whether Raman scattering is possibly enhanced by the QPM process in a $\chi^{(2)}$ modulated medium and what spectrum

---

[*] Phys. Rev. B, 2005, 72(6): 064307

characteristic may be revealed if this occurs. In this paper, we will report the experiment studies on QPM-enhanced phonon-polariton (P-P) Raman scattering in a manual microstructure—a HPLT crystal. The QPM approach was first operated to generate two laser beams to drive an intense coherent P-P field in the HPLT crystal and was then used to amplify the generated Raman signals by parametric amplification, which resulted in the generation of a stimulated P-P Raman scattering. The anti-Stokes and Stokes spectrum with very low Raman shift down to 2 cm^{-1} and very high scattering order up to the 11th rank was detected from the medium. The resulting spectrum exhibits a comb-shaped structure with an equal frequency interval tunable by changing the frequency of excited polariton. The QPM approach was successfully introduced into the $\chi^{(2)}$ medium for enhancement of inelastic scattering.

In experiment, the HPLT slice of 20 mm(x)×15 mm(y)×1 mm(z) was fabricated by a field poling technique.[15] The nearly circularly inverted domains have $-\chi^{(2)}$ distributed hexagonally in the $+\chi^{(2)}$ background, "lattice parameter" $a = 9.05$ μm, and a reversal factor $D = \sim 23\%$. This structure offers a reciprocal vector $|G| = 4\pi/\sqrt{3}\,a$ along the x axis, which is of sixfold symmetry. Figure 1 is the experimental setup. The laser source is a tunable optical parametric oscillator (OPO), which can supply tunable pair beams generated through down-conversion from a 355 nm laser. During this measurement, one beam was tuned at wavelength of around 1064 nm as signal, the other one dependently tuned around 532 nm as pump λ_0. The pair beams have nearly Gaussian profiles with a pulse width of ~ 5 ns, linewidth $\leqslant 0.1$ cm^{-1}, and repetition rate of 10 Hz, and were linearly polarized along the z axis of the crystal. Two lenses, of focal lengths $f = 40$ cm and $f = 20$ cm, were used to focus the pump and signal beams into the sample with beam divergences of 10 mrad and 16 mrad, respectively.

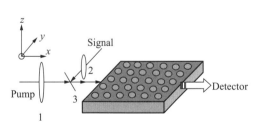

FIG. 1. Experimental setup in which 1 and 2 are focal lens and 3 is the mirror of high reflection for signal and high transmission for pump at 45°.

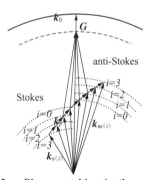

FIG. 2. Phase matching in the cascaded generation of anti-Stokes and Stokes Raman signals. Only the first three orders are sketched for simplification. The inner momentum parallelogram accounts for the initial QPM-DFG process and the outer one for the QPM parametric amplification for anti-Stokes and Stokes signals.

The pump and signal beams from the OPO were incident onto the HPLT sample and propagated along its x axis, which resulted in the generation of the third laser beam also at a wavelength of around 1064 nm by the difference frequency generation (DFG) process. The reciprocal G by $\chi^{(2)}$ modulation along the x axis made this process quasi-phase-matched. Conventionally, in LiTaO$_3$ crystal the lowest A$_1$ phonon mode (200 cm^{-1} or so) (Ref. [16]) and the corresponding phonon-polariton lying on its low-frequency wing can all be excited but rather weak if only a single laser beam is used. In the experiment, however, the excitation was greatly improved. The intense P-P field (k_p, ω_p) was driven coherently by two laser beams, k_1 and k_2, under the conservation conditions of wave vector and energy:

$$k_p = k_1 - k_2 \text{ and } \omega_p = \omega_1 - \omega_2. \qquad (1)$$

Henceforth we define $k_1(k_2)$ as the higher(lower) frequency beam between the signal and the difference frequency beams. Experimentally, $k_p = k_1 - k_2$ could be satisfied by making use of the divergence of k_1 and k_2.

The intense P-P fields then supplied the inelastic source and brought about new additional frequency $\omega_{as(n)} = \omega_1 + n\omega_p$ at anti-Stokes and $\omega_{s(n)} = \omega_2 - n\omega_p$ at Stokes, where $n = 1, 2, 3$, standing for the order of Raman scattering. Meanwhile the wave vector conservation demands

$$k_{as(n)} = k_{as(n-1)} + k_p \text{ and } k_{s(n)} = k_{s(n-1)} - k_p. \qquad (2)$$

Here $k_{as(n)}$ and $k_{s(n)}$ are the wave vectors of the nth-order anti-Stokes and Stokes, respectively, and for generation of the first-order Raman pair $k_{as(1)}$ and $k_{s(1)}$, we define $k_{as(0)} = k_1$ and $k_{s(0)} = k_2$. Equation (2) can be ensured since the crossing angle of $k_{as(n)}$ to $k_{as(n-1)}$ is rather small compared with the divergence of k_1 and k_2. Equations (1) and (2) can therefore be combined to be a cascaded automatic-phasematching (APM) two-step $\chi^{(2)}$ process. Figure 2 is the sketched momentum geometry of this cascaded process. Owing to an intense coherent P-P field being excited, the high-order scattering occurred. The Raman spectra exhibited a multi-peak structure with $2(n+1)$ separate frequencies with an equal frequency interval and the spatial distribution of the scattering light on the screen behind the sample presented a round spot or a round ring, depending on the phase-matching geometry of the foregoing QPM-DFG process. If the pump wavelength exceeds 531.64 nm, DFG must be noncollinear, i.e., the generated k_1 and k_2 are not along the direction of pump k_0 but construct a momentum triangle. Then a ring pattern can be expected on screen and the sequential APM two-step process occurred also within the ring. Experimentally the pattern developed from an obscured spot to a ring when the wavelength of pump k_0 varied from 531.64 nm to 532.39 nm, which was very consistent with theoretical estimation. In this experiment, the coherent P-P field was tuned in the range of 2 – 42 cm^{-1} by changing pump wavelength, which lay on the low-frequency wing of the lowest A$_1$-TO phonon mode. Although the P-P dispersion characteristic may still be disturbed by possible Debye relaxation mode, defect mode, or other low-frequency excitations in this crystal, the previous work gave that absorption coefficients of the polariton in this range were small, showing the polariton here to be photon-like.[17]

The Raman spectra corresponding P-P frequency $\omega_p = 19$ and 8.5 cm^{-1} are shown in Figs. 3(a) and 3(b), respectively. The insets are their output patterns distributed on the screen. The spectra have nearly symmetric distribution in which k_1, k_2 (the middle two peaks) exhibit almost identical intensity due to QPM interaction. The peaks are where the anti-Stokes or Stokes side has an order number of as many as 7 or 8. The intensity of high order Raman signal is relatively low but decreases more slowly with n getting larger. In Fig. 3(a) the anti-Stokes side, the intensity ratio of adjacent peaks $I_{as(n)}/I_{as(n-1)}$ differs greatly, from 14% ($n=1$) to 70% ($n=8$). The measured average power for the highest order of Raman shift was 100 nW or so under the pump power of 4 mW and the power of k_1 or k_2 of 10^{-2} mW. All peaks had almost the same bandwidth to be 0.15 nm limited by the spectrometer resolution of about 1.5 cm^{-1}.

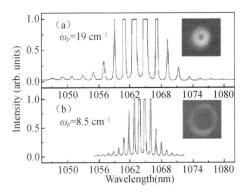

FIG. 3. (Color online) Raman spectra with (a) $\omega_p = 19$ cm^{-1} and $\lambda_0 = 531.82$ nm and (b) $\omega_p = 8.5$ cm^{-1} and $\lambda_0 = 531.92$ nm. The insets are the corresponding spatial distributions.

The intensity of these Raman peaks generated by the cascaded process was essentially related with the equivalent nonlinear susceptibility $\chi_p^{(3)}$ that was defined: $\chi_p^{(3)} = 4\pi\omega_p \chi_1^{(2)} \chi_2^{(2)} / c n_p \alpha_p^2$,[18] in which $\chi_1^{(2)} = (1-2D)\chi^{(2)}(-\omega_p;\omega_1,-\omega_2)$, $\chi_2^{(2)} = (1-2D)\chi^{(2)}(-\omega_{as(n)};\omega_{as(n-1)},\omega_p)$, α_p is the polariton absorption coefficient, n_p is the refraction index of polariton, and c is the light velocity in vacuum. Both $\chi_1^{(2)}$ and $\chi_2^{(2)}$ are in proportion to $\chi^{(2)}$ of crystal and structure-sensitive dependent on the reversal factor D of domain modulation, which distinguishes QPM material from homogeneous crystal. Experimentally, we did observe the dependence of Raman intensity on the parameter D. Besides, the direct four-wave-mixing (FWM) relating to third-order nonlinear susceptibility $\chi_d^{(3)}$ of the medium could also contribute to the multi-order Raman signal by fulfilling $k_{as(n)} = 2k_{as(n-1)} - k_{as(n-2)}$ and $k_{s(n)} = 2k_{s(n-1)} - k_{s(n-2)}$. But $\chi_d^{(3)}$ was much smaller than $\chi_p^{(3)}$ in our case because of the phonon-polariton excited here far from the resonance of A_1 mode.[18]

In a traditional P-P Raman experiment, Raman signal was very weak compared with incident beams and brought challenge in detection. Experiment with nanosecond pulse in GaP gave a relative intensity of $10^{-7} - 10^{-6}$,[19] 10^{-3} in calcite with picosecond pulse,[20] also 10^{-3} in LiNbO$_3$ even though femtosecond pulse was used.[21] Meanwhile a single $\chi^{(3)}$ process such as stimulated Raman scattering (SRS) and a direct FWM, which was realized in fiber[22] or cavity,[23] gave the prevalent value of $10^{-5} - 10^{-3}$ with pico – or nanosecond pulse; other cases gave a little larger conversion efficiency only if femtosecond pump was used[24] or the medium was of high Raman activity such as liquid or gas.[25] These experimental values were in good approximation with theoretical estimation by the relevant $\chi_p^{(3)}$ or $\chi_d^{(3)}$ mentioned above. So in our case with nanosecond pulse and performing at frequency far from phonon resonance, the relative intensity of 14% – 70% and the scattering order up to 8 should be attributed to the subsequent QPM parametric amplification process. It could bring the rapid growth of the seed signal.[26] Here the seed was the generated anti-Stokes and Stokes signals and k_0 serves as the pump. Then

$$k_0 = k_{as(n)} + k_{s(n)} + G \text{ and } \omega_0 = \omega_{as(n)} + \omega_{s(n)}, \tag{3}$$

$n = 1, 2, 3 \ldots$. In fact, setting $n = 0$ this equation can also describe the initial QPM-DFG process producing the two coherent fields k_1 and k_2, which is shown by the inner momentum parallelogram in Fig. 2 while the outer one stands for the parametric amplification process of $n \neq 0$. Therefore the amplitudes of each order anti-Stokes $E_{as(n)}$ and Stokes $E_{s(n)}$ should develop according to the coupling equations below:

$$\frac{dE_{as(n)}}{dx} = -i \frac{\omega_{as(n)}}{n_{as(n)} c} (f_{as(n)} \chi_p^{(3)} E_{as(n-1)} E_1 E_2^* + \chi_{eff}^{(2)} E_0 E_{s(n)}^*),$$

$$\frac{dE_{s(n)}}{dx} = -i \frac{\omega_{s(n)}}{n_{s(n)} c} (f_{s(n)} \chi_p^{(3)} E_{s(n-1)} E_2 E_1^* + \chi_{eff}^{(2)} E_0 E_{as(n)}^*), \tag{4}$$

in which E_0 was assumed to be constant since less than 5% of the pump was depleted experimentally. E_1 and E_2 were the amplitudes of k_1 and k_2, respectively; $\chi_{eff}^{(2)} = f\chi^{(2)}(-\omega_{as(n)}; \omega_0, \omega_{s(n)})/2$ was effective nonlinear susceptibility for this parametric amplification process. Here $f = 0.29$ was the Fourrier coefficient of reciprocal G; $n_{as(n)}$ and $n_{s(n)}$ were the refraction indexes; and $f_{as(n)}$ and $f_{s(n)}$ were spatial factors which were related with polariton's linear diffraction effect and its spatial overlap with related parametric waves.[26] The coupling equations are under proper approximation, such as the multi-polariton process, are not included because it involves higher order nonlinear susceptibility. So the APM two-step process expressed by the first term on the right-hand side of Eq. (4) was the main origination of the Raman signal. The second term described the amplification process of anti-Stokes and Stokes signals by pump. The traditional P-P Raman scattering through a two-step process can also be described by Eq. (4) but without the amplification term, so the intensity was rather weak. Exact numerical simulation of Eq. (4) was a little complicated due to the difficulty of the selection of several parameters, here for simplicity, we omitted the first term and substituted just a proper initial seed $I_{as(n)}(0)$ for the optical amplification process. Then we got the anti-Stokes intensity gain at its low limit, which is similar to the case in a traditional parametric process[27] and it is

$$G_{as(n)}(l) = \left(\frac{I_{as(n)}(l)}{I_{as(n)}(0)} - 1\right) \propto \frac{2\omega_{as(n)}\omega_{s(n)}\mu_0(\chi_{eff}^{(2)})^2}{n_0 n_{as(n)} n_{s(n)} \varepsilon_0 c^3} I_0 l^2, \tag{5}$$

where l is the interaction length, n_0 is the refraction index of pump wave, and ε_0 and μ_0 are the vacuum permittivity and susceptibility, respectively. $G_{as(n)}(l)$ is proportional to pump intensity I_0 and the square of l. In experiment, we measured the dependence of the Raman spectrum on the pump power, under the condition that k_1 and k_2 were adjusted to identical intensity, so the seed $I_{as(n)}(0)$ decided by the origination term in Eq. (4) was kept identical. Figures 4(a) and 4(b) reveal great differences both in scattering intensity and scattering order with moderate pump powers of 3.0 mW and 8.2 mW, respectively. The inset demonstrated that the intensity of the first-order anti-Stokes signal increased linearly with pump power, which can be deduced from Eq. (5). Moreover, an exponential growth can be expected if higher power pump is used.[26] Nevertheless, using the present model we cannot explain exactly the intensity ratio $I_{as(n)}/I_{as(n-1)}$, which increases with n from 14% ($n=1$) to 70% ($n=8$) in Fig. 3(a). We suggest a possible explanation that the high-order peaks can be produced through the several direct FWM processes. A richer variety of such phase-matching geometries appear for larger n, thus corrections should be made to the origination term in Eq. (4) especially when n gets much larger.

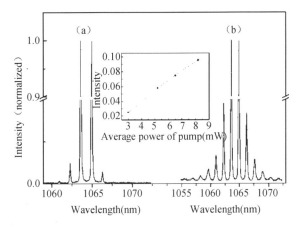

FIG.4. Raman spectra with different pump intensities at $\lambda_0 = 531.89$ nm. Average pump powers in (a) and (b) are 3.0 mW and 8.2 mW, respectively. The inset is the first anti-Stokes intensity going linearly with pump intensity.

We varied the wavelength of pump λ_0 in the range of 531.64 – 532.39 nm and observed the accompanied Raman spectra are similarly distributed but the intensity differs, especially at some pump wavelength the scattering order is up to 11. Figure 5 displays the first- and second-order anti-Stokes signal intensity dependent on λ_0 by plots I and II, respectively. Both exhibit two maximums. This can be attributed mainly to the difference in the process the Raman signal originated: as λ_0 tuned toward 531.64 nm (collinear QPM-

DFG) or neared 532.39 nm (maximal permissive wavelength for noncollinear QPM-DFG), the parametric interaction volume got to be smaller and this significantly decreased the DFG efficiency and accordingly decreased the generation of phonon-polariton and the intensity of the final Raman signal. When λ_0 neared 532 nm, the frequency of excited phonon-polariton was very low due to k_1 and k_2 being nearly degenerate. The diffraction effect got rather remarkable, which led $\chi_p^{(3)}$ contribution diminish sharply.[26] Thus plots in Fig. 5 should contain two maximums. All the higher-order peaks follow the same rule as shown in Fig. 5, except that they have lower intensity.

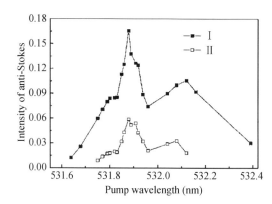

FIG. 5. The first-(I) and second-(II) order anti-Stokes intensity relative to the intensity of k_1 when pump wavelength varies.

In our experiment range, a low-frequency polariton in the range of 2 – 42 cm^{-1} was excited, which corresponded to a popular region of THz.[12] We did not find obvious evidences for low-frequency local excitation in this range, which consisted with most of the previous works.[14] Although the polariton in this range is photon-like, the two-step $\chi^{(2)}$ process, in some meaning, still differs from a general optical parametric process occurring in the visible and infrared ranges of medium. In fact, the polariton has a short free path varying from several millimeters to several centimeters due to damping,[19] which made the process not so efficient compared to a general cascaded $\chi^{(2)}$ process.

It is worth noting the P-P Raman signal here is crucially dependent on the reversal factor D. As $D \to 0.5$, $\chi_p^{(3)} \to 0$; the Raman signal only originates from direct FWM because $\chi_d^{(3)}$ always exists. In this case, the Raman signal can still be amplified by the later QPM parametric process. However, it is relatively weak compared with the present result because $\chi_d^{(3)} \ll \chi_p^{(3)}$ when ω_p is far away from that of the A_1 mode. We proved the presumption from a one-dimensionally periodically poled LiTaO$_3$ with $D \to 0.5$. In the present work we used a two-dimensional structure instead of a one-dimensional structure since the former has a better homogeneity of poling usually and can be well controlled for a required parameter D.

In summary, the P-P Raman scattering was experimentally studied in a HPLT crystal. In contrast to a single domain LiTaO$_3$ crystal, the HPLT crystal offers a valuable platform to perform QPM interaction. In this work, the QPM structure was used for a dual purpose: to assist in exciting intense polariton fields first and then to amplify the enhanced Raman signal. A simplified step-accomplished model has been proposed to describe the cascaded nonlinear processes and a set of coupling equations is developed to account for the experiment results. The resulting Raman spectrum exhibited a comb-shaped structure with high intensity, suggesting it may be of technological importance for developing a novel Raman laser with the multi-wavelength output and a tunable frequency interval. The work presents a novel and efficient method to study the inelastic scattering and the elementary excitation in a nonlinear medium.

References and Notes

[1] J. Armstrong, N. Bloembergen, J. Ducuing, and P. S. Pershan, Phys. Rev. **127**, 1918 (1962).

[2] P. A. Franken and J. F. Ward, Rev. Mod. Phys. **35**, 23 (1963).

[3] V. Berger, Phys. Rev. Lett. **81**, 4136 (1998).

[4] S. N. Zhu, Y. Y. Zhu, and N. B. Ming, Science **278**, 843 (1997).

[5] A. Chowdhury, S. C. Hagness, and L. McCaughan, Opt. Lett. **25**, 832 (2000).

[6] M. A. Arbore and M. M. Fejer, Opt. Lett. **22**, 151 (1997).

[7] D. K. Serland, P. Kumar, M. A. Arbore, and M. M. Fejer, Opt. Lett. **22**, 1497 (1997).

[8] A. M. Schober, G. Imeshev, and M. M. Fejer, Opt. Lett. **27**, 1129 (2002).

[9] C. B. Clausen, O. Bang, and Y. S. Kivshar, Phys. Rev. Lett. **78**, 4749 (1997).

[10] B. Xu and N. B. Ming, Phys. Rev. Lett. **71**, 1003 (1993).

[11] K. N. Alekseev and A. V. Ponomarev, JETP Lett. **75**, 174 (2002).

[12] Y. Sasaki and A. Yuri, Appl. Phys. Lett. **81**, 3323 (2002).

[13] M. J. A. de Dood, W. T. M. Irvine, and D. Bouwmeester, Phys. Rev. Lett. **93**, 040504 (2004).

[14] P. Xu, S. H. Ji, S. N. Zhu, X. Q. Yu, J. Sun, H. T. Wang, J. L. He, Y. Y. Zhu, and N. B. Ming, Phys. Rev. Lett. **93**, 133904 (2004).

[15] S. N. Zhu, Y. Y. Zhu, Z. Y. Zhang, H. Shu, H. F. Wang, J. F. Hong, and C. Z. Ge, J. Appl. Phys. **77**, 5481 (1995).

[16] A. F. Penna, A. Chaves, P. da R. Andrade, and S. P. S. Porto, Phys. Rev. B **13**, 4907 (1976).

[17] G. P. Wiederrecht, T. P. Dougherty, L. Dhar, D. E. Leaird, K. A. Nelson, and A. M. Weiner, Phys. Rev. B **51**, 916 (1995).

[18] G. Kh. Kitaeva, A. A. Mikhailovsky, P. S. Losevsky, and A. N. Penin, Opt. Commun. **138**, 242 (1997); G. Kh. Kitaeva, K. A. Kuznetsov, A. A. Mikhailovsky, and A. N. Penin, J. Raman Spectrosc. **31**, 767 (2000).

[19] J. P. Coffinet and F. De Martini, Phys. Rev. Lett. **22**, 60 (1969).

[20] R. R. Alfano and S. L. Shapiro, Phys. Rev. Lett. **26**, 1247 (1971).

[21] P. C. M. Planken, L. D. Noordam, J. T. M. Kennis, and A. Lagendijk, Phys. Rev. B **45**, 7106 (1992).

[22] A. V. Husakou and J. Herrmann, Appl. Phys. Lett. **83**, 3867 (2003).

[23] V. Pasiskevicius, A. Fragemann, and F. Laurell, Appl. Phys. Lett. **82**, 325 (2003).

[24] D. A. Akimov, E. E. Serebryannikov, and A. M. Zheltikov, Opt. Lett. **28**, 1948 (2003).

[25] H. Kawano, T. Mori, Y. Hirakawa, and T. Imasaka, Opt. Commun. **160**, 277 (1999).

[26] Y. R. Shen, *The Principles of Nonlinear Optics* (Wiley, New York, 1984).

[27] R. L. Byer, J. Nonlinear Opt. Phys. Mater. **6**, 549 (1997).

[28] We thank H. T. Wang, J. L. He, G. X. Chen, and A. Hu for valuable discussions and J. Sun for help during this experiment. This work is supported by grants from the National Advanced Materials Committee of China and the State Key Program for Basic Research of China (2004CB619003), and by the National Natural Science Foundation of China under Contract No. 90201008.

Nonlinear Čerenkov Radiation in Nonlinear Photonic Crystal Waveguides[*]

Y. Zhang, Z. D. Gao, Z. Qi, S. N. Zhu, and N. B. Ming

National Laboratory of Solid State Microstructures and Department of Physics,
Nanjing University, Nanjing, 210093, China

We study nonlinear Čerenkov radiation generated from a nonlinear photonic crystal waveguide where the nonlinear susceptibility tensor is modulated by the ferroelectric domain. Nonlinear polarization driven by an incident light field may emit coherently harmonic waves at new frequencies along the direction of Čerenkov angles. Multiple radiation spots with different azimuth angles are simultaneously exhibited from such a hexagonally poled waveguide. A scattering involved nonlinear Čerenkov arc is also observed for the first time. Čerenkov radiation associated with quasi-phase matching leads to these novel nonlinear phenomena.

A charged particle traveling faster than the speed of light can drive the medium to emit coherent light called Čerenkov radiation (CR)[1,2]. In such a process, the coherent radiation is observable at a conical wave front defined by the Čerenkov angle $\theta_c = \arccos(v'/v)$, where v is the speed of the moving charged particle and v' is the phase velocity of the radiation wave. CR occurs as $v > v'$. Light propagating in a nonlinear optical crystal or waveguide can also create such an emission via a second-order nonlinear process, such as Čerekov second harmonic generation (SHG)[3-8] and THz emission from optical beating[9-11]. Although closely resembling CR via relativistic charged particles, it has unique distinguishing features. Most importantly, it is a nonlinear optical process associated with the susceptibility tensor $\chi^{(2)}$ and the radiation source is not a point particle, but a spatially extended collection of dipoles driven by the incident light field, i. e., nonlinear polarization (P_{non}). Therefore, we call that nonlinear Čerenkov radiation(NCR). However, the phase velocity of the source does need exceed the radiation velocity. Comparing with a conversional $\chi^{(2)}$ process, NCR now can be realized only by extremely short optical pulses[9,10] or in the waveguide[3,4], which results in a relaxation of phase-matching requirements. This is relatively simple if the incident light is only monochromatic—for example, Čerenkov SHG from a planar waveguide[3]. In this case, the radiation source (second-order polarization) and incident light (guide modes of the

---

[*] Phys. Rev. Lett., 2008, 100(16):163904

waveguide) have the same phase velocity v. In a normally dispersive medium, it is faster than v' of SH and the harmonic radiation emits into the substrate. Because the geometry is planar and P_{non} is coherent over the full width of the beam, i.e., over many wavelengths, NCR emerges as a well-collimated beam with the Čerenkov angle relative to the direction of P_{non}.

We are concerned with two questions about NCR. Up to now, NCR has been usually generated from monochromatic light. If incident lights contain two or more than two frequency components, can they generate NCR? In principle the answer is yes, but it is not a simple accumulation of several Čerenkov SHG processes. For example, ω_1 and ω_2 for the simplest case P_{non} will include terms $2\omega_1$, $2\omega_2$, $\omega_1+\omega_2$, and $\omega_1-\omega_2$. Obviously, terms $2\omega_1$ and $2\omega_2$ can lead to Čerenkov SHG at frequency $2\omega_1$ and $2\omega_2$, as discussed above. If $\omega_1 \neq \omega_2$, phase velocity v_p of P_{non} for $\omega_3=\omega_1\pm\omega_2$ is equal to neither ω_1 nor ω_2. It is decided by

$$\frac{\omega_1}{v_1}x_1 \pm \frac{\omega_2}{v_2}x_2 = \frac{\omega_3}{v_p}x_p, \tag{1}$$

where v_i is phase velocity and x_i the direction of v_i ($i=1, 2, p$). In theory, P_{non} can emit NCR as long as v_p is faster than v', the phase velocity of ω_3, and Čerenkov radiation angle is defined by

$$\theta_c = \arccos(v'/v_p). \tag{2}$$

This process is more complicated but more general. From this point, we can analyze NCR with more frequency components. And if $\omega_1=\omega_2$, it becomes Čerenkov SHG.

It is easy to obtain from Eq. (1) that in a homogenous medium, NCR for frequency up-conversion, such as the generation of $2\omega_1$, $2\omega_2$, $\omega_1+\omega_2$, always satisfies $v_p > v'$ for normal dispersion, but does not for the process of $\omega_1-\omega_2$ because $v_p < v'$ at the case and Eq. (2) is violated. NCR is intrinsically determined by the medium's dielectric constants (or refractive index) and their dispersion; therefore, the Čerenkov angle is fixed. Can we change NCR's behavior by another physics process or microstructure? Several recent studies showed these could occur with phonon assistance[12] or using photonic crystal[13]. We provide a new approach by introducing nonlinear photonic crystal[14]. Differing from that in a homogeneous medium, the behavior of radiation source P_{non} in a nonlinear photonic crystal can be changed by the modulation of $\chi^{(2)}$ in terms of quasi-phase matching (QPM)[15—17]. P_{non} will change its phase π from positive domain to negative domain as $\chi^{(2)}$ alters its sign from $+1$ to -1, which leads to an effective change in v_p along the propagating direction of P_{non} (Fig. 1) and, therefore, changes the behavior of NCR, such as the threshold value and radiation angle, and even makes NCR appear in a frequency down-conversion process. These properties have not been studied in any detail until very recently.

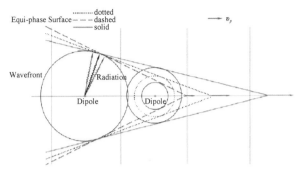

FIG. 1. (color online) A collection of dipoles (P_{non}) driven by incident light emits coherent radiation. In nonlinear photonic crystal P_{non} will change its phase $+\pi$ (dashed) or $-\pi$ (solid) as $\chi^{(2)}$ alters its sign compared with that in homogenous material(dotted), which leads to an effective change in the phase velocity of P_{non} as well as Čerenkov angle.

In this Letter, we study NCR generated from a nonlinear photonic crystal waveguide both experimentally and theoretically. For simplicity, here the fundamental beam only contains two frequency components, ω_1 and ω_2. The nonlinear photonic crystal provides a set of collinear and noncollinear reciprocals $G^{[14]}$ in this NCR process. In comparison with Eq. (1), the corresponding expression for sum-frequency generation (SFG) ($\omega_1 \neq \omega_2$) and SHG($\omega_1 = \omega_2, v_1 = v_2$) in such a structure is

$$\frac{\omega_1}{v_1} \boldsymbol{x}_1 + \frac{\omega_2}{v_2} \boldsymbol{x}_2 + \boldsymbol{G} = \frac{\omega_3}{v_p} \boldsymbol{x}_p. \tag{3}$$

The equation extends NCR from the homogeneous medium to nonlinear photonic crystal, showing that Čerenkov radiation can be realized in the QPM scheme. In the experiment, both ω_1 and ω_2 are guide modes and $\boldsymbol{x}_1$ is collinear with $\boldsymbol{x}_2$. $\boldsymbol{x}_p$ and $\boldsymbol{v}_p$ will depend on the reciprocal $\boldsymbol{G}$. According to Eqs. (2) and (3), in wave vector space, QPM-NCR can be written as

$$|\boldsymbol{\beta}(\omega_1) + \boldsymbol{\beta}(\omega_2) + \boldsymbol{G}| = k(\omega_3)\cos\theta_c, \tag{4}$$

where $\boldsymbol{\beta}(\omega_1)$ and $\boldsymbol{\beta}(\omega_2)$ are wave vectors of guided modes, $k(\omega_3) = \omega_3/v'$ is the magnitude of the radiation mode's wave vector in substrate. The schematically phase-matching geometry is shown in Fig. 2. The wave vector of P_{non} can make up a vector triangle together with $\boldsymbol{\beta}(\omega_1)$, $\boldsymbol{\beta}(\omega_2)$, and $\boldsymbol{G}$. When the Čerenkov condition [Eq.(2)] is satisfied, NCRs with frequency $\omega_3 = 2\omega_1$, $2\omega_2$, $\omega_1 + \omega_2$ can be simultaneously achieved and, respectively, emit along different azimuth angles.

In the experiment, the nonlinear photonic crystal was a hexagonally poled LiTaO$_3$ by the field poling technique with a domain interval $a = 9.0$ μm and reversal factor of $\sim 30\%$ [Fig. 2(a)]. The structure provides reciprocals of sixfold symmetry [Fig. 2(b)], defined

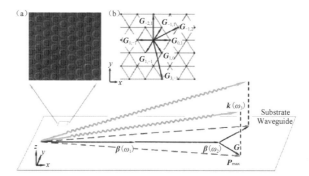

FIG. 2. (color online) Phase-matching process for NCR in nonlinear photonic crystal waveguide. Dashed lines represent polarization and curves are radiation waves. The insets are (a) domain structure and (b) reciprocal lattice of hexagonally poled LiTaO$_3$.

by $G_{m,n} = (4\pi/\sqrt{3}\, a)(\sqrt{m^2+n^2+m \cdot n})$ with m and n being indexes of reciprocals[14]. Then a planar waveguide was fabricated by a proton-exchange method. A laser-diode-pumped, Q-switched Nd: YAG dual-wavelength laser provided two z-polarized fundamental waves at 1.064 μm and 1.319 μm at the same time. The propagation constants of fundamental modes in the waveguide are $\beta(1.064\ \mu m) = 12.678\ \mu m^{-1}$ and $\beta(1.319\ \mu m) = 10.177\ \mu m^{-1}$. A screen was set at the position 15 cm away from the end face of the sample and no lens was used after the output end face. The operating temperature was kept at 25 ℃.

As these two fundamental beams were collinearly coupled into the waveguide by a cylindrical lens transmitting along the x axis of the crystal, a beautifully colorful pattern like a "Christmas tree," which was composed of spots and lines with red (R), green(G), and yellow(Y), was projected on the screen behind the crystal [Fig. 3(a)]. In the picture, five vertical lines, three with green and two with red, at the bottom of the picture, compose the "stems" of "tree." They correspond to the guided-to-guided QPM-SHG processes[15], but are not perfectly phase matched at 25 ℃ (or else, they will be brighter than that in the picture). The "crown" of the tree consists of colorful spots with red, yellow, and green. They distribute in mirror symmetry as shown in Fig. 3 (a). All the spots on the screen are Čerenkov radiation modes. They can be divided into groups and each contains three red, yellow, and green spots. The spots in a group associate the same $G_{m,n}$ as indexed in Fig. 3 (b). Red and green originate from SHG of two fundamental waves, respectively, and yellow spots come from their sum frequency. The brightest three spots locating in the center area of Fig. 3 (a) are a direct Čerenkov SHG or SFG in which no reciprocal vector is involved. These spots could appear in a homogenous LiTaO$_3$ planar waveguide as well. The yellow one in the center is located at an angle of 27.4° with the waveguide, which is in good agreement with the calculated 27.5° from Eq. (4). Two weaker green spots, which present just above the brightest green spot shown in Fig. 3(a), are SH of high-order guide modes of 1.064 μm. The lower spots associate with forward $G_{m,n}$ which retards the v_p of P_{non} wave, while the spots at the higher position involve backward $G_{m,n}$ that accelerates v_p. Especially, three spots in that group locating on the top of crown correspond to $G_{0,-1}$. It is antiparallel with the propagation direction of the fundamental waves. The exit angle of the yellow spot in this group is measured to be about 46.8°, which is much larger than that of direct Čerenkov SFG. In this case, the actual phase velocity v_p [$1.06v'$, calculated from Eq. (3)] goes faster than that ($1.02v'$) in the homogenous medium. The increase of v_p results in a larger Čerenkov angle. The result hints at the possibility of NCR in the frequency down-conversion process as long as actual v_p exceeds v' by introducing a backward reciprocal—for example, $v_p = 0.95v'$ in difference frequency generation with fundamental waves of 0.532 μm and 1.064 μm, which could be compensated using a reversed reciprocal.

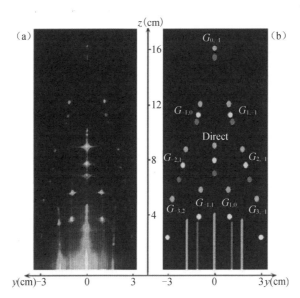

FIG. 3. (color) The pattern of NCR from fundamental waves of 1.064 μm and 1.319 μm projected on the screen behind the sample: (a) measured and (b) calculated.

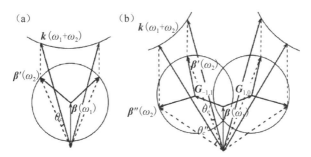

FIG. 4. Phase-matching processes for Čerenkov arcs of direct NCR (a) and QPM-NCR involving $G_{-1,1}$ and $G_{1,0}$ (b) with scattered $\beta'(\omega_2)$.

In Fig. 3 (a), we found a new phenomenon—an arc tangent with each Čerenkov spot, the brighter spot the brighter arc. According to analyses, these arcs are involved with elastic scattering of incident light in waveguide and emit in NCR by sum frequency of scattering and incident light. In a previous Letter[16], we reported conical SHG by conversional sum frequency of scattering and incident light in 2D nonlinear photonic crystal. In this experiment, such a sum-frequency process was achieved in Čerenkov radiation. In contrast to the bulk case, scattering light mainly exists in the x-y plane due to confinement of the waveguide. The Čerenkov condition may be understood in terms of Eq. (4), where one of two fundamental waves, such as $\boldsymbol{\beta}(\omega_2)$, is replaced by the scattering wave $\boldsymbol{\beta}'(\omega_2)$ that presents a continuously and symmetrically spatial distribution from $\boldsymbol{\beta}(\omega_2)$[16]. The phase-matching condition for this configuration can be written as

$$|\boldsymbol{\beta}(\omega_1) + \boldsymbol{\beta}'(\omega_2) + \boldsymbol{G}_{m,n}| = k(\omega_3)\cos\theta_c. \tag{5}$$

Phase-matching geometries are shown in Fig. 4, where Fig. 4 (a) corresponds to a direct NCR, and (b) correspond to QPM-NCRs being concerned with $\boldsymbol{G}_{-1,1}$ and $\boldsymbol{G}_{1,0}$, respectively. Obviously, one spot corresponds to one arc in SHG processes. If $\omega_1 \neq \omega_2$ in SFG processes, there will be two arcs with different curvatures to be tangent at the same spot because of $\boldsymbol{\beta}(\omega_1) + \boldsymbol{\beta}'(\omega_2) \neq \boldsymbol{\beta}'(\omega_1) + \boldsymbol{\beta}(\omega_2)$. However, the calculated result shows that the maximum interval for these two yellow arcs in our experiment is less than 0.1 mm, which is too small to be differentiated from the imaging on screen.

NCR can be analyzed by following the coupled-mode theory[7,8]. For the case of SHG, an approximately analytic expression of SH power can be deduced from the coupled wave equations by using a nondepletion approximation as below:

$$P(2\omega) = 2l\pi[P(\omega)]^2 |I_s + \alpha I_w|^2 \overline{d}_{33}^2 \frac{\left|2\frac{\omega}{v}x + \boldsymbol{G}_{m,n}\right|}{\rho} \left[\frac{n_e(2\omega)}{n_o(2\omega)}\right]^2, \tag{6}$$

where $P(\omega)$ and $P(2\omega)$ are powers of the fundamental wave and SH wave, l is the waveguide length, I_s and I_w represent the overlap of electric fields between fundamental and SH waves in substrate and in waveguide, $\overline{d}_{33}$ is an effective nonlinear coefficient, $\alpha < 1$ is due to the reduction of $\overline{d}_{33}$ in the proton-exchanged waveguide, ρ is the wave number of radiation mode, $n_e(2\omega)$ and $n_o(2\omega)$ are the extraordinary and ordinary refractive index in the medium. For the details of deductions, one can refer to [7,8]. Figure 5 shows the dependence of Čerenkov SH power on the input of fundamental waves at 1.064 μm. The harmonic power increased proportionally with the square of the fundamental power, for both direct Čerenkov SHG and QPM Čerenkov SHG, which is consistent with theoretical analysis. Assuming that the intensity of direct Čerenkov radiation is 1, we calculated relative intensity of QPM Čerenkov SHG with different $\boldsymbol{G}_{m,n}$ at the same input power. The result is well consistent with the measured data, as shown in Table 1. The coefficient α was fitted to be about 70% for our sample. It has been mentioned that Čerenkov radiation strongly depends on the discontinuity at the interfaces of waveguide[6]. The radiation in our experiment was weakened because of a graded index due to the annealing process in waveguide fabrication.

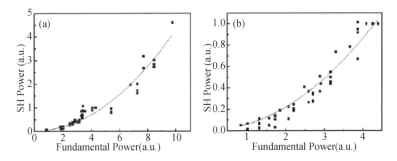

FIG. 5. (color online) Dependence of SH power on input power of fundamental 1.064 μm. (a) Direct Čerenkov SHG and (b) QPM Čerenkov SHG with $G_{1,-1}$. Dots are measured data which are well fitted to be quadratic (line).

TABLE 1. Measured and calculated relative intensity of Čerenkov SHG.

Intensity of SHG of 1.064 μm	Direct Čerenkov SHG	Čerenkov SHG with $G_{1,0}$	Čerenkov SHG with $G_{-1,1}$	Čerenkov SHG with $G_{1,-1}$	Čerenkov SHG with $G_{-1,0}$	Čerenkov SHG with $G_{0,-1}$
Measured (a.u.)	1	0.41	0.42	0.17	0.17	0.10
Calculated (a.u.)	1 (Assumed)	0.43	0.43	0.16	0.16	0.08

In general, the linear susceptibility tensor $\chi^{(1)}$ associates with the refractive index. It is a constant in a $\chi^{(2)}$ photonic crystal with inverse domain structure. $\chi^{(2)}$ photonic crystal is equivalent to a homogenous crystal in the linear optical regime. However, as incident light is strong enough to be able to drive an effective nonlinear polarization P_{non}, the situation will be greatly different. The modulation of $\chi^{(2)}$ can change the phase velocity of P_{non}. If $v_p = v'$, where v' is the phase velocity of free wave at the same frequency in the medium, a conventional nonlinear frequency conversion occurs, meanwhile, v' is collinear with v_p. As $v_p > v'$, although phase matching cannot achieve collinearly, effective frequency conversion can still be achieved by the NCR. As $v_p < v'$, no matching geometry exists; neither QPM parametric process nor QPM-NCR process can be observed. Compared with the conventional QPM parametric process, QPM-NCR can extend the tolerance of phase matching from $v_p = v'$ to $v_p \geqslant v'$, and multiple NCR processes can exist simultaneously because the Čerenkov phase-matching condition is automatically satisfied. Because of the fact that there are multiple modulation periods with hexagonal symmetry for a 2D hexagonal structure and every modulated period corresponds to one phase velocity, it is easily understood why NCRs can be divided into a group with the index of reciprocal m, n. And also it exhibits more plentiful radiation patterns compared with a direct NCR. These have been verified from theory to experiment in this work. In this sense, the action of $\chi^{(2)}$ modulation can be understood as an effective approach to control the phase velocity v_p of P_{non} to achieve the required nonlinear parametric process. For example, the backward reciprocal can be used to accelerate v_p of P_{non}, which provided a possibility to realize QPM-NCR in a parametric down-conversion process. A probable application for this process is to generate multiple entangled photon pairs for potential quantum communication and computation networks.

References and Notes

[1] P. A. Čerenkov, Dokl. Akad. Nauk SSSR **2**, 451 (1934).
[2] P. W. Gorham *et al.*, Phys. Rev. Lett. **99**, 171101 (2007).
[3] P. K. Tien, R. Ulrich, and R. J. Martin, Appl. Phys. Lett. **17**, 447 (1970).
[4] Y. Zhang *et al.*, Appl. Phys. Lett. **89**, 171113 (2006).
[5] N. A. Stanford and J. M. Connors, J. Appl. Phys. **65**, 1429 (1989).
[6] M.J. Li *et al.*, IEEE J. Quantum Electron. **26**, 1384 (1990).
[7] H. Tamada, IEEE J. Quantum Electron. **27**, 502 (1991).
[8] M. Vaya, K. Thyagarajan, and A. Kumar, J. Opt. Soc. Am. B **15**, 1322 (1998).
[9] C. D'Amico *et al.*, Phys. Rev. Lett. **98**, 235002 (2007).
[10] D. H. Auston *et al.*, Phys. Rev. Lett. **53**, 1555 (1984).
[11] G. A. Askar'yan, Phys. Rev. Lett. **57**, 2470 (1986).
[12] T. E. Stevens *et al.*, Science **291**, 627 (2001).
[13] C. Luo *et al.*, Science **299**, 368 (2003).
[14] V. Berger, Phys. Rev. Lett. **81**, 4136 (1998).
[15] K. Gallo *et al.*, Opt. Lett. **31**, 1232 (2006).
[16] P. Xu *et al.*, Phys. Rev. Lett. **93**, 133904 (2004).
[17] S. Sinha *et al.*, Opt. Lett. **31**, 347 (2006).
[18] This work is supported by the National Natural Science Foundation of China (No. 60578034 and No. 10534020), and by the National Key Projects for Basic Researches of China (No. 2006CB921804 and No. 2004CB619003).

Nonlinear Volume Holography for Wave-Front Engineering*

Xu-Hao Hong,[1,2] Bo Yang,[1,3] Chao Zhang,[1,3] Yi-Qiang Qin,[1,3] and Yong-Yuan Zhu[1,2]

[1] *Key Laboratory of Modern Acoustics, National Laboratory of Solid State Microstructures and Collaborative Innovation Center of Advanced Microstructures, Nanjing University, Nanjing 210093, People's Republic of China*

[2] *School of Physics, Nanjing University, Nanjing 210093, China*

[3] *School of Modern Engineering and Applied Science, Nanjing University, Nanjing 210093, China*

The concept of volume holography is applied to the design of an optical superlattice for the nonlinear harmonic generation. The generated harmonic wave can be considered as a holographic image caused by the incident fundamental wave. Compared with the conventional quasi-phase-matching method, this new method has significant advantages when applied to complicated nonlinear processes such as the nonlinear generation of special beams. As an example, we experimentally realized a second-harmonic Airy beam, and the results are found to agree well with numerical simulations.

Nonlinear optics has been developed rapidly in recent decades, opening up new vistas in many classical fields. Because of the ability of generating special beams (such as the nonlinear the Airy beam or vortex beam), the study on two-dimensional optical superlattices(OSL) with the quasi-phase-matching method has become a hot topic[1-10]. A reciprocal space phase-matching technique has been developed to deal with these new topics[11-14]. The key of this method contains two main parts, that is, the transverse axes forming a certain type of curved domain configuration for the phase modulation, and the propagation axis providing an appropriate poled period for compensating the phase mismatching. This technique, combined with the traditional quasi-phase-matching method and the wave-front engineering, has brought great possibilities to manipulate harmonics in many different fields[15-18].

The phase-matching problem can be also analyzed in real space. In our previous work, we developed a phase-matching technique to control the nonlinear wave front[19]. In this method, each area of the nonlinear medium is treated as a source of the nonlinear secondary wave, radiating harmonic waves outward, and the excited harmonic wave at a given area can be determined locally by the nonlinear coefficient and the fundamental wave (FW). The

---

* Phys. Rev. Lett., 2014,113(16):163902

wave front at any position is related to the sum of the previously generated harmonic waves. Using this principle we have realized several nonlinear beam manipulations[19,20]. Furthermore, some nonlinear phenomenon such as the nonlinear Cherenkov radiation and the Raman-Nath diffraction[21—23] can be also well explained in this regard. However, this method also has its limitations. In some more complicated cases, such as nonlinear generation of special beams, this theory is found difficult to work with practically. Recently, a two-stage approach has been proposed which is able to transfer the fundamental wave into an arbitrary designed second-harmonic beam[24], but the wave-front shaping is performed by a linear optical process rather than a nonlinear process. And to date, none of the above wave-front engineering theories could provide a universal method to deal with arbitrary special beams all in the nonlinear regime. In this Letter, we show that this problem can be solved by introducing the nonlinear volume holography technique, which extends the concept of holography from linear optics to nonlinear optics.

Holography, being different from photography, is a technique that can store and reconstruct both the amplitude and phase of an object wave[25—29]. There are basically two categories of holography based on the difference of the recording medium: plane holography and volume holography. The volume holography not only inherits the specialty of the plane holography, but also has some additional features such as Bragg selectivity[30—33]. Conventionally, the holography is based on the interference fringes of the reference beam and the object beam recorded on a photographic film through corrugated surface gratings. Or in a photorefractive crystal the fringes can be imprinted as refractive index variation. Here, by transferring the fringes into the modulation of the nonlinear coefficient in a ferroelectric crystal, the nonlinear hologram can be realized.

Based on the secondary wave principle, in a second-harmonic generation (which acts as object beam) process the harmonic wave satisfies the brief expression $dE_{2\omega} \propto \chi^{(2)} E_\omega^2 dx$, here $\chi^{(2)}$ is the nonlinear coefficient. Considering the pump undepletion approximation and the plane wave form of FW, E_ω could be written as $E_\omega = A_\omega \exp(-ik_1 x)$, where k_1 is the wave vector of FW. Then the excited second-harmonic generation is depicted as follows:

$$dE_{2\omega} = C\chi^{(2)} A_\omega^2 \exp(-i2k_1 x) dx \qquad (1)$$

In this equation the initial phase $2k_1 x$ can be considered as the spatial phase shift caused by a wave with the wave vector $2k_1$ before the excitation of the harmonic wave. In order to develop the nonlinear interference theory, a special light with wave vector of $2k_1$ and frequency of $\omega_2 = 2\omega_1$ should be introduced to interfere with the harmonic waves. It is interesting to find that the nonlinear polarization wave (NPW) meets all the demands. The NPW is generated by the FW, and then it radiates an optical field with the frequency doubled. For the sake of simplicity, here the NPW was set to be a plane wave [$E_p = A_p \exp(-i2k_1 x)$, which acts as the reference beam] and all the discussions are in a two dimensional system. By interfering with the NPW, the harmonic wave $[E_h(x,y)]$ can be stored in the form of interference fringe as Fig. 1(a) shows. The schematic diagram shows the general situation where the SH beam is noncollinear with the NPW. While in our experiment, we choose in-line (collinear) holography to generate the SH beam for the purpose of gaining higher nonlinear efficiency. The orange and green beams stand for the NPW and the harmonic wave, respectively, and they interfere with each other in the overlapped area to record both the amplitude and phase information of the harmonic wave. The light intensity distribution of the interference area is

$$I(x,y) = |E_p + E_h|^2 = |E_p|^2 + |E_h|^2 + E_p^* \cdot E_h + E_p \cdot E_h^*$$
$$= |E_p|^2 + |E_h|^2 + 2E_p E_h \cos\phi. \tag{2}$$

Here ϕ is the phase difference between these two waves. It is worth noting that because of the law of energy conservation ($\omega_2 = 2\omega_1$), the interference pattern between the NPW and the harmonic wave is independent of the time. That is to say the interference pattern in real space is stable without any beating frequency or vibration phenomenon.

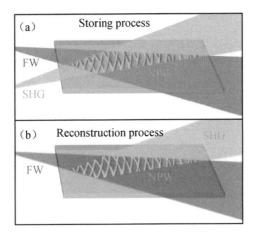

FIG.1. (color online) Schematics of (a) the storing and (b) reconstruction processes in nonlinear volume holography.

Ignoring the background signal $I_0 = |E_p|^2 + |E_h|^2$, the normalized light field is $I' = \cos\phi$. Here the range of I' is between -1 and 1.

In order to record the "light distribution" into a nonlinear crystal as a hologram, the controllable quadratic nonlinear coefficient is introduced, which is defined as

$$\chi^{(2)} = d_{ij}\{I'\}, \tag{3}$$

where d_{ij} is the element of the quadratic susceptibility used in the nonlinear process. In commonly used nonlinear crystal, the $\chi^{(2)}$ cannot be tuned in the full range from $-d_{ij}$ to d_{ij}. Taking the congruent lithium tantalate as an example, with the electric-field poling method only 180° antiparallel ferroelectric domain can be produced. The coefficient $\chi^{(2)}$

becomes binary format with two potential values only, either d_{ij} or $-d_{ij}=d_{ij} \cdot e^{i\pi}$. Thus, the structure function $f(x,y)$ (binary modulation function) required by nonlinear volume hologram can be simply determined by

$$f(x,y) = \text{sgn}\{I'\}. \tag{4}$$

Considering that the binarization of the structure function might bring some error to the generated SH wave, we simulated the Airy beam generated in a binary structure and in an ideal continuous structure. The difference is small see [34]. In addition, there already exist some methods which can be used to mimic the structure function, such as the detour-phase encoding technique, where both the amplitude and phase information can be recorded by tuning the duty circle and the position of each domain[35].

The reconstruction process shown in Fig. 1 (b) is the reverse process of the storing one. When the developed nonlinear hologram is illuminated by the FW(shown in red), the FW propagates through the hologram and excites harmonic waves (shown in green) with a different initial phase which is determined by the binary nonlinear optical coefficient. Then the harmonic waves diffract and interfere to form a certain pattern. The final field intensity E is equal to I' multiplied by the reference beam E_p, giving

$$E = E_p \alpha (E_p^* E_h + E_p E_h^*) = \alpha |E_p|^2 E_h + \alpha E_p^2 E_h^* = E' + E'', \tag{5}$$

where α is a normalized coefficient. The first term E' is proportional to the initial SH wave E_h, and the other is the phase-conjugated one which can be ignored. Except for a constant factor $\alpha |E_p|^2$, the reconstructed wave E' is exactly the same as the expected harmonic wave E_h.

In single sentences, the reference wave in nonlinear holography is the NPW, while the object wave is the SH wave. The hologram pattern is determined by the interference fringe, and the reconstruction method is the reverse of the restoring process. For demonstration of the idea, we manipulated a wave with a complicated wave front, the Airy beam. According to Eq. (4), the OSL structure can be obtained as follows:

$$f(x,y) = \text{sgn}\{\text{Ai}[s-(2/\xi)^2]\exp[i\Delta kx + i(s\xi/2) - i(\xi^3/12)] + \text{c.c.}\}, \tag{6}$$

where Ai is the Airy function, $s=y/y_0$ represents a dimensionless traverse coordinate, y_0 is an arbitrary traverse scale, $\xi=x/k_2 y_0^2$ is a normalized propagation distance, and $\Delta k=k_2-2k_1$ is the wave vector mismatch in the second harmonic generation process[36,37].

We chose the $ee-e$ nonlinear process with the congruent LiTaO$_3$. In this case $d_{ij}=d_{33}$. The ones before the minus symbol represent the polarization of FW, after it represents the SH wave. Here the wavelengths of FW and the SH are 1064 nm and 532 nm, respectively, and T is set to be 150 ℃ to avoid the photorefractive effect and $y_0=20$ μm. The OSL structure thus designed is no longer periodic along the propagation direction. Because of the domain limitation in our traditional poling process (several microns), the beam "acceleration" cannot be set too large, otherwise the structure is too small to be fabricated. The numerical simulation of the SH field intensity distribution and the designed structure are shown in Figs. 2(a)-2(b).

The OSL structure was fabricated by two-dimensional poling of a 500 μm thick z-cut congruent lithium tantalate and then the x faces were polished. The length of the sample is 2 cm. Figure 2 (c) shows the optical microscopic image of the sample surface after HF acid etching. Because of the practical poling technique, the duty cycle of the experimental structure is different from the calculated one. However, it will only lead to a slight loss to the conversion efficiency of the SH wave. Figure 3(a) shows the experimental layout. In the experiment a cw laser (LE-LS-1064-500TA) was used with the wavelength 1064 nm. An Airy beam was generated at the second-harmonic frequency. The power of FW was fixed at 500 mW with the peak intensity about 50 W/cm^2 and 1 mm beam diameter. There was no other optical element between the laser and the sample. A filter was fixed closely to the sample to block the infrared light while transmitting the green harmonic wave. The spacing between Airy lobes was designed to be dozens of microns. In order to amplify the details of Airy lobes, an imaging system(including lens, screen, and a camera) was used. The distance between the lens and the screen(image distance) was set at 500 mm, which is more than 16 times longer than the focal length (30 mm). Calculated by the lens imaging formula, the projection on the screen was the real image of the beam cross profile at the position of 32 mm (the distance between the object plane and the lens). Figure 3(a) shows the imaging system, which is inside the dashed box. When moving the entire imaging system the distances of each element are kept constant. Figure 3 (b) shows a series of the photographs with spacing of 40 mm to display the Airy beam's most exotic features. It exhibits the propagation dynamics of a typical Airy beam which is nondiffraction and acceleration forming a parabolic arc.

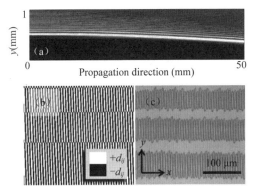

FIG. 2. (color online) (a) Numerical simulation of a propagation nonlinear Airy beam. (b) Calculated and (c) optical microscopic images of OSL.

To illustrate the self-healing character, a needle was inserted into the light path in the experiment and it blocks the second and the third lobes [the dashed curve in Fig. 3 (c)]. Likewise, photographs were shot and the propagation trajectory is shown in this figure. Using the imaging system, the light distribution nearby the obstacle has been detected. At

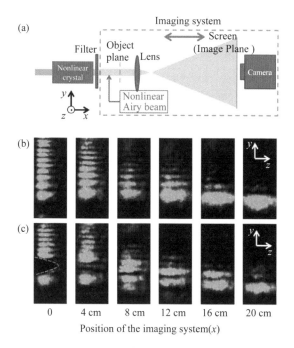

FIG. 3. (color online) (a) Experimental setup for the generation of the nonlinear Airy beam. The imaging system is moving along the x axis during the measurement. (b) Transverse patterns of the generated Airy beam and (c) the self-healing property of the nonlinear Airy beam.

the position of $x=40$ mm the lost lobes start to reappear, and at about $x=80$ mm the image recovers the normal Airy wave profile [shown in Fig. 3(c)]. However, some weak diffraction light was inevitably generated below the main lobe.

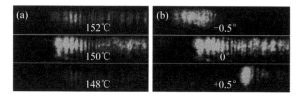

FIG. 4. (color online) Transverse patterns showing (a) temperature selectivity and (b) angular selectivity of the nonlinear volume holography.

Just like the linear volume holography, there are also strong wavelength, temperature, and angular selectivities in the nonlinear volume holography. For example, the temperature and angular dependence properties in the experiment are shown in Figs. 4 (a) and 4(b), respectively. When the temperature was shifted 2℃ away from 150℃, the main lobe of the Airy beam disappeared and the cross profile no longer kept the Airy beam wave profile. Likewise, when the incidence angle varied 0.5°, the SH distribution changed significantly. These selectivities impose restrictions on the light path, that is, no optical element is needed to be inserted between the laser and the sample, thus avoiding the distortion of the FW.

The method can be even applied to the case where the harmonic wave doesn't have an analytical form. In this situation, in order to engineer the wave front of an arbitrary harmonic wave in far field, the Fresnel diffraction formula should be adopted to calculate the propagation field in the recording area (interference region):

$$E_h(x,y) = \frac{C}{\sqrt{\lambda x}} \exp(ikx) \int_{-\infty}^{\infty} E_h(0,y_0) \exp\left[i\frac{k}{2x}(y-y_0)^2\right] dy_0. \qquad (7)$$

Using this equation, we can numerically calculate the field distribution $E_h(x,y)$ provided that the initial wave front $E_h(x,y)\big|_{x=0}$ is given; thus, the required OSL structure can be designed accordingly by inserting the calculated field distribution E_h into Eqs. (2) and (4).

It is interesting to compare the present work with the approach used in Ref. [11], which is more similar to the Fourier holography, while our method belongs to the Fresnel holography. The Fourier holography deals with the wave information in the reciprocal space. Since no lens is needed to realize the optical Fourier transform in the reconstruction process, the nonlinear Fresnel holography might be more useful than the Fourier holography for nanophotonics and integrated photonics where a smaller and more compact device is preferred.

In conclusion, by introducing the concept of volume holography into the nonlinear optics we have developed a new method for nonlinear wave-front engineering. It has significant advantages when dealing with complicated nonlinear generation processes involving special beams[34]. As a demonstration, we have experimentally generated a second-harmonic Airy beam. The method is universal and could be applied to other nonlinear optical processes. In principle, it might even be applied to some nonoptical processes that require complicated wave-front modulation, such as the Airy beam generation of surface plasmon polaritons and electron beams.

References and Notes

[1] V. Berger, Phys. Rev. Lett. **81**, 4136 (1998).

[2] Y. Zhang, J. M. Wen, S. N. Zhu, and M. Xiao, Phys. Rev. Lett. **104**, 183901 (2010).

[3] J. M. Wen, Y. Zhang, S. N. Zhu, and M. Xiao, J. Opt. Soc. Am. B **28**, 275 (2011).

[4] N. G. R. Broderick, G. W. Ross, H. L. Offerhaus, D. J. Richardson, and D. C. Hanna, Phys. Rev. Lett. **84**, 4345 (2000).

[5] J. R. Kurz, A. M. Schober, D. S. Hum, A. J. Saltzman, and M. M. Fejer, IEEE J. Sel. Top. Quantum Electron. **8**, 660 (2002).

[6] J. J. Chen and X. F. Chen, Phys. Rev. A **80**, 013801 (2009).

[7] A. Arie and N. Voloch, Laser Photonics Rev. **4**, 355 (2010).

[8] P. Zhang, Y. Hu, D. Cannan, A. Salandrino, T. C. Li, R. Morandotti, X. Zhang, and Z. G. Chen, Opt. Lett. **37**, 2820 (2012).

[9] P. Zhang, Y. Hu, T. C. Li, D. Cannan, X. B. Yin, R. Morandotti, Z. G. Chen, and X. Zhang, Phys. Rev. Lett. **109**, 193901 (2012).

[10] N. V. Bloch, K. Shemer, A. Shapira, R. Shiloh, I. Juwiler, and A. Arie, Phys. Rev. Lett. **108**, 233902 (2012).

[11] T. Ellenbogen, N. V. Bloch, A. G. Padowicz, and A. Arie, Nat. Photonics **3**, 395 (2009).

[12] I. Doleva and A. Arie, Appl. Phys. Lett. **97**, 171102 (2010).

[13] A. Shapira, I. Juwiler, and A. Arie, Opt. Lett. **36**, 3015 (2011).

[14] A. Shapira, R. Shiloh, I. Juwiler, and A. Arie, Opt. Lett. **37**, 2136 (2012).

[15] L. Li, T. Li, S. M. Wang, C. Zhang, and S. N. Zhu, Phys. Rev. Lett. **107**, 126804 (2011).

[16] I. Dolev, I. Epstein, and A. Arie, Phys. Rev. Lett. **109**, 203903 (2012).

[17] L. Li, T. Li, S. M. Wang, and S. N. Zhu, Phys. Rev. Lett. **110**, 046807 (2013).

[18] N. V. Bloch, Y. Lereah, Y. Lilach, A. Gover, and A. Arie, Nature(London) **494**, 331 (2013).

[19] Y. Q. Qin, C. Zhang, Y. Y. Zhu, X. P. Hu, and G. Zhao, Phys. Rev. Lett. **100**, 063902 (2008).

[20] X. H. Hong, B. Yang, D. Zhu, C. Zhang, H. Huang, Y. Q. Qin, and Y. Y. Zhu, Opt. Lett. **38**, 1793 (2013).

[21] S. M. Saltiel, Y. Sheng, N. Bloch, D. N. Neshev, W. Krolikowski, A. Arie, K. Koynov, and Y. S. Kivshar, IEEE J. Quantum Electron. **45**, 1465 (2009).

[22] Y. Zhang, Z. D. Gao, Z. Qi, S. N. Zhu, and N. B. Ming, Phys. Rev. Lett. **100**, 163904 (2008).

[23] Y. Sheng, W. Wang, R. Shiloh, V. Roppo, A. Arie, and W. Krolikowski, Opt. Lett. **36**, 3266 (2011).

[24] A. Shapira, A. Libster, Y. Lilach, and A. Arie, Opt. Commun. **300**, 244 (2013).

[25] D. Gabor, Nature(London) **161**, 777 (1948).

[26] T. Dresel, M. Beyerlein, and J. Schwider, Appl. Opt. **35**, 6865 (1996).

[27] T. Dresel, M. Beyerlein, and J. Schwider, Appl. Opt. **35**, 4615 (1996).

[28] D. N. Christodoulides and T. H. Coskun, Opt. Lett. **21**, 1460 (1996).

[29] E. N. Leith and J. Upatnieks, J. Opt. Soc. Am. **52**, 1123 (1962).

[30] G. A. Rakuljic, V. Leyva, and A. Yariv, Opt. Lett. **17**, 1471 (1992).

[31] M. C. Bashaw, J. F. Heanue, A. Aharoni, J. F. Walkup, and L. Hesselink, J. Opt. Soc. Am. B **11**, 1820 (1994).

[32] G. A. Rakuljic and V. Leyva, Opt. Lett. **18**, 459 (1993).

[33] F. H. Mok, Opt. Lett. **18**, 915 (1993).

[34] See Supplemental Material at http://link.aps.org/supplemental/10.1103/PhysRevLett.113.163902 for more examples and analysis.

[35] A. W. Lohmann and D. P. Paris, Appl. Opt. **6**, 1739 (1967).

[36] M. V. Berry and N. L. Balazs, Am. J. Phys. **47**, 264 (1979).

[37] G. A. Siviloglou, J. Broky, A. Dogariu, and D. N. Christodoulides, Phys. Rev. Lett. **99**, 213901 (2007).

[38] This work is supported by the State Key Program for Basic Research of China (Grant No. 2010CB630703), the National Natural Science Foundation of China (Grants No. 11374150, No. 11074120, No. 11274163, and No. 11274164) and Priority Academic Program Development of Jiangsu Higher Education Institutions of China(PAPD).

Nonlinear Talbot Effect[*]

Yong Zhang,[1,2,3] Jianming Wen,[1,2] S. N. Zhu,[1] and Min Xiao[1,2,3]

[1] *National Laboratory of Solid State Microstructures and Department of Physics,
Nanjing University, Nanjing 210093, China*

[2] *Department of Physics, University of Arkansas, Fayetteville, Arkansas 72701, USA*

[3] *School of Modern Engineering and Applied Science, Nanjing University, Nanjing 210093, China*

    We propose and experimentally demonstrate the nonlinear Talbot effect from nonlinear photonic crystals. The nonlinear Talbot effect results from self-imaging of the generated periodic intensity pattern at the output surface of the crystal. To illustrate the effect, we experimentally observed second-harmonic Talbot self-imaging from 1D and 2D periodically poled LiTaO$_3$ crystals. Both integer and fractional nonlinear Talbot effects were investigated. The observation not only conceptually extends the conventional Talbot effect, but also opens the door for a variety of new applications in imaging technologies.

    The Talbot effect[1,2], a near-field diffraction phenomenon in which self-imaging of a grating or other periodic structure replicates at certain imaging planes, holds a variety of applications in imaging processing and synthesis, photolithography, optical testing, optical metrology, spectrometry, optical computing[3], as well as in electron optics and microscopy[4]. Recent progress has been made in areas such as atomic waves[5,6], nonclassical light[7], waveguide arrays[8], and X-ray phase imaging[9]. However, all the above achievements are limited in studying properties of input fundamental signals. In this Letter, we demonstrate, for the first time, the nonlinear Talbot effect, i.e., the formation of second-harmonic (SH) self-imaging instead of the fundamental one from periodically poled LiTaO$_3$ (PPLT) crystals. This demonstration not only maintains all characteristics of the conventional Talbot effect, but offers a new way to image objects with periodic structures with higher spatial resolution. The conceptual extension achieved here thus opens a door for broader scopes of applications in imaging techniques.

    The conventional Talbot effect is well understood by the Fresnel-Kirchhoff diffraction theory, as first explained analytically by Lord Rayleigh in 1881[2], attributing its origin to the interference of diffracted beams. The simplicity and beauty of such Talbot self-imaging have since then attracted many researchers and resulted in numerous interesting and original applications that represent competitive solutions to various scientific and technological problems. The optical self-imaging phenomenon usually requires a highly

---

[*] Phys. Rev.Lett., 2010, 104(18):183901

spatially coherent illumination. The self-imaging disappears when the lateral dimensions of the light source are increased. As noted a long time ago, when the source is made spatially periodic and it is placed at the proper distance in front of a periodic structure, a fringe pattern is formed in the space behind the structure. The first example of this type was performed by Lau[10], who used two amplitude gratings of the same spatial period illuminated incoherently. However, up to today, all the research on the self-imaging has been limited in studying the properties of the input beams and using real gratings for imaging. Bypassing these limitations will not only enrich the conventional self-imaging research, but also offer new methods for imaging technologies. Here, we present the first experimental demonstration beyond the conventional Talbot effect, in which the observed self-images are not produced by the input fundamental beam but by the SH field generated in PPLT nonlinear crystals. In our experiment, the grating is the periodic intensity patterns appearing on the output surface of the crystal due to the periodic domain structures, i.e., the modulated second-order nonlinear susceptibility. This difference thus distinguishes our results from the conventional self-imaging research.

PPLT crystals have been extensively used as a workhorse for laser frequency conversion[11], optical switching[12], wave-front engineering[13], and quantum information processing[14]. Numerous interesting phenomena have been discovered in both 1D and 2D nonlinear photonic crystals, such as solitons[15], entangled photons[16], conical second-harmonic generation (SHG)[17,18], and nonlinear Čerenkov radiation[19]. Benefitting from their modified nonlinear properties, here we report a novel SH Talbot self-imaging demonstrated with the use of such crystals. Different from the conventional Talbot effect[1,3], the observed self-imaging is a consequence of the $\chi^{(2)}$ nonlinear optical process, which we call the nonlinear Talbot effect. As mentioned before, the Talbot effect is attributed to the interference of diffracted beams from periodic structures, a prerequisite condition to realize such an effect[1-3]. In PPLT crystals, this condition is fulfilled by the periodic patterns of the SH intensity difference distributed on the output surface. The SH intensity generated from the domain walls is different from that inside domains because the nonlinear coefficients near the nonideal domain walls will be changed due to the crystal lattice distortion after the poling process[20]. In the experiment, we have directly observed periodically distributed SH intensity patterns in recorded self-images. Although the observed phenomenon well resembles the conventional Talbot effect, several interesting features distinguish this demonstration from the previous observations. Besides no real grating used in the experiment, spatial resolution improvement by a factor of 2, due to frequency doubling, is powerful to high-resolution imaging, compared with the simple input-pump imaging. In principle, if the material allows the Nth-order harmonic generation, spatial resolving power can be enhanced by N times using the nonlinear Talbot effect reported here. Moreover, this experiment conceptually extends the conventional

Talbot effect and thus paves a way for new applications.

The 1D and 2D PPLT slices were fabricated through an electric-field poling technique at room temperature[21]. In the experiments, we used a femtosecond mode-locked Ti:sapphire laser operating at a wavelength of 800 nm as the fundamental input field. The pulse width is ~100 fs with a repetition rate of 82 MHz. As shown in Fig. 1, the fundamental beam from the Ti:sapphire laser was first reshaped by a telescope device, composed of two focusing lenses, to achieve a near-parallel beam with a spot size of ~100 μm, which propagates along the z axis of the PPLT sample and whose polarization is parallel to the x axis of the crystal. Although LiTaO$_3$ crystal has a space group of $3m$ (C_{3v}), only the d_{21} component contributes to the SHG process in our experimental configuration[22]. After the sample, a bandpass filter was used to filter out the near-infrared fundamental field. The generated SH intensity pattern was magnified by an objective lens and projected onto a CCD camera. The SH patterns at different imaging planes were recorded by moving the objective lens along the SH propagation direction, which was controlled by a precision translation stage. We emphasize that the nonlinear Talbot effect is a lensless imaging process. The objective used here is to magnify the self-images for easy observations.

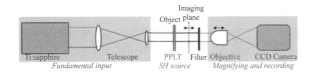

FIG. 1. (color online) Experimental setup. The PPLT sample is placed at the waist plane of the fundamental wave. The patterns on different imaging planes are recorded by a CCD camera through moving the objective.

Figs. 2 and 3, respectively, show the integer SH self-images from 1D and 2D PPLT crystals. In the experiment, we have sequentially recorded such integer Talbot self-images at about several Talbot lengths. For simplicity, we arranged d_{21} as the only nonzero nonlinear coefficient in the tensor to contribute to the collinear SHG process. To verify this, we measured the polarization of the generated SH wave and found that its polarization was indeed along the y axis of the crystal. As described above, d_{21} at the domain walls is different from that inside the domains. Such periodic domain structures result in the generated SH intensity patterns displaying the same periodicity at the sample output surface. An imaginary grating is thus formed for self-imaging. From the Fresnel-Kirchhoff diffraction theory, the diffracted field amplitude $A(\bm{r}_1)$ is defined in terms of the aperture function of the object $t(\bm{r})$ and the coherent amplitude of the source $S(\bm{r}_s)$. Here $\bm{r}_1$, $\bm{r}$, and $\bm{r}_s$ are located at the observation, object, and source planes, respectively. The diffracted amplitude $A(\bm{r}_1)$ is given by

$$A(\mathbf{r}_1) = \frac{\exp[2i\pi(d_1+d_2)/\lambda]}{i\lambda d_1 d_2} \int d\mathbf{r}_s S(\mathbf{r}_s) \int d\mathbf{r} t(\mathbf{r}) \times$$
$$\exp\left[\frac{i\pi |\mathbf{r}-\mathbf{r}_s|^2}{\lambda d_1}\right] \exp\left[\frac{i\pi |\mathbf{r}_1-\mathbf{r}|^2}{\lambda d_2}\right], \quad (1)$$

where d_1 is the propagation distance between the object and the source, and d_2 is the distance from the object to the observation plane. In our experiment, the SH source may be treated as a plane wave or a Gaussian beam. The aperture function comes from the spatially periodic domain structures, i.e., the periodically engineered $\chi^{(2)}$. One notable difference from conventional self-imaging is that λ in Eq. (1) is the SH wavelength, which is half of the wavelength λ_p of the fundamental input beam. This difference leads to a factor of 2 in imaging resolution improvement, compared with the traditional direct input-output measurement.

In our first experiment, a 1D PPLT sample with a domain period of $a = 8.0$ μm and a duty cycle of $\sim 50\%$ [see the SEM picture shown in Fig. 2(a)] was chosen to illustrate the effect. Figs. 2(b) and 2(c) are the recorded self-images on the first and third Talbot planes, respectively. For a plane-wave illumination, the self-images repeat at multiples of the SH Talbot length $z_T = 4a^2/\lambda_p$, which gives 320 μm and 960 μm for the first and third Talbot imaging planes. The experimentally measured lengths were 330 μm±5 μm and 1020 μm±10 μm, respectively.

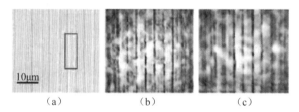

FIG. 2. (color online) 1D SH Talbot self-imaging. (a) The domain structure of the 1D PPLT crystal; (b) the SH self-image at the first Talbot plane; (c) the SH self-images at the third Talbot plane. The marked domain in (a) is a little narrower than the others.

The discrepancy between the experiment and theory can be readily corrected by taking into account the Gaussian profile of the beam, which reinterprets the SH Talbot length as $z_T = 4(a^2/\lambda_p)(r_z/r_0)^2$, where r_z and r_0 represent the Gaussian beam radii at the imaging plane and the waist plane. In the experiments, the PPLT sample was placed at the waist plane of the Gaussian beam. The corrected theoretical lengths are now 320 μm with $r_z/r_0 = 1.00$ μm and 1018 μm with $r_z/r_0 = 1.03$, consistent with the experimental data. Periodic SH interference fringes are clearly observable on the first Talbot plane [Fig. 2(b)] and the period coincides with the periodicity of the domain structure. The bright fringes correspond to the domains exactly. The intensities from positive or negative domains have no appreciable difference because the value of the nonlinear coefficient inside the domains does not change during the poling process. The dark fringes correspond to the domain walls. In the self-imaging, the SH intensity produced at the domain walls is much weaker than that inside the domains. From Fig. 2(b), the width of the domain walls is estimated to be 0.5 μm, which agrees with the reported range from 100 nm to few microns[23]. One difference between the SEM image [Fig. 2(a)] and the self-image [Fig. 2(b)] is that the SEM imaging is sensitive to the imperfections in the domain structure (the marked square) while the self-imaging produces a uniform interference pattern. It is because only the periodically distributed SH light is self-imaged at the Talbot planes while the nonperiodic part is not reproduced. This could be useful for the application of optical lithography. We also experimentally found that the image qualities after the third Talbot plane became worse than those at the first and second planes. One reason is that higher-order diffraction fields cannot be totally collected by the CCD camera with a limited numerical aperture.

To illustrate the integer SH Talbot effect from a 2D modulated nonlinear crystal, a hexagonally-poled LiTaO$_3$ slice, was adopted in the second experiment. The period of the domain structure of the sample is $a=9.0$ μm and the duty cycle is ~30% as shown in the SEM image in Fig. 3(a). The SH self-images were also successively observed at several Talbot imaging planes. For the ideal plane-wave illumination, the Talbot length is deduced to be $z_T = 3a^2/\lambda_p$. Taking into account the Gaussian profile, the SH Talbot length is corrected to be $z_T = 3(a^2/\lambda_p) \times (r_z/r_0)^2$. Figs. 3(b) and 3(c) show the images recorded at the first and third SH Talbot planes, respectively. Same as for the 1D case, the SH waves generated at the domain walls are weaker than that created inside the domains. The domain wall in the 2D structure exhibits a ring-shaped structure [Fig. 3(b)], while it appears as a dark line in the 1D PPLT [Fig. 2(b)]. Similar as the 1D structure, the qualities of the self-images after the third Talbot plane [Fig. 3(c)] become worse due to the loss of higher-order SH diffractions. The measured first and third SH Talbot lengths were 315 μm±5 μm and 970 μm±10 μm, which agree well with the calculated 304 μm($r_z/r_0=1.00$) and 966 μm ($r_z/r_0=1.03$), respectively.

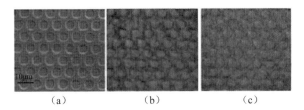

(a)   (b)   (c)

FIG.3. (color online) 2D SH Talbot self-imaging. (a) The domain structure of the hexagonally poled LiTaO$_3$ crystal; (b) the SH self-image at the first Talbot plane; (c) the SH self-image at the third Talbot plane.

Besides these integer SH Talbot effect, in the third experiment we have also investigated the fractional SH self-images occurring at the intermediate Talbot distances, $z=(p/q)z_T$, where p and q are integers with no common factor. It is well known that fractional Talbot effect[24—26] has a close connection with fractional revivals, quantum carpet[27], and Gauss sums[28]. Compared with the integer self-imaging, fractional SH

Talbot effect exhibits more interesting and complicated interference patterns. We have experimentally observed such patterns for both 1D and 2D PPLT crystals. In the 1D case, because of the limited illumination area on the sample, the quality of fractional self-images is greatly reduced and becomes blurry in contrast with the integer cases [Figs. 2(b) and 2(c)]. The situation is dramatically changed for the 2D case, where the SH fractional Talbot images have high quality and show number of unique properties in comparison with the integer case. In Figs. 4(a)-4(f), we present several representative fractional self-images with different fractional (p/q) parameters. For instance, the SH interference pattern at the $z_T/7$ plane [Fig. 4(a)] is periodically distributed bright spots with the same lattice structure as in the integer case. In Figs. 4(b) and 4(c), the period of the hexagonal structure is halved. In Figs. 4(c)-4(f), a 30° rotation of the structure is obviously observable and even a π-phase shift can be deduced from Figs. 4(d)-4(f). We found that the fractional Talbot images are very sensitive to the precision positions of the imaging planes. Further detailed studies of these fractional SH Talbot effects will be presented elsewhere.

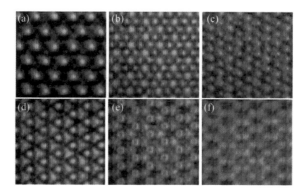

FIG. 4. (color online) The SH patterns (a)-(f) are corresponding to fractional Talbot images at different fractional Talbot planes.

In recent years, SH imaging technique has been developed into a powerful near-field imaging tool for visualizing various ferroic domains[29], carrier motion[30], biomolecular array[31], and collagen modulation[32]. Comparing with the conventional SH imaging, such SH Talbot self-imaging does not need an imaging lens and no reference SH wave is needed, which greatly simplifies the experimental setup. For periodically-poled ferroelectric domains, although chemical etching method is adopted as a standard procedure to look at the domain structures[33], one major disadvantage of this method is that the sample surface is damaged. The newly observed nonlinear Talbot effect provides a better (optical) way to easily check the domains without damaging the sample surfaces. This will be very helpful for inspecting integrated nonlinear optical devices such as nonlinear photonic waveguides and wavelength conversion devices, which cannot be done by the conventional Talbot effect. More importantly, the conceptual generalization demonstrated here is not limited to the optical signals, and could also apply to other research fields where similar situations may exist. This effect can also be further considered with nonclassical light states[7] to achieve sub-Rayleigh images. Generally speaking, the demonstrated effect not only enriches conventional imaging techniques, but also offers a new method for imaging in broad applications.

References

[1] H. F. Talbot, Philos. Mag. **9**, 401 (1836).
[2] L. Rayleigh, Philos. Mag. **11**, 196 (1881).
[3] K. Patorski, Prog. Opt. **27**, 1 (1989).
[4] J. M. Cowley, *Diffraction Physics* (North-Holland, Amsterdam, 1995).
[5] M. S. Chapman et al., Phys. Rev. A **51**, R14 (1995).
[6] C. Ryu et al., Phys. Rev. Lett. **96**, 160403 (2006).
[7] K.-H. Luo et al., Phys. Rev. A **80**, 043820 (2009).
[8] R. Iwanow et al., Phys. Rev. Lett. **95**, 053902 (2005).
[9] F. Pfeiffer et al., Nature Mater. **7**, 134 (2008).
[10] E. Lau, Ann. Phys. (Leipzig) **437**, 417 (1948).
[11] S. N. Zhu, Y. Y. Zhu, and N. B. Ming, Science **278**, 843 (1997).
[12] C. Langrock et al., J. Lightwave Technol. **24**, 2579 (2006).
[13] Y. Q. Qin et al., Phys. Rev. Lett. **100**, 063902 (2008).
[14] P. Kumar et al., Quant. Info. Proc. **3**, 215 (2004).
[15] H. Kim et al., Opt. Lett. **28**, 640 (2003).
[16] X. Q. Yu et al., Phys. Rev. Lett. **101**, 233601 (2008).
[17] P. Xu et al., Phys. Rev. Lett. **93**, 133904 (2004).
[18] S. M. Saltiel et al., Phys. Rev. Lett. **100**, 103902 (2008).
[19] Y. Zhang et al., Phys. Rev. Lett. **100**, 163904 (2008).
[20] M. Bazzan et al., in *Ferroelectric Crystals for Photonic Applications*, edited by P. Ferraro, S. Grilli, and P. De Natale, (Springer-Verlag, Berlin, Heidelberg, 2009).
[21] S. N. Zhu et al., J. Appl. Phys. **77**, 5481 (1995).
[22] R. W. Boyd, *Nonlinear Optics* (Academic, New York, 2003), 2nd ed.
[23] T. J. Yang et al., Phys. Rev. Lett. **82**, 4106 (1999).
[24] J. T. Winthrop and C. R. Worthington, J. Opt. Soc. Am. **55**, 373 (1965).
[25] M. Paturzo et al., Opt. Lett. **31**, 3164 (2006).
[26] C.-S. Guo et al., Opt. Lett. **32**, 2079 (2007).
[27] M. V. Berry, I. Marzoli, and W. P. Schleich, Phys. World **14**, 39 (2001).
[28] D. Bigourd et al., Phys. Rev. Lett. **100**, 030202 (2008).
[29] M. Fiebig et al., Nature (London) **419**, 818 (2002).
[30] T. Manaka et al., Nat. Photon. **1**, 581 (2007).
[31] P. J. Campagnola and L. M. Loew, Nat. Biotechnol. **21**, 1356 (2003).
[32] E. Brown et al., Nature Med. **9**, 796 (2003).
[33] V. Bermudez et al., J. Cryst. Growth **191**, 589 (1998).
[34] We acknowledge partial support from the National Science Foundation (U.S.A.) and partial support by the 111 Project B07026 (China).

Diffraction Interference Induced Superfocusing in Nonlinear Talbot Effect[*]

Dongmei Liu[1], Yong Zhang[1], Jianming Wen[2], Zhenhua Chen[1], Dunzhao Wei[1], Xiaopeng Hu[1], Gang Zhao[1], S. N. Zhu[1] & Min Xiao[1,3]

[1] *National Laboratory of Solid State Microstructures, College of Engineering and Applied Sciences, School of Physics, Nanjing University, Nanjing 210093, China*, [2] *Department of Applied Physics, Yale University, New Haven, Connecticut 06511, USA*, [3] *Department of Physics, University of Arkansas, Fayetteville, Arkansas 72701, USA*

We report a simple, novel subdiffraction method, i.e. diffraction interference induced superfocusing in second-harmonic (SH) Talbot effect, to achieve focusing size of less than $\lambda_{SH}/4$ (or $\lambda_{pump}/8$) without involving evanescent waves or subwavelength apertures. By tailoring point spread functions with Fresnel diffraction interference, we observe periodic SH subdiffracted spots over a hundred of micrometers away from the sample. Our demonstration is the first experimental realization of the Toraldo di Francia's proposal pioneered 62 years ago for superresolution imaging.

Focusing of a light beam into an extremely small spot with a high energy density plays an important role in key technologies for miniaturized structures, such as lithography, optical data storage, laser material nanoprocessing and nanophotonics in confocal microscopy and superresolution imaging. Because of the wave nature of light, however, Abbe[1] discovered at the end of 19th century that diffraction prohibits the visualization of features smaller than half of the wavelength of light (also known as the Rayleigh diffraction limit) with optical instruments. Since then, many efforts have been made to improve the resolving power of optical imaging systems, and the research on overcoming the Abbe-Rayleigh diffraction limit has become an energetic topic (for recent reviews see Refs. [2]–[4]).

Historically, an early attempt to combat the diffraction limit can be traced back to the work by Ossen[5] in 1922, in which he proved that a substantial fraction of emitted electromagnetic energy can be squeezed into an arbitrarily small solid angle. Inspired by the concept of super-directivity[6], in his seminal 1952 paper Toraldo di Francia[7] suggested that a pupil design provide an accurately tailored subdiffracted spot by using a series of concentric apertures with different phases. Based on the mathematical prediction that band-limited functions are capable of oscillating faster than the highest Fourier components

---

[*] Sci.Rep., 2013, 4:6134

carried by them (a phenomenon now known as superoscillation[8]), Berry and Popescu[9] in their recent theoretical analysis pointed out that subwavelength localizations of light could be obtained in Talbot self-imaging[10,11] under certain conditions. With use of a nanohole array, Zheludev's group demonstrated the possibility to focus the light below the diffraction limit[12,13]. By using a sequence of metal concentric rings with subwavelength separations, they further reported well-defined, sparsely distributed subdiffracted light localizations in a recent optical superoscillating experiment[14]. Despite the newly developments of quantum imaging[15] and quantum lithography[16] allow the formation of sub-Rayleigh diffracted spots, the severe reliance on specific quantum entangled states and sophisticated measurement devices limits their practical applications.

By exploring evanescent components containing fine details of an electromagnetic field distribution, researchers working in near-field optics have invented powerful concepts, such as total internal reflectance microscopy[17] and metamaterial-based superlens[18,19], to overcome the barrier of the diffraction limit. Most near-field techniques operate at a distance extremely close (typically hundreds of nanometers) to the object in order to obtain substantial subdiffracted spots. Since these techniques cannot image an object beyond one wavelength, they are not applicable to image into objects thicker than one wavelength, which greatly limits their applicability in many situations. There also exists a broad category of functional super-resolution imaging techniques which use clever experimental tools and known limitations on the matter being imaged to reconstruct the super-resolution images. The representative ones include stimulated emission depletion[20], spatially-structured illumination microscopy[21], stochastic optical reconstruction microscopy[22], and super-resolution optical fluctuation imaging[23].

Principle and experimental scheme Here we introduce an alternative scheme, i. e. diffraction interference induced superfocusing in nonlinear Talbot effect[24,25], to achieve subdiffraction by exploiting the phases of the second-harmonic (SH)[24—27] fields generated from a periodically-poled LiTaO$_3$ (PPLT) crystal. The poling inversions in the PPLT crystal, typically with a period of few micrometers, make the SH waves generated in the negative domains possess a π phase shift relative to those in the positive domains. The destructive interference between these two SH waves in the Fresnel diffraction region shrinks the point spread functions below the diffraction limit and leads to subwavelength focused spots, resembling a similar idea as suggested by Toraldo di Francia sixty-two years ago[7]. The essential physics behind this experiment is schematically illustrated in Fig. 1. From college textbook *Optics*, we know that as a circular aperture with a diameter of d [Fig. 1(a)] is uniformly illuminated, the focused light intensity has a spatial profile of the Airy disk with a radius of $\lambda/2d$, first noticed by Abbe as the spatial resolution limit. To beat this limit, in 1952 Toraldo di Francia theoretically discovered that if the object consists of a series of concentric apertures with different transmission amplitudes and/or phases, for example, a circular aperture with α phase and its neighboring ring with β phase

[Fig. 1(b)], due to diffraction interference the Airy disk can be shrunk below the Abbe limit with the cost of pushing the energy into the fringe wings [Fig. 1(d)]. In our experiments, the "object" is the generated SH waves with a periodic binary phase distribution ($\alpha = 0$ and $\beta = \pi$) from a PPLT crystal [Fig. 1(c)], which is designed to produce periodic superfocused spots. To the best of our knowledge, our observation of superfocusing in nonlinear Talbot self-imaging is the first realization of Toraldo di Francia's proposal.

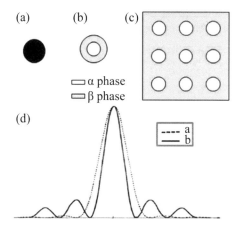

FIG. 1. Theoretical model. (a) A circular aperture. (b) A donut-like aperture with an α-phase inside circle and a β-phase outside ring. (c) The "object" in nonlinear Talbot effect, i.e., the generated SH waves with binary phases ($\alpha = 0$ and $\beta = \pi$) at the output surface of a squarely poled LiTaO$_3$ crystal. (d) Diffracted intensity distributions with apertures of (a) and (b) in the Fresnel near field. Obviously, the aperture (b) gives a smaller Airy disk than the aperture (a). With use of the aperture (c), periodic superfocused spots are expected to be observed in the Fresnel near field.

Besides, because of the phase matching the generated SH signals are automatically band-limited, a key ingredient for superoscillations[9]. These two unique and coexisting features distinguish the current scheme from previous works that involve either evanescent waves, metal nanostructures, luminescent objects or quantum states. Our demonstration can be considered as the first experimental realization of the Toraldo di Francia's proposal for subdiffraction[7] and supperresolution with superoscillations[9] in nonlinear optics. This method allows to produce subdiffracted SH spots over 100 μm easily, and has, in principle, no fundamental lower bound to limit the focusing ability. As such, we have observed superfocused SH spots with the size of less than one quarter of the SH wavelength ($\lambda_{SH}/4$) at the distance of tens of micrometers away, which is comparable to the superoscillation experiment[14], but without employing subwavelength metal nanoholes. We thus expect our imaging technique to provide a super-resolution alternative for various applications in photolithography, medical imaging, molecular imaging, as well as bioimaging.

In our proof-of-principle experiment, the periodic domain structures of the PPLT crystal help create subwavelength foci with prescribed sizes and shapes, as the SH waves with different phases propagate freely and interfere destructively. The achievable subdiffraction patterns depend on parameters such as the periodicity of domain structures, sizes of the domain structures, and the propagation distance. Experimentally, superfocused SH spots with sizes of less than $\lambda_{pump}/8$ have been recorded at 27.5 μm away from the sample. In comparison with the superoscillatary experiment[14] using a binary-amplitude metal mask, the current scheme explores the π phase difference between the SH fields generated from inside and outside of the domains, respectively. Note that such a π-phase shift does not exist in grating-based Talbot effect and linear superfocusing systems. Besides, the structure of the PPLT crystal does not involve complicated nano-fabrications and has large-scale parameters than light wavelengths.

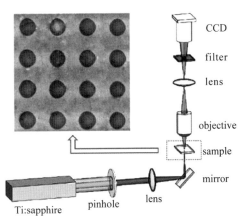

FIG. 2. Schematic diagram of the experimental setup. The sample is placed at the focal plane of the lens. The SH patterns at different imaging planes are recorded by a CCD camera mounted on a nanopositioning precision translation stage. Inset is the SEM image of the 2D squarely poled LiTaO$_3$ slice with a period of 5.5 μm and duty cycle of ∼35%.

Similar to our previous studies on SH Talbot effects[24,25], the super-focusing setup is schematically shown in Fig. 2. A femtosecond mode-locked Ti:sapphire laser was operated at a wavelength of $\lambda_{pump}=900$ nm as the fundamental input field. The pulse width was about 75 fs with a repetition rate of 80 MHz. As illustrated in Fig. 2, the fundamental pump laser was first shaped by a pinhole and focusing lens to produce a near-parallel beam with a beam size of ∼100 μm, and then directed into a 2D squarely-poled LiTaO$_3$ slice along the z axis with its polarization parallel to the x axis of the crystal. The sample with the size of 20 mm (x)×20 mm (y)×0.5 mm (z) was placed on a nanopositioning precision translation stage, and the SEM image of its domain structures (with a period of a =5.5 μm and a duty cycle of ∼35%) is depicted in the inset of Fig. 2. Despite the LiTaO$_3$

crystal has a space group of 3m (C_{3v}), mainly the d_{21} component contributes to the SH generation in the current experimental configuration. After the sample, an objective lens ($\times 100$) with a high numerical aperture of $NA=0.7$ was used to magnify the SH intensity patterns ($\lambda_{SH}=450$ nm). With this secondary imaging process, the magnified SH intensity distributions were recorded by the CCD camera. To remove the near-infrared fundamental field, a bandpass filter was placed between the objective lens and the CCD camera. The SH patterns at different imaging planes were recorded along the SH propagation direction by moving the microscope stage, which was controlled by the precision translation stage.

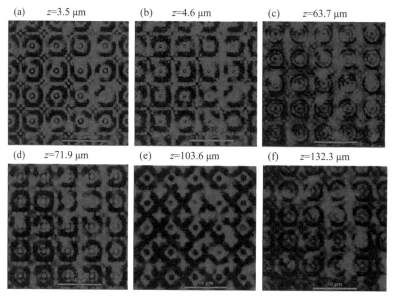

FIG 3. Recorded images of the SH patterns with a conventional optical microscope at different Talbot planes. The images shown here are all formed within one Talbot length. (a) and (b) are close to the sample surface. (c) is the SH pattern at 1/2 Talbot plane, which is laterally shifted by half a period in comparison with the primary SH self-image. (f), at the first Talbot plane. Periodic focusing spots can be observed in most of the imaging planes. At certain planes (e), however, the focusing spots disappear because of destructive interference of the SH waves.

Experimental results To ensure the good alignment requested in superfocusing, we begin with the observation of SH Talbot self-imaging in comparison with theoretical predictions. Characteristic SH field patterns recorded at different distances (z) away from the output surface of the sample are presented in Fig. 3, representing a variety of "photonic carpets" in the Fresnel diffraction region. As indicated in Figs. 3(a)–(f), the diffraction patterns change dramatically along with the focus being moved away from the crystal. The primary Talbot self-imaging was observed at $z=132.3$ μm [Fig. 3(f)], which is well consistent with the theoretically calculated SH Talbot length[24,25] of $z_t=4a^2/\lambda_{pump}=134.4$ μm. One previously unconfirmed feature appears at about 1/2 Talbot length where a square array-like SH self-image is laterally shifted by half the width of the domain period [comparing Fig. 3(c) with Fig. 3(f)]. In fractional Talbot planes, one can see complicated diffraction patterns, which result from the Fresnel diffraction interference of the SH waves. In proximity to the end face of the sample [Figs. 3(a)-(d)], the SH waves form periodic focused spots at the center of each unit. The focusing size varies with the propagation distance z and superfocusing occurs at certain planes. At some other planes [e. g., 4/5 fractional Talbot plane as shown in Fig. 3(e)], the focusing spots disappear due to destructive interference. By carefully examining the patterns, one can find that the detailed structures in every single unit are very sensitive to the observation distance, especially when close to the sample. For example, as the observation plane moves from $z=3.5$ μm [Fig. 3(a)] to $z=4.6$ μm [Fig. 3(b)], the rings at the center shrink and the fractal array at the corner evolves. The key factor is that the phases of the SH waves develop sensitively along the propagation distance.

After confirming the alignment, we are ready to look for the sub-diffracted SH spots at different propagation distances. According to the Abby resolution limit, the diffraction lower bound is about 321 nm in the current system. However, in the experiment we have recorded series of subwavelength focused spots at different propagation distances. Figure 4 shows some typical images with subdiffracted focused spots. For $\lambda_{SH}=450$ nm, at a distance of $z=27.5$ μm a subdiffracted spot was identified with a full-width-at-the-halfmaximum (FWHM) of 106 nm [see the enlarged area in Fig. 4(a)], i.e. $0.117\times\lambda_{pump}$ or $0.235\times\lambda_{SH}$, which surpasses that of the super-oscillatory lens ($0.29\times\lambda_{pump}$; a focal spot of 185 nm in diameter for a wavelength of 640 nm, Ref. [14]). The recorded cross-section of the spot shown in Fig. 4(e), without any data post-processing, is well fitted with a Lorentzian lineshape. The background is resulted from the imperfect domain structures. Due to the imperfections such as defects in the crystal and the nonuniform domain structures, not all the theoretically predicted superfocused spots were observable in the experiment. At $z=31.5$ μm the selected subdiffraction spot [Fig. 4(b)] has a FWHM of 187 nm [Fig. 4(f)]. We notice that in the current scheme, the cross-section profile of the measured subdiffracted spot fits better with a Lorentzian curve rather than with a Gaussian shape, which is in contrast to those focusing spots obeying the diffraction limit. For instance, at the distance of $z=91.4$ μm, a focusing spot with a size of 227 nm (close to the diffraction limit) is well fitted by a Gaussian curve [Figs. 4(d) and 4(h)]. In the present experiment, the largest distance where we can still find superfocusing spots is at $z=133.4$ μm [almost at the primary SH Talbot plane, Fig. 4(c)] and these subwavelength spots with a FWHM down to 168 nm [Fig. 4(g)] well follow a Lorentzian profile. We also carefully analyzed the power distribution in superfocusing. For the input pump power setting at 590 mW, the total power of the generated SH field is $\sim$4.64 mW. The power of each central superfocused spot in Fig. 4(a)-(d) is about (0.227, 1.236, 0.713, 1.269) nW, respectively. As expected, most of the energy is distributed in side fringes while the center spot contains less than one thousandth of the total SH power.

Simulations To theoretically verify that the superfocusing feature in our experiment is

134 介电体超晶格
Superlattices and Microstructures of Dielectric Materials

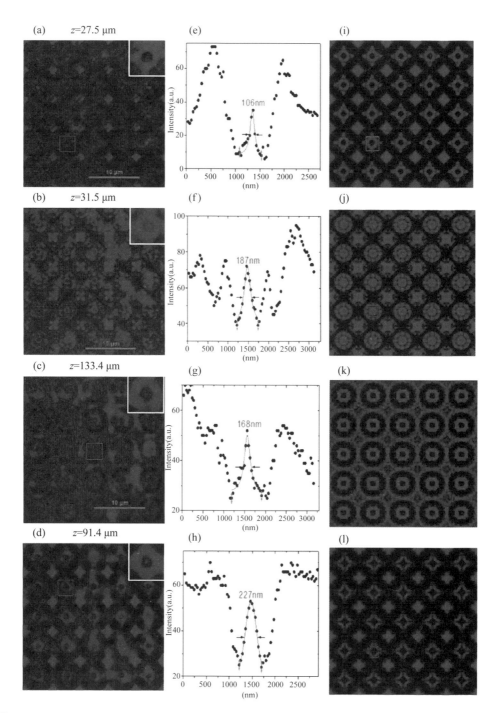

FIG. 4. Typical measured results of superfocusing. (a) - (d) are experimentally recorded SH patterns at different observation distances, where the insets are the enlarged images of the selected focused spots. The cross-sections of the selected focused spots in (a) - (d) are given, respectively, in (e) - (h), whose centers are fitted with a Lorentzian (e) - (g) or Gaussian (h) lineshape. The black dots in (e) - (h) are experimental data. (i) - (l) are theoretical simulations corresponding to (a) - (d), respectively.

indeed formed by the constructive/destructive interference of propagating SH waves, we performed numerical simulations using the angular spectrum method. For the self-images shown in Figs. 4(a)-(d), the simulations predict the same patterns as depicted in Figs. 4(i)-(l) by using the experimental parameters. For Figs. 4(e)-(h), the simulations yield the focused spot sizes of 118 nm, 191 nm, 185 nm, and 225 nm, respectively, which indicate good agreements with the experimental data. Our simulations also reveal that the width of the domain walls is a very important parameter for the imaging pattern. We find, both theoretically and experimentally, that wide domain walls (>1 μm) may completely change the image patterns and eliminate the superfocusing phenomenon at those spatial planes. Thanks to the high-quality sample used in the experiment, the computed patterns by using a model without considering domain walls well match the experimental results. This implies that the domain walls here are narrow enough to be negligible in the process of image formations. To further examine the observed subdiffraction effect, we have chosen another hexagonally-poled PPLT structure with a period of 9 μm and the duty cycle~30% (previously used for the illustration of SH Talbot effect[24,25]). The input pump laser was still operated at the 900 nm wavelength. In the Fresnel near field, subwavelength focusing has also been observed (not shown here).

Discussion

One may notice that the CCD camera used in the experiment has pixels with the size of 6.45×6.45 μm^2. College textbook *Optics* tells us that a passive Gaussian thin lens won't lead to a subdiffracted focused spot for a plane-wave illumination. One interesting question then arises: how can one image the subwavelength spots from the focusing plane to the CCD camera? It turns out that in the spirit of superoscillations[9], the superposed waves at the center superfocused spot can be effectively approximated as a new (pseudo-)wave with an effective wavenumber N times larger than the original SH wavenumber, where N is the resolution enhancement over the Abbe limit achieved in the experiment. The center focused spots with these effective wavenumbers therefore can be imaged in the CCD camera through the traditional imaging process as described above. Experimentally, we have further verified this analysis through looking at the size of a standard focused spot (following the Abbe diffraction limit, obtained after a Gaussian thin lens) magnified by passing another Gaussian lens. Different from this latter case, in our experiments we have observed different size amplification, which is a strong evidence of superoscillations. We notice that in the previous superoscillatory experiments[13,14], no such analysis has been implemented.

Subwavelength focusing with such a square array or hexagonal array PPLT could allow light to be squeezed into the scale of less than a quarter of the SH wavelength (i.e., less than one eighth of the fundamental wavelength!), thereby opening new avenues of studying light-matter interactions, single-molecule sensing, nanolithography, and nanoscale imaging. By optimizing the parameters (such as the periodicity, domain structure, and propagation distance), it is possible to shrink the focused spot size down to tens of nanometers, which would be comparable with those functional super-resolution imaging techniques[20-23]. Moreover, thanks to the excellent electrooptic characteristics of PPLT, one may continuously tune the phase of the SH waves produced in the crystal, and control the interference and focusing in the far field by applying an electric field[28]. To effectively remove the background noise, we have recently used concentrically poled nonlinear crystals[26,27] by following the original proposal proposed by Toraldo di Francia[7]. The preliminary experimental results (not shown here) clearly show substantial improvement on the quality of the superfocused light spots.

Subdiffraction imaging holds many exciting promises in various areas of science and technologies. The extension of the current method to optical microscope may improve the resolving power down to nanometer scale, which would be very useful for non-invasive subwavelength biomedical imaging. Another potential application is in optical lithography at ultra-small scales, which is the key to scaling down integrated circuits for high-performance optoelectronics. Optical data storage and biosensing may also benefit from this promising scheme to process information within an ultra-small volume, thereby increasing storage densities or sensing resolution.

In summary, we have proposed and demonstrated an easy way to reduce the point spread function below $\lambda_{SH}/4$ in SH Talbot effect with the periodically-poled LiTaO$_3$ crystal. The method does not involve evanescent waves nor subwavelength structures in the object. Through the destructive/constructive interference, the subdiffracted SH spots can be observed up to 133.4 μm away from the sample for $\lambda_{pump}=900$ nm. The numerical simulations have confirmed the experimental results with excellent agreements. Our work can be considered as the first realization of the proposals made by Toraldo di Francia[7] and Berry & Popescu[9]. Furthermore, our investigation can potentially have a wide range of applications including subwavelength imaging, as a mask for biological molecule imaging, optical lithography and focus devices.

References and Notes

[1] Abbe, E. Beitrage zur theorie des mikroskops und der mikroskopischen wahrnehmung. *Arch. Mikrosk. Anat.* **9**, 413—468 (1873).

[2] Serrels, K. A., Ramsay, E., Warburton, R. J. & Reid, D. T. Nanoscale optical microscopy in the vectorial focusing regime. *Nature Photon.* **2**, 311—314 (2008).

[3] Kawata, S., Inouye, Y. & Verma, P. Plasmonics for near-field nano-imaging and superlensing. *Nature Photon.* **3**, 388—394 (2009).

[4] Mosk, A. P., Lagendijk, A., Lerosey, G. & Fink, M. Controlling waves in space and time for imaging and focusing in complex media. *Nature Photon.* **6**, 283—292 (2012).

[5] Oseen, C. W. Die Einsteinsche Nadelstichstrahlung und die Maxwellschen Gleichungen. *Ann. Phys.*

(*Leipzig*) **374**, 202—204 (1922).

[6] Schelkunoff, S. A. A Mathematical Theory of Linear Arrays. *Bell Syst. Tech. J.* **22**, 80—107 (1943).

[7] Toraldo di Francia, G. Super-gain antennas and optical resolving power. *Suppl. Nuovo Cim.* **9**, 426—438 (1952).

[8] Aharonov, Y., Anandan, J., Popescu, S. & Vaidman, L. Superpositions of time evolutions of a quantum system and a quantum time-translation machine. *Phys. Rev. Lett.* **64**, 2965—2968 (1990).

[9] Berry, M. V. & Popescu, S. Evolution of quantum superoscillations and optical superresolution without evanescent waves. *J. Phys. A* **39**, 6965—6977 (2006).

[10] Patorski, K. The self-imaging phenomenon and its applications. *Prog. Opt.* **27**, 1—108 (1989).

[11] Wen, J.-M., Zhang, Y. & Xiao, M. The Talbot effect: Recent advances in classical optics, nonlinear optics, and quantum optics. *Adv. Opt. Photon.* **5**, 83—130 (2013).

[12] Huang, F. M, Zheludev, N. I., Chen, Y. & Javier Garcia de Abajo, F. Focusing of light by a nanohole array. *Appl. Phys. Lett.* **90**, 091119 (2007).

[13] Huang, F. M. & Zheludev, N. I. Super-resolution without evanescent waves. *Nano Lett.* **9**, 1249—1254 (2009).

[14] Rogers, E. T. F. *et al.* A super-oscillatory lens optical microscope for subwavelength imaging. *Nature Mater.* **11**, 432—435 (2012).

[15] Wen, J.-M., Du, S. & Xiao, M. Improving spatial resolution in quantum imaging beyond the Rayleigh diffraction limit using multiphoton W states. *Phys. Lett. A* **374**, 3908—3911 (2010).

[16] Boto, A. N. *et al.* Quantum interferometric optical lithography: Exploring entanglement to beat the diffraction limit. *Phys. Rev. Lett.* **85**, 2733—2736 (2000).

[17] Ambrose, E. J. A surface contact microscope for the study of cell movements. *Nature* **178**, 1194 (1956).

[18] Pendry, J. B. Negative refraction makes a perfect lens. *Phys. Rev. Lett.* **85**, 3966—3969 (2000).

[19] Zhang, X. & Liu, Z. Superlenses to overcome the diffraction limit. *Nature Mater.* **7**, 435—441 (2009).

[20] Hell, S. W. Far-field optical nanoscopy. *Science* **316**, 1153—1158 (2007).

[21] Gustafsson, M. G. Nonlinear structured-illumination microscopy: Wide-field fluorescence imaging with theoretically unlimited resolution. *Proc. Natl. Acad. Sci. U.S.A.* **102**, 13801—13806 (2005).

[22] Rust, M., Bates, M. & Zhuang, X. Sub-diffraction-limit imaging by stochastic optical reconstruction microscopy (STORM). *Nature Methods* **3**, 793—796 (2006).

[23] Dertinger, T. *et al.* Superresolution optical fluctuation imaging (SOFI). *Adv. Exp. Med. Biol.* **733**, 17—21 (2012).

[24] Zhang, Y., Wen, J.-M., Zhu, S. N. & Xiao, M. Nonlinear Talbot effect. *Phys. Rev. Lett.* **104**, 183901 (2010).

[25] Chen, Z.-H. *et al.* Fractional second-harmonic Talbot effect. *Opt. Lett.* **37**, 689—691 (2012).

[26] Kasimov, D. *et al.* Annular symmetry nonlinear frequency converters. *Opt. Express* **14**, 9371—9376 (2006).

[27] Saltiel, S. M. *et al.* Generation of second-harmonic conical waves via nonlinear Bragg diffraction. *Phys. Rev. Lett.* **100**, 103902 (2008).

[28] Liu, D.-M., Zhang, Y., Chen, Z.-H., Wen, J.-M. & Xiao, M. Acoustic-optic tunable second-harmonic Talbot effect based on periodically-poled $LiNbO_3$ crystals. *J. Opt. Soc. Am. B* **29**, 3325—3329 (2012).

[29] Aharonov, Y., Albert, D. Z. & Vaidman, L. How the result of a measurement of a component of the spin of a spin-1/2 particle can turn out to be 100. *Phys. Rev. Lett.* **60**, 1351—1354 (1988).

[30] This work was supported by the National Basic Research Program of China (Nos. 2012CB921804 and 2011CBA00205), the National Science Foundation of China (Nos. 11274162, 61222503, 11274165 and 11021403), and the Science Foundation of Jiangsu Province (No. BK20140590), the New Century Excellent Talents in University, and the Priority Academic Program Development of Jiangsu Higher Education Institutions (PAPD).

[31] The principle of superfocusing in nonlinear Talbot effect was suggested by J.W. and Y.Z. with contribution from M.X. D.L., D.W. and Z.C. performed the superfocusing experiments and data post-processing. X.H., G.Z. and S.N.Z. fabricated the 2D PPLT crystals. Optical characterization and modeling of the superfocusing were undertaken by D.L., D.W. and Z.C. under the guidance of Y.Z. and J.W. J.W., Y.Z. and M.X. supervised the project and wrote the manuscript with contributions from all co-authors. J.W., Y.Z., M.X., D.L., Z.C. and D.W. contributed to the discussions of results and planning of experiments.

[32] Methods: Model of the simulations. We used the same parameters of the sample to model the "aperture function", and also took into account the π phase shift of the SH waves in negative domains. After propagation distance z, the diffracted SH field is computed by the Rayleigh-Sommerfeld diffraction formula[25]: $U(x,y,z) = \int_{-f_m}^{f_m}\int_{-f_m}^{f_m} A_0(f_x,f_y)\exp[ikz\sqrt{1-(\lambda f_x)^2-(\lambda f_y)^2}]\exp[i2\pi(f_x x + f_y y)]df_x df_y$, where $A_0(f_x,f_y)$ is the angular-spectrum representation of the sample aperture function at $z=0$. In the model, the SH field is simplified to be a plane wave, and the integration limits are bounded by the phase-matching condition in the range of $[-f_m, f_m]$. Despite this method allows accurate calculations on the evolution of intensity distribution recorded in the CCD camera, in light of weak measurement[29] superoscillations offer an alternative but interesting interpretation on the phenomenon.

Cavity Phase Matching via an Optical Parametric Oscillator Consisting of a Dielectric Nonlinear Crystal Sheet[*]

Z.D. Xie, X.J. Lv, Y.H. Liu, W. Ling, Z.L. Wang, Y.X. Fan, and S.N. Zhu

*National Laboratory of Solid State Microstructures and Department of Physics,
Nanjing University, Nanjing 210093, China*

We experimentally demonstrate cavity phase matching for the first time using a sheet optical parametric oscillator which is made of an x-cut $KTiOPO_4$ crystal sheet. This microcavity presents 220 kW peak power capability for near-frequency-degenerate parametric outputs with up to 23.8% slope efficiency. It also features unique spectral characteristics such as single-longitudinal-mode and narrow linewidth. These attractive properties predict broad applications of such a mini-device, such as terahertz generation, photonic integration, spectroscopy, and quantum information, etc.

In the recent years, optical microcavities have received considerable attention as integrated light sources and devices for the future applications in optical sensing, computing, and communication. The wave-guide-style microcavities, especially the whispering-gallery-mode microcavities are attractive due to their ultrahigh optical quality factors Q and thereby the unique spectral features. They not only work as laser active devices such as the microdisk laser[1], but also have been demonstrated to be useful for parametric processes, for example, second-harmonic generation[2-4], tunable frequency comb generation[5], parametric oscillation[6-8], and Raman lasing[9], etc. On the other hand, as proven by the fast-developing vertical-cavity surface-emitting lasers, the Fabry-Perot microcavities (FPMCs) have great advantages for high-power operation, while at the same time maintaining good beam qualities. Despite the lack of experimental realizations, theoretical studies[10-12] have shown potential applications of FPMC in nonlinear optics, which can be dated back to the early work by Armstrong *et al.* in 1962[13]. Among these studies, the subcoherence length FPMC is of special interest, where the resonance also serves to compensate for the phase mismatching and efficient frequency conversion is expected. The principle of this so-called cavity phase matching (CPM) can be interpreted in comparison with the well-known quasi-phase-matching (QPM) method[14-18]. As shown in Fig. 1(a), QPM requires engineering microstructure inside the nonlinear medium to compensate the phase mismatch by reversing orientation of the nonlinear polarization;

---

[*] Phys.Rev.Lett., 2011,106(8):083901

while in the case of CPM, the resonance recirculation in cavity can ensure that the traveling light and reflected light are exactly in phase in every circling. This is equivalent to extend the effective nonlinear interaction path, and therefore, increase the frequency conversion efficiency by a factor of Q[13].

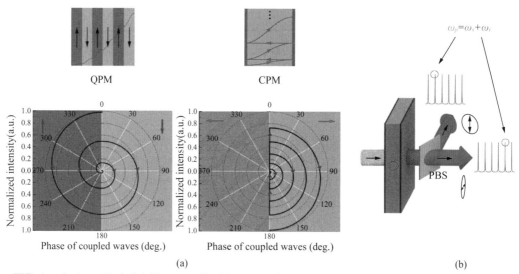

FIG. 1. (color online) (a) The normalized intensity of the converted light as a function of the phase of the three coupled light waves in a QPM material (left) and a sheet nonlinear cavity (right). In QPM material, the nonlinear polarization of the crystal is reversed every coherence length, and therefore, the light waves always have the correct phase for nonlinear amplification. The phase correction is also the key issue for CPM; however, it is achieved by the total reflection on the cavity mirrors. (b) In the case of type-II optical parametric oscillation, the reflection circulation in the CPM also requires a cavity mode matching for signal and idler with the incidence pump. Only one pair of the signal and idler modes can be excited simultaneously in this study, which ensures single-longitudinal-mode oscillation.

In this Letter, we revisit the concept of CPM, and have realized it experimentally for the first time, by using a sheet optical parametric oscillator (SOPO). The SOPO consists of a dielectric nonlinear crystal sheet, whose two end-faces are high-reflectance coated, forming a high-Q FPMC that can provide double resonances for both signal and idler. Although the thickness of the sheet is less than one coherence length, such a sheet oscillator exhibits high slope efficiency and peak power output, as well as near-transform-limited spectral and near-diffraction-limited spatial features. These novel features are of wide interest in the nonlinear optics, laser physics, and optical technology communities.

Similar to the case of QPM, the CPM allows a certain amount of phase mismatch, and therefore it is necessary to introduce an effective nonlinear coefficient d_{eff} to describe the strength of the nonlinear coupling, instead of the nonlinear coefficient d of the material. As shown in Fig. 2, d_{eff} can be expressed in the form of the following sinc function:

$$d_{\text{eff}} = d \left| \text{sinc}\left(\frac{\pi l_{\text{cav}}}{2 l_{\text{coh}}}\right) \right|, \tag{1}$$

where[13,19]

$$l_{coh} = \pi/(k_p - k_s - k_i), \qquad (2)$$

and l_{cav} is the cavity length; $\boldsymbol{k}_p$, $\boldsymbol{k}_s$, and $\boldsymbol{k}_i$ are the wave vectors of pump, signal, and idler, respectively.

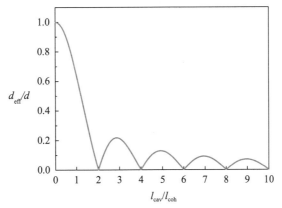

FIG. 2. (color online) When $l_{cav} \leqslant l_{coh}$, unidirectional conversion is ensured which yields to a higher d_{eff} than the first-order QPM. When l_{cav} is even times of l_{coh}, a complete destructive interference occur and $d_{eff} = 0$. When l_{cav} approaches odd times of l_{coh}, d_{eff} reaches peak value which may be used for high order CPM.

The most efficient CPM is achieved when $l_{cav} \leqslant l_{coh}$, otherwise destructive interference occurs and d_{eff} drops significantly. When l_{cav} is around odd times of l_{coh}, the high order CPM takes place, however, with smaller d_{eff}. In our study, we always keep $l_{cav} \leqslant l_{coh}$, i.e., the first-order CPM.

For the x-cut $KTiOPO_4$ (KTP) crystal, we choose type-II matching for the OPO system, where the idler has the same polarization as the pump along the y axis, while the signal beam has an orthogonal polarization along the z axis. The SOPO we use is fabricated from a monolithic crystal after fine polishing, down to 217 μm in thickness and with the surface area of 2 mm × 2 mm in the yz plane. The two end faces of the crystal sheet are both antireflection coated for 532 nm (reflectivity $R < 1\%$), and high-reflection coated ranging from 1000 nm to 1100 nm with $R > 99.8\%$ for the input surface and $R = 98.2\%$ for the output surface, respectively. The SOPO shows a high Q value up to 1.6×10^5 (or a finesse F of 292) for signal or idler in terms of the measurement, which means the 217 μm thick crystal sheet due to the resonance in cavity has an equivalent interaction length of $l_{Equ} = 63$ mm.

Considering the dispersion of x-cut KTP crystal[20], the first-order CPM allows large bandwidths of the signal and idler beams at 532 nm pump, which corresponds to a tunable signal output with a wavelength range from 1016.4 nm to 1069.4 nm. Meanwhile, a noncritical phase matching occurs:

$$532 \text{ nm (pump)} \rightarrow 1041.4 \text{ nm (signal)} + 1087.6 \text{ nm (idler)}. \qquad (3)$$

This indicates the coherence length $l_{coh} \rightarrow \infty$, and the phase matching is always satisfied at the point. As deviating from this point, the CPM plays a particular role in extending the bandwidth of the noncritical phase matching and making an effective interaction length F times of l_{cav}.

In our experiment, the SOPO was pumped by a single-longitudinal-mode frequency-doubled yttrium-aluminum-garnet laser, which was the pump source of a commercial tunable OPO system (Sunlite, Continuum, Santa Clara, CA), with pulse duration of 5 ns and repetition rate of 10 Hz. The pump beam was first focused onto a small pinhole for selecting TEM_{00} mode, and then imaged onto the SOPO. The SOPO worked in a temperature-controllable oven with an accuracy of 0.01℃. The doubly resonance configuration required precise longitudinal mode matching for signal and idler simultaneously [Fig. 1(b)]. When a proper temperature was set, the signal and idler beams could be generated in pairs with near TEM_{00} mode. As shown in Fig. 3, the spectra of the signal and idler captured by a spectral meter presented a single-longitudinal-mode oscillation. Since the resolution of spectra were limited by the spectra meter to about 60 GHz, which is sufficient to resolve the longitudinal mode spacing (~390 GHz) of the SOPO but not enough to measure the linewidth of the outputs, we had to do further measurement of the signal (or idler) using a Mach-Zehnder interferometer. The

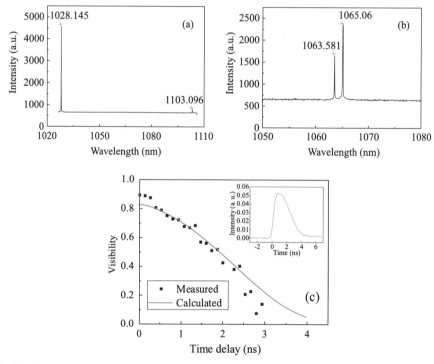

FIG. 3. (color online)Captured spectra of signal and idler beams at $t=54.00℃$ (a) and $t=113.50℃$ (b). The low strength of signal at 1100 nm originates from the lower efficiency of the silicon CCD of spectral meter at this wavelength. (c) The visibility of interference as a function of the relative time delay of two arms of the interferometer. The solid curve is the Fourier transform of captured pulse waveform (see insertion). It fits well with the measured data which show the signal light is nearly transform-limited.

interference visibility was recorded by changing the relative time delay between the two arms. As shown in Fig. 3(b), the linewidth of the signal (or idler) was around 440 MHz, indicating the output pulses were nearly transform limited. This is very important for spectroscopic applications.

Interestingly, the signal and idler were tuned in a quasi-continuous way by changing operating temperature or pump wavelength. The longitudinal mode hopping occurred when the operating temperature changed around every 2.5℃. The hopping originated from the doubly resonance configuration of the cavity as discussed previously. The experimental data exhibited a 47.8 nm tuning range (for signal) when the operating temperature varied from 32.00℃ to 115.60℃ [Fig. 4(a)]. This tuning corresponded to a coherence length change from $1.02 l_{cav}$ to $96.8 l_{cav}$, which was covered by the first-order CPM. The tuning was achieved by the change of pump wavelength as well. The tunable pump beam was provided by the output of the Sunlite OPO system. This pump had a pulse width of 4 ns, a repatriation rate of 10 Hz and a narrow linewidth of less than 0.075 cm^{-1} ($\sim$2.2 GHz), which was comparable with the FWHM of the transmission peaks of the sheet cavity. As shown in Fig. 4(b), broad tuning up to 38.8 nm for the signal was achieved while the pump's wavelength only changed 0.45 nm. Many diode lasers can be tuned in such a narrow range and may be used as a pump for such a compact tunable SOPO device.

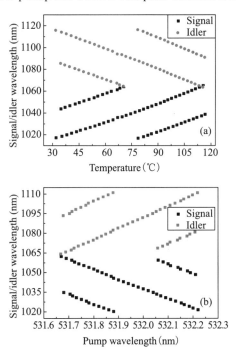

FIG. 4. (color online) Wavelengths of signal and idler lights as a function of temperature (a) or the pump wavelength (b). It worth noting that there is only one pair of signal and idler beams at one temperature or pump wavelength because of the small bandwidth of the longitudinal mode matching, although the matching conditions seem to be close in the above figures.

We measured the output power dependence on the pump at the operating temperature of 54.00 ℃, when the signal and idler located at 1028.14 nm and 1103.10 nm, respectively. The corresponding $l_{coh} = 1.98 l_{cav}$, i.e., $d_{eff} = 0.9d$, according to Eq. (1). As shown in Fig. 5(a), the oscillation started with a threshold of about 50 μJ and the total output (signal and idler) exceeded 51 μJ at the pump of 295 μJ, with a peak conversion efficiency of 18.6%. The measured slope efficiency was around 21.2%. Considering the 10% loss of the high-pass filters placed in front of the detectors to block the pump light, the actual conversion efficiency and slope efficiency can be calibrated to be 21.7% and 23.6%, respectively. In the above measurement the waist of pump beam was 300 μm, and an actual output energy density of 72 mJ/cm^2 was achieved from this thin sheet. Enlarging the optical aperture can result in further increase of the output power. We increased the beam size to 1.3 mm in the following measurement. The total output of signal and idler raised up to a maximum of 778 μJ, which corresponded to a peak power of 220 kW [Fig. 5(b)]. The conversion efficiency was still as high as 17.3% at 4.5 mJ pump energy without noticeable damage to this SOPO. The sub-milli-joule level output was sufficient for the study of nonlinear effects in a number of different applications, and watt-level average power could be expected if using a quasicontinuous laser pump for such a SOPO. We noticed slight drop in the efficiency and beam quality for the large pump beam size, which was caused by the imperfection on the cavity surfaces. By improving the polishing quality, higher efficiency and better beam quality can be achieved, and the output power is scalable by further enlarging the surface area of the SOPO.

Like the conventional QPM method, the CPM extends the equivalent interaction length of the nonlinear interaction without the requirement of nature phase matching, and is universal to realize phase matching for the entire transparent range for the nonlinear medium. Moreover, the SOPO does not need the poling technique which is usually tricky for fabricating the QPM materials, which means it is applicable to nonferroelectric crystals, such as the famous BBO, LBO, etc.; a large optical aperture is easy to be achieved which allows considerable power output from such a monolithic microcavity. In this work, we have shown an example of SOPO with submillimeter thickness, where a small dispersion of the nonlinear crystal is required to provide a relatively long coherence length. However, it is not a prerequisite for the application of the CPM technique. With the modern fine polishing technique, it is possible to manufacture thin nonlinear crystal sheets down to the ~10 μm level, which has been reported for the frequency doubling of ultrafast lasers[21]. Thus the CPM technique may be adapted to more general cases, that is, the thinner crystal sheet with smaller coherence length and higher Q value. CPM will offer an efficient, high-power solution for nonlinear optical frequency conversion in a thin, monolithic nonlinear device.

In the area of quantum optics, the optical parametric oscillator is widely used to generate squeezing[22] or entanglement[23,24]. As an integrated device, the SOPO features a

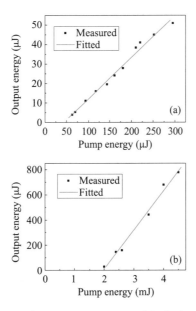

FIG. 5. (color online) Output pulse energy of both signal and idler light as a function of pump pulse energy with pump light waist of 300 μm (a) or 1.3 mm (b).

large optical aperture with single-longitudinal-mode and TEM$_{00}$ output, and therefore, high spectral and spatial brightness nonclassical light can be obtained at both above and below threshold, which is preferred for high fidelity quantum communication and computation.

In our experiment, the SOPO emits near-frequency-degenerate signal and idler beams in pairs, with a quasi-continuously tunable frequency difference ranging from 0.35 THz to 26.1 THz. By the means of difference frequency generation and simultaneous resonance of signal and idler, terahertz (THz) radiation should exist inside the cavity considerably. In this work, we measured the near infrared output and did not detect the THz signal from the SOPO. In theory, one can roughly estimate THz output to be of tens of milliwatts peak power according to the ∼1 MW intracavity peak power of signal and idler, under our experimental condition. The study for the generation of effective THz radiation from such a SOPO is under way.

References and Teose

[1] S. L. McCall et al., Appl. Phys. Lett. **60**, 289 (1992).
[2] V. S. Ilchenko et al., Phys. Rev. Lett. **92**, 043903 (2004).
[3] J. U. Fürst et al., Phys. Rev. Lett. **104**, 153901 (2010).
[4] T. J. Kippenberg, Physics **3**, 32 (2010).
[5] P. Del'Haye et al., Nature (London) **450**, 1214 (2007).
[6] A. A. Savchenkov et al., Opt. Lett. **32**, 157 (2007).

[7] J. U. Fürst *et al.*, Phys. Rev. Lett. **105**, 263904 (2010).

[8] T. Beckmann *et al.*, arXiv:1012.0801v1.

[9] S. M. Spillane, T. J. Kippenberg, and K. J. Vahala, Nature (London) **415**, 621 (2002).

[10] E. Rosencher, B. Vinter, and V. Berger, J. Appl. Phys. **78**, 6042 (1995).

[11] V. Berger, X. Marcadet, and J. Nagle, Pure Appl. Opt. **7**, 319 (1998).

[12] R. Haïdar, N. Forget, and E. Rosencher, IEEE J. Quantum Electron. **39**, 569 (2003).

[13] J. A. Armstrong *et al.*, Phys. Rev. **127**, 1918 (1962).

[14] D. Feng *et al.*, Appl. Phys. Lett. **37**, 607 (1980).

[15] M. M. Fejer *et al.*, IEEE J. Quantum Electron. **28**, 2631 (1992).

[16] V. Pruneri *et al.*, Appl. Phys. Lett. **67**, 2126 (1995).

[17] S.N. Zhu, Y.Y. Zhu, and N.B. Ming, Science **278**, 843 (1997).

[18] J.E. Schaar, K.L. Vodopyanov, and M.M. Fejer, Opt. Lett. **32**, 1284 (2007).

[19] Y. R. Shen, *The Principles of Nonlinear Optics* (Wiley, New York, 1984).

[20] S. Emanueli and A. Arie, Appl. Opt. **42**, 6661 (2003).

[21] A. Shirakawa *et al.*, Appl. Phys. Lett. **74**, 2268 (1999).

[22] Ling-An Wu *et al.*, Phys. Rev. Lett. **57**, 2520 (1986).

[23] Z.Y. Ou and Y. J. Lu, Phys. Rev. Lett. **83**, 2556 (1999).

[24] X. H. Bao *et al.*, Phys. Rev. Lett. **101**, 190501 (2008).

[25] The research was supported by the National Natural Science Foundation of China (No. 11021403 and No. 10734010). Z. D. Xie acknowledges his support from the Scientific Research Foundation of the Graduate School of Nanjing University. The authors thank P. Xu and Y.Q. Lu for valuable discussions.

第三章 克尔非线性光学超晶格与光子晶体
Chapter 3 Kerr-type Nonlinear Optical Superlattices and Photonic Crystals

3.1 Light Transmission in Two-dimensional Optical Superlattices

3.2 Optical Bistability in a Two-Dimensional Nonlinear Superlattice

3.3 Experimental Observations of Bistability and Instability in a Two-Dimensional Nonlinear Optical Superlattice

3.4 Gap Shift and Bistability in Two-dimensional Nonlinear Optical Superlattices

3.5 Optical Bistability in Two-dimensional Nonlinear Optical Superlattice with Two Incident Waves

3.6 Three-dimensional Self-assembly of Metal Nanoparticles: Possible Photonic Crystal with a Complete Gap Below the Plasma Frequency

3.7 Parity-time Electromagnetic Diodes in a Two-dimensional Nonreciprocal Photonic Crystal

3.8 Nonreciprocal Light Propagation in a Silicon Photonic Circuit

3.9 Experimental Demonstration of a Unidirectional Reflectionless Parity-time Metamaterial at Optical Frequencies

3.10 Plasmonic Airy Beam Generated by In-Plane Diffraction

3.11 Collimated Plasmon Beam: Nondiffracting versus Linearly Focused

3.12 The Anomalous Infrared Transmission of Gold Films on Two-Dimensional Colloidal Crystals

3.13 Localized and Delocalized Surface-plasmon-mediated Light Tunneling Through Monolayer Hexagonal-close-packed Metallic Nanoshells

3.14 Experimental Observation of Sharp Cavity Plasmon Resonances in Dielectric-metal Core-shell Resonators

3.15 Magnetic Field Enhancement at Optical Frequencies Through Diffraction Coupling of Magnetic Plasmon Resonances in Metamaterials

第三章　克尔非线性光学超晶格与光子晶体

王振林

　　采用激光干涉并结合其他物理效应在介质中刻写正弦调制光栅结构,从而对材料线性光学系数——折射率产生周期性调制,形成光学超晶格;由于折射率的调制幅度可达到 10^{-2} 量级,比在晶体中传播 X 射线的折射率调制幅度高两个数量级,因此光在这种线性光学系数周期性调制的结构中传播时能够产生新颖的光学效应。论文(3.1~3.5)介绍了我们在克尔(Kerr)非线性二维光学超晶格方面的研究成果,目的是通过在光学超晶格结构中引入克尔非线性,从而达到多通道的光控制的光开关效应。我们首先发展了四波近似线性衍射动力学(3.1)和四波近似非线性衍射动力学理论(3.2),研究了布拉格条件下光在二维光学超晶格中的传播和光响应特性,预见了一种新型光学双稳机制——折射率调制光学双稳机制。由于这一双稳特性是在严格相位匹配条件下得到的,因此有别于非线性一维周期体系中的相位失配双稳机制。随后我们在掺铁铌酸锂光折变晶体中通过光折变效应写入二维光学超晶格,并在其中观测到光响应的迟滞迴线现象(3.3)。实验结果尽管从定量上与理论预测不完全吻合,这是由于光折变材料本身的性质与理论假设的克尔非线性在机理和响应速度上存在差异所致。我们进一步给出了非布拉格条件下光学超晶格的出射光谱特性,得出了二维光学超晶格光子带隙出现的条件与带隙宽度的解析表达式,预见了克尔非线性二维光学超晶格还存在一种透射态与非透射态之间的双稳态开关效应(3.4),本质上这是一种相位失配双稳机制。此外,我们讨论了通过双束光输入以降低光学超晶格光学双稳态开关阈值的可行性(3.5)。

　　在我们开展二维光学超晶格非线性效应研究的同时,Yablonovitch[E. Yablonovitch, Phys. Rev. Lett. 58, 2059 (1987)]和John[S. John, Phys. Rev. Lett. 58, 2486 (1987)]开启了以折射率强调制的三维周期结构——光子晶体的研究领域。对于三维光子晶体而言,由于光沿着三维方向都会受到强烈的布拉格散射,有可能在三维布里渊空间出现带隙重叠区域——形成光子全带隙,波长位于全带隙内的光波会被禁止在光子晶体内沿任何方向传播。因此,全带隙光子晶体为人们控制原子自发辐射等光学过程提供了新途径。我们在研究折射率弱调制的非线性光学超晶格光响应特性的同时,也在研究折射率强调制的光子晶体,并首先提出了由贵金属微球或金属微球壳自组织构成三维面心立方结构,来获得可见/近红外光谱区光子全带隙的理论设想,这一全带隙光子晶体设计方案不仅借助了微球自组织易于构造出三维周期结构的优点,同时也解决了由纯介电微球自组织形成的面心立方结构不会产生全带隙的缺点(3.6)。在实验上,我们以介质微球自组织形成的有序结构——胶体晶体为模版,通过无电镀工艺在微球表面形成银纳米包裹层,从而形成金属包裹介质球的有序结构[Z. Chen et al., Adv. Mater. 16, 417 (2004)]。这一材料复制技术克服了之前人们采取先化学包裹形成核壳结构微球,然后试图通过微球自组织获取有序结构样品的困难。

为了开展这类微结构功能材料未来的应用研究与探索,我们发展了单分散、尺寸可控二氧化硅胶体微球生长与形貌控制技术[J. H. Zhang, et al., J. Mater. Res. 18, 649 (2003)],不仅能有效抑制二氧化硅微球生长过程中新核的形成和颗粒之间的聚集,而且无需对正硅酸乙酯的蒸馏过程。此外,我们发展的聚苯乙烯微球合成技术克服了传统两步种子生长技术中需要采取对苯乙烯溶胀的过程,大大提高了合成的效率[J. H. Zhang et al., Mater. Lett. 57, 4466 (2003)]。

最近电子拓扑序的研究启发人们对于同样具有波粒二象性的光子实施量子调控的新思路,而光子晶体为在较大尺度上研究光子的量子行为及其新效应提供了便利。论文(3.7)介绍了在外磁场下磁光光子晶体由于时间反演破缺所引起的单通边界态,特别是利用镜面对称破缺单元结构,打破宇称对称性并保持时空对称,实现体光波的非互易传播,从而设计出光的体波二极管。基于点缺陷磁光光子晶体微腔并结合电光调制效应,还可以设计出外加电场和外加磁场调控的单通滤波器和非互易的边界态光二极管器件[C. He et al., Appl. Phys. Lett. 96, 111111 (2010)]。而通过一种非均匀空间的变换,将该磁光介质光子晶体作为虚拟空间进行坐标变换,可产生一种单向隐身状态[C. He et al., Appl. Phys. Lett. 99, 151112 (2011)]。

然而,在芯片上如何实现光的单向传输仍是一个很大的挑战。论文(3.8,3.9)介绍了我们在折射率含有虚部的非厄米量子开放系统中实现光的有效调控的新思想,特别是提出在折射率的复数空间域调控光子的势函数,从而为在开放非厄米光学系统研究新型光子材料和量子力学基本对称性问题提供了新的范式。我们利用时空对称性破缺的临界现象,在硅基光子芯片中巧妙地引入实部和虚部的相位匹配调制,实现了光波单向模式的转换(3.8)。进一步,我们结合耦合模理论和散射矩阵本征值分析方法,在纯损耗硅基波导中实现了Bragg(布拉格)反射的剪裁,进而实现了光的单向无反射传播(3.9)。

以光作为信息载体,由于受到衍射极限的限制,在未来光回路和光集成方面存在很大的瓶颈制约。然而,沿金属表面传播、场强沿界面的法线方向迅速衰减的表面等离极化激元(surface plasmon polariton,SPP)以其特有的亚波长局域特性而受到人们关注。作为面内调控的一种方式,SPP波在平整金属平面传播受结构引入而引起的散射及衍射特性,为人们在金属表面引入类似光子晶体结构来操纵光子的传播开拓了新思路。论文(3.10,3.11)介绍了利用有限周期结构阵列中倒格点延展特性形成非严格的波矢匹配,实现SPP波的定向衍射,乃至形成一些特殊的波束,从而达到对原入射波束调控的功能。我们提出了调控波束的"非完美匹配布拉格衍射法"(3.10),针对3/2次型相位分布艾里波束,在金属银膜上制备非周期纳米孔阵结构,成功观测到具有无衍射自弯曲自修复的SPP艾里波束、宽带SPP波聚焦和波分复用功能[L. Li et al., Nano Lett. 11, 4357 (2011)]。我们进一步发展了通过相位分布来调控波束强度的方法,实现了在一定区域内由相位补偿的"无损耗"准直SPP波束(3.11)。这些研究结果不仅揭示SPP波束的形成机制,同时也为亚波长光传输、信号处理和光子集成等拓展了新的可能途径。

除了金属膜表面传播的SPP波之外,贵金属颗粒表面还支持局域表面等离激元共振(Localized Surface Plasmon Resonances,LSPRs),当颗粒之间相互靠近时,LSPRs会通过近场耦合而发生相互作用产生更强的局域场。因此,通过改变贵金属颗粒的形状、结构、空间上的排列方式,可以对LSPRs的频率、电磁场分布、响应特性进行调控。论文(3.12,3.13,

3.14)则介绍了围绕金属纳米球壳这类颗粒体系 LSPRs 特性所开展的理论与实验研究。通过由介质微球自组织形成的二维胶体晶体为衬底,通过物理沉积制备出空间上相互链接的金属纳米半球壳阵列,不仅观测到增强光透射现象,并且揭示共振透射峰是由于局域在介质球内低品质因子的腔模激发所引起的(3.12)。通过去除二维胶体晶体的支撑而降低介质衬底效应,可以将局域电场释放到金属与空气的分界面,从而使得共振峰位置对环境周围折射率变化的敏感度提高 4 倍[Y. Li et al., Opt. Express. 18, 3546 (2010)]。我们预见了完整金属纳米球壳所支持一类新的 LSPRs 模式——腔模共振,这些模式不仅由于其局域场被完全限制在腔内而具有很高的品质因子;当金属球壳组成二维周期阵列结构后,借助这些模式的激发可产生窄线宽的共振透射现象(3.13)。在实验上,我们巧妙地发展了完整且厚度均匀的金属纳米球壳制备新方法,成功观测具有不同角动量量子数的腔模式,并观测到品质因子高达到 100、线宽为 12nm 的磁偶极子 LSPR 腔模(3.14),实验测量与理论计算极好地吻合。这些研究成果为利用 LSPR 腔模实现三维空间电磁场局域增强,研究光与物质的相互作用提供了一个可靠的实验体系。

与等离激元密切相关的另一个领域是超构材料。超构材料可以产生天然材料所不具有的介电响应性质。在超构材料中,一种不可缺少的构造单元是能产生高频磁响应的"磁原子",一般采用特定金属微结构来产生,其尺寸远小于其 LSPR 共振波长。如何来进一步提高局域磁场,有可能为人们研究光与物质的相互作用提供一个新的调控手段。论文(3.15)介绍了我们针对由磁原子构成的二维阵列中沿着面内传播的集体模与"磁原子"产生耦合条件和耦合特性的研究,预测在 Wood 反常情况下发生强耦合并形成一种杂化模式;杂化模的局域磁场比入射电磁波的磁场高出 450 倍,比远离耦合情况下"磁原子"的磁场提高了 5 倍,为提高"磁原子"中磁场强度提供了新的方案。

Chapter 3　Kerr-type Nonlinear Optical Superlattices and Photonic Crystals

Zhenlin Wang

By using laser interference in combination with other physical effects, sinusoidally modulated gratings can be written in media to form optical superlattices (OSLs). The modulation depth of refractive index can be as high as 10^{-2}, which is two orders of magnitude higher than the perturbation in a crystal for X-ray, the propagation and interference of light in OSLs can lead to novel optical properties. Our research efforts on two-dimensional (2D) Kerr-type nonlinear OSLs are reported in papers (3.1~3.5). Our research motivation is the realization of multiple output switches by introducing 2D OSL structures into Kerr-type nonlinear media. We firstly developed linear (3.1) and nonlinear (3.2) four-wave dynamical diffraction theories, studied the wave propagation in 2D OSLs and optical responses under Bragg conditions, and predicted a novel index modulation bistability mechanism. This mechanism validates in the rigorous phase matching condition in two orthogonal directions and thus is different from the bistablity mechanism in 1D nonlinear periodic medium, the later requires a certain content of phase mismatch. Subsequently, we fabricated the 2D OSLs in $LiNbO_3$:Fe crystals using the photorefractive effect and observed optical responses with hysteresis (3.3). The experimental results give a qualitative support of the theoretical predictions because the nonlinearity of the photorefractive crystal has a different nature from the Kerr nonlinearity assumed in the theory. We further presented the condition for the photonic band gap of 2D OSLs as well as the analytical expression of the gap width, and predicted an optical switching between the transmitting on-state and non-transmitting off-state by tuning the incident wavelength near the band gap edge of the 2D OSLs (3.4). In addition, we discussed the possibility to reduce the threshold for bistable output switches by illuminating the 2D nonlinear OSLs with two incident beams that both satisfy the Bragg condition (3.5).

While we studied the nonlinear effects in 2D OSLs, Yablonovitch [E. Yablonovitch, Phys. Rev. Lett. 58, 2059 (1987)] and John [S. John, Phys. Rev. Lett. 58, 2486 (1987)] stimulated a new line of research on micro-structured materials where the refractive index is modulated periodically in three dimensions with much stronger amplitudes, termed as photonic crystals (PhCs). For three-dimensional (3D) PhCs, light can Bragg-scatter strongly in all three dimensions, which makes it possible to open up a complete photonic bandgap (PBG) that is the overlapping part of the stop gaps in 3D Brillouin zone. The PhCs with a complete PBG thereby provide a novel means for control of spontaneous emission processes of atoms and molecules. In addition to the studies of OSLs with weakly perturbed refractive index, we investigated the PhCs with strongly modulated refractive index. We firstly proposed that complete PBGs in near infrared and optical frequencies can be realized in a 3D self-assembled face-centered cubic (FCC) lattice of noble-metal spheres or dielectric spheres coated with a thin layer of noble metal (3.6). Such a novel design scheme of PhCs with complete PBGs not only takes advantage of the fact that 3D periodic microstructures can be easily constructed by self-assembly of spherical particles, but also alleviates the restriction that the self-assembled FCC structures of dielectric spheres do not support a complete PBG. Experimentally, we developed a method for preparing 2D and 3D close-packed dielectric-metal core-shell resonators by using dielectric colloidal crystals as templates and seeding/electroless deposition of silver nanoparticles on dielectric microspheres [Z. Chen et al., Adv. Mater. 16, 417 (2004)]. The above template replication technique overcomes the difficulties of obtaining the highly ordered structures by self-organization of the as-prepared metal-coated composite particles. In order to facilitate future applications of the microstructured functional materials mentioned above, we developed a seeded growth technique to synthesize monodispersed silica nanospheres with a controllable and tunable size [J. H. Zhang, et al., J. Mater. Res. 18, 649 (2003)], which can depress the formation of new nuclei and the aggregation of adhesion of spheres and do not need the distillation of tetraethyl orthosilicate. In addition, we developed a method for preparing monodispersed polystyrene nanospheres, which can shorten the growth period by removing the swelling process of styrene in traditional two-step seeded growth techniques [J. H. Zhang et al., Mater. Lett. 57, 4466 (2003)].

Recently, the investigations of electronic topological states have inspired a novel quantum approach to manipulating photons and phonons, which also behave the wave-particle duality. Photonic crystals provide a macroscopic platform to investigate and manipulate the quantum properties and unique functionalities of light. We addressed the edge dependence of the unidirectional light transport in magneto-optical PhCs, and synergized time-reversal breaking by applying magnetic field and parity breaking of unit cells in space to realize nonreciprocal transport of bulk optical waves in magneto-optical PhCs under the conserved parity-time symmetry (3.7). For example, a unidirectional photonic edge state by the time-reversal breaking in magneto-optical PhCs under an external magnetic bias have been proposed and demonstrated. We have designed a point defect-enabled micro-cavity and demonstrated a unidirectional optical filter and a nonreciprocal optical diode based on magneto-optical PhCs under external electric and magnetic fields [C. He et al., Appl. Phys. Lett. 96, 111111 (2010)]. Moreover, we adopted a method based on inhomogeneous virtual coordinate-space transformation and proposed magneto-optical PhCs for the demonstration of unidirectional cloaking [C. He et al., Appl. Phys. Lett. 99, 151112 (2011)].

Recently, a great challenge remains towards realizing unidirectional light transport on-a-chip. Collaborating with the group from Institute of Caltech, we have (3.8, 3.9) provided a new paradigm to study the exceptional point and symmetry breaking as well as their related quantum phase transition by exploiting the spontaneous breaking of parity-time symmetry, and by judiciously designing a in-phase separation of real and imaginary refractive index modulations. This could open a door to investigate the fundamental symmetry in quantum mechanics and light transport for Non-Hermitian photonics, in which optical properties of new kinds of materials can be explored. We have successfully constructed artificial materials with controllable dielectric properties in the entire complex permittivity or permeability plane for the first time and achieved unidirectional photonic mode conversion in a silicon integrated photonic circuit (3.8). We further developed a coupled mode theory and analysis method of eigenvalue of scattering matrix, to engineer the Bragg reflection and demonstrate unidirectional reflectionless light transport in a pure passive silicon waveguide with the modulation of complex dielectric constants (3.9).

As an information carrier, light faces a big bottleneck in compact integration for optical and photonic chip due to the diffraction limit. However, surface plasmonic polariton (SPP), as a kind of surface wave that propagates along the metal surface with the field rapidly decaying from interface, has provided a possible solution due to its subwavelength localization properties and thus received great attentions. Being capable of manipulating the SPP in a planar dimension, the scattering and diffraction of SPP waves by PhC-like structures on metal surface open a new avenue in steering photon in-plane propagation. We carried on systematic studies that utilize a non-perfect phase matching process of the scattered SPP wave to form directional coherent diffractions, and thus to construct a number of special SPP beams, so as to reach an at-will manipulation of the SPPs (3.10, 3.11). In detail, we first proposed a strategy called "non-perfectly matched Bragg diffraction" to manipulate the phase and wave front of SPP waves (3.10). Based on this method, we designed and fabricated nanohole arrays on a silver surface with the 3/2 type phase modulation according to the Airy beam requirement, successfully observed the SPP Airy beams with non-diffractive, self-bending and self-healing properties in experiments. Next, a broadband SPP focusing and wavelength division multiplexing (WDM) were achieved as well [L. Li et al., Nano Lett. 11, 4357 (2011)]. Moreover, based on the phase design, an intensity controlled SPP beam engineering was further developed, and was successfully used to realize a "lossless" collimated SPP beam (3.11). Our works not only provide an in-depth understanding of the SPP beam formation, and but also illuminate new possibilities in sub-wavelength optical propagation, information processing and photonic integrations.

In addition to the SPPs that can propagate along the metal/dielectric interface, they can be confined to noble metal nanoparticles to form the so-called localized surface plasmon resonances (LSPRs). When two or more metal nanoparticles are in close vicinity to each other, LSPRs of the individual nanoparticles can interact via their optical near fields. Therefore, the resonance frequencies and field distributions of LSPRs on noble metal nanoparticles can be tuned by varying the size, morphology, metal composition, local dielectric environment and their spatial arrangement. Our theoretical and experimental investigations of LSPRs supported by the dielectric-metal core-shell architectures are demonstrated in refs. 3.12~3.14. By physically depositing thin metal films onto a monolayer colloidal crystal substrate, we obtained interconnected periodic metal half-shells and observed extraordinary optical transmission (EOT) phenomena that are mediated by the excitations of low-quality cavity modes localized within the dielectric cores (3.12). After removing the supporting dielectric cores, the electric fields can be released to the metal/air interface, which is accessible to the detected species, leading to a four-fold enhancement in the dielectric environment sensitivity as compared to that of the as-prepared metal half-shells with supporting dielectric cores [Y. Li et al., Opt. Express 18, 3546 (2010)]. We also theoretically predicted that complete metal nanoshells are capable of supporting a new class of LSPRs, called cavity plasmon resonances, which not only have high quality factors as a result of the strong field confinement within the dielectric cores, but also can mediate narrow-band resonant transmission when a 2D array of metal nanoshells is formed (3.13). Very recently, we developed a novel approach for preparing uniform and nearly complete metal nnaoshells, and experimentally observed a series of multipolar cavity plasmon resonances including a magnetic dipolar cavity resonance with a remarkably high quality factor of ~100 and a narrow linewidth of ~12 nm (3.14). Our experimental findings are supported by excellent agreement with theoretical calculations, and may offer a good experimental system for studying light-matter interactions using the strongly enhanced localized electromagnetic fields via the excitations of cavity plasmons.

Metamaterials also represent an exciting emerging research area closely related to SPPs. Artificially engineered metamaterials can enable unprecedented electromagnetic properties that cannot be obtained with naturally occurring materials. A subwavelength resonator that exhibits high frequency magnetism and is customarily acted by a kind of metallic nanostructure, called "artificial magnetic atom" is an indispensable element for the construction of metamaterials. Achieving large enhancement on the localized magnetic field may open up new possibilities for studying light-matter interactions. Our theoretical investigations of the diffraction coupling between magnetic SPRs and in-plane propagating collective surface modes in a 2D array of magnetic atoms are demonstrated in(3.15). We predicted that at the optical resonance of the hybrid mode arising from the strong diffraction coupling, the magnetic field intensity is about 450 times of the incident field, which is enhanced to nearly 5 times larger than that at the magnetic resonance of individual magnetic atom, providing a new mechanism to enhance the local magnetic field.

Light Transmission in Two-dimensional Optical Superlattices[*]

Jing Feng

Laboratory of Solid State Microstructures, Nanjing University, Nanjing, People's Republic of China

Nai-Ben Ming

Center for Condensed Matter Physics and Radiation Physics, Chinese Center of Advanced Science and Technology (World Laboratory), Beijing, People's Republic of China

The propagation and interaction of light in the two-dimensional optical superlattice, a two-dimensional periodically modulated medium, were investigated by the dynamical theory with the four-wave approximation, which is based on an incident beam that satisfies Bragg conditions in two orthogonal directions. The excited fields and the intensities of exiting waves in the two-dimensional optical superlattice were calculated. The relation between relative intensities of the exiting waves and angular deviation of the incident beam as well as the index-modulated parameters of the superlattice was obtained. Also discussed are possible applications of the superlattices in optical communications.

1. Introduction

In a one-dimensional periodically modulated medium, the coupled-mode theory[1] is used to solve the problems of the propagation and interaction of light. In this medium two waves viewed as independent in the unperturbed structure are coupled to each other by perturbation in some way. The interference in one-dimensional modulated structure was analyzed by Russell with aid of the two wave approximation dynamical theory similar to the theory of X-ray diffraction. His results were consistent with experimental results.[2,3]

In our paper we present a four-wave dynamical theory in two-dimensional periodically modulated media referred to as two-dimensional (2D) optical superlattices. In such 2D optical superlattices, the refractive index is no longer a constant, but rather a periodic function of spatial coordinates. In our paper we consider only a case of a 2D orthogonal optical superlattice similar to the sinusoidal cross gratings, as shown in Fig. 1. We assume that light propagates in the plane of the cross gratings; this is a dynamical process. The behavior of light through the cross gratings, as is well-known, is involved in a kinematical process. The modern technology of crystal growth,[4,5] molecular-beam epitaxy,[6] and the use of acousto-optic effects[7] make the construction of 2D optical superlattices possible.

---

[*] Phys.Rev.A, 1989, 40(12): 7047

When the incident beam satisfies Bragg conditions in two orthogonal directions and then falls on the 2D optical superlattice, four propagating waves will be excited in the medium; the momentum conservation relation of the light in the 2D optical superlattice is shown in Fig. 2. Such a four-wave problem cannot be explained by either the coupled-mode theory[1] or the two-wave dynamical theory.[2,3] A four-wave dynamical theory which can deal with interaction and propagation of light in both 1D and 2D periodically modulated media is needed. The four-wave dynamical theory is used in our paper in order to solve the propagation equations of the Floquet-Bloch wave that consists of four waves traveling in different directions. With the aid of dispersion surfaces in the X-ray diffraction theory,[8] all the wave fields inside and outside of the medium have analytical expressions under several different forms of the dispersion surface due to the index-modulated parameters in two modulation directions of the 2D optical superlattice. Setting the modulated parameter in one of two modulation directions to zero, the 2D optical superlattices turn into 1D periodically modulated media and the results obtained by the four-wave dynamical theory are in good agreement with those by the coupled-mode theory[1] and two-wave dynamical theory.[2,3] Additional information will be given by considering the interaction and propagation of four waves in the 2D optical superlattices. Such superlattices are expected to have potential applications in optical communications.

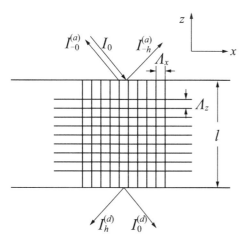

FIG.1. Schematic diagram of the propagation of light in the 2D optical superlattice. Four exiting waves are diffracted from the 2D optical superlattice.

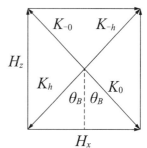

FIG.2. Momentum conservation relations (3a) and (3b) for deriving the Bragg conditions (4a) and (4b).

2. Four-wave Approximation Dynamical Theory

Let us consider light propagating in an isotropic loss-less two-dimensional rectangular optical superlattice, assuming that the electrical conductivity is zero and the magnetic permeability is unity. The propagation constant of the 2D optical superlattice is defined by the following equation:

$$\beta^2(\mathbf{r}) = k_m^2 [1 + M_x \cos(H_x x) + M_z \cos(H_z z)], \quad (1)$$

where M_x, M_z are modulating strength and H_x, H_z are reciprocal vectors of the 2D optical superlattice along the x and z axes, which are related to spatial periods Λ_x, Λ_z, that is

$$H_x = \frac{2\pi}{\Lambda_x}, H_z = \frac{2\pi}{\Lambda_z}. \quad (2)$$

$k_m = N\omega/c$ in Eq. (1) is the wave number of light in the 2D optical superlattice with an effective mean index N. Our theoretical construction is similar to the theory of X-ray diffraction.[8] The difference between the propagation and interaction of light in the 2D optical superlattices and X-ray diffraction in crystals is that there are no extra free charges in the superlattices. In the former case, we can set $\nabla \cdot \mathbf{E} = 0$. In X-ray diffraction, however, $\nabla \cdot \mathbf{E} = \nabla \cdot (\mathbf{D} - 4\pi \mathbf{P}) = -4\pi \nabla \cdot \mathbf{P}$ is normally nonzero. So the propagation equation of light in 2D optical superlattices is scalar and polarization is not considered. In our theory, the Floquet-Bloch wave consists of four waves; the wave vectors of the four waves are $\mathbf{K}_0$, $\mathbf{K}_h$, $\mathbf{K}_{-0}$, and $\mathbf{K}_{-h}$, respectively. There are two momentum conservation relations shown in Fig. 2, which are

$$\mathbf{K}_0 - \mathbf{K}_h = \mathbf{H}_x, \quad (3a)$$

$$\mathbf{K}_{-h} - \mathbf{K}_0 = \mathbf{H}_z. \quad (3b)$$

From above equations, we get Bragg conditions:

$$2K_0 \sin\theta_B = H_x \text{ or } 2\Lambda_x \sin\theta_B = \lambda, \quad (4a)$$

$$2K_0 \cos\theta_B = H_z \text{ or } 2\Lambda_z \cos\theta_B = \lambda, \quad (4b)$$

where $\lambda = 2\pi c/N\omega$. It is worthy to point out that an X ray can hardly satisfy the two Bragg conditions (4a) and (4b) at the same time in a crystal because the lattice periods are nonadjustable; in our discussion, however, the parameters Λ_x, Λ_z of the 2D optical superlattice can be adjusted to make the incident beam satisfy equations (4a) and (4b) in the 2D optical superlattice.

If the effective mean index N of the 2D optical superlattice is not the same as the index of the space outside it, reflection and refraction will occur on the boundary. The refracted angle and transmission coefficient of light are determined by the Snell law and the Fresnel formulas. The reflection and refraction on the boundary of the 2D optical superlattice will not be considered here; we assume that the effective mean index of the 2D optical superlattices equals the index of the surrounding medium.

In reciprocal space, four-wave vectors $\mathbf{K}_0$, $\mathbf{K}_h$, $\mathbf{K}_{-0}$, and $\mathbf{K}_{-h}$ are on the reflection sphere as shown in Fig. 3. Comparing with other reciprocal lattice points, the fields of the four waves are much larger than all other fields corresponding to the other nodes of reciprocal lattices. So the four-wave approximation is valid under our condition.[8]

The scalar Floquet-Bloch wave (polarized in the y axis) can be written as

$$E(\boldsymbol{r}) = \sum_n E_n \exp(i\boldsymbol{K}_n \cdot \boldsymbol{r}), n = 0, h, -0, -h \tag{5}$$

where E_n is the amplitude of the plane wave with wave vector $\boldsymbol{K}_n$ ($n = 0, -0, h,$ and $-h$), and we put (5) into the propagation equation:

$$[\nabla^2 + \beta^2(\boldsymbol{r})]E(\boldsymbol{r}) = 0. \tag{6}$$

Setting the coefficients of like exponentials to zero, we get following matrix equation:

$$\begin{pmatrix} -(\boldsymbol{K}_0^2 - k_m^2) & M_x k_m^2/2 & 0 & M_z k_m^2/2 \\ M_x k_m^2/2 & -(\boldsymbol{K}_h^2 - k_m^2) & M_z k_m^2/2 & 0 \\ 0 & M_z k_m^2/2 & -(\boldsymbol{K}_{-0}^2 - k_m^2) & M_x k_m^2/2 \\ M_z k_m^2/2 & 0 & M_x k_m^2/2 & -(\boldsymbol{K}_{-h}^2 - k_m^2) \end{pmatrix} \begin{pmatrix} E_0 \\ E_h \\ E_{-0} \\ E_{-h} \end{pmatrix} = 0. \tag{7}$$

From Fig. 3, we can see that $\boldsymbol{K}_{-0} = -\boldsymbol{K}_0, \boldsymbol{K}_{-h} = -\boldsymbol{K}_h$. At a sufficient approximation, we assume that

$$(\boldsymbol{K}_0^2 - k_m^2) \approx 2k_m(K_0 - k_m), \tag{8}$$

$$(\boldsymbol{K}_h^2 - k_m^2) \approx 2k_m(K_h - k_m). \tag{9}$$

Let us then introduce the segment parameters ξ_0 and ξ_h, which are defined $\xi_0 = K_0 - k_m$, and $\xi_h = K_h - k_m$, respectively. Now we can rewrite (7) as

$$\begin{pmatrix} -\xi_0 & M_x k_m/4 & 0 & M_z k_m/4 \\ M_x k_m/4 & -\xi_h & M_z k_m/4 & 0 \\ 0 & M_z k_m/4 & \xi_0 & M_x k_m/4 \\ M_z k_m/4 & 0 & M_x k_m/4 & \xi_h \end{pmatrix} \begin{pmatrix} E_0 \\ E_h \\ E_{-0} \\ E_{-h} \end{pmatrix} = 0. \tag{10}$$

In order to get nonzero solutions of the fields E_0, E_h, E_{-0}, and E_{-h}, we set the determinant of this system to zero and get the following equation:

$$\xi_0 \xi_h = (k_m/4)^2 (M_x^2 - M_z^2). \tag{11}$$

The dispersion surface is composed of two hyperbolas $S^{(1)}$ and $S^{(2)}$, as seen in Figs. 4 and 5. The dispersion surface is the locus of the excitation points. An incident beam in the 2D optical superlattice with vector $\boldsymbol{K}_0$ will excite two points $A^{(1)}$ and $A^{(2)}$, so the segments ξ_0 and ξ_h have two values:

$$\xi_0^{(j)} = K_0^{(j)} - k_m, \xi_h^{(j)} = K_h^{(j)} - k_m, j = 1, 2 \tag{12}$$

$\boldsymbol{K}_0^{(j)}$ and $\boldsymbol{K}_h^{(j)}$ are the vectors starting from point $A^{(j)}$ to points 0 and H, and $\boldsymbol{K}_{-0}^{(j)}$ and $\boldsymbol{K}_{-h}^{(j)}$ are the vectors from point $A^{(j)}$ to points -0 and $-H$.

Combining Eqs. (10) with (11), we obtain

$$E_h^{(j)} = k_m (M_z E_{-0}^{(j)} + M_x E_0^{(j)}) / 4\xi_h^{(j)}, j = 1, 2 \tag{13a}$$

$$E_{-h}^{(j)} = -k_m (M_x E_{-0}^{(j)} + M_z E_0^{(j)}) / 4\xi_h^{(j)}, j = 1, 2 \tag{13b}$$

From Eq. (11) it can be seen that the dispersion surface is determined by strength M_x, M_z. When M_x is larger than M_z, i.e., $M_z/M_x < 1$, the dispersion surface has the form of Fig. 4. In the case of the incident angle slightly deviating from the Bragg angle, i.e., with wave vector $\boldsymbol{K}_0$ near the midpoint of maximum, there are four waves excited, each of which is in fact combined by two plane waves in one direction. So four direction exiting waves are expected from the 2D optical superlattice. If $M_z/M_x > 1$, the form of the dispersion surface is shown in Fig. 5. With K_0 near the midpoint of maximum, there is no intersection between n_0, the normal of surface of the 2D optical superlattice, and hyperbolas $S^{(1)}$, $S^{(2)}$. If $M_z/M_x = 1$, the hyperbolas $S^{(1)}$, $S^{(2)}$ vanish and the dispersion surface becomes two intersecting lines T_0 and T_h as in Figs. 4 and 5. In this case only wave K_0 couples with K_{-0}, and K_h with K_{-h}; the exciting waves are independent of the angular deviation of the incident beam from the Bragg angle.

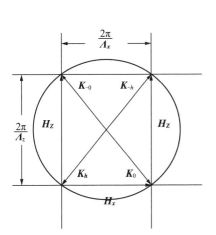

FIG. 3. Reflection sphere in the reciprocal space for the kinematical approximation.

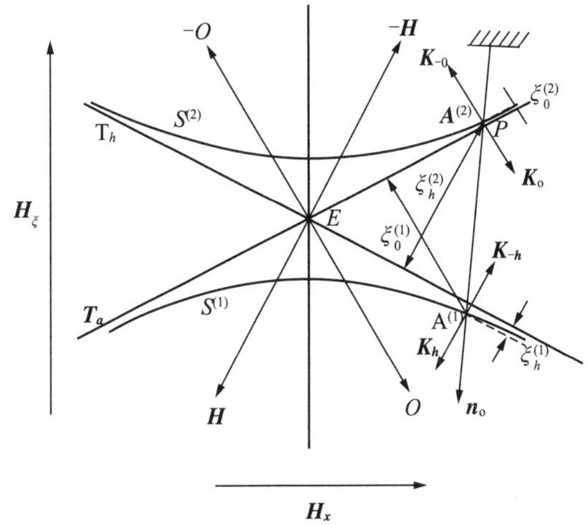

FIG. 4. Dispersion surface in reciprocal space for the strength ratio $m < 1$ in the four-wave approximation.

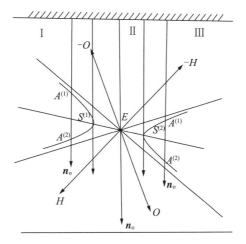

FIG. 5. Dispersion surface in reciprocal space for the strength ratio $m > 1$ in the four-wave approximation.

There are three different cases to present according to the ratio M_z/M_x, which we present below.

2.1 $M_z/M_x < 1$

If a light E_0 falls on the 2D optical superlattice, the boundary conditions are

$$E_0^{(1)} + E_0^{(2)} = E_0, \tag{14a}$$

$$E_h^{(1)} + E_h^{(2)} = 0, \tag{14b}$$

$$E_{-h}^{(1)} \exp(iK_{-hz}^{(1)}l) + E_{-h}^{(2)} \exp(iK_{-hz}^{(2)}l) = 0, \tag{14c}$$

$$E_{-0}^{(1)} \exp(iK_{-0z}^{(1)}l) + E_{-0}^{(2)} \exp(iK_{-0z}^{(2)}l) = 0, \tag{14d}$$

where $E_n^{(j)}$ ($j=1,2$; $n=0, h, -0$, and $-h$) are the excited fields in the 2D optical superlattices, $K_{nz}^{(j)}$ are the z components of wave vectors $K_n^{(j)}$ ($j=1, 2$ $n=0, h, -0$, and $-h$), and l is the length of the medium along the z axis. In Fig. 4, if $\eta = \overline{PE}$, which represents derivation of the incident beam from the Bragg angle in Eqs. (3) and (4), we have the equation

$$\overline{PA}^{(j)} = \gamma^{-1} \xi_0^{(j)} = \gamma^{-1}(\xi_h^{(j)} - k_m \eta \sin 2\theta_B). \tag{15}$$

Here $\gamma = \cos\theta_B$.

With Eqs. (11) and (15), we obtain

$$\xi_0^{(j)} = -\frac{k_m \eta \sin 2\theta_B}{2} \mp \left(\frac{k_m^2}{4}\eta^2 \sin^2 2\theta_B + \frac{1}{16}k_m^2 M^2\right)^{1/2}, \tag{16a}$$

$$\xi_h^{(j)} = \frac{k_m \eta \sin 2\theta_B}{2} \mp \left(\frac{k_m^2}{4}\eta^2 \sin^2 2\theta_B + \frac{1}{16}k_m^2 M^2\right)^{1/2}, \tag{16b}$$

Where $j=1,2$ and $M^2 = M_x^2 - M_z^2$.

We now introduce the angular coordinates

$$\omega = \sinh(v) = 2\eta \sin 2\theta_B/M. \tag{17}$$

We can rewrite $\xi_h^{(j)}$ with the new coordinate v,

$$\xi_h^{(j)} = \mp k_m M \exp(\mp v)/4. \quad j=1,2 \tag{18}$$

As in the theory of X-ray diffraction, we have the relation

$$K_{nz}^{(1)} - K_{nz}^{(2)} = -\Delta, \tag{19}$$

with

$$\Delta = k_m M (1+w^2)^{1/2}/\cos\theta_B. \tag{20}$$

Here $n=0, h, -0$, and $-h$.

Using the relation $m = M_z/M_x$, $M^2 = M_x^2 - M_z^2$, we can write Eq. (19) as follows:

$$\Delta = k_m M_x (1-m^2)^{1/2}(1+w^2)^{1/2}/\cos\theta_B. \tag{21}$$

With Eqs. (13), (14), (18), (19), and (20), the wave amplitudes $E_0^{(j)}$, $E_{-0}^{(j)}$ can be obtained,

$$E_0^{(j)} = \frac{E_0 \exp(\mp v)[-M_x^2 \cosh(v) + M_z^2 \exp(\pm i\Delta l/2)\cosh(v + i\Delta l/2)]}{2[M_z^2 \cosh(v + i\Delta l/2)\cosh(v - i\Delta l/2) - M_x^2(\cosh(v))^2]}, \tag{22a}$$

$$E_{-0}^{(j)} = \pm \frac{E_0 \exp(-v)\exp(\pm i\Delta l/2)[-M_x M_z \cosh(v)\exp(i\Delta l/2) + M_x M_z \cosh(v - i\Delta l/2)]}{2[M_z^2 \cosh(v + i\Delta l/2)\cosh(v - i\Delta l/2) - M_x^2(\cosh(v))^2]}. \quad j=1,2$$
$$\tag{22b}$$

With Eqs. (13) and (22), the intensities of exiting waves $I_0^{(d)}, I_{-0}^{(a)}, I_h^{(d)}$, and $I_{-0}^{(a)}$ can be calculated. They are illustrated in Fig. 1:

$$E_0^{(d)} \exp(iK_{0z}^{(d)}l) = E_0^{(1)} \exp(iK_{0z}^{(1)}l) + E_0^{(2)} \exp(iK_{0z}^{(2)}l), \tag{23a}$$

$$I_0^{(d)} = E_0^{(d)}(E_0^{(d)})^* = \frac{(m^2-1)^2(w^2+1)\{w^2+\cos^2[B(1+\omega^2)^{1/2}]\}}{(m^2\{w^2+\cos^2[B(1+w^2)^{1/2}]\}-(1+w^2))^2}I_0, \tag{23b}$$

where $m = M_z/M_x$, $B = \pi l M/\lambda \cos\theta_B$,

$$E_{-0}^{(a)} = E_{-0}^{(1)} + E_{-0}^{(2)}, \tag{24a}$$

$$I_{-0}^{(a)} = E_{-0}^{(a)}(E_{-0}^{(a)})^* = \frac{m^2 \sin^4[B(1+w^2)^{1/2}]}{(m^2\{w^2+\cos^2[B(1+w^2)^{1/2}]\}-(1+w^2))^2}I_0, \tag{24b}$$

$$E_h^{(d)} \exp(iK_{hz}^{(d)}l) = E_h^{(1)} \exp(iK_{hz}^{(1)}l) + E_h^{(2)} \exp(iK_{hz}^{(2)}l). \tag{25a}$$

Solving Eqs. (13), (17), (20), (22), and (25), we get

$$I_h^{(d)} = E_h^{(d)}(E_h^{(d)})^* = \frac{(1+w^2)[\sin B(1+w^2)^{1/2}]^2(1-m^2)}{(m^2\{w^2+\cos^2[B(1+w^2)^{1/2}]\}-(1+w^2))^2}I_0. \tag{25b}$$

As in Eq. (25),

$$E_{-h}^{(a)} = E_{-h}^{(1)} + E_{-h}^{(2)}, \tag{26a}$$

$$I_{-h}^{(a)} = \frac{m^2(1-m^2)[\sin B(1+w^2)^{1/2}]^2\{w^2+\cos^2[B(1+w^2)^{1/2}]\}}{(m^2\{w^2+\cos^2[B(1+w^2)^{1/2}]\}-(1+w^2))^2}I_0, \tag{26b}$$

with

$$B = A(1-m^2)^{1/2} \tag{27}$$

and

$$A = \pi l M_x/\lambda \cos\theta_B. \tag{28}$$

If we choose the ratio $l/\lambda \sim 10^4$, $M_x \sim 10^{-4}$ and 10^{-3}, A are in the order of 1 and 10, respectively. The relative intensities of the forward propagation wave $I_0^{(d)}/I_0$ versus angular coordinate w for $A = 3.14$ and 31.4, $m = 0.5$ have been plotted in Figs. 6(a) and 6(b). When l and m are fixed, from Eq. (28), the strength M_x is proportional to A. In (21), the wave-vector difference in the z direction of the two excited plane waves propagating along the same direction in the 2D optical superlattice is determined by M_x; in Figs. 6(a) and 6(b), it is proportional to A. Because the oscillation in the angular distribution of the intensity of the exiting wave in one propagation direction is caused by the wave-vector difference in z axis of the two excited plane waves in that direction, the oscillation in Fig. 6(b) is faster than that in Fig. 6(a).

Figures 7 – 9 are the intensities in the exit lights in other propagation directions. They are all scaled in I_0, the intensity of incident beam intensity. In Eqs. (23)–(26), the intensities of exiting waves are a periodic function of the variable B or superlattice thickness l and index strength M_x, M_z. Thus the periodic variation indicated in the expression of exiting waves will be observed directly on the exit face of a wedge-shaped plate. We can see the pendulum solutions in Figs. 6 – 9, similar to those in the X-ray diffraction theory. With the pendulum solutions, we can study the subsidiary maxima or fine structure of the angular distribution of the exiting waves out from the 2D optical superlattice. In our case, the strength M_x and M_z could be much larger than electric susceptibility of crystals in the X-ray diffraction theory, so that the coordinate w in Figs. 6 - 9 is in a larger unit compared with that in the X-ray diffraction theory. The angular distribution of the pendulum solution would possibly be observed in the experiment.

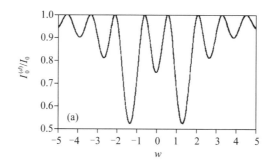

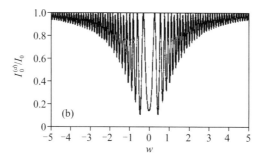

FIG. 6. Ratio of the intensity of wave $I_0^{(d)}$ to that of the incident beam I_0 at various coordinates w: (a) $A = 3.14$, $m = 0.5$; (b) $A = 31.4$, $m = 0.5$.

2.2 $M_z/M_x > 1$

When $m > 1$, M is complex; we rewrite (17) as

$$w = 2\eta \sin 2\theta_B / |M| = 2\eta \sin 2\theta_B / (M_z^2 - M_x^2)^{1/2}; \tag{29}$$

from (16b),

$$\xi_h^{(j)} = w \mp (w^2 - 1)^{1/2} = k_m |M| \exp(\mp v)/4. \tag{30}$$

From boundary conditions (13) and (14), excited fields $E_0^{(j)}$ and $E_{-0}^{(j)}$ in the medium are

$$E_0^{(j)} = \mp \frac{E_0 \exp(\mp v)[-M_x^2 \sinh(v) + M_z^2 \exp(\mp i\Delta l/2)\sinh(v + i\Delta l/2)]}{2[M_z^2 \sinh(v + i\Delta l/2)\sinh(v - i\Delta l/2) - M_x^2 (\sinh(v))^2]}, \tag{31a}$$

$$E_{-0}^{(j)} = \mp \frac{E_0 \exp(-v)\exp(\pm i\Delta l/2)[-M_x M_z \sinh(v)\exp(i\Delta l/2) + M_x M_z \sinh(v - i\Delta l/2)]}{2[M_z^2 \sinh(v + i\Delta l/2)\sinh(v - i\Delta l/2) - M_x^2 (\sinh(v))^2]}. \tag{31b}$$

Similar to Eqs. (23)-(26), the electric fields are

$$E_0^{(d)} = \frac{E_0[-M_x^2 \sinh(v + i\Delta l/2)\sinh(v) + M_z^2 \sinh(v)\sinh(v + i\Delta l/2)]}{M_z^2 \sinh(v + i\Delta l/2)\sinh(v - i\Delta l/2) - M_x^2 (\sinh(v))^2}, \tag{32}$$

$$E_{-0}^{(a)} = \frac{E_0 M_x M_z \sin^2(\Delta l/2)}{M_z^2 \sinh(v+\mathrm{i}\Delta l/2)\sinh(v-\mathrm{i}\Delta l/2) - M_x^2(\sinh(v))^2}, \tag{33}$$

$$E_h^{(d)} = \mathrm{i}\frac{E_0}{|M|} \frac{M_x \sinh(v)\sin(\Delta l/2)(M_z^2 - M_x^2)}{M_z^2 \sinh(v+\mathrm{i}\Delta l/2)\sinh(v-\mathrm{i}\Delta l/2) - M_x^2(\sinh(v))^2}, \tag{34}$$

$$E_{-h}^{(a)} = \mathrm{i}\frac{E_0}{|M|} \frac{M_z \sin(\Delta l/2)\sinh(v+\mathrm{i}\Delta l/2)(M_z^2 - M_x^2)}{M_z^2 \sinh(v+\mathrm{i}\Delta l/2)\sinh(v-\mathrm{i}\Delta l/2) - M_x^2(\sinh(v))^2}. \tag{35}$$

We divide the range of maximum into three regions in Fig. 5: region Ⅰ, $w < -1$; region Ⅲ, $w > 1$; and region Ⅱ, $-1 \leqslant w \leqslant 1$. In regions Ⅰ and Ⅲ the normal of the surface of the 2D optical superlattice n_0 intersects with the corresponding hyperbolas branching at two points $A^{(1)}$, $A^{(2)}$; the corresponding wave vectors $\boldsymbol{K}_n^{(j)}$ of $A^{(1)}$, $A^{(2)}$ are real. The intensities of exiting waves are

$$I_0^{(d)} = \frac{(m^2-1)^2(w^2-1)\{w^2 - \cos^2[B'(w^2-1)^{1/2}]\}}{((w^2-1) - m^2\{w^2 - \cos^2[B'(w^2-1)^{1/2}]\})^2} I_0, \tag{32'}$$

$$I_{-0}^{(a)} = \frac{m^2 \sin^4[B'(w^2-1)^{1/2}]}{((w^2-1) - m^2\{w^2 - \cos^2[B'(w^2-1)^{1/2}]\})^2} I_0, \tag{33'}$$

$$I_h^{(d)} = \frac{(m^2-1)(w^2-1)\sin^2[B'(w^2-1)^{1/2}]}{((w^2-1) - m^2\{w^2 - \cos^2[B'(w^2-1)^{1/2}]\})^2} I_0, \tag{34'}$$

$$I_{-h}^{(a)} = \frac{(m^2-1)m^2 \sin^2[B'(w^2-1)^{1/2}]\{w^2 - \cos^2[B'(w^2-1)^{1/2}]\}}{((w^2-1) - m^2\{w^2 - \cos^2[B'(w^2-1)^{1/2}]\})^2} I_0, \tag{35'}$$

where $B' = A(m^2-1)^{1/2}$.

It can be seen in Fig. 5 that in region Ⅱ, $-1 \leqslant w \leqslant 1$, the normal n_0 has no intersection with two hyperbolas branches, so $(w^2-1)^{1/2} = \mathrm{i}(1-w^2)^{1/2}$, and the wave vectors $\boldsymbol{K}_n^{(j)}$ become complex. The intensities of four exiting waves are

$$I_0^{(d)} = \frac{(m^2-1)^2(1-w^2)\{1-w^2 + \sinh^2[B'(1-w^2)^{1/2}]\}}{((1-w^2) - m^2\{1-w^2 + \sinh^2[B'(1-w^2)^{1/2}]\})^2} I_0, \tag{32''}$$

$$I_{-0}^{(a)} = \frac{m^2 \sinh^4[B'(1-w^2)^{1/2}]}{((1-w^2) - m^2\{1-w^2 + \sinh^2[B'(1-w^2)^{1/2}]\})^2} I_0, \tag{33''}$$

$$I_h^{(d)} = \frac{(m^2-1)\sinh^2[B'(1-w^2)^{1/2}](1-w^2)}{((1-w^2) - m^2\{1-w^2 + \sinh^2[B'(1-w^2)^{1/2}]\})^2} I_0, \tag{34''}$$

$$I_{-h}^{(a)} = \frac{(m^2-1)m^2 \sinh^2[B'(1-w^2)^{1/2}]\{1-w^2 + \sinh^2[B'(1-w^2)^{1/2}]\}}{((1-w^2) - m^2\{1-w^2 + \sinh^2[B'(1-w^2)^{1/2}]\})^2}, \tag{35''}$$

$$B' = A(m^2-1)^{1/2} = \pi l M_x(m^2-1)^{1/2}/\lambda \cos\theta_B. \tag{36}$$

Equations (32)-(35) are plotted in Figs. 10-13, where w ranges from -5 to 5. In Figs. 10-13, we can see that at the midpoint of maximum, i.e., $w = 0$, and when $A = 3.14$, the transmitting beams are very weak, $I_0^{(d)} \approx I_h^{(d)} \approx 0$. Most of the energy of the incident beam has changed into the backward wave $I_{-0}^{(a)}$ and $I_{-h}^{(a)}$ in the range $-1 \leqslant w \leqslant 1$. The ratio $I_{-0}^{(a)}/I_{-h}^{(a)}$ at $w = 0$ is controlled by the strength M_x, M_z and thickness of the 2D optical superlattice along the z axis.

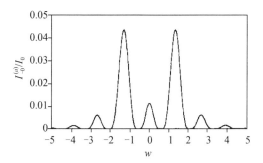

FIG. 7. Ratio of the intensity of wave $I_{-0}^{(a)}$ to that of the incident beam I_0 at various coordinates w with $A=3.14$, $m=0.5$.

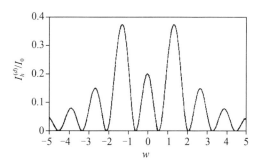

FIG. 8. Ratio of the intensity of wave $I_h^{(d)}$ to that of the incident beam I_0 at various coordinates w with $A=3.14$, $m=0.5$.

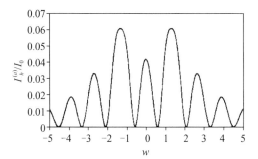

FIG. 9. Ratio of the intensity of wave $I_{-h}^{(a)}$ to that of the incident beam I_0 at various coordinates w with $A=3.14$, $m=0.5$.

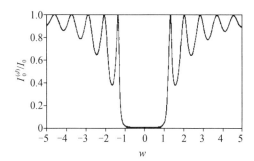

FIG. 10. Ratio of the intensity of wave $I_0^{(d)}$ to that of the incident beam I_0 at various coordinates w with $A=3.14$, $m=1.5$.

2.3 $M_z/M_x \to 1$

When $m \to 1$, causing $\xi_0^{(j)} \to 0$, $\xi_h^{(j)} \to 0$, the dispersion surface turns into the two intersecting lines T_0 and T_h. There exists coupling only between waves $\boldsymbol{K}_0$ and $\boldsymbol{K}_{-0}$, or between $\boldsymbol{K}_h$ and $\boldsymbol{K}_{-h}$. The four waves $\boldsymbol{K}_0$, $\boldsymbol{K}_h$, $\boldsymbol{K}_{-0}$, and $\boldsymbol{K}_{-h}$ do not couple with each other except for the above cases. The exiting waves tend to the certain values, possibly with oscillation when $m \to 1$. We can calculate these values from Eqs. (23)–(25). For any values of w, the following equations always hold:

$$\lim_{m \to 1} I_0^{(d)} = \left(\frac{1}{A^2+1}\right)^2 I_0, \tag{37}$$

$$\lim_{m \to 1} I_0^{(a)} = \left(\frac{A^2}{A^2+1}\right)^2 I_0, \tag{38}$$

$$\lim_{m \to 1} I_h^{(d)} = \left(\frac{A}{A^2+1}\right)^2 I_0, \tag{39}$$

$$\lim_{m \to 1} I_{-h}^{(a)} = \left(\frac{A}{A^2+1}\right)^2 I_0, \tag{40}$$

where A was expressed in Eq. (28).

Normally, A is much larger than unity, so most of the energy of the incident beam turns into $I_{-0}^{(a)}$. What interests us is that all intensities of waves in Eqs. (37)–(40) are independent of the angular coordinate w. We can imagine that when a light propagates with the incident angle near the Bragg angle in the 2D optical superlattice, which has an index strength ratio equal to unity, the incident beam will reflect back in the direction opposite the incident direction and other exiting waves are very weak, according to Eqs. (37)–(40).

3. Discussion

The intensity of each exiting wave can be controlled by modulating the index strength M_x, M_z of the 2D optical superlattice or by changing its thickness l. There are four waves propagating in different directions and our exit waves in the 2D optical superlattice. The interaction and propagation of light in the 2D optical superlattices are more plentiful than those in the 1D periodically modulated index structures; the former affords a new means for optical communications. If the acoustic wave is used to produce the periodically modulated index in the 2D optical superlattices, the acoustic-optic interaction provides a new way to probe the acoustic field in a solid and to manipulate the optical radiation. Such effects in 2D optical superlattices can be used to construct optical devices in areas such as light modulation, beam deflection, and signal processing.

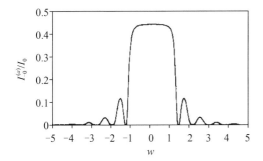
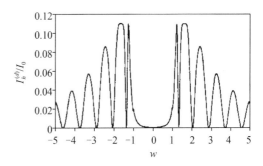

FIG. 11. Ratio of the intensity of wave $I_{-0}^{(a)}$ to that of the incident beam I_0 at various coordinates w with $A = 3.14$, $m = 1.5$.

FIG. 12. Ratio of the intensity of wave $I_h^{(d)}$ to that of the incident beam I_0 at various coordinates w with $A = 3.14$, $m = 1.5$.

Let us look at a simple example of using 2D the optical superlattice to design an optical switch with modulating strength ratio m. Generally, there are four lights from the 2D optical superlattice. However, the incident beam will be diffracted into one of the diffraction directions as m takes some particular values. If we choose $A = 3\pi/2 = 4.71$, $w = 0$, and the acoustic field is applied only along the x axis, i.e., $m = M_z/M_x = 0$, the intensities of two exiting waves are $I_0^{(d)} = 0$, $I_h^{(d)} = I_0$, respectively, where I_0 is the intensity of the incident beam. Now another acoustic field is applied in the z direction, and a suitable intensity of acoustic field is chosen to make the strength ratio $m=\sqrt{5}/3$; then $I_0^{(d)}$ reaches its maximum, $I_0^{(d)}=I_0$, $I_h^{(d)}=0$. The intensities $I_0^{(d)}$, $I_h^{(d)}$ at two values of m are shown in Fig. 14. According to the discussion in Sec. II C, when the strength ratio m is changed from $\sqrt{5}/3$ to 1, the intensities become $I_{-0}^{(a)} \approx I_0$, $I_0^{(d)} \approx I_h^{(d)} \approx I_{-h}^{(a)} \approx 0$. The incident beam is diffracted into the $I_{-0}^{(a)}$ direction. With this optical switch, three diffraction states of the incident beam can be realized. In principle, after setting proper parameters, we could have the incident beam diffracted into one of the four propagating directions with wave vectors $\boldsymbol{K}_0$, $\boldsymbol{K}_h$, $\boldsymbol{K}_{-0}$, and $\boldsymbol{K}_{-h}$.

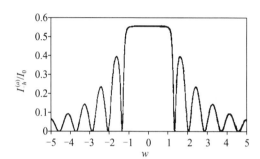

FIG. 13. Ratio of the intensity of wave $I_{-h}^{(a)}$ to that of the incident beam I_0 at various coordinates w with $A=3.14$, $m=1.5$.

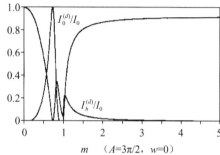

FIG. 14. Relationship between the intensities of two exiting waves $I_0^{(d)}$, $I_h^{(d)}$ and the strength ratio m when the incident beam is in the exact Bragg angle.

4. Conclusion

Light propagation in the 2D optical superlattice has been treated in detail. In Figs. 6 – 13, the intensities of the exiting waves from the 2D optical superlattice have the forms of the pendulum solution, and the pendulum solutions here are different from those in X-ray diffraction theory with a two-wave approximation. Because the strength ratio m determines the dispersion surface form, in three ranges of m, three different expressions of the electric fields in and out of the 2D optical superlattices are given. Replacing the values of sine and cosine squared in Eqs. (23)–(26) and (32)–(40) by their values over the full oscillation period, average values of intensities can be obtained. From the average intensities, we can get their angular half-width. With I_0 fixed, the half-width is determined by index strength M_x, M_z. Because the index strength can be as large as the order of 10^{-2} (see Russell[2,3]), the angular half-width of the average intensities of waves in the 2D optical superlattices is much larger than that of the X-ray intensities in crystals which normally have an electric susceptibility under the order of 10^{-4}. Therefore the angular half-width of the average intensities of waves in 2D optical superlattices is expected to be larger than the angular divergence of lasers, and plane-wave approximation is suitable to solve the problems of the propagation and interaction of light in 2D optical superlattices.

References and Notes

[1] A. Yariv, IEEE J. Quantum Electron. **QE-9**, 919 (1973).

[2] P.St. J. Russell, Phys. Rev. A **33**, 3232 (1986).

[3] P.St. J.Russell, Phys. Rev. Lett. **56**, 596 (1986).

[4] N. B. Ming, J. F. Hong, and D. Feng, J. Mater. Sci. **27**, 1663 (1982).

[5] T. H. Xue, N. B. Ming, J. S. Zhu, and D. Feng, Chin. Phys. **4**, 554 (1982).

[6] A. Y. Cho and J. R. Arthur, *Progress in Solid State Chemistry* (Pergamon, New York, 1975), Vol. 10, pp. 157—191.

[7] Robert Adler, IEEE Spectrum **4**, 42 (1967).

[8] Z. Pinsker, *Dynamical Scattering of X-rays in Crystals* (Springer-Verlag, Berlin, 1978).

[9] The authors are grateful to Professor R. M. Sanford for his great assistance. We also thank Professor ShuSheng Jiang and Ming-Sheng Zhang for valuable discussions; and Yiang Qu, Jiu-Fang Chao, and Yon Chen for preparing some figures. This project was supported by the National Science Foundation of China.

Optical Bistability in a Two-Dimensional Nonlinear Superlattice[*]

Bin Xu

National Laboratory of Solid State Microstructures, Nanjing University,
Nanjing 210008, People's Republic of China

Nai-Ben Ming

Center for Condensed Matter Physics and Radiation Physics,
China Center of Advanced Science and Technology (World Laboratory),
P. O. Box 8730, Beijing 100080, People's Republic of China
and National Laboratory of Solid State Microstructures, Nanjing University,
Nanjing 210008, People's Republic of China

Based on our theoretical results of dynamical behaviors of light transmission in optical superlattices, we show that there is a new type of optical bistable mechanism, i.e., the index-modulation mechanism, existing in a two-dimensional superlattice that contains dielectric nonlinearity of the Kerr form. This bistable mechanism is not exhibited in one-dimensional superlattices.

A nonlinear system with positive feedback will exhibit bistable behaviors under suitable conditions. That is to say, two necessary elements are expected to coexist in a dielectric system that possesses optical bistability[1]: One is the nonlinear response element, requiring that the transmission light field respond in nonlinear form to parameters that are related to the transmission process; the other is the positive feedback element, requiring that the values of these parameters change with the transmission field. In dispersive bistability in a Fabry-Pérot etalon[2] or in a one-dimensional (1D) superlattice[3—9], the related parameter is the mismatch of the propagating wave vector that detunes from the cavity-matching condition or the Bragg condition. The average energy density of the transmission field in the cavity or in the superlattice is an oscillation function of the parameter[7,9]. This nonlinear response element, coupled with the feedback element that the wave vector's mismatch is established by the field's average energy density in the medium via the Kerr form dielectric nonlinearity, will produce bistability in the system's input/output relation. We may attribute the same phase-mismatch mechanism to the two kinds of bistabilities existing in the Fabry-Pérot etalon and the 1D superlattice. Normally in the all-optical bistable systems of nonabsorptive type so far put forward, the feedback

---

[*] Phys.Rev.Lett.,1993,71(7):1003

element is provided by the dielectric nonlinearity of the Kerr form. Novel optical bistable mechanisms can be proposed by searching for new types of nonlinear response elements. In fact, if a nonlinear response element exists in a system, people can find it by checking the dynamics of light transmission in the corresponding linear dielectric system. In this letter, after examining the dynamics of multiwave diffraction in a two-dimensional (2D) linear optical superlattice, we show that a new type of bistable mechanism, i.e., the index-modulation mechanism, exists in a 2D superlattice containing the Kerr form dielectric nonlinearity. Compared with the phase-mismatch mechanism, the related parameter in the index-modulation mechanism is not the wave vector's mismatch, but the index-modulation strength[10] of the 2D superlattice.

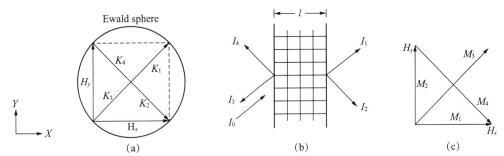

FIG. 1. Four-wave diffraction in the two-dimensional periodic superlattice. (a) Bragg condition with four reciprocal points located on the Ewald sphere. (b) Schematic diagram of four-wave diffraction in real space. (c) Four index-modulation strengths M_1, M_2, M_3, and M_4, as Fourier components corresponding to reciprocal vectors H_x, H_y, H_x+H_y, and H_x-H_y, respectively.

We first consider a linear model of an isotropic lossless medium with its refractive index modulated periodically in two dimensions. The periodicity is expressed by vectors $\boldsymbol{H}_x$ and $\boldsymbol{H}_y$ in reciprocal space. The incident wave vector satisfies the Bragg condition that four reciprocal points are located on the Ewald sphere[11]. Thus four diffracted waves will be excited as Fig. 1 shows. Feng and Ming[10] first pointed out that in this multiwave diffraction process, the intensities of the diffracted waves are oscillation functions of the index-modulation strengths, which are the Fourier components of the periodic refractive index. Here in this work, with the aid of numerical computation, we use the dispersion equation which is similar to that of the multiwave diffraction in X-ray scattering theory[11] to accurately solve the transmission problem in this 2D structure. The following scalar propagation equation is used to describe the transmission process:

$$[\nabla^2 + \beta_0^2(\boldsymbol{r})]E(\boldsymbol{r}) = 0. \tag{1}$$

The β_0 is the propagation constant expressed by

$$\beta_0^2(\boldsymbol{r}) = k_v^2[1 + M_1\cos(\boldsymbol{H}_x \cdot \boldsymbol{r}) + M_2\cos(\boldsymbol{H}_y \cdot \boldsymbol{r}) + M_3\cos((\boldsymbol{H}_x + \boldsymbol{H}_y) \cdot \boldsymbol{r}) + M_4\cos((\boldsymbol{H}_x - \boldsymbol{H}_y) \cdot \boldsymbol{r})], \tag{2}$$

where k_v is the wave number in the average refractive index and $M_i(i=1, 2, 3, 4)$ are the Fourier components by expanding the 2D periodic refractive index along the reciprocal vectors lying on the Ewald sphere [see Fig. 1(c)]. With the multiwave approximation in dynamical diffraction theory[11], the solution of Eq. (1) in such a periodic structure is the following Bloch wave:

$$E(\boldsymbol{r}) = \exp(i\boldsymbol{K}_1 \cdot \boldsymbol{r}) \sum_{i=1}^{4} E_i \exp(i\boldsymbol{H}_i \cdot \boldsymbol{r}) = \sum_{i=1}^{4} E_i \exp(i\boldsymbol{K}_i \cdot \boldsymbol{r}). \tag{3}$$

As Fig. 1(a) shows, $\boldsymbol{K}_1$ is the incident wave vector satisfying the exact Bragg condition. $\boldsymbol{K}_i$ ($=\boldsymbol{K}_1 + \boldsymbol{H}_i$) is the excited diffracted wave vector that locates on the Ewald sphere. Considering the dispersive effect caused by the multiwave interaction[11], we replace wave vector $\boldsymbol{K}_i$ with $\boldsymbol{k}_i = \boldsymbol{K}_i + \boldsymbol{\delta}$ in Eq. (3), where $\boldsymbol{\delta} = \delta \hat{\boldsymbol{s}}$ and $\hat{\boldsymbol{s}}$ is a unit vector normal to the entrance surface. As a sufficient approximation, we assume $1 - (k_i/k_v)^2 = -(2\boldsymbol{K}_i \cdot \hat{\boldsymbol{s}}/k_v^2)\delta$. Then the following dispersion equation is derived by putting Eqs. (2) and (3) into Eq. (1):

$$\begin{vmatrix} -\sqrt{2}\delta/k_v & \frac{1}{2}M_2 & \frac{1}{2}M_3 & \frac{1}{2}M_1 \\ \frac{1}{2}M_2 & -\sqrt{2}\delta/k_v & \frac{1}{2}M_1 & \frac{1}{2}M_4 \\ \frac{1}{2}M_3 & \frac{1}{2}M_1 & \sqrt{2}\delta/k_v & \frac{1}{2}M_2 \\ \frac{1}{2}M_1 & \frac{1}{2}M_4 & \frac{1}{2}M_2 & \sqrt{2}\delta/k_v \end{vmatrix} \begin{vmatrix} E_1 \\ E_2 \\ E_3 \\ E_4 \end{vmatrix} = 0. \tag{4}$$

We assume the eigenmodes and the eigenvalues of the dispersion equation are $\{E_i^{(j)}\}$ ($i=1,2,3,4$) and $\delta^{(j)}$, respectively ($j=1,2,3,4$). Then the transmission field $E(\boldsymbol{r})$ is expressed as

$$E(\boldsymbol{r}) = \sum_{i=1}^{4} E_i(\boldsymbol{r}) = \sum_{i=1}^{4} \sum_{j=1}^{4} \lambda^{(j)} E_i^{(j)} \exp[i(\boldsymbol{K}_i + \boldsymbol{\delta}^{(j)}) \cdot \boldsymbol{r}], \tag{5}$$

where the coefficient $\lambda^{(j)}$ is determined by the boundary conditions: at $x=0$, $E_1(\boldsymbol{r}) = E_0$ and $E_2(\boldsymbol{r}) = 0$; at $x=1$, $E_3(\boldsymbol{r}) = E_4(\boldsymbol{r}) = 0$. The relative intensities of the four diffracted waves $I_i = |E_i|^2/|E_0|^2$ at the exit boundaries can thus be obtained with numerical computation. As our results of the 3D plots in Fig. 2 show, all four diffracted intensities are oscillation functions of the index-modulation strengths M_i. According to the above outline for bistability, such a dynamical behavior revealed in a 2D superlattice provides a new type of nonlinear response element. The related parameters here are the four index-modulation strengths. It is expected that if a dielectric nonlinearity of the Kerr form is considered, the values of the index-modulation strengths will be perturbed by the interference of the four diffracted waves in the transmission field. This is the feedback element. We naturally expect that the bistability may be exhibited in the incident-diffracted relations of a 2D superlattice containing Kerr form dielectric nonlinearity.

To investigate such possible bistability of the index-modulation mechanism, here we apply a kind of selfconsistent method to achieve the incident-diffracted relations. The

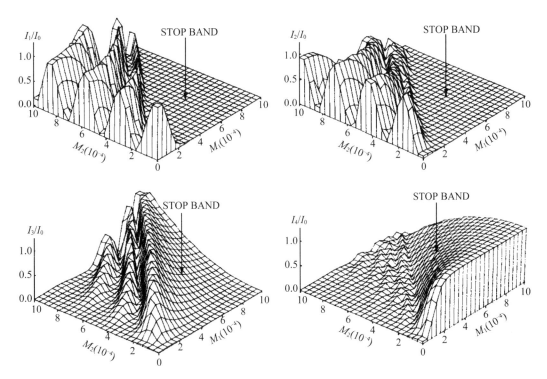

FIG. 2. When incident wave satisfies the Bragg condition, plots of the intensities of the four diffracted waves at exit boundaries, as functions of the index modulation strengths M_1 and M_2, with $M_3 = M_4 = 10^{-5}$, and $k_v l = 3 \times 10^4$. These functions are in oscillation forms in the allowed band.

distribution of the refractive index that affects the transmission process is indicated by the four indexmodulation strengths M_i ($i = 1, 2, 3, 4$). When the incident wave satisfying the Bragg condition excites four diffracted waves, the interference of the field will give perturbations to the values of the four index-modulation strengths via the Kerr form nonlinear term. The perturbed index-modulation strengths M_i' ($i = 1, 2, 3, 4$) will then return to affect the transmission field. Eventually this dynamical interaction between the transmission field and the index modulation strengths will reach a stable state, i.e., the two are in a self-consistent manner. Such a stable convergent self-consistent solution can be easily obtained with the numerical computation. According to the multiwave approximation in dynamical diffraction theory[11], only the index modulation strengths with reciprocal vectors lying on the Ewald sphere will effectively affect the light transmission process; other terms about the refractive index distribution in the medium are discarded. In dealing with the perturbations of the index modulation strengths brought by the field interference, as an approximation, we only consider the terms with reciprocal vectors lying on the Ewald sphere. The Kerr form nonlinear term is assumed to be $\Delta \varepsilon = \alpha |E|^2$. With Eq. (5), we express the perturbations as

$$\Delta M_1 = \alpha \left(\sum_{j=1}^{4} |\lambda^{(j)}|^2 (E_1^{(j)} E_4^{(j)*} + E_2^{(j)} E_3^{(j)*}) \right) + \text{c.c.}, \tag{6a}$$

$$\Delta M_2 = \alpha \Big(\sum_{j=1}^{4} |\lambda^{(j)}|^2 (E_1^{(j)} E_2^{(j)*} + E_3^{(j)} E_4^{(j)*}) \Big) + \text{c.c.}, \tag{6b}$$

$$\Delta M_3 = \alpha \Big(\sum_{j=1}^{4} |\lambda^{(j)}|^2 E_1^{(j)} E_3^{(j)*} \Big) + \text{c.c.}, \tag{6c}$$

$$\Delta M_4 = \alpha \Big(\sum_{j=1}^{4} |\lambda^{(j)}|^2 E_2^{(j)} E_4^{(j)*} \Big) + \text{c.c.}. \tag{6d}$$

With the numerical computation, given M_i, we have a set of $\{E_i^{(j)}\}$; then with Eq. (6), we have $M_i' = M_i + \Delta M_i$ ($i=1,2,3,4$) to replace the values M_i; then a new set of $\{E_i^{(j)}\}$ is obtained to determine the next new values $M_i' \cdots$, until a stable convergent solution is achieved. If the stable self-consistent solution converges on one value, it means that the incident-diffracted relations are single valued at this incident intensity I_0; if the solution converges on two or more stable values, the incident-diffracted relations are bivalued or multivalued, and bistability or multistability is thus exhibited. Figure 3 is one of our results that shows bistability (multistability). The shapes of the incident-diffracted curves are determined by the superlattice's original index-modulation strengths. There always appear discontinuous jumps of intensity in the bistable (multistable) regions and the jumps can be either from the higher values to the lower or from the lower to the higher. The threshold for bistability here is comparable with that of the dispersive bistability in a Fabry-Pérot etalon.

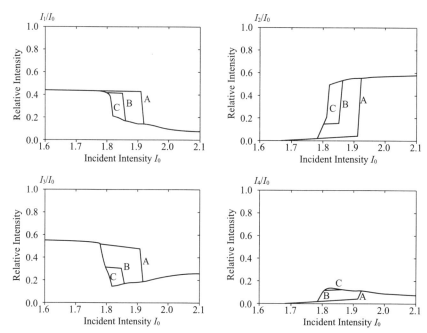

FIG. 3. With dielectric nonlinearity of the Kerr form considered, the intensities of the four diffracted waves as functions of the incident intensity I_0, with $M_1 = 4.5 \times 10^{-4}$, $M_2 = 6.0 \times 10^{-4}$, $M_3 = M_4 = 10^{-5}$, and $k_v l = 8 \times 10^4$. The unit of I_0 corresponds to $\alpha |E_0|^2 |_{I_0=1} = 10^{-4}$. Curves A, B, and C are three different stable convergent self-consistent solutions obtained by numerical computation.

This index-modulation mechanism for bistability, however, is characteristic of multiwave diffraction cases in 2D superlattices. It is not exhibited in a 1D superlattice because there is only one parameter of index-modulation strength in a 1D superlattice; when the incident wave satisfies the Bragg condition, the transmissivity is a monotonous function of the parameter[10].

The bistability of the index-modulation mechanism is only exhibited in the allowed band of the transmission spectrum[12], for, as Fig. 2 shows, the oscillations only appear in this region. The peak and the trough of an oscillation correspond to the high and low transmission states of a diffractive wave, respectively.

In the allowed band of the 2D superlattice, we do not consider the bistability of the phase-mismatch mechanism. Figure 4 is our numerical result of the average energy density of the transmission field in the 2D linear superlattice as a function of the propagating wave vector's mismatch from the Bragg condition. Because of the multiwave diffraction in four directions of two-dimensional superlattice, the oscillations in the allowed band are unlike those in a 1D superlattice[7,9], but in a flattened form. Thus it cannot provide the nonlinear response element for effective bistability of the phase-mismatch mechanism.

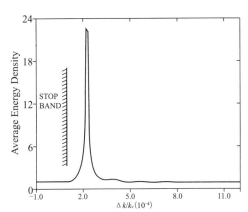

FIG. 4. The average energy density in the two-dimensional superlattice as a function of the mismatch of the propagating wave vector from the Bragg condition, in ratio with that of the incident wave, with $M_1 = 4.5 \times 10^{-4}$, $M_2 = 6.0 \times 10^{-4}$, $M_3 = M_4 = 10^{-5}$, and $k_v l = 3 \times 10^4$.

In the region near the transmission transition from the forbidden (stop) state to the allowed, as Fig. 4 shows, there is always a sharp resonance peak, and this peak corresponds to transmissivity of unity. The phase-mismatch mechanism may be exhibited here for the existence of such a resonance peak. Then the two mechanisms will coexist to lead to bistability with switching jumps from the forbidden transmission state to the allowed, and owing to the closely spaced sharp forms of the oscillations here (see Fig. 2), the bistable threshold is lower than that in the allowed band[9].

At present, there are several techniques available to fabricate a 2D nonlinear superlattice, such as microstructural etching techniques, the growth of colloidal crystal, the use of the acousto-optic effect, and the use of the photorefractive effect, etc.

In photorefractive materials, the net refractive index change is only produced by field interference, and has expression similar to that of the Kerr form dielectric nonlinearity. As the perturbations of the index-modulation strengths we discussed above are just resulting from interference, the bistability of only the index-modulation mechanism will be exhibited in such a medium, where a 2D periodic refractive index grating can be constructed beforehand by the volume holographic recording method and can be thermally fixed in some way[13]. We suggest this way to be the most convenient method to experimentally demonstrate the index-modulation mechanism.

This 2D superlattice can be used to construct more compact bistable devices in integrated optics for its advantages in optical geometrical arrangements.

References and Notes

[1] See, for example, H. M. Gibbs, *Optical Bistability: Controlling Light with Light* (Academic, New York, 1985).
[2] H. M. Gibbs, S. L. McCall, and T. N. C. Venkatesan, Phys. Rev. Lett. **36**, 1135 (1976).
[3] H. G. Winful, J. H. Marburger, and E. Garmire, Appl. Phys. Lett. **35**, 379 (1979).
[4] F. Delyon, Y. E. Lévy, and B. Souillard, Phys. Rev. Lett. **57**, 2010 (1986).
[5] W. Chen and D. L. Mills, Phys. Rev. B **35**, 524 (1987).
[6] L. Kahn, N. S. Almeida, and D. L. Mills, Phys. Rev. B **37**, 8072 (1988).
[7] J. He and M. Cada, IEEE J. Quantum Electron. **27**, 1182 (1991).
[8] J. Danckaert *et al.*, Phys. Rev. B **44**, 8214 (1991).
[9] V. M. Agranovich, S. A. Kiselev, and D. L. Mills, Phys. Rev. B **44**, 10917 (1991).
[10] J. Feng and N. B. Ming, Phys. Rev. A **40**, 7047 (1989), and references therein.
[11] Z. G. Pinsker, *Dynamical Scattering of X-rays in Crystal* (Springer-Verlag, Berlin, 1978).
[12] P. Yeh, A. Yariv, and C. S. Hong, J. Opt. Soc. Am. **67**, 423 (1977).
[13] P. Günter, Phys. Rep. **93**, 199—299 (1982).
[14] This work is supported by a grant for the Key Research Project in Climbing Program from the National Science and Technology Commission of China.

Experimental Observations of Bistability and Instability in a Two-Dimensional Nonlinear Optical Superlattice[*]

Bin Xu

National Laboratory of Solid State Microstructures, Nanjing University,
Nanjing 210008, People's Republic of China

Nai-Ben Ming

Center for Condensed Matter Physics and Radiation Physics,
China Center of Advanced Science and Technology (World Laboratory),
P.O. Box 8730, Beijing 100080, People's Republic of China
and National Laboratory of Solid State Microstructures, Nanjing University,
Nanjing 210008, People's Republic of China

Optical bistability in a two-dimensional nonlinear superlattice, resulting from the index-modulation mechanism, was observed in experiment for the first time. The two-dimensional nonlinear superlattice was constructed by recording a two-dimensional refractive index grating into a photorefractive material, a Fe-doped $LiNbO_3$ single crystal. Optical instability was also observed in this experiment.

Optical bistability in a one-dimensional (1D) superlattice containing Kerr-form dielectric nonlinearity was first proposed by the pioneering work of Winful, Marburger, and Garmire[1] more than ten years ago. This idea had then been developed in a number of theoretical papers[2—7]. Recently, Cada et al.[8] carried out an experiment using GaAs/AlAs periodically layered structures and observed the all-optical logic operations. This may have been the first experimental attempt to demonstrate the bistable behavior in a 1D superlattice. The bistability in such a 1D periodic medium and that in a Fabry-Pérot etalon are potentially analogous because of their dispersive natures in these two types of resonant structures, respectively. In a 1D superlattice, it is the change of the phase mismatch between the propagating wave vector and the periodic structure that brings the incident wave from a forbidden transmission state to an allowed state, or from a low-transmission state to a high-transmission state in the allowed band[4—7]. The bistability in a 1D superlattice is thus attributed to the phase-mismatch mechanism[9]. A recent theoretical work by the present authors has revealed that a novel bistable mechanism, i.e., the index-modulation mechanism, which is characteristically different from the phase-mismatch

---

[*] Phys. Rev. Lett., 1993, 71(24): 3959

mechanism, exists in a two-dimensional (2D) superlattice containing Kerr-form dielectric nonlinearity[9]. It is known that in 1D superlattices the light's transmission in the exact phase-matched condition (the incident wave vector satisfies the exact Bragg condition) corresponds to the forbidden state[6,7]. However, in 2D superlattices, when the transmitting light satisfies the exact Bragg condition of multiwave diffraction, whether the transmission state is located in the forbidden band or in the allowed band, is determined by the values of a set of parameters which are defined as the index-modulation strengths of the 2D periodic structure[9,10]. If the values of these parameters are arranged suitably, the light will propagate in the allowed band. Changes of these values will lead to high or low-transmission states for each diffracted wave, since in the allowed band in the exact Bragg diffraction condition, the intensities of the multidiffracted waves are oscillation functions of these parameters[9,10]. This element provides the basis of the index-modulation bistable mechanism, which is exhibited in 2D superlattices but not in 1D cases. When the Kerr-form dielectric nonlinearity is considered, as the theory in Ref.[9] predicts, the values of the index-modulation strengths will be perturbed by the interference of the transmission-diffraction light field in the medium, forming a positive feedback element to establish bistable behaviors in the incident-diffracted relations of the 2D superlattice[9].

To experimentally demonstrate this index-modulation mechanism, the photorefractive materials are ideal candidates. In these materials, the net refractive index change, caused by only the interference fringe of even a weak cw light field, has an expression similar to the Kerr-form dielectric nonlinearity. It will be potentially the same for the existence of the index-modulation bistable mechanism in photorefractive materials as in the Kerr-form nonlinear media[9]. The 2D superlattice in a photorefractive material can be constructed beforehand by recording a 2D refractive index grating into the medium with the volume holographic recording method. This is a convenient way to qualitatively, though not quantitatively, demonstrate the existence of the index-modulation mechanism in a 2D nonlinear superlattice. In this letter, we report the first experimental observations of optical bistability, as well as instability, in such a 2D nonlinear superlattice.

The photorefractive material used in the experiment is an oxidized Fe-doped $LiNbO_3$ (0.1 wt.% Fe) single domain crystal. A beam of the blue line (488 nm, TEM_{00} mode) of an argon-ion laser was split into three nearly-equal-intensity ones, S_1, S_2, and S_3, which were recombined and interfered to cause spatial variations of the refractive index. Thus a 2D square periodic grating was recorded into the crystal as Fig. 1(a) shows. (The index-modulation strengths are Fourier components of the grating strength) The intensities of the beams S_i ($i=1,2,3$) were set to be about 2.5 W/cm^2. The diameter of the beams, stipulating the size of the grating, was 0.2 cm. Despite the difficulty in determining precisely the grating strength in this holographic writing process, after an exposure time of about 30 s to 40 s, the grating's diffraction efficiency reached a value in the range from 10% to 15% (the material's absorption is high). We should point out that sufficient high diffraction efficiency is a prerequisite for the following experiments. This oxidized Fe:LiNbO$_3$ sample showed high resistance to relax and the grating remained stable in the sample after the writing beams were moved. Then a beam I_0, with the same wavelength and the same incident direction as the beam S_1, was incident on this 2D grating with a much weaker intensity which was adjusted by an attenuator from 0 to 0.8 W/cm^2. Because this incident beam satisfied the exact Bragg condition, four diffracted waves were excited as Fig. 1 (b) shows. This is the multiwave diffraction model the present authors used in the theoretical work[9] and here the allowed transmission is permitted. The photodiodes were used for measuring the diffracted intensities. With slowly adjusting the incident intensity I_0 and then recording the changes of the diffracted intensities with I_0 by the X-Y recorder, the bistable behaviors would be clearly plotted if they appeared in the incident-diffracted relations.

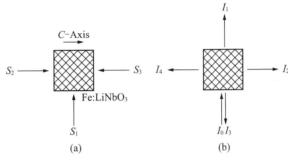

FIG. 1. (a) Optical geometry for recording the 2D refractive index grating into a Fe-doped LiNbO$_3$ single domain crystal. (b) Schematic diagram of four-wave diffraction. I_0 is the incident wave satisfying the exact Bragg condition of four-wave diffraction.

In the illumination of the incident wave, the trapped electrons in the space charge pattern of the original grating are excited and drift. Because there is an interference field existing in the medium due to the incident wave diffracted into four directions, the space charge pattern that the drifting of the free carriers is constructing coincides with the original one. By the electric field induced modulations on the refractive index, and because of the low illuminating intensity and the low erase sensitivity in this oxidized sample[11], this redistribution of the trapped electrons relative to the former actually gives perturbations to the values of the original index-modulation strengths of the superlattice. The strengths of the perturbations depend on the incident intensity and they may be positive or negative, respectively, with different interference field. This fact provides the feedback element for the index-modulation mechanism which is expected to exist in such a 2D superlattice. The change of the values of the perturbed index-modulation strengths causes a change of the transmission-diffraction field, and this changed field will return to affect the perturbations of the index-modulation strengths. Diffracted intensities will thus change nonlinearly with the incident intensity. Figures 2 and 3 are two of the results

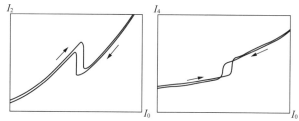

FIG. 2. With incident wave I_0 satisfying the Bragg condition of four-wave diffraction, and its intensity adjusted to form a cycle, the changings of the diffracted intensities I_2 and I_4 recorded. Intensity jumps occur simultaneously in the diffracted waves.

recorded in experiments that demonstrate bistable behaviors with discontinuous jumps of the diffracted intensities. The hysteresis loops appearing here in the cycles of the figures are somewhat complicated to identify, for the redistribution of the trapped electrons in the space charge pattern is irreversible with the changing of the incident intensity and erasure occurs in this incident-diffracted process[12]. The experimental results may be improved by means of the fixing methods to fix the first-constructed 2D grating into the photorefractive sample[13,14]. In Fig. 3 it is noticed that the lower branch of the hysteresis loop develops into instability. As we analyzed above, the transmissiondiffraction field and the perturbed index-modulation strengths are interacting with each other. The observed bistable region in Fig. 2 indicates that this kind of interaction reached a stable state, and the stable state could be more than one; if this interaction did not reach a stable state, instability was exhibited as shown in Fig. 3.

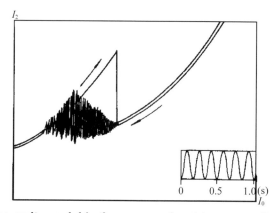

FIG. 3. One result recorded in the same experimental process as that of Fig. 2, showing not only the intensity jump but also the behavior of instability. Inset: The intensity pulses periodically with time when I_0 is fixed in the region of instability. The pulsation period is of the order of 0.1 s, corresponding to the response time of the photorefractive effect in the sample.

To check the results further, in an experiment we improved the incident intensity and made it fixed at about 2 W/cm^2. The sample's erase sensitivity increased a lot under such a high illuminating intensity. The 2D grating would be erased gradually and steadily with time when it diffracted the incident light. Then we recorded how the diffracted intensities were changed during this erasure process.

Figure 4 shows the experimental results observed in this erasure process. It is seen that the diffracted intensities are not always diminished continuously with time but with abrupt change or discontinuity, and behavior of instability was observed (see Fig. 5). This system can enter the regions of bistability or instability by means of either adjusting the incident intensity or changing the values of the index-modulation strengths.

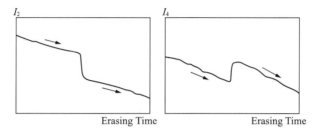

FIG. 4. During the process of erasing the 2D grating by the illumination of the incident beam, discontinuous jumps of the diffracted intensities recorded in I_2 and I_4. Notice that I_2 and I_4 jump simultaneously and in opposite directions.

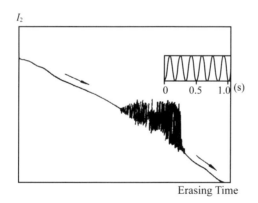

FIG. 5. During the erasure process, instability observed in diffracted wave I_2. Inset: Sine-wave periodic pulsations in the region of instability.

The behaviors of instability result from the fact that the transmission-diffraction field in the medium and the perturbed index-modulation strengths cannot evolve to a convergent and self-consistent state with time. This happens when the parameters are in certain ranges. To verify that the index-modulation mechanism may lead to instability, here we borrow the model that we have used in the preceding theoretical work[9]. In the model, a self-consistent solution of the iteration method is applied to achieve the stable state between the field and the perturbed index-modulation strengths; i.e., with numerical computation, the solution converges in a finite number of iterations. However, we find that when the incident intensity increases to certain values, the solution does not converge, but either pulses periodically [see Fig. 6(a)] or shows chaos [see Fig. 6(b)] with infinite numbers of iterations. Details about the instability in a 2D nonlinear superlattice can be investigated by solving a set of coupled partial differential equations that involve the field and the material's parameters, which we plan to discuss in a later paper.

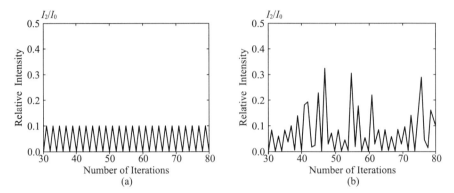

FIG. 6. In the theoretical model of four-wave diffraction in a 2D square superlattice containing Kerr-form dielectric nonlinearity (see Ref.[9]) at constant incident intensity I_0, the varying of the diffracted intensity I_2 obtained numerically in iterative processes. Divergent solutions appear instead of convergent self-consistent solutions between the perturbed index-modulation strengths and the transmission-diffraction field, such as (a) periodic pulsations at $I_0 = 2.23$ and (b) chaos at $I_0 = 2.66$. The setting of other parameters about the 2D superlattice follows exactly that in Ref.[9].

In conclusion, we have observed experimentally for the first time the bistability and the instability of the index-modulation mechanism in a 2D nonlinear superlattice that was constructed by recording a 2D refractive index grating into a photorefractive material, a Fe-doped $LiNbO_3$ crystal. The experimental results here are just qualitative, and we find that the bistable hysteresis loops and the discontinuous jumps of the diffracted intensities are not as large as those of the theoretical calculations. These discrepancies may be due to the deviations of the experimental conditions from those in the theoretical model in these facts: (i) The writing and the incident beams are of nonuniform Gaussian profile, not the plane waves in the theoretical consideration. (ii) The nonlinear dielectricity in photorefractive materials is not exactly the same as that in the Kerr-form media. (iii) The volume holographic writing of a grating into a photorefractive material is in fact a dynamical process, the 2D grating deviated from an exact periodic one[13]. (iv) The material's absorption, effectively affecting the light transmission process, is not considered. We recommend that further experimental investigations to verify the theory quantitatively may be carried out by using etching techniques, patterned epitaxy, etc., to fabricate 2D periodic structures in dielectric media with Kerr-form nonlinearity. This is also where the interest of such practical bistable devices lies.

References and Notes

[1] H. G. Winful, J. H. Marburger, and E. Garmire, Appl. Phys. Lett. **35**, 379 (1979).

[2] F. Delyon, Y. E. Lévy, and B. Souillard, Phys. Rev. Lett. **57**, 2010 (1986).

[3] W. Chen and D. L. Mills, Phys. Rev. B **35**, 524 (1987).

[4] L. Kahn, N. S. Almeida, and D. L. Mills, Phys. Rev. B **37**, 8072 (1988).

[5] J. Danckaert et al., Phys. Rev. B **44**, 8214 (1991).

[6] V. M. Agranovich, S. A. Kiselev, and D. L. Mills, Phys. Rev. B **44**, 10917 (1991).

[7] J. He and M. Cada, IEEE J. Quantum Electron. **27**, 1182 (1991).

[8] M. Cada et al., Appl. Phys. Lett. **60**, 404 (1992).

[9] B. Xu and N. B. Ming, Phys. Rev. Lett. **71**, 1003 (1993).

[10] J. Feng and N. B. Ming, Phys. Rev. A **40**, 7047 (1989).

[11] R. Orlowski, E. Kratzig, and H. Kurz, Opt. Commun. **20**, 171 (1977).

[12] T. K. Gaylord et al., J. Appl. Phys. **44**, 896 (1973).

[13] P. Günter, Phys. Rep. **93**, 199—299 (1982).

[14] If the grating is thermally fixed into the photorefractive sample, the electronic space charge pattern will be transformed into an ionic pattern that will not be affected by the light's illumination. If the incident power is strong and the sample's erase sensitivity is high enough, the perturbations of the index-modulation strengths of the electron originally brought by interference will change instantaneously with the changing of the field. Such reversible devices are practical for bistability. However, efforts should be made to achieve sufficient high diffraction efficiency after the thermal fixing.

[15] We are glad to thank Professor D. Feng for stimulating discussions. Valuable discussions with Professor Z. J. Yang and Dr. Y. Y. Zhu are also greatly acknowledged. This work is supported by a grant for the Key Research Project in Climbing Program from the National Science and Technology Commission of China.

Gap Shift and Bistability in Two-dimensional Nonlinear Optical Superlattices[*]

Zhen-Lin Wang, Yong-Yuan Zhu, and Zhen-Ju Yang
National Laboratory of Solid State Microstructures, Nanjing University, Nanjing 210093, China
Nai-Ben Ming
National Laboratory of Solid State Microstructures, Nanjing University, Nanjing 210093, China and China Center of Advanced Science and Technology (World Laboratory), P.O. Box 8730, Beijing 100080, China

We numerically investigate electromagnetic wave propagation in a two-dimensional optical superlattice with Kerr-type nonlinearity. We show that, with proper modulation depths of the refractive index, a stop gap in the spectrum is possible for a wave propagating at the Bragg angle of the structure. The location of the stop gap depends critically on the incident wave power. We demonstrate that this gap-shift effect can induce an intensity-dependent transmission. This property has important applications in optical bistable switching.

The dispersion relation provides the key to understanding electromagnetic (e.m.) wave propagation through periodic dielectric structure.[1] The solution to the dispersion relation may contain stop gaps within which the superlattice becomes perfectly reflecting. The gaps separate bands within which propagating wave solutions are allowed. In the search for photonic band-gap (PBG) materials,[2] two-dimensional (2D) periodic structures have received theoretical[3–5] and experimental[4–8] attention because superlattices of this type are relatively easy to fabricate. In this paper, we propose a 2D PBG structure, of which its refractive index is continuously modulated in two directions. We show that with the inclusion of nonlinearity of the medium the incident wave power leads to a shift in the location of the stop gap. This nonlinear mechanism can be used to construct a different class of bistable optical devices.

We consider a lossless 2D optical superlattice (OSL) with a simple Kerr-type nonlinearity shown in Fig. 1(a). We assume that the linear refractive index of a 2D nonlinear OSL is weakly, sinusoidally modulated in the $\hat{x}$ and $\hat{y}$ directions. Thus the refractive index of a 2D nonlinear OSL is given by

$$n = n_0 + n_x \cos(H_x x) + n_y \cos(H_y y) + n_a |E|^2/4\pi, \tag{1}$$

where n_0 is the average effective index, n_x and n_y are the index modulation depths and are taken to be rather weak, H_x and H_y are the reciprocal lattice vectors, n_a is the nonlinear

---

[*] Phys. Rev. B, 1996, 53(11):6984

coefficient of the medium, and E is the electric field of the optical wave. In what follows we will call the ratio $m = n_x/n_y$ the modulation ratio of a 2D nonlinear OSL. The 2D periodic structures described by Eq. (1) can be fabricated by using a holographic recording technique.[9] Another such structure would be a doubly periodic planar waveguide where Eq. (1) describes the distribution of an effective mode index.[10] The existence of solitary waves in the 2D nonlinear periodic medium with strong modulation depths has been demonstrated by John and Aközbek by using a variational method.[11]

As is known, the phenomenon of dynamical diffraction of an e.m. wave in a spatially periodic medium occurs when the wave vectors of incoming $\boldsymbol{k}$ and diffracted $\boldsymbol{k} + \boldsymbol{H}_\sigma$ waves fulfill the Bragg condition $|\boldsymbol{k} + \boldsymbol{H}_\sigma| \approx |\boldsymbol{k}|$, where $\boldsymbol{H}_\sigma$ is the reciprocal lattice vector. For a 2D OSL, it is possible for the incoming wave vector $\boldsymbol{k}$ to satisfy simultaneously two exact equalities for $\boldsymbol{H}_\sigma = -\boldsymbol{H}_x$ and $\boldsymbol{H}_\sigma = \boldsymbol{H}_y$. We define this situation to be the Bragg case of a 2D OSL,

$$2k_B \sin\theta_B = H_x, \quad 2k_B \cos\theta_B = H_y, \tag{2}$$

where θ_B is the Bragg angle and $k_B = n_0 \omega_B/c$ the Bragg wave number with ω_B being the Bragg frequency. As it will be shown, in a 2D OSL, whether or not a frequency stop gap for a wave propagating in the vicinity of the Bragg angle occurs depends on the modulation ratio m in contrast to the properties of two-wave Bragg diffraction in one-dimensional linear periodic media.[1]

In the four-wave approximation, the Bloch waves in the periodic structure can be decomposed into

$$E(\boldsymbol{r}) = \sum_\sigma E_\sigma \exp(\mathrm{i}\boldsymbol{K}_\sigma \cdot \boldsymbol{r}), \tag{3}$$

where E_σ is the amplitude of the partial mode with wave vector $\boldsymbol{K}_\sigma$ and σ denotes $o, h, -o$, and $-h$, which are the nodes of the reciprocal lattice [Fig. 1(b)]. A useful representation of the dispersion effect caused by multiwave interaction is to introduce a new set of axes (ξ_o, ξ_h), which are defined through $\xi_o = (K_o - k_B)/k_B$, $\xi_h = (K_h - k_B)/k_B$. Then straightforward manipulations similar to the ones used in the linear multiwave diffraction dynamics lead to the nonlinear matrix equation for E_σ,

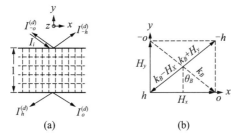

FIG. 1. (a) Schematic of a two-dimensional optical superlattice and (b) exact Bragg conditions in a two-dimensional optical superlattice.

$$\begin{vmatrix} 4(\delta-\xi_o)+\Delta M_0 & M_x+\Delta M_x & \Delta M_{x-y} & M_y+\Delta M_y^* \\ M_x+\Delta M_x^* & 4(\delta-\xi_h)+\Delta M_0 & M_y+\Delta M_y^* & \Delta M_{x+y}^* \\ \Delta M_{x-y}^* & M_y+\Delta M_y & 4(\delta+\xi_o)+\Delta M_0 & M_x+\Delta M_x^* \\ M_y+\Delta M_y & \Delta M_{x+y} & M_x+\Delta M_x & 4(\delta+\xi_h)+\Delta M_0 \end{vmatrix} \begin{vmatrix} E_o \\ E_h \\ E_{-o} \\ E_{-h} \end{vmatrix} = 0, \tag{4}$$

where $M_x = 2n_x/n_0$ and $M_y = 2n_y/n_0$ are the linear index modulation strengths and $\delta = (\omega - \omega_B)/\omega_B$ is a frequency dephasing parameter of the operating frequency ω from the Bragg frequency ω_B and is assumed to be small. The field-dependent index modulation strengths in Eq. (4) are expressed as

$$\Delta M_x = M_a(E_o E_h^* + E_{-h} E_{-o}^*),$$
$$\Delta M_y = M_a(E_{-o} E_h^* + E_{-h} E_o^*),$$
$$\Delta M_{x+y} = M_a E_{-h} E_h^*,$$
$$\Delta M_{x-y} = M_a E_o E_{-o}^*,$$
$$\Delta M_0 = M_a \sum_\sigma |E_\sigma|^2,$$

where $M_a = n_a/n_0$ and σ takes $o, h, -o,$ and $-h$. To obtain Eq. (4), the approximations $1 - K_o^2/k^2 \approx 2(\delta - \xi_o)$, $1 - K_h^2/k^2 \approx 2(\delta - \xi_h)$, $1 - K_{-o}^2/k^2 \approx 2(\delta + \xi_o)$, and $1 - K_{-h}^2/k^2 \approx 2(\delta + \xi_h)$ were used, which follow from the inequalities $n_x \ll n_0$, $n_y \ll n_0$, and $\delta \ll 1$. The relation between ξ_o and ξ_h is, in the geometry of Fig. 1, $\xi_o = \xi_h - \eta \sin 2\theta_B$, where η is the angular deviation from the Bragg angle. Though a 2D nonlinear OSL can be studied by using the formalism proposed above for any arbitrary angular deviation, we shall treat only the simple case of Bragg incidence.

In the limit of low power, the field-dependent terms are negligible. In this case, setting the determinant of coefficients to zero, we obtained the dispersion equation:

$$\xi_o^4 + B\xi_o^2 + C = 0, \tag{5}$$

where

$$B = \left(\frac{M_y}{2}\right)^2 \left[-\delta^2 + \frac{(1-m^2)}{2}\right],$$
$$C = \left(\frac{M_y}{2}\right)^4 \left[\frac{1}{16}(1-m^2)^2 + \delta^4 - \frac{\delta^2}{2}(m^2+1)\right].$$

It is readily obtained that Eq. (5) yields real values for ξ_o for all δ when $m > 1$. For a 2D OSL with $m < 1$, however, the regime where $|\delta| < (1-m)M_y/4$ corresponds to all ξ_o's having an imaginary part and thus to evanescent Bloch waves. Outside the regime, ξ_o has real-value solutions indicating propagating Bloch waves. So a frequency stop gap centered at the Bragg frequency appears if $m < 1$. Under the Bragg incidence, the width of the stop gap is given by $\Delta\omega = (1-m)M_y\omega_B/4$. For convenience, we adopt two variables $\Omega = 2(\omega - \omega_B)/\Delta\omega$ and $L = k_B l/\cos\theta_B$ in the following discussion, where l is the thickness of the 2D nonlinear OSL. As an example, we choose a 2D nonlinear OSL with structure parameters $M_a = 10^{-4}$, $L = 3.5 \times 10^4$, $M_x = 4.5 \times 10^{-5}$, and $M_y = 4.5 \times 10^{-4}$, thus with a modulation ratio $m=0.1$. The incoming wave intensity I_i is normalized to $I_{i,c}$ with $M_a I_{i,c} = 10^{-4}$.

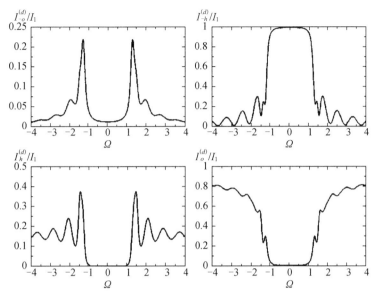

FIG. 2. Linear transmission coefficients of four diffracted waves of the 2D nonlinear OSL as a function of the frequency detuning parameter Ω.

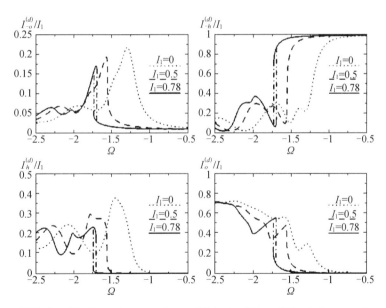

FIG. 3. Nonlinear transmission coefficients of four diffracted waves of the 2D nonlinear OSL as a function of the frequency detuning parameter Ω with different input intensities.

In Fig. 2, we plot the linear transmission coefficients as functions of the frequency detuning parameter Ω of the incident wave. Clearly, total reflection (i.e., a stop gap) occurs whenever $|\Omega|<1$.

Figure 3 shows the nonlinear transmission coefficients on the low-frequency side of the stop gap of our interest. It is seen that due to the nonlinearity of the medium two quantitative changes in the stop gap occur. First, the position of the stop gap shifts to the low-frequency side as the input power increases. This effect may be used to design an intensity-driven optical limiter as suggested by Scalora et $al.$ for 1D nonlinear PBG materials.[12] Second, with increasing input wave energy the width of the stop gap also expands. In this example, the intensity for bistability is found to be $M_a I_i = 5 \times 10^{-5}$, a value that is comparable to a nonlinear distributed feedback structure[13] where for operating frequency within the gap bistability is mediated by the excitation of gap solitons.[14—17] For an input intensity of $M_a I_i = 7.8 \times 10^{-5}$, the coefficients exhibit clear bivalued features at the low-frequency side near the stop gap.

The relations between the relative outputs and the input are shown in Fig. 4 for two beams of different frequencies. Both frequencies are tuned in the passband below the stop gap (linear case) so that the nonlinear effect can shift the gap towards the operating frequency as the intensity is increased. The critical detuning for which the bistability just starts to occur is -1.55. For beams with frequency detuning larger than 1.55 below the gap, it will initially be transmitted. As the beam intensity increases, its effect is to widen and shift the gap towards the operating frequency; it will do so to such extent that the operating frequency will find itself inside the gap. Diffracted beams $I_o^{(d)}$ and $I_h^{(d)}$ then shut off. With decreasing input, the relative outputs do not retrace their original paths and output-versus-input functions exhibit typical bistable hysteresis.

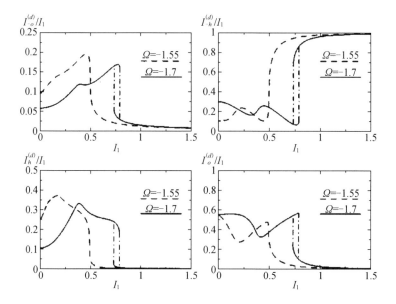

FIG. 4. Relative outputs as a function of input of the 2D nonlinear OSL for two different frequency detunings.

It is important to note that the bistable switching behaviors shown in Fig. 4 are different from that of the index-modulation bistability mechanism in a 2D nonlinear OSL previously proposed by us[18] in that the latter is, in effect, a switching between different transmitting states (TS's), rather than between a nontransmitting state (NTS) and a TS because the modulation ratio of the 2D nonlinear OSL discussed there is larger than unity, thus with no stop gap as specified above. The bistability revealed here, however, is based on the occurrence of the stop gap and the nonlinear effect on it. Therefore the optical response exhibits bistable switching between TS's and NTS's (at least for exiting beams $I_o^{(d)}$ and $I_h^{(d)}$ for a light beam with its frequency operating in the band below the gap. With optimizing structure parameters, we can also achieve a high contrast at the switching on and switching off for the beam $I_{-h}^{(d)}$.

The results we presented here for the case of $M_a > 0$ are also sustained for a 2D OSL with a negative nonlinear coefficient. In this case we select the operating frequency in the passband but above the gap because the nonlinear effect now is to shift the stop gap to the high-frequency side of the gap.

In summary, we have shown that for a 2D OSL with proper index modulation depths, a stop gap in the linear spectra is possible for light propagating in the vicinity of the Bragg angle. The inclusion of nonlinearity in the superlattice alters substantially the position and width of the stop gap. For a light beam with its frequency in the band, an appropriate change of the input power can switch the exiting beams between TS's and NTS's. These properties may have applications in designing actual 2D nonlinear OSL bistable devices.

References and Notes

[1] H. Kogelnik and C. V. Shank, J. Appl. Phys. **43**, 2327 (1972).
[2] E. Yablonovitch, Phys. Rev. Lett. **58**, 2059 (1987). For a recent review of photonic band structures, see special issue on photonic band structures, edited by C. M. Bowden, J. P. Dowling, and H. O. Everitt [J. Opt. Soc. Am. B **10** (2)(1993)].
[3] M. Phihal, A. Shambrook, A. A. Maradudin, and P. Sheng, Opt. Commun. **80**, 199 (1991).
[4] S. L. McCall, P. M. Platzman, R. Dalichaouch, D. Smith, and S. Schultz, Phys. Rev. Lett. **67**, 2017 (1991).
[5] R. D. Meade, K. D. Brommer, A. M. Rappe, and J. D. Joannopoulos, Appl. Phys. Lett. **61**, 495 (1992).
[6] W. M. Robertson, G. Arjavalingam, R. D. Meade, K. D. Brommer, A. M. Rappe, and J. D. Joannopoulos, Phys. Rev. Lett. **68**, 2023 (1992).
[7] P. L. Gourley, J. R. Wendt, G. A. Vawter, T. M. Brennau, and B. E. Hammous, Appl. Phys. Lett. **64**, 687 (1994).
[8] S. Y. Lin and G. Arjavalingam, Opt. Lett. **18**, 1666 (1993).
[9] B. Xu and N. B. Ming, Phys. Rev. Lett. **71**, 3959 (1993).
[10] R. Zengerle, J. Mod. Opt. **34**, 1589 (1987).
[11] S. John and N. Aközbek, Phys. Rev. Lett. **71**, 1168 (1993).

[12] M. Scalora, J. P. Dowling, C. M. Bowden, and M. J. Bloemer, Phys. Rev. Lett. **73**, 1368 (1994).

[13] H. G. Winful, J. H. Maburger, and E. Garmire, Appl. Phys. Lett. **35**, 379 (1979).

[14] W. Chen and D. L. Mills, Phys. Rev. Lett. **58**, 160 (1987).

[15] D. L. Mills and S. E. Trullinger, Phys. Rev. B **36**, 947 (1987).

[16] C. M. de Sterke and J. E. Sipe, Phys. Rev. A **38**, 5149 (1989).

[17] L. M. Kahn, K. Huang, and D. L. Mills, Phys. Rev. B **39**, 12 449 (1989).

[18] B. Xu and N. B. Ming, Phys. Rev. Lett. **71**, 1003 (1993).

[19] This work was supported by a grant for the Key Research Project in Climbing Program from the National Science and Technology Commission of China.

Optical Bistability in Two-dimensional Nonlinear Optical Superlattice with Two Incident Waves[*]

Xiang-fei Chen, Ya-lin Lu, and Zhen-lin Wang

*National Laboratory of Solid State Microstructures, Nanjing University, Nanjing 210093,
and Center for Advanced Studies in Science and Technology of Microstructures,
Nanjing 210093, People's Republic of China*

Nai-ben Ming

*National Laboratory of Solid State Microstructures, Nanjing University, Nanjing 210093,
and Center for Condensed Matter Physics and Radiation Physics, China Center of Advanced Science
and Technology (World Laboratory), P.O. Box 8730, Beijing 100080,
and Center for Advanced Studies in Science and Technology of Microstructures,
Nanjing 210093, People's Republic of China*

With two coherent waves incident to a two-dimensional (2D) nonlinear optical superlattice containing Kerr form dielectric nonlinearity, the optical bistability related to transition between a forbidden transmission state and an allowed transmission state can occur through the index-modulation mechanism. This kind of bistability is difficult to obtain in the case of only one incident wave. The power cost for the bistability might be very low.

A nonlinear optical system with positive feedback will exhibit bistability. Optical bistability in one-dimensional (1D) nonlinear superlattices, which was first proposed by the pioneering work of Winful et al.[1] more than 10 years ago, is one example.[2-10] The bistability is related to the transition between a forbidden transmission state (FTS) and an allowed transmission state (ATS). In the presence of gap soliton, an intense wave, whose frequency lies within the forbidden gap, can bring itself into the allowed band of transmission spectrum through the Kerr form nonlinearity.[3,4] Recently, wave propagation in a two-dimensional (2D) optical superlattice, i.e., with its refractive index periodically modulated in two dimensions, has been studied by Feng and Ming.[11] They have shown that, near the Bragg condition, the light transmission strongly depends upon the index-modulation (IM) strengths. In the presence of Kerr form nonlinearity, the interference in the transmission field will perturb the IM strengths. Therefore, a positive feedback is formed.[12,14] This new type of bistable mechanism, named IM mechanism, has been demonstrated in 2D nonlinear superlattices, both theoretically and experimentally.[12-14] The available results are only related to the transition of transmission in the allowed band.[14] However, to our knowledge, the optical bistability related to transition between a

---

[*] Appl. Phys. Lett., 1995, 67(24):3538

FTS and an ATS has not been discovered in 2D nonlinear superlattices up to now. In addition, in the light of our careful numerical investigations, it is also difficult to obtain this kind of optical bistability through the IM mechanism in the case of only one incident wave in 2D nonlinear superlattices. In this letter, we propose a new scheme in which two coherent Bragg incident waves have been used. We have found that the optical bistability related to transition between a FTS and an ATS can be obtained through the IM mechanism in 2D nonlinear superlattices by this scheme. This kind of 2D nonlinear superlattice bistable device might have potential application in integrated optics.

The 2D nonlinear superlattice which is considered here is an isotropic lossless Kerr form nonlinear medium. Its linear periodically modulated refractive index is expressed as follows:

$$n = n_0 + n_z \cos(H_z z) + n_x \cos(H_x x), \tag{1}$$

where n_z and n_x are IM strengths along z and x direction, respectively, and satisfy n_z, n_x $\ll n_0$; H_z and H_x denote the periodicity in reciprocal space. Two coherent incident waves E_{in1} and E_{in2}, with the same incident Bragg angle, symmetrically fall down the 2D superlattices [see Fig. 1(a)]. The wave vectors satisfy the exact Bragg condition where four reciprocal points are located on the Ewald sphere. Thus, four Bloch diffracted waves will be excited in the medium [see Fig. 1(b)]. In the linear theory, the IM diffracted relations have been presented in a detailed way by Feng and Ming.[11] In terms of their results, when $n_z > n_x$, the wave vectors of Bloch waves have an imaginary part so that the Bloch waves are evanescent. This is very similar to the situation in which a wave located in the forbidden gap propagates down a 1D superlattice, and then are so-called forbidden light transmissions in this letter. On the contrary, when $n_z < n_x$, the wave vectors of Bloch waves are real, and then the Bloch waves will propagate through the medium unimpeded. These are so-called allowed light transmissions. When the exact Bragg condition is satisfied, one can write the field in the medium as a sum of two forward and two backward diffracted waves:

$$E(z) = E_0(z)\exp(i\mathbf{K}_0 \cdot \mathbf{r}) + E_h(z)\exp(i\mathbf{K}_h \cdot \mathbf{r}) + E_{-0}(z)\exp(-i\mathbf{K}_0 \cdot \mathbf{r}) + \\ E_{-h}(z)\exp(-i\mathbf{K}_h \cdot \mathbf{r}), \tag{2}$$

where $K_0 = K_h = k_m$. k_m is the wave number in the average refractive index of the medium. The relation between transmittance and IM strengths is plotted in Fig. 2. The transmittance is defined as $(I_0 + I_h)/I_{in}$, where I_{in} is the incident intensity. The similar results can be found in Ref. [12]. It is seen from Fig. 2 that two distinct regions obviously exist. One is the region in which $n_z > n_x$. The amplitudes of the Bloch waves decay exponentially with propagation distance into the medium, and then the transmittance is very weak. These are related to the FTS. The other is the region in which $n_z < n_x$. The amplitudes of the Bloch waves will oscillate with the IM strengths. These are related to the ATS. For convenience, we define the parameter m as n_x/n_z. Thus, the case of $m < 1$ is related to the FTS, and that of $m > 1$ is related to the ATS.

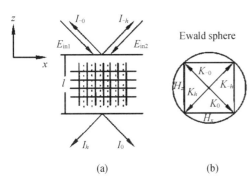
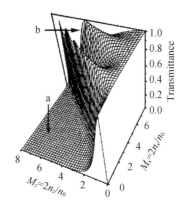

FIG. 1. Four-wave diffraction in the two-dimensional periodic superlattice. (a) Schematic diagram of four-wave diffraction in real space. The dotted lines denote the incident-dependent IM along x-direction formed by incident fields E_{in1} and E_{in2}. (b) Bragg condition with four reciprocal points located on the Ewald sphere.

FIG. 2. The relation between transmittance and index-modulation strengths under the exact Bragg condition. There exist two distinct regions. Region a is related to the forbidden transmission state. Region b is related to the allowed transmission state. The structure parameters $k_m l/\cos\theta_B = 4\times 10^4$.

In the presence of two incident waves, the interference formed by two incident waves in the medium is characterized by a periodic spatial variation of the intensity. When Kerr form nonlinearity is considered, as shown in Fig. 1(a), the nonlinear response in the medium leads to the formation of incident-dependent periodic IM along the x direction with is periodicity characterized by H_x in reciprocal space and with its strength proportional to the incident intensities. In the figure, the dotted lines denote the incident-dependent IM along the x direction. This kind of process is similar to that of volume grating formation. When the incident-dependent periodic IM matches the preconstructed IM [namely the dotted and the solid lines in Fig. 1(a) are overlapped], the total IM strength along the x direction, and thus the effective m, is enhanced in proportion to the incident intensities. Therefore, in the case of two incident waves, for a 2D superlattice with its m less than 1, the sufficient increase of the two incident intensities will lead to the value of effective m greater than 1. That is to say, the two intense waves can bring their transmission from FTS into the ATS. We expect that the optical bistability can occur through this transition. Furthermore, for a 2D superlattice whose value of m is near but smaller than 1, a bit of enhancement of IM strength along the x direction can lead to this kind of transition. So, a very low input power will be required to achieve such a transition. This advantage might be beneficial for constructing low power cost 2D optical bistable devices.

The linear formalism of refractive IM is shown in Eq. (1), and the Kerr form nonlinearity can be described by the nonlinear polarization term:

$$P_{NL}(r) = n_0 n_a |E|^2 E/4\pi, \qquad (3)$$

where n_a is the nonlinear index. Using Eqs. (2) and (3) in Maxwell's wave equation, we obtain in the slowly varying envelope approximation:

$$\frac{dE_0}{dz} = -i\frac{k_m}{4\cos\theta_B}[2\delta n_0 E_0 + (M_x + \delta n_x)E_h + \delta n_{x-z}E_{-0} + (M_z + \delta n_z^*)E_{-h}], \quad (4a)$$

$$\frac{dE_h}{dz} = -i\frac{k_m}{4\cos\theta_B}[(M_x + \delta n_x^*)E_0 + 2\delta n_0 E_h + (M_z + \delta n_z^*)E_{-0} + \delta n_{x+y}^*E_{-h}], \quad (4b)$$

$$\frac{dE_{-0}}{dz} = -i\frac{k_m}{4\cos\theta_B}[\delta n_{x-z}^* E_0 + (M_z + \delta n_z)E_h + 2\delta n_0 E_{-0} + (M_x + \delta n_x^*)E_{-h}], \quad (4c)$$

$$\frac{dE_{-h}}{dz} = i\frac{k_m}{4\cos\theta_B}[(M_z + \delta n_z)E_0 + \delta n_{x+z}E_h + (M_x + \delta n_x)E_{-0} + 2\delta n_0 E_{-h}], \quad (4d)$$

where $M_z = 2n_z/n_0$, $M_x = 2n_x/n_0$, $\delta n_0 = 2\alpha(|E_0|^2 + |E_h|^2 + |E_{-0}|^2 + |E_{-h}|^2)$, $\delta n_z = 2\alpha(E_h^* E_{-0} + E_0^* E_{-h})$, and $\delta n_x = 2\alpha(E_0 E_h^* + E_{-0}^* E_{-h})$, $\delta n_{x+z} = 2\alpha(E_h^* E_{-h})$, $\delta n_{x-z} = 2\alpha(E_0 E_{-0}^*)$, and θ_B is the Bragg angle. Here, α is defined as n_a/n_0.

The four diffracted waves at their exit boundaries are determined by the field equations, the material equations, and the boundary conditions $E_0(0) = E_{in1}$, $E_h(0) = E_{in2}$, $E_{-0}(-l) = 0$, $E_{-h}(-l) = 0$. For convenience, we define the parameter γ as $|E_{in1}|^2/|E_{in2}|^2$, the parameter φ as the relative phase between E_{in1} and E_{in2}, and I_{in} as $|E_{in1}|^2 + |E_{in2}|^2$. I_{in} is total incident intensity. Because of the complexity of these equations, Eqs. (4) can only be solved by numerical methods.

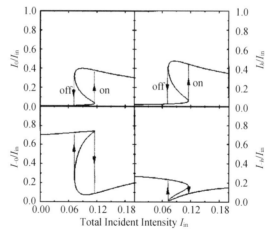

FIG. 3. Bistable behavior in the incident-diffracted relations obtained by numerical solution with two waves incident to 2D nonlinear superlattices. The structure parameters $k_m l/\cos\theta_B = 4\times 10^4$, $M_z = 4\times 10^{-4}$, $M_x = 3.7\times 10^{-4}$, $\gamma = 0.2$, and $\varphi = 0$. I_{in} is measured I_{unit}, with $\alpha|E_{unit}|^2 = 10^{-4}$.

We use the standard finite-difference approximation to integrate the Eqs. (4a)–(4d). Here, we only consider the case in which the incident-dependent IM along the x direction matches the original one, i.e., the case in which the relative phase φ between two incident waves is equal to zero. For a 2D superlattice with its m less than 1, the relation between the output intensities and the total incident intensity is plotted in Fig. 3. There are four output intensities, i.e., two transmitted intensities I_0 an I_h, and two reflected intensities

I_{-0} and I_{-h}. As in all bistable devices, the negative slope regions are unstable. Therefore, as the input is slowly varied, the output shows discontinuous jumps at certain critical input intensities, and then a hysteresis loop is traced out. The hysteresis loop width depends on M_z, M_x and $k_m l/\cos\theta_B$. The figure shows that, at lower input laser power, the transmitted intensities I_0 and I_h are very weak and related to the FTS. Once above the switch-on input intensity, the two transmitted waves will propagate through the medium unimpeded. Then I_0 and I_h become strong and related to the ATS. So, optical bistability related to transition between a FTS and an ATS is obtained with two incident waves. In the example with m near 1 as shown in Fig. 3, the switch-on intensity is the order of 0.1. The unit of I_{in} corresponds to $\alpha|E_{unit}|^2=10^{-4}$. The relative variation of index of the medium induced by incident intensities is the order of 10^{-5}. Thus, transmissivity becomes bivalued at rather low laser powers.

In conclusion, we have numerically demonstrated that optical bistability related to transition between a FTS and an ATS can occur when two incident waves are considered. The optical bistability originates from the IM mechanism, and the very low power cost is required in an appropriate structure parameter of a 2D nonlinear superlattice. This might be useful to construct a practical low power cost 2D optical bistable device, although the case with two incident waves is less convenient than that with one incident wave. In further work, we will propose a novel structure, which can be used to generate similar optical bistability with the usage of only one incident wave.

References

[1] H. G. Winful, J. H. Marburger, and E. Garmire, Appl. Phys. Lett. **35**, 379 (1979).
[2] F. Delyon, Y. E. Levy, and B. Souillard, Phys. Rev. Lett. **57**, 2010 (1986).
[3] W. Chen and D. L. Mills, Phys. Rev. Lett. **58**, 160 (1987).
[4] W. Chen and D. L. Mills, Phys. Rev. B **35**, 524 (1987).
[5] L. Kahn, N. S. Almeida, and D. L. Mills, Phys. Rev. B **37**, 8072 (1988).
[6] J. Danckaert et al., Phys. Rev. B **44**, 8214 (1991).
[7] V. M. Agranovich, S. A. Kiselev, and D. L. Mills, Phys. Rev. B **44**, 10917 (1991).
[8] J. He and M. Cada, IEEE J. Quantum Electron. **27**, 1182 (1991).
[9] M. Cada et al. (unpublished).
[10] C. J. Herbert and M. S. Malcuit, Opt. Lett. **18**, 1783 (1993).
[11] J. Feng and N. B. Ming, Phys. Rev. A **40**, 7047 (1989).
[12] B. Xu and N. B. Ming, Phys. Rev. Lett. **71**, 1003 (1993).
[13] B. Xu and N. B. Ming, Phys. Rev. Lett. **71**, 3959 (1993).
[14] B. Xu and N. B. Ming, Phys. Rev. A **50**, 5197 (1994).
[15] We thank Dr. Yan-Qing Lu and Dr. Gui-Peng Luo for useful discussions. This work is supported by a grant for the Key Research Project in Climbing Program from the State Science and Technology Commission of China.

Three-dimensional Self-assembly of Metal Nanoparticles: Possible Photonic Crystal with a Complete Gap Below the Plasma Frequency[*]

Zhenlin Wang,[1,2] C. T. Chan,[1] Weiyi Zhang,[2] Naiben Ming,[2] and Ping Sheng[1]

[1] Department of Physics, Hong Kong University of Science and Technology, Clear Water Bay, Kowloon, Hong Kong, China

[2] National Laboratory of Solid State Microstructures, Nanjing University, Nanjing 210093, China

We study theoretically the optical properties of a three-dimensional self-assembly of spherical-metal nanoparticles. Our band-structure calculations show that photonic band gaps in near infrared and optical frequencies can be realized in a fcc lattice of metal particles with radii of approximately 160 nm. When absorption is taken into account, we found that the gap is preserved in good metals like silver. The metal nanoparticles can be replaced by dielectric spheres coated with a layer of good metal.

Photonic crystals are three-dimensional periodic dielectric structures, with lattice constants of the order of the desired electromagnetic (EM) wavelength of operation. The free-EM wave dispersion in such crystals is greatly modified due to the presence of periodicity and refractive-index contrast. Under favorable circumstance, a photonic band gap (PBG) in the dispersion relation can open up, in which light cannot propagate in any direction.[1] Such a structure is believed to have a deep impact on a wide range of photonic applications.[2-5]

Recently, considerable progress has been made in constructing three-dimensional photonic crystals with submicrometer periodicity, using the sophisticated "top-down" fabrication techniques,[6] the self-assembly of monodispersed microspheres and related infiltration method,[7] and the holographic lithography.[8] Compared to the micromachining methods for the production of PBG crystal for visible wavelengths, the self-organization process is a very cheap and relatively easy method too.

On the other hand, the interaction of light with nanoscale metal particles has been the subject of extensive research.[9] Researchers have used metal nanoparticles as building blocks for novel materials.[10-14] We will show in this paper that metallic nanoparticles can be used as building blocks for photonic crystals operating at near IR and optical frequencies. It is not intuitively obvious that metallic nanoparticles can serve the purpose. First, metals are dispersive and can be rather absorbing at IR and optical frequencies.

---

[*] Phys. Rev. B, 2001, 64(11):113108

Second, it is very difficult to fabricate metal nanoparticles that are monodispersed spheres with radii of the order of 150 nm, the scale length needed to realize photonic gaps at optical frequencies. We are going to show that both of these problems can be solved. First, our calculations will show that the absorption problem can be alleviated by a careful choice of the metal component. In order to build a convincing case, we will use experimentally measured dielectric constants for the frequency range of interest (in contract to Drude or Drude-like models that are used frequently in the literature)[15] in all of our calculations. We will also suggest a simple and practical way (using metal coated silica spheres) to bypass the need of using metal nanoparticles of large (150 nm) radius.

We will draw our conclusions from results that are based on band-structure calculations employing periodic boundary conditions and multiple-scattering technique,[16] as well as an "on-shell" method[17] that gives directly transmission/reflection/absorption coefficients for a slab of finite thickness. The band-structure code sets the imaginary part of the dielectric constants to zero, which allows us to determine whether a photonic band gap exists or not in the absence of absorption. The slab calculations consider explicitly both the real and imaginary parts of the dielectric constants, modeling the metal faithfully according to published complex dielectric constants, and will help us to see what absorption will do to the band gap. We will consider three common metals in our calculations: silver, copper, and nickel. Our band-structure calculations show that fairly large photonic band gaps are obtained in optical frequencies for all the metals we have considered. The band gaps are the consequence of a large and negative real part of the dielectric constant, a condition that is easily satisfied by most metals up to IR and even optical frequencies. However, when absorption is taken into account, we found that the gap survives in good metals like silver, but are seriously compromised in nickel. We emphasize again that the complex dielectric constants are taken from experimental measurements[18] for all the metals in all of our calculations. We shall focus on the face-centered-cubic(fcc) arrangement of spheres, which is the structure usually obtained in self-assembly processes of spheres. We will see later (in Fig. 2) that the size of the photonic gaps increases with the filling ratio of metal nanospheres, but we are going to assume a 66% filling ratio for almost all of the calculations although the filling ratio can reach 74% (close-packing case) for a fcc structure. This is because in experimental implementations, we expect that the metal spheres would be coated with protective inert coatings to avoid oxidation. A lower-filling ratio of the metal spheres also cuts down absorption.

From tabulated values in Ref. [19], the plasma frequencies for silver, copper, and nickel are found to be 9.0 eV, 7.4 eV, and 4.9 eV, respectively, so that the real part of the dielectric constant of metals has the property of being negative and large in magnitude for wavelengths in near IR and optical frequencies that lie well below the plasma energies. This large dielectric contrast should be favorable for the opening of a PBG in these frequencies. This is illustrated in Fig. 1, where we display the band structures of fcc arrays of sphericalmetal nanoparticles in air with particle radius $r = 160$ nm. The filling ratio of the metal component is $f = 0.66$ for all metals. We found that all the metallo-dielectric (MD) nanostructures possess a large PBG between the fifth and sixth bands. As the PBG opens as a result of the real part of dielectric constant ε_r being large and negative, we expect similar effect in self-organizations of other metal nanoparticles such as gold and aluminum. We found that the gap edges show some variations for silver, copper, and nickel as a consequence of their different magnitude and dispersion relations of the dielectric constant in these frequencies. A small region of distorted dispersion in the lower part of the figure is observed for copper and nickel due to the mismatch of dielectric constants in the optical and IR frequencies tabulated from separate experiments.[18]

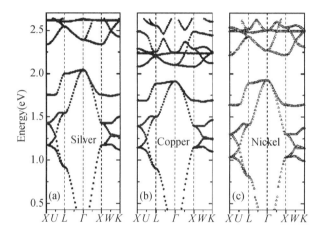

FIG. 1. Calculated band structure for fcc photonic crystals consisting of spherical-metal nanoparticles with radius $r = 160$ nm for (a) silver, (b) copper, and (c) nickel. All structures have the same filling ratio $f = 0.66$. In the band-structure calculations, the imaginary part of the dielectric constant is set to zero.

The frequency and size of the photonic gap depend on both the filling ratio and the radius of the spheres. We found that silver spheres with $r = 160$ nm has the largest gap in optical frequencies for a filling ratio of $f = 0.66$. We now consider the dependence of PBG position and gap-width/midgap frequency as functions of filling ratio of the metal spheres with a fixed radius $r = 160$ nm. Results are shown in Figs. 2(a) and 2(b). A PBG emerges when the volume fraction of the metal component exceeds a certain critical value, and increases as a function of the filling ratio. The critical volume fraction is seen to be nearly the same for all metals. The gap width reaches its maximum for the close-packing case ($f = 0.74$). Although we get bigger gaps for higher-filling fractions, it is most likely that a lower-filling ratio (e.g., 66%) will be used experimentally for aforementioned reasons.

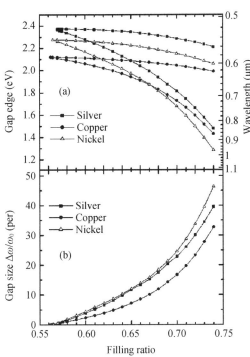

FIG. 2. Dependence of the gap position and gap-width/midgap frequency as functions of filling ratio for fcc photonic crystals for silver, copper, and nickel nanoparticles with radius $r=160$ nm.

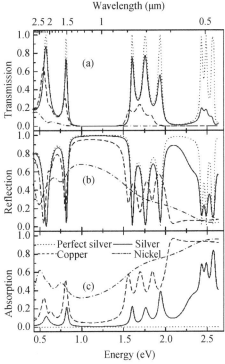

FIG. 3. Calculated transmission (a), reflection (b), and absorption (c) by four layers of (111) orientated fcc photonic-crystal slabs for silver, copper, and nickel nanoparticles with radius $r=160$ nm. Experimental complex dielectric constants are used. For comparison, the spectra of a photonic-crystal slab composed of "nonabsorbing" silver spheres [Imag(ε)=0] are also shown.

In band-structure calculations, only the real part of the dielectric constant is used since the size of the photonic band gap can be precisely defined only in the absence of absorption. It is important to know how the features and the properties of the PBG would be affected by material absorption. In the microwave and far-infrared region, metal behaves almost like a perfect conductor, and photonic crystals incorporated with metal scatterers have been studied at these low frequencies.[20] From near IR down to ultraviolet, absorption must be considered carefully. In the slab calculations we are going to describe, which give transmission/reflection/absorption spectra, the material absorption is included naturally by turning on the imaginary part of the complex (experimentally measured) dielectric constant ε.

We have calculated the transmission, reflection, and absorption spectra through a slab of four layers of a (111) oriented MD photonic crystal, which displays a PBG in the visible region from the band-structure calculation as shown in Fig. 1. Results for silver, copper, and nickel are compared and the transmission, reflection, and absorption of light by the thin slabs are plotted in Figs. 3(a),3(b), and 3(c), respectively. For comparison with the band structure, the spectra for the nonabsorbing silver (imaginary part of ε set to zero) particles are also shown. Band structure shows there is a directional gap from 0.92 eV to 1.54 eV and an absolute gap from 2.04 eV to 2.40 eV along the ΓL direction. This is almost perfect correspondence between the band-structure result and the calculated transmission spectrum for "nonabsorbing" silver-photonic crystal, which shows two pronounced dips (almost 100% reflectance) centered at 1.21 eV and 2.19 eV, respectively. When the effect of absorption is included by turning on the imaginary part of ε, noticeable changes in the transmission/reflection spectra are observed, especially for nickel in the high-energy region. Generally speaking, absorption is small for longer wavelengths and increases with frequency. For silver particles, the reflection within the directional gap remains almost the same as that of the "nonabsorbing" case; while it shows a notable decrease near the blue edge of the higher stop gap. Nevertheless, a maximum reflection of 90% can still be achieved in the stop gap. An increase in the slab thickness can lead to a small increase in the reflection within the stop gap, but the reflection coefficient saturates quickly for further increase in the number of layers.

For nickel, the pass bands at higher frequencies predicted by the band structure calculations for "nonabsorbing" nickel spheres become a region of strong absorption when the imaginary part of the dielectric constant is taken into account. The absorption is strong enough so that the edges of the photonic band gap are smeared out. The higher-frequency stop gap is totally destroyed, and even the lower-frequency directional gap is hardly discernible as the band edges are smoothed by the absorption. Since photonic band-gap originates from the collective scattering effect of many spheres, the gap cannot be well established if light cannot reach the deeper layers due to strong absorption. Thus it would be difficult to observe any photonic-gap effect in nanostructured crystals made of nickel nanoparticles. For MD crystals consisting of copper particles, the directional gap is still fairly well-defined both in the reflection and transmission spectra; while the reflection within the stop gap reduces to a small peak adjacent to the peaks resulting from the interlayer interference. Our calculations suggest that both the directional gap in the infrared and the PBG in the visible region are observable for MD photonic-crystal built with good-metal nanoparticles.

We note that a PBG around the plasma frequency has been found recently for a fcc lattice of spherical scatterers with a real Drude-like dielectric function by Moroz.[21] As the dielectric constant changes sign across the plasma frequency in a real-Drude model, it can be arbitrarily small near the plasma frequency, producing a large dielectric contrast relative to the embedding dielectric medium, which results in the PBG. Compared to Moroz's PBG, our predicted PBG opens in the frequencies where the real part of the dielectric constant ε_r is large and negative.

Current technology produces metal nanoparticles that are typically smaller than

150 nm in radius.[9,14] These are too small for the construction of a MD PBG in the visible region. However, we can bypass this hurdle by using metal-coated spheres instead of solid-metal nanoparticles. As the penetration depth of a good metal is of the order of several tens of nanometer, coating a thin metal layer on monodispersed dielectric particles should work as solid-metal nanoparticles do. This is illustrated in Fig. 4, where we show the transmission, reflection, and absorption spectra of a fcc array of silver-coated dielectric nanoparticles. The coated particle consists of a silver shell with a thickness of 40 nm and a silica core with a radius of 120 nm. The dielectric constant of silica is chosen to be $\varepsilon =$ 2.1 as dispersion and absorption of silica at optical frequencies can be neglected. We found that the spectra are almost the same as solid-silver nanoparticle. Several methods have been developed to synthesize these composite nanoparticles that consist of a dielectric-core coated with a metallic shell.[22—24] In particular, the thickness of the metal shell can be controlled fairly, precisely.[22]

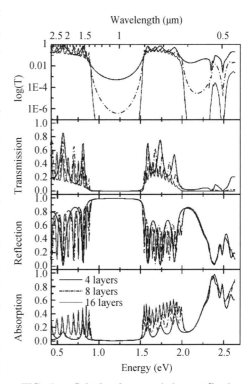

FIG. 4. Calculated transmission, reflection, and absorption by four, eight, and 16 layers of a (111) orientated slab of a fcc lattice of silver-coated silica nanoparticle. The silver shell has a thickness of 40 nm and the silica core has a radius of 120 nm.

In Fig. 4, we also compare the spectra for photonic crystal slabs with four, eight, and 16 layers of coated spheres. We see from this figure that the photonic gaps are already very well established for as little as four layers. The transmission in the pass bands diminishes as the number of layer increases, as it should be, but the transmission/reflection/absorption within the photonic gap quickly saturates with the slab thickness. These metal nanoparticle derived systems are such potent photonic band-gap systems that a very thin slab (four to eight layers) should be good for most purposes.

In short, we showed that spherical-silver nanoparticles of radii of about 160 nm are plausible building blocks for self-assembled photonic crystals operating at optical frequencies. The difficulty of fabricating such particles can be bypassed by using silver-coated silica spheres, with the added advantage that the frequency of the PBG can be tuned simply by changing the size of the silica core.

References and Notes

[1] E. Yablonovitch, Phys. Rev. Lett. **58**, 2059 (1987).

[2] J.D. Joannopoulos, P.R. Villeneuve, and S. Fan, Nature (London) **386**, 143 (1997).

[3] S. John and J. Wang, Phys. Rev. B **43**, 12 772 (1991).

[4] S.-Y. Lin *et al.*, Science **282**, 274 (1998).

[5] T.F. Krauss and R.M. De La Rue, Prog. Quantum Electron. **23**, 51 (1999).

[6] S. Noda and A. Sasaki, Jpn. J. Appl. Phys. **36**, 1907 (1997); S.-Y. Lin *et al.*, Nature (London) **394**, 251 (1998).

[7] B.T. Holland, C.F. Blanford, and A. Stein, Science **281**, 538 (1998); J.E.G. Wijnhoven and W.L. Vos, *ibid.* **281**, 802 (1998); A.A. Zakhidov *et al.*, *ibid.* **282**, 897 (1998); S.H. Park and Y. Xia, Adv. Mater. **10**, 1045 (1998); P.V. Braun and P. Wiltzius, Nature (London) **402**, 603 (1999); J.F. Bertone *et al.*, Phys. Rev. Lett. **83**, 300 (1999); A. Blanco *et al.*, Nature (London) **405**, 437 (2000).

[8] M. Campbell *et al.*, Nature (London) **404**, 53 (2000).

[9] U. Kreibig and M. Vollmer, *Optical Properties of Metal Clusters* (Springer, Berlin, 1995).

[10] A. Taleb, V. Russier, A. Courty, and M.P. Pileni, Phys. Rev. B **59**, 13 350 (1999).

[11] Z.L. Wang *et al.*, Adv. Mater. **10**, 808 (1998).

[12] M. Brust *et al.*, Langmuir **14**, 5425 (1998).

[13] M. Aindow *et al.*, Philos. Mag. Lett. **79**, 569 (1999).

[14] A.K. Boal *et al.*, Nature (London) **404**, 746 (2000).

[15] A.R. McGurn and A.A. Maradudin, Phys. Rev. B **48**, 17 576 (1993); V. Yannopapas, A. Modinos, and N. Stefanou, *ibid.* **60**, 5359 (1999); I. El-Kady *et al.*, *ibid.* **62**, 15 299 (2000).

[16] J. Korringa, Physica (Utrecht) **13**, 392 (1947); W. Kohn and N. Rostoker, Phys. Rev. **94**, 1111 (1954).

[17] J.B. Pendry, J. Mod. Opt. **41**, 209 (1993); N. Stefanou, V. Karathanos, and A. Modinos, J. Phys.: Condens. Matter **4**, 7389 (1992); K. Ohtaka and Y. Tanabe, J. Phys. Soc. Jpn. **65**, 2276 (1996).

[18] Optical freqencies data from ultraviolet to near IR (0.64 eV) are taken from P.B. Johnson and R.W. Christy, Phys. Rev. B **6**, 4370 (1972); for Ag and Cu; **9**, 5056 (1974); for Ni; IR data below 0.64 eV from Ref. 19.

[19] M.A. Ordal *et al.*, Appl. Opt. **22**, 1099 (1983).

[20] D.R. Smith *et al.*, Appl. Phys. Lett. **65**, 645(1994); E.R. Brown and O.B. McMahon, *ibid.* **67**, 2138 (1995); E. Ozbay *et al.*, *ibid.* **69**, 3797 (1996); K.A. McIntosh *et al.*, *ibid.* **70**, 2937 (1997); D.F. Sievenpiper, M.E. Sickmiller, and E. Yablonovitch, Phys. Rev. Lett. **76**, 2480 (1996); S. Fan, P.R. Villeneuve, and J.D. Joannopoulos, Phys. Rev. B **54**, 11 245 (1996); J.B. Pendry, J. Phys.: Condens. Matter **8**, 1085 (1996); W.Y. Zhang *et al.*, Phys. Rev. Lett. **84**, 2853 (2000).

[21] A. Moroz, Phys. Rev. Lett. **83**, 5274 (1999); Europhys. Lett. **50**, 466 (2000).

[22] S.J. Oldenberg *et al.*, Chem. Phys. Lett. **288**, 243 (1998).

[23] C.-W. Chen, T. Serizawa, and M. Akashi, Langmuir **15**, 7998 (1999).

[24] W. Wen *et al.*, Phys. Rev. Lett. **82**, 4248 (1999).

[25] This work is supported in part by RGC through HKUST6145/99P and the key research project in "Climbing Program" by the National Science and Technology Commission of China. Z. W. acknowledges partial financial support from the Natural Science Foundation of China (NSFC). W. Z. also acknowledges partial support from NSFC under "Excellent Youth Foundation."

Parity-time Electromagnetic Diodes in a Two-dimensional Nonreciprocal Photonic Crystal[*]

Cheng He,[1] Ming-Hui Lu,[1] Xin Heng,[2] Liang Feng,[3] and Yan-Feng Chen[1]

[1] *National Laboratory of Solid State Microstructures & Department of Materials Science and Engineering, Nanjing University, Nanjing 210093, People's Republic of China*

[2] *Bio-Rad Laboratories, Hercules, California 94547, USA*

[3] *Department of Electrical Engineering, California Institute of Technology, Pasadena, California 91125, USA*

We propose a kind of electromagnetic (EM) diode based on a two-dimensional nonreciprocal gyrotropic photonic crystal. This periodic microstructure has separately broken symmetries in both parity (P) and timereversal (T) but obeys parity-time (PT) symmetry. This kind of diode could support bulk one-way propagating modes either for group velocity or phase velocity with various types of negative and positive refraction. This symmetry-broken system could be a platform to realize abnormal photoelectronic devices, and it may be analogous to an electron counterpart with one-way features.

1. Introduction

The principle of symmetry is very important in the field of fundamental physical science.[1-14] For example, in optics, the one-way edge mode, as an optical counterpart to the quantum Hall effect, was reported in magneto-optical photonic crystals (MOPC's) with periodic and parity (P) symmetries[1-3] but broken time-reversal (T) symmetry.[4-9] In condensed-matter physics, a single surface Dirac cone, due to spin-orbit coupling, results in a topological insulator with certain topological symmetry and unbroken T symmetry.[10,11] Recently, a new class of optical models constructed by gain and loss materials, which have non-Hermitian Hamiltonians with both broken P and T symmetries but which obey PT symmetry, was demonstrated with some intriguing light-propagation properties.[12-14] In all of these cases, the existence of off-diagonal components in the Hamiltonian matrix is an essential and common mathematic structure in determining the unpaired eigenvalues. In the context of the MOPC, it has imaginary off-diagonal elements in its permeability matrix, thus it could be considered to be an excellent candidate to explore the anomalous behaviors stemming from broken symmetries in optics.

In this work, we propose a two-dimensional (2D) MOPC whose bulk modes support one-way propagation for either group velocity (V_g) or phase velocity (V_p) due to

---

[*] Phys.Rev.B, 2011, 83(7):075117

asymmetry of its photonic band structure. Contrary to the well-discussed PT symmetric models constructed by a periodic gain-loss medium,[12—14] this 2D MOPC separately breaks both P and T symmetries but obeys PT symmetry, which comes from the imaginary off-diagonal elements in its permeability matrix. We call this periodic microstructure a PT EM diode, which means that the transmission is unidirectional. We also show that besides the traditional one-way propagation, all four types of refraction,[15—19] that is, right-handed positive, left-handed negative, right-handed negative, and left-handed positive refraction, could be realized in a one-way feature in one MOPC for either V_g or V_p.

2. Models and Methods

The MOPC's[20,21] have demonstrated some remarkable optical performance, such as enhancement of the Faraday effect,[22,23] nonreciprocal superprisms,[24] ultracompact isolators and circulators,[25,26] and one-way chiral edge modes.[4—9] All of these effects have made use of the broken T symmetry. Here, we focus on a unique photonic crystal (PC) structure, namely a nonreciprocal photonic crystal (NRPC), in which both P and T symmetries are broken.[27,28] The 2D NRPC of interest is composed of a square lattice of yttrium-iron-garnet (YIG) cylinders (permittivity $\varepsilon = 15\varepsilon_0$, permeability $\mu = \mu_0$, and radius $r = 0.3a$, where a is the lattice constant) embedded in an air medium.

The one-way properties of NRPC's occur when both P and T symmetries are broken with

$$\omega(k) \neq \omega(-k), \tag{1}$$

where k and $-k$ represent a pair of counterpropagating wave vectors. To break T symmetry, we apply an external dc magnetic field (out of the plane along the $+z$ direction) of 1600 G to the 2D NRPC, which induces the permeability matrix to become a gyromagnetic form,

$$\overset{\leftrightarrow}{\mu} = \begin{pmatrix} \mu & i\kappa & 0 \\ -i\kappa & \mu & 0 \\ 0 & 0 & \mu_0 \end{pmatrix} = \overset{\leftrightarrow}{\mu}_d - i\kappa z \times \overset{\leftrightarrow}{I}, \tag{2}$$

where $\mu = 14\mu_0$ and $\kappa = 12.4\mu_0$ at 4.28 GHz.[29] $\overset{\leftrightarrow}{\mu}_d$ is the diagonal part. By reversing the external magnetic field, the off-diagonal part would flip its signs too. The gyromagnetic permeability tensor in the xy plane shows broken T symmetry,

$$\pi^{-1} \overset{\leftrightarrow}{\mu} \pi = \overset{\leftrightarrow}{\mu}_d + i\kappa z \times \overset{\leftrightarrow}{I} \neq \overset{\leftrightarrow}{\mu}, \tag{3}$$

where π is the mirror reflection transform operator. On the other hand, $\overset{\leftrightarrow}{\mu}$ has P symmetry, that is

$$\sigma^{-1} \overset{\leftrightarrow}{\mu} \sigma = \overset{\leftrightarrow}{\mu}, \tag{4}$$

where σ is the spatial inversion operator. However, the P symmetry can be broken by using semicylinders along the y axis in each prime cell [Fig. 1(a)]. Then, the real space symmetry is broken relative to the y axis but remains intact relative to the x axis. As a result, both P and T symmetries are broken, which would result in the one-way characteristics.

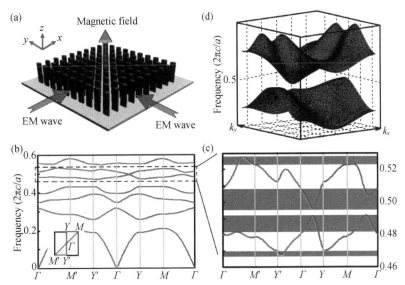

FIG. 1. (color online) (a) Schematic of the square lattice of YIG semicylinders (black) on a copper plate (yellow) in air. (b) Band structure of YIG PC, where $\varepsilon=15\varepsilon_0$ and $r=0.3a$ with 1600 G+z dc external applied magnetic field ($\mu=14\mu_0$, $\kappa=12.4\mu_0$). The inset indicates two typical opposite high-symmetry directions in reciprocal space. (c) Zoom-in band structure of the fifth and sixth bands. The shadow regions indicate the typical one-way frequency region. The red and blue lines represent the opposite high-symmetry directions, respectively, as shown in (c). (d) Whole band structure of the fifth and sixth bands reduced in the first BZ.

For transverse magnetic (TM) modes (electric field out of plane), Maxwell's equation can be written as

$$(\Theta_0 + V) | E \rangle = \omega^2 | E \rangle, \tag{5}$$

where $\Theta_0 = \varepsilon^{-1}(r) \nabla \times \mu^{-1}(r) \nabla \times$ is the nonmagneto part of the Hamiltonian while $V = \varepsilon^{-1}(r) \nabla \times [\overset{\leftrightarrow}{\mu}^{-1}(r) - I\mu^{-1}(r)] \nabla \times$ represents the gyrotropic perturbation. The imaginary off-diagonal part is entirely described by V. Both Θ_0 and V are the functions of position. Under the $\hat{P}$ operation, $\hat{k} \to -\hat{k}, \hat{r} \to -\hat{r}$ ($\hat{r}$ and $\hat{k}$ are the position and momentum operators, respectively), the right semicylinders change to the left semicylinders, which means the incident direction of the wave is transformed to the opposite side (σ operation). Under the $\hat{T}$ operation, $\hat{k} \to -\hat{k}, \hat{r} \to \hat{r}, i \to -i$, the imaginary off-diagonal part changes sign while the position of $\hat{r}$ is kept unchanged, which is equivalent to reversing the external applied magnetic field or transforming by a mirror reflection relative to the x axis (π_x operation). Analogous to the well-discussed PT symmetric systems, which require the real part of the potential to be an even function of position and the imaginary part to be odd,[12,13] in this case the real part of perturbation V with a pair of counterpropagating EM waves satisfies the even function, while the imaginary part with opposite magnetic fields or with a pair of mirror-symmetric (π_x operation) incident EM waves is an odd function.[12-14]

In other words, in this NRPC there are two cases for PT symmetry: (i) a pair of counterincident EM waves with opposite external magnetic fields; (ii) a pair of symmetric incident EM waves relative to the y axis ($\sigma + \pi_x = \pi_y$) operation, which satisfies the even function real part of the potential and the odd imaginary part. Under this condition, the transmissions would not be one-way because the eigenvalues of Eq. (5) would come out in pairs. Furthermore, there are two cases for separately broken P and T symmetries: (i) a pair of counterincident EM waves with the same external magnetic fields; (ii) a pair of symmetric incident EM waves relative to the x axis with the same external magnetic fields. Due to this symmetry-broken system, the eigenvalues of Eq. (5) would not be in pairs anymore. By fixing a frequency ω, if the Bloch vectors k [satisfying $e^{ika} \boldsymbol{E}(r) = \boldsymbol{E}(r+a)$ for the Bloch-Floquet theorem], which indicate the direction of the incident wave, have unpaired eigenvalues, the one-way group velocity would be realized from some incident directions. Otherwise, if the Bloch vectors k have the same sign (only positive or minus), the one-way phase velocity would be realized.

The following numerical results are calculated with the finite-element method (FEM) by the COMSOL MULTIPHYSICS 3.3 program. We used its radiofrequency (rf) module in the 2D TM mode with periodic boundary conditions, and we thoroughly reduced the adverse influence from meshing errors and outer boundaries.

3. Results and Discussion

3.1 Asymmetric Band Structure

When both P and T symmetries are simultaneously broken, the asymmetric band structure can be derived, as shown in Fig. 1(b). Due to the asymmetric structure, we should choose two opposite high-symmetry directions to show the one-way characters [the inset in Fig. 1(b) indicates two typical opposite high-symmetry directions in reciprocal space]. Three things should be noticed in this dispersion relation: (i) the two sides of the Brillouin zone (BZ) center in the first four bands are almost the same; (ii) the band structures remain symmetric in YM and $Y'M'$ but are asymmetric in other directions corresponding to the center of the BZ, which stems from the fact that the k space remains symmetric relative to the k_y direction; and (iii) a pair of high-symmetry points are still equivalent points, which must have the same eigenfrequencies according to the Bloch-Floquet theorem. An enlarged picture of the fifth and sixth bands is shown in Fig. 1(c). The strongly bended dispersion curves are due to the broken symmetries, which then result in the band structures upshifting. By comparing the two sides of the center of the

BZ, we identify some bump-shape and pit-shape areas in either side, which indicate the different one-way characters as shown in the shadow regions of Fig. 1(c). The whole band structure of the fifth and sixth bands in the reduced BZ is shown in Fig. 1(d). In bump-shape and pit-shape areas, the Bloch vectors are totally shifted to one side with the same sign, which means that the frequency in these regions would be one-way frequency. For example, the incident wave with some angles from the left would transmit but be completely reflected from the right independent of the angles. Large regions of the incident angle and frequency would be realized in one direction. Therefore, this 2D NRPC can provide much flexibility for the device design in terms of operating frequencies and incident directions.

Near the frequency $0.495(2\pi c/a)$ in the ΓY direction [Fig. 1(c)], the maximum and minimum values in the fifth and sixth bands, respectively, are very close. This suggests that some Dirac points may not necessarily be at the high-symmetry points.[30] Generally, in a symmetric system, the Dirac point is always located at the high-symmetry point, which indicates the linear degeneracy of two bands originating from the symmetry of the eigenfunction, that is, the symmetry of the component and lattice. In the NRPC, due to the broken symmetry of the component medium YIG, some occasional linear degeneracy might occur, causing a shift in the position away from the high-symmetry point. Actually, the one-way character with the Dirac point may make it useful and convenient to study some pseudospin and spin characters in the PC for photons and graphene for electrons, respectively.[31—33]

3.2 One-way Group Velocity

To realize the one-way group velocity, we choose the incident direction along the x direction in the y-cut NRPC. We study the one-way bulk modes of V_g in the sixth band. The equifrequency surface (EFS) is used to analyze the one-way propagation of the EM waves. As shown in Fig. 2(a), the EFS is asymmetric, in which a pit frequency ranging from about $0.495(2\pi c/a)$ to $0.505(2\pi c/a)$ is enclosed with a white solid line corresponding to $k_x \in 2\pi/a \cdot [-0.15, 0.15]$ and $k_y \in 2\pi/a \cdot [0.1, 0.38]$, respectively. As an example, we choose a pair of counterpropagating incident waves with $\omega = 0.5(2\pi c/a)$ at an incident angle of $+/-30°$ from the lower left and upper right, respectively, shown as blue (denoted as k) and red (k') arrows in Fig. 2(a). The white and black circles represent the EFS at a frequency of $0.5(2\pi c/a)$ in NRPC and in air, respectively. Only then can the k wave vector of the incident EM waves from $+30°$ intersect the EFS of the NRPC. The k' ($-30°$ incident) wave vector is not able to intersect the EFS, which means that due to the mismatch of the phase condition, the incident wave with $-30°$ incident angle is totally reflected, as shown in Figs. 2(b) and 2(c).

Moreover, a pair of incident EM waves that are symmetric relative to the x axis is also nonreciprocal, while a pair of incident EM waves that are symmetric relative to the y axis

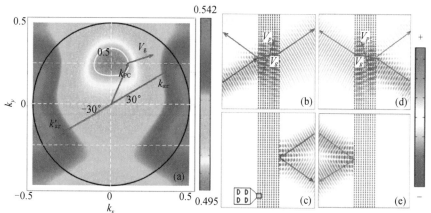

FIG. 2. (color online) One-way propagation for group velocity with the y-cut NRPC. (a) Contour plot for the sixth band, where the white and black circles are the EFS at a frequency of $0.5(2\pi c/a)$ in the NRPC and in air, respectively. V_g and V_p with a pair of counterincident waves along the x direction are indicated by blue ($+30°$) and red ($-30°$) arrows, respectively. The corresponding electric-field distribution is shown with (b) $+30°$ from the lower left and (c) $-30°$ from the upper right. Parts (d) and (e) are the PT symmetric cases corresponding to (b) and (c), respectively.

can be treated as a PT symmetric case. Figures 2(d) and 2(e) are the PT symmetric cases corresponding to Figs. 2(b) and 2(c), respectively, and vice versa. In other words, the NRPC can be a lens for the upper part of the xy plane since it allows the incident waves from the lower part to transmit, while it turns out to be a mirror for the lower part, which can only get the waves to reflect from the same side of the upper incident waves. Furthermore, if the frequency is chosen in the bump frequency range of the fifth band, for example, $0.48(2\pi c/a)$, the opposite one-way direction can be realized.

We plot the angle-dependent transmission and refractive spectrum with the two opposite incident waves along the x direction in the y-cut NRPC in Fig. 3. Figure 3(a) clearly shows that $0.5(2\pi c/a)$ and $0.48(2\pi c/a)$ ($+/-30°$ incident) are two prominent frequencies having different one-way features. Within a relatively broad bandwidth of $\Delta\omega/\omega \approx 5\%$, our NRPC structure shows an excellent transmission-reflection ratio, greater than 40 dB. In addition, large incident angles are allowed for one-way refraction in this case: about $20°-40°$ at a frequency of $0.5(2\pi c/a)$, and about $10°-50°$ at a frequency of $0.48(2\pi c/a)$ [Fig. 3(b)]. Figures 3(c) and 3(d) show the angles of refraction of V_g and V_p, respectively, with various incident angles, where negative V_g represents negative refraction and negative V_p indicates backward phase propagation. Due to the asymmetric EFS, various refraction types could occur in a one-way feature in this model. For instance, at a frequency of $0.48(2\pi c/a)$ with $\theta_{in}=-30°$, the $V_g<0$ means that it is the right-handed negative refraction [red solid dots in Fig. 3(d)]. More interestingly, "birefraction" would occur, that is, the coexistence of negative and positive V_p with $\theta_{in}=53°$

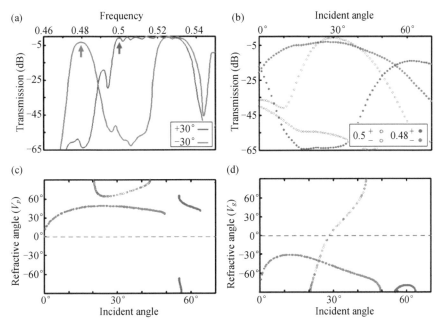

FIG. 3. (color online) Transmission vs refraction angles with the *y*-cut NRPC. Blue and red indicate incident waves from the lower left (+) and the upper right (−), respectively. (a) Transmission spectra (10 layers) of +/− 30° incident. (b) Transmission at a frequency of $0.5(2\pi c/a)$ (open circles) and 0.48 (solid dots) with various incident angles. Refractive angles of (c) V_p and (d) V_g at a frequency of 0.5 $(2\pi c/a)$ (open circles) and $0.48(2\pi c/a)$ (solid dots) with various incident angles.

[blue solid dots in Fig. 3(c)].

We also calculate three typical propagating cases, as shown in Mov. S1.[34] At a frequency of $0.50(2\pi c/a)$ with incident angle $\theta_{in} = +/-23°$, the propagation is the one-way right-handed negative refraction. At a frequency of $0.48(2\pi c/a)$ with incident angle $\theta_{in} = +/-30°$, the propagation is also right-handed negative refraction, but with different one-way characters.

In other words, the case in Fig. 2 is the one-way right-handed positive refraction. On the other hand, the one-way right-handed negative refraction can be achieved at the same frequency with a smaller incident angle, that is, $\theta_{in} < 28°$ [blue open circles in Figs. 3(c) and 3(d)]. If we choose the operating frequency region in the fifth band, a bump-shaped region would allow us to construct one-way left-handed positive and one-way left-handed negative refraction, respectively.

3.3 Refractive Types

In general, there are two key factors to determine the optical refraction: one is the conservation of the wave vector, $k_\parallel^{PC} = k_\parallel^{incident}$, the other is the causality of the wave propagation, $V_{g\perp}^{PC} \cdot V_{g\perp}^{incident} > 0$, where $\perp$ and $\parallel$ represent the directions perpendicular to and parallel to the boundary of the NRPC, respectively. For these reasons, only the right intersecting point in Fig. 2(a) could be considered as a physical reality. It should be noticed that the directions of $k_\perp^{PC}$ (indicating the forward or backward phase propagation) and $V_{g\parallel}^{PC}$ (indicating the positive or negative refraction) could not be restricted. Therefore, there are a total of four types of refraction in this system. Here, a general table was used to categorize these four types of refraction, as shown in Fig. 4. In addition to the well-known right-handed positive (upper left part of Fig. 4) and left-handed negative (lower right part of Fig. 4) refractions, in the symmetry-broken system the group velocity and phase velocity were often separated into different directions to form the left-handed positive (lower left part of Fig. 4) and right-handed negative (upper right part of Fig. 4) refractions. The advantage of this category is that we can clearly distinguish that the negative refraction is decided by the sign of $V_{g\parallel}^{incident} V_{g\parallel}^{PC}$ while the backward phase propagation is decided by the sign of $V_{p\perp}^{incident} V_{p\perp}^{PC}$. This category can help us to understand that positive refraction can also amplify the evanescent wave, while the forward phase propagation can also realize the negative refraction.[18]

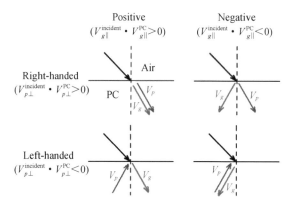

FIG. 4. (color online) Category of four types of refraction. $\perp$ and $\parallel$ represent the directions normal to and along the boundary between air and the NRPC, respectively.

By using these criteria, the refractive waves in Fig. 2(b) are cataloged as the right-handed and positive refraction. All four types of refraction can be achieved with the one-way character for V_g in this model too. Due to the strong bending of EFS, the refractive angle would vary rapidly when sweeping the incident angle. Moreover, flipping the external applied magnetic field could reverse the direction of one-way propagation. This property would have great potential in preparing some new optoelectronic devices, such as superprisms and optical switches.

3.4 One-Way Phase Velocity

To discuss the one-way phase propagation, we change the incident direction along the y direction in the x-cut NRPC. The waves operating in the same pit frequency domain as in the previous discussion would propagate with the one-way features not for V_g but for V_p.

As shown in Fig. 5(a), we choose a pair of counterincident waves with angle $+/-10°$ along the y direction at a frequency of $0.5(2\pi c/a)$. The blue and red arrows correspond to $+10°$ incident and $-10°$ incident waves from the lower left and upper right, respectively. Each of the counterincident waves would have the same two intersecting points. Due to the causality, these counterincident waves would choose different intersecting points to propagate. The EM waves show right-handed refraction for a $+10°$ incident wave and left-handed refraction for a $-10°$ incident wave, as shown in Figs. 5(b) and 5(c), respectively. When $\theta_{in} < 12°$, V_p stays positive along the $+y$ axis. To realize the one-way phase velocity in the NRPC, two conditions should be satisfied: (i) both P and T symmetries must be broken to cause the unpaired Bloch vectors; (ii) these unpaired Bloch vectors should have the same sign (i.e., the EFS should be totally shifted to one side), which means that the symmetry breaking should be large enough.

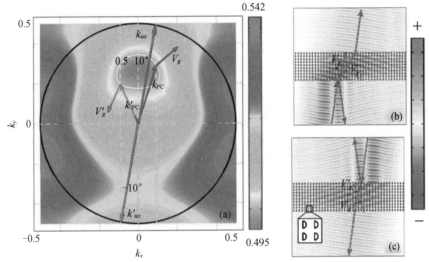

FIG. 5. (color online) One-way propagation for phase velocity with the x-cut NRPC. (a) Contour plot for the sixth bands with $+/-10°$ incident angle along the y direction at a frequency of $0.5(2\pi c/a)$. The corresponding electric-field distribution is shown with (b) $+10°$ from the lower left and (c) $-10°$ from the upper right.

For the one-way property of V_p, we plot the angles of refraction of V_g and V_p, respectively, versus various incident angles along the y axis in the x-cut NRPC in Fig. 6. The different signs of V_p represent different handedness (left-handed or right-handed), while the different signs of V_g indicate refraction types (negative or positive). Frequency $0.5(2\pi c/a)$ with less than $12°$ incident angle is the typical one-way range for V_p. At a frequency of $0.48(2\pi c/a)$ with $\theta_{in} < 20°$, a pair of counterincident EM waves would both propagate left-handed but with different refractive types [solid line in Figs. 6(a) and 6(b)]: positive refraction for the incident wave from the lower left, negative refraction for the incident wave from the upper right. Moreover, all four types of refraction could be realized with one-way character for V_p. By reversing the external magnetic field, the one-way V_p would be reversed, which also obeys PT symmetry. These properties could be applied in some photoelectronic devices such as a phase filter and a one-way phase compensator.

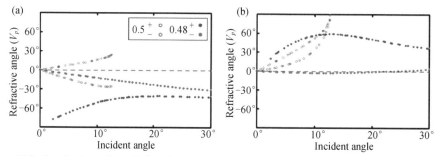

FIG. 6. (color online) Refractive angles with the x-cut NRPC. Blue and red indicate incident waves from the lower left (+) and upper right (−), respectively. Refractive angles of (a) V_p and (b) V_g at a frequency $0.5(2\pi c/a)$ (open circles) and $0.48(2\pi c/a)$ (solid dots) with various incident angles.

Similarly, we choose three typical propagating cases as shown in Mov. S2.[35] At a frequency of $0.50(2\pi c/a)$ with normal incidence, the propagations of the counterincident waves have the same direction of V_p but with opposite handedness: right-handed for the incident wave from the lower part, left-handed for the incident wave from the upper part. At a frequency $0.48(2\pi c/a)$ with incident angle $\theta_{in}=+/-30°$, the one-way direction of V_p is reversed: left-handed positive refraction for the incident wave from the lower part, right-handed positive refraction for the incident wave from the upper part.

Recently, a time-reversal method was proposed to realize a perfect lens with negative refraction by four-wave mixing.[19] The evanescent wave without carrying energy can be collected for perfect imaging due to the backward phase propagation. In the NRPC (Fig. 5), the counterincident waves would have opposite handedness, which would have a one-way phase-compensation effect in only one direction. This means that the amplification of the evanescent wave in a PC may not necessarily be associated with T symmetry. Therefore, the NRPC provides an alternative device model to realize the one-way perfect lens in a T symmetry-broken system.

3.5 Composite Structure

As mentioned earlier, the one-way operating frequency is about $0.5(2\pi c/a)$. From an experimental point of view, the lattice constant should be chosen about 3.5 cm. The one-way frequency is near 4.28 GHz. Although the structure under consideration has one-way features in the flat fifth and sixth bands, the wavelength is about double the lattice constant, which may not be difficult to measure in experiment. On the other hand, a composite model could lower the one-way bands, as shown in Fig. 7. The band structure of a composite NRPC is shown in the inset [Fig. 7(a)], where the left semicylinder is a dielectric medium ($\varepsilon = 16\varepsilon_0$), the right semicylinder is YIG ($\varepsilon = 15\varepsilon_0$, $\mu = 14\mu_0$, $\kappa = 12.4\mu_0$), and $r = 0.3a$. The one-way features occur in the second and third bands. The enlarged band structure is shown in Fig. 7(b). The shadow regions indicate the typical one-way frequency region. The red and blue lines represent the opposite high-symmetry directions. Insets are the EFS of the second and third bands. By enlarging the condition of the asymmetry, the region of one-way frequency would be enlarged correspondingly. With regard to the measurement of one-way V_p, the changes of phase can be determined by increasing or decreasing the number of periods of NRPC's. The one-way phase propagation may be achieved by comparing the different phase changes of two counterincident waves.

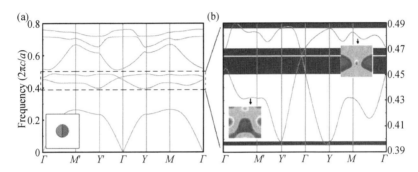

FIG. 7. (color online) (a) Band structure of composite PC (inset), where the left semicylinder is the dielectric medium ($\varepsilon = 16\ \varepsilon_0$), the right semicylinder is YIG ($\varepsilon = 15\ \varepsilon_0$, $\mu = 14\ \mu_0$, $\kappa = 12.4\ \mu_0$), and $r = 0.3a$. (b) Zoom-in band structure of the second and third bands. The shadow regions indicate the typical one-way frequency region. The red and blue lines represent the opposite high-symmetry directions. Insets are the EFS of the second and third bands.

4. Conclusions

These effects can be attributed to the non-Hermitian eigenequation. With broken P and T symmetries, the Bloch vectors could have unpaired real values for the same frequency. Compared to the one-way edge mode with P symmetry,[4−9] the eigenequation is Hermitian, and the real parts of the Bloch vectors are always equal with opposite signs.[9] This means that they cannot support the one-way bulk mode. On the other hand, due to the periodicity of the NRPC, where the eigenvalues of high-symmetry points remain the same, the intersections of the normal incident wave vector would result in a pair corresponding to forward and backward energy flow, respectively. Therefore, one-way propagation of V_g would never happen for normal incident waves in periodic cases. However, the V_p can be one-way for normal incidence, which can be treated as another novel phenomenon (phase diode) stemming from the broken P and T symmetries.

In summary, we proposed a 2D NRPC with broken P and T symmetries simultaneously. The asymmetry of P and T would bring more parameters in the band engineering to control and realize the PT EM diodes, which could show one-way features either for V_g or V_p due to the NRPC's asymmetric band structure. Four types of refraction cases could be realized in one NRPC, depending on the operating frequency and incident angle. These properties may have great potential applications in optoelectronic phase-control devices, such as optical isolators, optical switches, and phase compensators. Using such a macrosystem of NRPC's with PT symmetry, we have the potential to realize some exotic quantum phenomena that can mimic electronic quantum effects, such as the optical counterpart of quantum spin Hall effects or the optical counterpart of topological insulators.

References and Notes

[1] S. John, Phys. Rev. Lett. **58**, 2486 (1987).
[2] E. Yablonovitch, Phys. Rev. Lett. **58**, 2059 (1987).
[3] J. D. Joannopoulos, R. D. Meade, and J. N. Winn, *Photonic Crystals: Modeling the Flow of Light* (Princeton University Press, Princeton, NJ, 1995).
[4] F. D. M. Haldane and S. Raghu, Phys. Rev. Lett. **100**, 013904 (2008).
[5] Z. Wang, Y. D. Chong, J. D. Joannopoulos, and M. Soljačić, Phys. Rev. Lett. **100**, 013905 (2008).
[6] Z. Wang, Y. D. Chong, J. D. Joannopoulos, and M. Soljačić, Nature (London) **461**, 772 (2009).
[7] X. Ao, Z. Lin, and C. T. Chan, Phys. Rev. B **80**, 033105 (2009).
[8] C. He, X. L. Chen, M. H. Lu, X. F. Li, W. W. Wan, X. S. Qian, R. C. Yin, and Y. F. Chen, Appl. Phys. Lett. **96**, 111111 (2010).
[9] C. He, X. L. Chen, M. H. Lu, X. F. Li, W. W. Wan, X. S. Qian, R. C. Yin, and Y. F. Chen, J. Appl. Phys. **107**, 123117 (2010).
[10] H. Zhang, C. X. Liu, X. L. Qi, X. Dai, Z. Fang, and S. C. Zhang, Nature Phys. **5**, 438 (2009).
[11] Y. Xia, D. Qian, D. Hsieh, L. Wray, A. Apl, H. Lin, A. Bansil, D. Grauer, Y. S. Hor, R. J. Cava, and M. Z. Hasan, Nature Phys. **5**, 398 (2009).
[12] K. G. Makris, R. El-Ganainy, D. N. Christodoulides, and Z. H. Musslimani, Phys. Rev. Lett. **100**, 103904 (2008).
[13] C. E. Rüter, K. G. Makris, R. El-Ganainy, D. N. Christodoulides, M. Segev, and D. Kip, Nature Phys. **6**, 192 (2010).
[14] P. M. Rinard and J. W. Calvert, Am. J. Phys. **39**, 753 (1971).
[15] V. G. Vesselago, Sov. Phys. Usp. **10**, 509 (1968).
[16] L. Feng, X. P. Liu, Y. F. Tang, Y. F. Chen, J. Zi, S. N. Zhu, and Y. Y. Zhu, Phys. Rev. B **71**, 195106 (2005).
[17] N. H. Shen, Q. Wang, J. Chen, Y. X. Fan, J. P. Ding, H. T. Wang, Y. Tian, and N. B. Ming, Phys. Rev. B **72**, 153104 (2005).
[18] L. Feng, X. P. Liu, M. H. Lu, Y. B. Chen, Y. F. Chen, Y. W. Mao, J. Zi, Y. Y. Zhu, S. N. Zhu, and N. B. Ming, Phys. Rev. Lett. **96**, 014301 (2006).
[19] J. B. Pendry, Science **322**, 71 (2008).
[20] I. L. Lyubchanskii, N. N. Dadoenkova, M. I. Lyubchanskii, E. A. Shapovalov, and Th. Rasing, J. Phys. D **36**, R277 (2003).
[21] M. Inoue, R. Fujikawa, A. Baryshev, A. Khanikaev, P. B. Lim, H. Uchida, O. Aktsipetrov, A.

Fedyanin, T. Murzina, and A. Granobsky, J. Phys. D **39**, R151 (2006).

[22] R. Wolfe, R. A. Lieberman, V. J. Frantello, R. E. Scotti, and N. Kopylov, Appl. Phys. Lett. **56**, 426 (1990).

[23] M. Inoue, K. Arai, T. Fujii, and M. Abe, J. Appl. Phys. **83**, 6768 (1998).

[24] I. Bita and E. L. Thomas, J. Opt. Soc. Am. B **22**, 1199 (2005).

[25] Z. Wang and S. Fan, Opt. Lett. **30**, 1989 (2005).

[26] Z. Wang and S. Fan, Appl. Phys. B **81**, 369 (2005).

[27] Z. Yu, Z. Wang, and S. Fan, Appl. Phys. Lett. **90**, 121133 (2007).

[28] M. Vanwolleghem, X. Checoury, W. Smigaj, B. Gralak, L. Magdenko, K. Postava, B. Dagens, P. Beauvillain, and J. M. Lourtioz, Phys. Rev. B **80**, 121102(R) (2009).

[29] D. M. Pozar, *Microwave Engineering*, 2nd ed. (Wiley, New York, 1998).

[30] S. L. Yu, J. X. Li, and L. Sheng, Phys. Rev. B **80**, 193304 (2009).

[31] R. A. Sepkhanov, J. Nilsson, and C. W. J. Beenakker, Phys. Rev. B **78**, 045122 (2008).

[32] K. S. Novoselov, A. K. Geim, S. V. Morozov, D. Jiang, M. I. Katsnelson, I. V. Grigorieva, S. V. Dubonos, and A. Firsov, Nature (London) **438**, 197 (2005).

[33] Y. Zhang, Y. W. Tan, H. L. Stormer, and P. Kim, Nature (London) **438**, 201 (2005).

[34] See supplemental material at [http://link.aps.org/supplemental/ 10.1103/PhysRevB.83.075117] Movies of three typical onw-way V_g cases.

[35] See supplemental material at [http://link.aps.org/supplemental/ 10.1103/PhysRevB.83.075117] Movies of three typical onw-way V_p cases.

[36] The work was supported jointly by the National Basic Research Program of China (Grant No. 2007CB613202) and the National Nature Science Foundation of China (Grant No. 50632030). We also acknowledge the support from the Nature Science Foundation of China (Grant No. 10874080) and the Nature Science Foundation of Jiangsu Province (Grant No. BK2009007).

Nonreciprocal Light Propagation in a Silicon Photonic Circuit*

Liang Feng,[1,2,4] Maurice Ayache,[3] Jingqing Huang,[1,4] Ye-Long Xu,[2] Ming-Hui Lu,[2]
Yan-Feng Chen,[2] Yeshaiahu Fainman,[3] Axel Scherer[1,4]

[1] *Department of Electrical Engineering, California Institute of Technology, Pasadena, CA 91125, USA*
[2] *Nanjing National Laboratory of Microstructures, Nanjing University, Nanjing, Jiangsu 210093, China*
[3] *Department of Electrical and Computer Engineering, University of California, San Diego, La Jolla, CA 92093, USA*
[4] *Kavli Nanoscience Institute, California Institute of Technology, Pasadena, CA 91125, USA*

Optical communications and computing require on-chip nonreciprocal light propagation to isolate and stabilize different chip-scale optical components. We have designed and fabricated a metallic-silicon waveguide system in which the optical potential is modulated along the length of the waveguide such that nonreciprocal light propagation is obtained on a silicon photonic chip. Nonreciprocal light transport and one-way photonic mode conversion are demonstrated at the wavelength of 1.55 micrometers in both simulations and experiments. Our system is compatible with conventional complementary metal-oxide-semiconductor processing, providing a way to chip-scale optical isolators for optical communications and computing.

An example of nonreciprocal physical response, associated with the breaking of time-reversal symmetry, is the electrical diode[1]. Stimulated by the vast application of this one-way propagation of electric current, considerable effort has been dedicated to the study of nonreciprocal propagation of light. The breaking of time-reversal symmetry of light is typically achieved with magneto-optical materials that introduce a set of antisymmetric off-diagonal dielectric tensor elements[2-4] or by involving nonlinear optical activities[5,6]. However, practical applications of these approaches are limited for the rapidly growing field of silicon (Si) photonics because of their incompatibility with conventional complementary metal-oxide-semiconductor (CMOS) processing. Si optical chips have demonstrated integrated capabilities of generating[7-11], modulating[12], processing[13] and detecting[14] light signals for next-generation optical communications but require on-chip nonreciprocal light propagation to enable optical isolation in Si photonics.

Parity-time (PT) symmetry is crucial in quantum mechanics. In contrast to conventional quantum mechanics, it has been proposed that non-Hermitian Hamiltonians where $\hat{H}^+ \neq \hat{H}$ can still have an entirely real spectrum with respect to the PT symmetry[15,16].

---

* Science, 2011, 333:729

Due to the equivalence between the Schrödinger equation in quantum mechanics and the wave equation in optics, PT symmetry has been studied in the realm of optics with non-Hermitian optical potentials[17—19]. The breaking of PT symmetry has recently been experimentally observed, showing asymmetric characteristics transverse to light propagation above the PT threshold[20,21]. Here, we have designed a Si waveguide integrated with complex optical potentials that have a thresholdless broken PT symmetry along the direction of light propagation, thus creating on-chip nonreciprocal light propagation.

On a Si-on-insulator (SOI) platform, the designed two-mode Si waveguide is 200 nm thick and 800 nm wide, allowing a fundamental symmetric quasi-TE mode with a wave vector of $k_1 = 2.59 k_0$ and a higher-order antisymmetric mode with a wave vector of $k_2 = 2k_0$ at the wavelength of 1.55 μm. Periodically arranged optical potentials are implemented in the Si waveguide and occupy half of the waveguide width in the x direction [Fig. 1(a)]. The optical potentials have a complex modulation in their dielectric constants along light propagation in the z direction compared with the Si waveguide background ($\varepsilon_{Si} = 12.11$), as shown in Fig. 1(b)

$$\Delta\varepsilon = \exp[iq(z - z_0)] \tag{1}$$

where $q = k_1 - k_2$, and z_0 is the starting point of the first modulation region. This complex exponential variation of $\Delta\varepsilon$ along the z direction introduces a one-way wave vector that is intrinsically not reciprocal because its corresponding Fourier transform is one-sided to the guided light inside the Si waveguide. These complex optical potentials are located in phase with each other with a spacing of $2\pi/q$ (or multiples of $2\pi/q$) in between, such that light modulation always remains in phase with and is consistently applied to guided light. We chose the dielectric constant modulation to be completely passive in order to make the

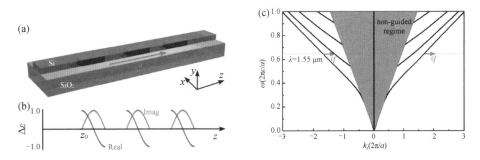

FIG. 1. (a) Nonreciprocal light propagation in a Si photonic circuit. Based on a SOI platform, PT optical potentials with exponentially modulated dielectric constants, as depicted in (b) where blue and red curves represent the real and imaginary parts of $\Delta\varepsilon$, respectively, are embedded in the Si waveguide to introduce an additional wave vector q to guided light. (c) Band diagram for TE-like polarization of the Si waveguide, where the frequency and wave vector are normalized with $a = 1$ μm. At the wavelength of 1.55 μm, if incoming light is a fundamental symmetric mode, one-way mode conversion is only expected for backward propagation where the phase-matching condition is satisfied as indicated by arrows.

experiment easier to perform, meaning that the modulation length of each optical potential is π/q. Therefore, no gain is required to construct these optical potentials. From a quantum mechanics analysis, these optical potentials have a spontaneously broken PT symmetry with a non-Hermitian Hamiltonian $\hat{H}^+(-x,z) \neq \hat{H}(x,z)$ or $\hat{H}^+(x,-z) \neq \hat{H}(x,z)$[22], suggesting noncommutative binary operations to the Hamiltonian $PT\hat{H}^+ \neq \hat{H}PT$. In our system, this is observed as nonreciprocal light propagation through the optical potentials in the Si waveguide.

More intuitively, the introduced one-way wave vector q shifts the incoming photons of the symmetric mode with an additional spatial frequency: k_1+q for forward propagation and $-k_1+q$ for backward propagation. The mode transition between the symmetric mode and the antisymmetric mode can happen only when the phase-matching condition is approximately satisfied $\Delta k = \pm(k_1-k_2)+q \approx 0$, where + and − represent forward and backward propagation, respectively. In our case, for an incoming symmetric mode the phase-matching condition is only valid in the backward direction, supporting a one-way mode conversion from k_1 to k_2 [Fig. 1(c)]. In the modulated regime, the electric field of light is given by

$$E(x,z,t) = A_1(z)E_1(x)e^{i(k_1 z - \omega t)} + A_2(z)E_2(x)e^{i(k_2 z - \omega t)} \qquad (2)$$

where $E_{1,2}(x)$ are normalized mode profiles of two different modes, and $A_{1,2}(z)$ are the corresponding normalized amplitudes of two modes, respectively. Assuming a slowly varying approximation, the coupled-mode equations can be expressed as follows:

$$\frac{d}{dz}A_1(z) = iB_1 \exp(iqz)A_1(z) + iC_1 A_2(z)$$
$$\frac{d}{dz}A_2(z) = iC_2 \exp(i2qz)A_1(z) + iB_2 \exp(iqz)A_2(z) \qquad (3)$$

for forward propagation and

$$\frac{d}{dz}A_1(z) = -iB_1 \exp(iqz)A_1(z) - iC_1 \exp(i2qz)A_2(z)$$
$$\frac{d}{dz}A_2(z) = -iC_2 A_1(z) - iB_2 \exp(iqz)A_2(z) \qquad (4)$$

for backward propagation, where $B_1 = \frac{1}{2k_1}\frac{\omega^2}{c^2} \times \frac{\int E_1^*(x)E_1(x)dx}{\int |E_1(x)|^2 dx}$, $C_1 = \frac{1}{2k_1}\frac{\omega^2}{c^2}\frac{\int E_1^*(x)E_2(x)dx}{\int |E_1(x)|^2 dx}$, $C_2 = \frac{1}{2k_2}\frac{\omega^2}{c^2}\frac{\int E_2^*(x)E_1(x)dx}{\int |E_2(x)|^2 dx}$, and $B_2 = \frac{1}{2k_2}\frac{\omega^2}{c^2}\frac{\int E_2^*(x)E_2(x)dx}{\int |E_2(x)|^2 dx}$.

The mode transition can happen only when the phase-matching condition is satisfied as the exponential term disappears because $\exp(i\Delta kz) = 1$. Therefore, it is evident that with an initial condition of $A_1=1$ and $A_2=0$, photons from the symmetric mode can be converted to the antisymmetric mode only for backward propagation, whereas A_2 remains 0 for forward propagation, indicating negligible mode conversion. The nonreciprocity of the mode transition here results from the spontaneous breaking of the PT symmetry of guided light by the engineered complex optical potentials. It is worth emphasizing that this nonreciprocal unidirectional mode transition is always valid with any modulation intensity, indicating a completely thresholdless breaking of PT symmetry[22], in stark contrast to previous work on threshold PT symmetry breaking[20,21].

Fully vectorial three-dimensional (3D) finite element method simulations have been performed to validate the proposed nonreciprocal propagation of guided light at the telecom wavelength of 1.55 μm. With a TE-like symmetric incident mode, after forward propagating through the PT optical potentials where $\Delta\varepsilon$ follows the exponential modulation, guided light does not meet any phase-matching condition with the additional wave vector q and therefore retains the same symmetric mode profile. However, for backward propagation, it is evident that the antisymmetric mode is converted from the incoming symmetric mode due to the phase matching with the additional wave vector q, showing a one-way mode transition [Fig. 2(a)]. The nonreciprocal light propagation can also be analytically calculated using the coupled-mode theory from Eqs. 2 to 4 (fig. S1), consistent with the simulated results.

However, the approach so far demonstrated to create the exponentially modulated $\Delta\varepsilon$[21] is difficult to integrate with Si photonics. It is therefore necessary to design an equivalent guidedmode modulation that at a macroscopic scale mimics the intrinsically microscopic exponential modulation of the PT optical potentials. To simplify fabrication, each complex exponential modulation is separated into two different modulation regions: one providing only the imaginary sinusoidal modulation of the dielectric constant covering one transverse half space (bottom) of the waveguide, and the other creating the real cosine modulation occupying the other transverse half space (top) of the waveguide [Fig. 2(b)], as follows:

$$\Delta\varepsilon_{real} = -\cos[q(z-z_0)]$$
$$\Delta\varepsilon_{imag} = i\sin[q(z-z_0)] \quad (5)$$

Although individual sinusoidal or cosine modulation does not contribute to the breaking of PT symmetry, simultaneous modulations of both cause an equivalent nonreciprocal one-way mode transition. Guided light in different half spaces experiences complementary mode modulation from each other and therefore behaves as if the PT optical potentials do exist. Moreover, to have only the positive $\Delta\varepsilon_{real}$ of the modulations for ease of fabrication, regions of $\Delta\varepsilon_{real}$ are shifted $\pi/2q$ in the z direction: $\Delta\varepsilon_{real} = \sin[q(z-(z_0+\pi/2q))]$ [Fig. 2(c)]. The resulting one-way mode transition of guided light consequently remains the same.

Finally, to achieve sinusoidal optical potentials using microscopically homogeneous materials, sinusoidal-shaped structures are adopted on top of the Si waveguide for both real

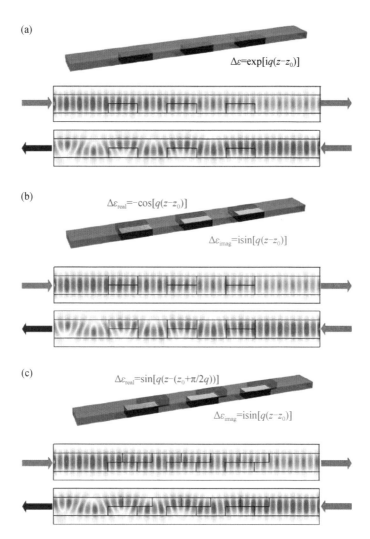

FIG. 2. Evolution of PT optical potentials in the Si waveguide (top) and their corresponding field distributions of E_x for forward (middle) and backward (bottom) propagation with an incoming symmetric mode. (a) Original PT optical potentials with exponentially modulated dielectric constant. (b) Two different kinds of optical potentials with real cosine and imaginary sinusoidal modulated dielectric constants. (c) Optical potentials with the real part modulation in (b) are shifted $\pi/2q$ in the z direction.

and imaginary modulations to mimic the modulations described in Eq. 5 [Fig. 3(a)]. An 11-nm germanium (Ge)/18-nm chrome (Cr) bilayer structure is applied for the imaginary modulation $\Delta\varepsilon_{imag}$ as guided modes have the same effective indices as $\Delta\varepsilon = i$. For the real modulation $\Delta\varepsilon_{real}$, additional 40-nm Si on top of the original Si waveguide achieves the same effective indices of guided modes as $\Delta\varepsilon = 1$. The length, period, and locations of these sinusoidal shaped structures follow those in Fig. 2(c). The designed sinusoidal-shaped structures have almost the same effective indices of the waveguide modes, as if the real and imaginary function-like modulations exist in the waveguide [Fig. 3, (b) and (c)], such that the same unidirectional wave vector q can be introduced. Therefore, an equivalent one-way mode transition is realized, as shown in Fig. 3(d): Forward propagating light remains in the symmetric profile, whereas mode conversion from the symmetric mode to the antisymmetric mode exists for backward propagation. It is thus evident that our classical waveguide system successfully mimics the quantum effect inherently associated with a broken PT symmetry. Overall, at different steps of the evolution of PT optical potentials from Fig. 2 to Fig. 3, guided light exhibits almost identical phase and intensity for both forward and backward propagation, further proving the equivalence of our classical design to the quantum PT potentials for guided light. To confirm the thresholdless condition of the breaking of PT symmetry in our system, we also designed and simulated different Si and Ge/Cr combinations corresponding to different modulation intensities from $\Delta\epsilon = 0.25\exp[iq(z-z_0)]$ to $\Delta\epsilon = 0.75\exp[iq(z-z_0)]$ (fig. S2). Although conversion efficiencies reduce as modulation intensities decrease, one-way mode conversion always exists, indicating that the breaking of PT symmetry in our system is spontaneous without any threshold.

A picture of the fabricated device is shown in Fig. 4(a). Nonreciprocal light propagation in the Si waveguide was observed using a near-field scanning optical microscope[23]. In experiments, a tapered fiber was used to couple light into the waveguide. Although the fundamental symmetric mode is dominant in incidence, there also exists some power coupled to the antisymmetric mode as shown in Fig. 4(b). Consistent with simulations, light remains predominantly the fundamental symmetric mode after propagating through the optical potentials for forward propagation. However, the symmetric-mode-dominant incoming light in backward propagation clearly shows mode conversion to the antisymmetric mode after the device. With phase information of guided light simultaneously obtained using our heterodyne system[22,23], we applied the Fourier transform analysis to the measured field distribution. The results further confirm that the breaking of PT symmetry in our system as the conversion from the symmetric mode to the antisymmetric mode can be observed only in the backward direction (fig. S3). It is therefore evident that one-way mode conversion and nonreciprocal light propagation have been successfully realized in CMOS-compatible Si photonics, paving the way to on-chip optical isolation. Although the insertion loss of about 7 dB is observed through the optical potentials, it can be completely compensated by incorporating gain into the imaginary part modulation of the PT optical potentials in Eqs. 1 and 5. One can easily envision constructing chip-scale isolators by extending the demonstrated nonreciprocal light propagation. The excited antisymmetric mode can be removed in transmitted fields by implementing, next to the PT optical potentials, an optical mode filter that completely reflects the antisymmetric mode but allows the symmetric mode to transmit. Optical isolation with large extinction ratio can be achieved by controlling interference between incidence and reflection.

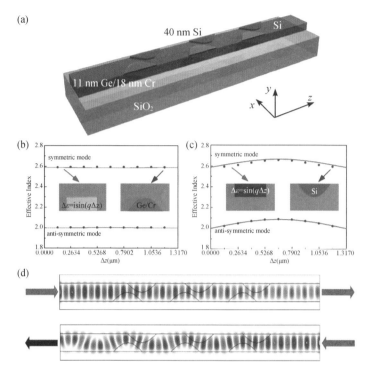

FIG. 3. (a) Design of the metallic-Si waveguide to mimic the light modulation of PT optical potentials. (b) Effective indices of symmetric and antisymmetric modes with the imaginary part sinusoidal-modulated optical potential (red lines) and the sinusoidal-shaped Ge/Cr bilayer structure (blue dots). (c) Modes' effective indices with the real part sinusoidal-modulated optical potential (red lines) and the sinusoidal-shaped Si structure (blue dots). Insets in (b) and (c) show the considered waveguide, and Δz starts from where modulation begins. (d) Numerical mappings of E_x for forward (upper) and backward (lower) propagation with an incoming symmetric mode.

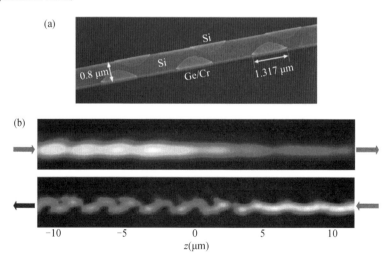

FIG. 4. (a) Scanning electron microscope image of the fabricated device. (b) Measured near-field amplitude distribution of light in the one-way mode converter for both forward (upper) and backward (lower) light propagation.

The nonreciprocal light propagation we have accomplished in a Si photonic circuit is expected to have strong impacts in both fundamental physics and device applications. The feasibility of mimicking complicated quantum phenomena in classical systems promises completely chip-scale optical isolation for rapidly growing Si photonics and optical communications. The proposed one-way system is completely linear and expected to have higher efficiencies and broader operation bandwidths than nonlinear strategies. Analogously the concept of this one-way propagation can be applied to other classical waves, such as sound, as a promising method to drastically increase the rectification efficiency over current nonlinearity-induced isolation[24].

References and Notes

[1] S.M. Sze, *Semiconductor Devices: Physics and Technology* (Wiley, New York, ed. 2, 2001).
[2] J. D. Jackson, *Classical Electrodynamics* (Wiley, New York, ed. 3, 1998).
[3] T. Amemiya, K. Abe, T. Tanemura, T. Mizumoto, Y. Nakano, *IEEE J. Quantum Electron.* **46**, 1662 (2010).
[4] Z. Wang, Y. Chong, J. D. Joannopoulos, M. Soljacic, *Nature* **461**, 772 (2009).
[5] K. Gallo, G. Assanto, K. R. Parameswaran, M. M. Fejer, *Appl. Phys. Lett.* **79**, 314 (2001).
[6] Z. Yu, S. Fan, *Nat. Photonics* **3**, 91 (2009).
[7] D. Liang, J. E. Bowers, *Nat. Photonics* **4**, 511 (2010).
[8] R. Chen *et al.*, *Nat. Photonics* **5**, 170 (2011).
[9] O. Painter *et al.*, *Science* **284**, 1819 (1999).
[10] M. P. Nezhad *et al.*, *Nat. Photonics* **4**, 395 (2010).
[11] X. Sun *et al.*, *Opt. Lett.* **34**, 1345 (2009).
[12] M. Hochberg *et al.*, *Nat. Mater.* **5**, 703 (2006).
[13] M. A. Foster *et al.*, *Nature* **456**, 81 (2008).
[14] J. Michel, J. Liu, L. C. Kimerling, *Nat. Photonics* **4**, 527 (2010).
[15] C. M. Bender, S. Boettcher, *Phys. Rev. Lett.* **80**, 5243(1998).
[16] C. M. Bender, *Rep. Prog. Phys.* **70**, 947 (2007).
[17] K. G. Makris, R. El-Ganainy, D. N. Christodoulides, Z. H. Musslimani, *Phys. Rev. Lett.* **100**, 103904 (2008).
[18] R. El-Ganainy, K. G. Makris, D. N. Christodoulides, Z. H. Musslimani, *Opt. Lett.* **32**, 2632 (2007).
[19] Z. H. Musslimani, K. G. Makris, R. El-Ganainy, D. N. Christodoulides, *Phys. Rev. Lett.* **100**, 030402 (2008).
[20] A. Guo *et al.*, *Phys. Rev. Lett.* **103**, 093902 (2009).
[21] C. E. Rüter *et al.*, *Nat. Phys.* **6**, 192 (2010).
[22] Materials and methods are available as supporting material on *Science* Online.
[23] M. Abashin *et al.*, *Opt. Express* **14**, 1643 (2006).
[24] B. Liang, X. S. Guo, J. Tu, D. Zhang, J. C. Cheng, *Nat. Mater.* **9**, 989 (2010).
[25] Support by NSF and NSF ERC Center for Integrated Access Networks grant EEC-0812072, Defense Advanced Research Projects Agency under the Nanoscale Architecture for Coherent Hyperoptical Sources program grant W911NF-07-1-0277, National Basic Research Program of China grant 2007CB613202, National Nature Science Foundation of China grants 50632030 and 10874080, and

Nature Science Foundation of Jiangsu Province grant BK2007712. L. F., M. A., J. H., these authors contributed equally to this work. We thank Nanonics, Ltd., for extensive training and support in near-field scanning optical microscopy. M. A. acknowledges support of a Cymer Corp. graduate fellowship.

[26] See Supplemental material at www.sciencemag.org/cgi/content/full/333/6043/729/DC1, Materials and Methods, Figs. S1 to S3, References.

Experimental Demonstration of a Unidirectional Reflectionless Parity-time Metamaterial at Optical Frequencies*

Liang Feng[1], Ye-Long Xu[2], William S. Fegadolli[1,3,4], Ming-Hui Lu[2], José E. B. Oliveira[3], Vilson R. Almeida[3,4], Yan-Feng Chen[2] and Axel Scherer[1]

[1] *Department of Electrical Engineering and Kavli Nanoscience Institute, California Institute of Technology, Pasadena, California 91125, USA*

[2] *National Laboratory of Solid State Microstructures and Department of Materials Science and Engineering, Nanjing University, Nanjing, Jiangsu 210093, China*

[3] *Department of Electronic Engineering, Instituto Tecnológico de Aeronáutica, São José dos Campos, São Paulo 12229-900, Brazil*

[4] *Division of Photonics, Instituto de Estudos Avançados, São José dos Campos, São Paulo 12229-900, Brazil*

Invisibility by metamaterials is of great interest, where optical properties are manipulated in the real permittivity-permeability plane[1,2]. However, the most effective approach to achieving invisibility in various military applications is to absorb the electromagnetic waves emitted from radar to minimize the corresponding reflection and scattering, such that no signal gets bounced back. Here, we show the experimental realization of chip-scale unidirectional reflectionless optical metamaterials near the spontaneous parity-time symmetry phase transition point where reflection from one side is significantly suppressed. This is enabled by engineering the corresponding optical properties of the designed parity-time metamaterial in the complex dielectric permittivity plane. Numerical simulations and experimental verification consistently exhibit asymmetric reflection with high contrast ratios around a wavelength of of 1,550 nm. The demonstrated unidirectional phenomenon at the corresponding parity-time exceptional point on-a-chip confirms the feasibility of creating complicated on-chip parity-time metamaterials and optical devices based on their properties.

Parity-time-symmetric material, also called synthetic matter, is a new class of metamaterials in which parity and time reversal symmetries are combined in such a way that non-Hermitian Hamiltonians may still possess real energy spectra[3—5], creating physical properties of quantum mechanics and quantum field theories in classical systems[6]. Owing to the equivalence between the Schrödinger equation in quantum mechanics and the wave equation in optics, the combined parity-time symmetry has been

---

* Nat.Mater.,2013,12(2):108

intensively studied in classical optical systems with non-Hermitian optical potentials. By exploiting optical modulation of the refractive index in the complex dielectric permittivity plane and engineering both optical absorption and amplification, parity-time metamaterials can lead to a series of intriguing optical phenomena and devices, such as dynamic power oscillations of light propagation[7—13] and coherent perfect absorber-lasers[14—16].

Another important aspect of such parity-time optical materials is that the evolution of parity-time symmetry becomes measurable through the quantum-optical analogue[17—19]. The threshold of parity-time symmetry breaking was clearly visualized as the corresponding energy spectra change dramatically from real to complex after this phase transition point. This threshold is therefore called an exceptional point or spontaneous parity-time symmetry breaking point, where amplitudes of the real and imaginary parts of the modulated refractive index are identical. An interesting phenomenon named unidirectional invisibility was theoretically proposed at this exceptional point in parity-time metamaterials, where reflection from one direction is diminished[20—23]. With the analysis using the scattering matrix (S-matrix)[15,24—26], in the combined parity-time system, the eigenvalues of the S-matrix are all unimodular in the exact parity-time phase but become a pair of reciprocal moduli in the parity-time symmetry broken phase. In particular, the eigenstates in the parity-time metamaterials become degenerate at the exceptional point, causing unidirectional invisibility where it is reflectionless when probed from one side. Most recently, it has been experimentally demonstrated using gain/loss balanced parity-time optical fibre networks in the temporal domain[27], but the realization of the spatial analogue still remains challenging[28] and its on-chip implementation is expected to underpin a new generation of photonic devices[29]. However, optical gain is difficult to achieve using conventional complementary metal-oxide-semiconductor silicon (Si) technology. In this Letter, therefore, in contrast to previous studies[20—23,27], we theoretically propose and experimentally realize an Si-based optical non-Hermitian parity-time system with only absorptive media on the Si-on-insulator (SOI) platform. Because gain and loss are no longer compensated, the corresponding S-matrix is always not unitary. However, a similar exceptional point still exists where the parity-time phase becomes degenerate as the loss in the system is represented by an additional term of attenuation[30]. Remarkably, unidirectional reflection can still be expected at this exceptional point of the proposed passive parity-time metamaterial. In other words, the implemented passive parity-time metamaterial can be viewed as creating a normal parity-time-symmetric potential in a lossy background whose spectra are invariant in the corresponding Fourier space.

As depicted in Fig. 1a, the studied passive parity-time metamaterial embedded inside SiO_2 is an 800-nm-wide and 220-nm-thick Si waveguide with periodic modulation of its dielectric permittivity. The waveguide supports a fundamental mode with a wave vector of $k_1 = 2.69k_0$ (k_0 is the wave vector in air) at a wavelength of 1,550 nm. Therefore to form a

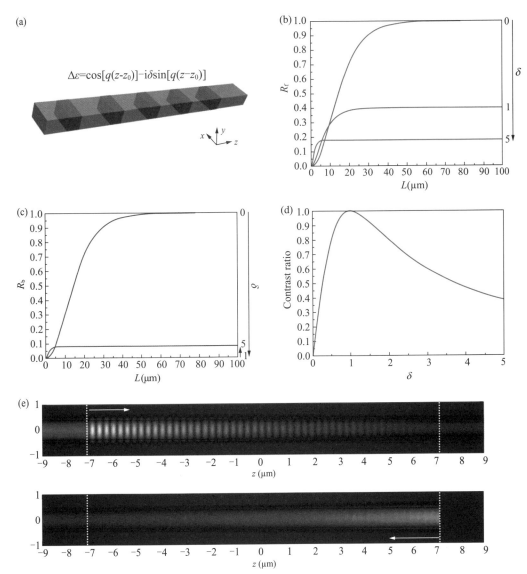

FIG.1. Characteristics of evolution of parity-time symmetry in the proposed passive parity-time metamaterial. a, Schematic of the passive parity-time metamaterial on an SOI platform. Periodically arranged parity-time optical potentials with the modulated dielectric permittivity of $\Delta\varepsilon = \cos(qz) - i\delta\sin(qz)$ ($4n\pi/q + \pi/q \leqslant z \leqslant 4n\pi/q + 2\pi/q$) are introduced into an 800-nm-wide and 220-nm-thick Si waveguide embedded in SiO_2. The period is $4\pi/q = 575.5$ nm. b, c, Dependences of reflectance of the passive parity-time metamaterial on the modulation length with different values of δ at the wavelength of 1,550 nm in forward (b) and backward (c) directions, respectively. In contrast to the monotonic decrease in the forward direction, reflection in the backward direction first reaches a minimum value of $R_b = 0$ at $\delta = 1$ and then increases as δ becomes larger. d, The corresponding contrast ratio of reflectivities in both directions at different values of δ. e, Electric field amplitude distribution of light in the parity-time metamaterial at its exceptional point, where $\delta = 1$ for both forward (upper) and backward (lower) light propagation at a wavelength of 1,550 nm. The 3D FDTD simulation (FDTD Solutions 7.5, Lumerical) consists of 25 parity-time optical potentials with a period of 575.5 nm. Guided light is set to be incident along $+z/-z$ in the forward/backward direction at the boundaries of the parity-time metamaterial (marked with dashed white lines). Therefore light behind the lines is attributed only to reflection.

Bragg grating to reflect the fundamental mode of guided light, the introduced modulation of the dielectric permittivity is $\Delta\varepsilon = \cos(qz) - i\delta\sin(qz)$, where $q = 2k_1$ and $4n\pi/q + \pi/q \leqslant z \leqslant 4n\pi/q + 2\pi/q$. The period is chosen to be $4\pi/q$, corresponding to the second Bragg order, for ease of experimental implementation as described below, but it is worth noting that our structure contains the same parity-time characteristics as the first Bragg order with a period of $2\pi/q$. In the modulated regime, the electric field can be written as $E(x,y,z) = A(z)E(x,y)e^{ik_1 z} + B(z)E(x,y)e^{-ik_1 z}$, where $A(z)$ and $B(z)$ are the amplitudes of forward and backward fundamental modes, respectively. With slowly varying approximation, the coupled-mode equations can be derived as

$$\begin{cases} \dfrac{dA(z)}{dz} = -\dfrac{\delta}{2\pi}\alpha A(z) + i\dfrac{1-\delta}{8}\kappa B(z) \\ \dfrac{dB(z)}{dz} = -i\dfrac{1+\delta}{8}\kappa A(z) + \dfrac{\delta}{2\pi}\alpha B(z) \end{cases}$$

where α and κ denote attenuation and mode coupling between forward and backward fundamental modes (see Methods and Supplementary Information). The transfer matrix for the optical modulation from $z=0$ to $z=L$ can be written as

$$\begin{pmatrix} A(L) \\ B(L) \end{pmatrix} = \begin{bmatrix} M_{11} & M_{12} \\ M_{21} & M_{22} \end{bmatrix} \begin{pmatrix} A(0) \\ B(0) \end{pmatrix},$$

where

$$M_{11} = \cosh(\gamma L) - \frac{\delta}{2\pi}\frac{\alpha}{\gamma}\sinh(\gamma L), \quad M_{12} = i\frac{1-\delta}{8}\frac{\kappa}{\gamma}\sinh(\gamma L),$$

$$M_{21} = -i\frac{1+\delta}{8}\frac{\kappa}{\gamma}\sinh(\gamma L), \quad M_{22} = \cosh(\gamma L) + \frac{\delta}{2\pi}\frac{\alpha}{\gamma}\sinh(\gamma L)$$

and

$$\gamma = \sqrt{(\delta\alpha/2\pi)^2 + (1-\delta^2)\kappa^2/64}.$$

The corresponding **S**-matrix of the parity-time system (notice that the **S**-matrix here is not the same as the convention in electromagnetics)[23—26] is

$$\mathbf{S} = \begin{bmatrix} t & r_b \\ r_f & t \end{bmatrix} = a\mathbf{S}' = \frac{\sqrt{1+M_{12}M_{21}}}{M_{22}} \times \begin{pmatrix} \dfrac{1}{\sqrt{1+M_{12}M_{21}}} & \dfrac{M_{12}}{\sqrt{1+M_{12}M_{21}}} \\ \dfrac{-M_{21}}{\sqrt{1+M_{12}M_{21}}} & \dfrac{1}{\sqrt{1+M_{12}M_{21}}} \end{pmatrix}$$

where $a = \sqrt{1+M_{12}M_{21}}/M_{22}$, t is the transmission amplitude, and r_f and r_b are the amplitudes of reflection in forward and backward directions, respectively. Because the eigenvalues of $\mathbf{S}(s_n = t \pm \sqrt{r_f r_b})$ correspond to different phases of parity-time, the parity-time phase and its dependence on δ are analysed accordingly. When $0 \leqslant \delta < 1$, the eigenvalues

$$s_n = as_n' = a\frac{1 \pm i\dfrac{\kappa}{8\gamma}\sinh(\gamma L)\sqrt{1-\delta^2}}{\sqrt{1+M_{12}M_{21}}}$$

where s'_n is unimodular. It is therefore evident that the parity-time symmetry of the studied passive system is similar to the parity-time symmetric phase in the balanced gain/loss system but with an additional attenuation term $|a|$ (see Supplementary Information for detailed comparison with typical parity-time systems with balanced gain/loss). When $\delta > 1$,

$$s_n = as'_n = a \frac{1 \pm \frac{\kappa}{8\gamma}\sinh(\gamma L)\sqrt{\delta^2 - 1}}{\sqrt{1 + M_{12}M_{21}}}$$

corresponds to the broken parity-time phase in the gain/loss system. Remarkably, at $\delta = 1$, $s_n = as'_n = \exp(-\alpha L/2\pi)$, where $s'_n = 1$, and therefore the eigenvalues of the **S**-matrix become degenerate, indicating that the phase transition takes place at this exceptional point. The evolution of parity-time symmetry can be macroscopically observed via forward and backward reflection from the Bragg grating. The corresponding reflectivities of the Bragg grating for forward and backward directions are, respectively,

$$R_f = \left|\frac{M_{21}}{M_{22}}\right|^2, R_b = \left|\frac{M_{12}}{M_{22}}\right|^2.$$

Figure 1b,c shows the calculated reflection spectra as a function of modulation lengths at different values of δ for forward and backward directions, respectively. Because the parity-time metamaterial is only absorptive, the reflectivity reaches an asymptotic value after enough modulation. However, forward and backward directions show significantly distinguishable tendencies of the corresponding maximum reflection when δ increases. It is evident that $\delta = 1$ is the exceptional point corresponding to the phase transition of parity-time symmetry where the Bragg grating is unidirectional reflectionless. The exceptional characteristics owing to the parity-time phase transition can be clearly visualized from the contrast ratio between the forward and backward reflectivities ($C = |(R_f - R_b)/(R_f + R_b)| = (2\delta/(1+\delta^2))$), shown in Fig. 1d, where the Bragg grating is assumed long enough such that the reflectivities in both directions approach their corresponding asymptotic values. At $\delta = 1$, the extreme of the dispersion is reached, which is direct evidence of the parity-time phase transition. Even with the frequency detuning, the system is unidirectional reflectionless in a broad band, with a contrast ratio of 1 at the exceptional point (see Supplementary Information). Consistent with the above theoretical analysis, mappings of light intensity in the waveguide using three-dimensional (3D) finite difference time domain (FDTD) simulations only show a reflected field in the forward direction (Fig. 1e). Moreover, the interference pattern between incidence and reflection can be observed only in the forward direction, further confirming that the system is unidirectional reflectionless at its exceptional point.

To realize the proposed unidirectional reflectionless parity-time metamaterial at its exceptional point ($\delta = 1$), an equivalent guided-mode modulation is designed using additional structures on top of the waveguide to mimic the microscopic parity-time modulation on a macroscopic scale. However, the balance of the modulation amplitude in the real part $\Delta\varepsilon_{rea}=\cos(qz)$ in an entire period is broken, because additional structures on top can only result in a higher effective index. To be consistent with the final design, regions of $\Delta\varepsilon_{real}$ are extracted from original parity-time optical potentials and the corresponding cosine modulation is shifted $5\pi/2q$ in the z direction to become a positive sinusoidal modulation, whereas the imaginary part modulation remains at the same location to provide in-phase modulation to guided light together with the shifted real part. Although the real and imaginary part modulations are now separated, our in-phase arrangement creates an equivalent optical modulation to guided light in both amplitude and phase as if the original complex parity-time optical potential still existed (see Supplementary Information). Finally, to achieve these sinusoidal-function modulated optical potentials using microscopically homogeneous materials, sinusoidal-shaped combo structures (combination of a sinusoidal shaped structure and its mirror image to the transverse direction) are adopted on top of the Si waveguide (Fig. 2a). Mode effective indices with additional Si and germanium (Ge)/chrome (Cr) bilayer combo structures are consistent with the real and imaginary part modulations, respectively, showing the equivalence of the designed structure to the parity-time metamaterial (Fig. 2b). The characteristics of this parity-time metamaterial have also been verified using FDTD simulations. The reflectivities are significantly distinguished in the forward and backward directions, with about 11 dB of extinction ratio in the studied wavelength range from 1,520 to 1,580 nm (Fig. 2c). It is worth noting that the resonance peak of the Bragg grating moves to around 1,560 nm, because the modulation in the real part is only positive, which creates a higher effective index and thus red-shifts the resonance peak compared with the case in Fig. 1 where the real part modulation has balanced positive and negative amplitudes (see Supplementary Information). The contrast ratio between the forward and backward directions is plotted in Fig. 2d, showing that reflection in the backward direction is significantly suppressed. The asymmetry in reflection from the designed parity-time metamaterial can also be visualized from mappings of light propagating inside the waveguide (Fig. 2e): forward propagating light and its reflection form strong interference, whereas reflection is barely seen with backward incidence. It is thus evident that the designed on-chip waveguide system successfully mimics the unidirectional effect inherently associated with the exceptional point.

In experiments, however, measuring in-line reflection would require the use of external components such as optical circulators or directional couplers, which in turn cause additional insertion loss and extra noise in the measurements, making it difficult to perform. Therefore, we propose a strategy using on-chip waveguide directional couplers to measure the corresponding reflectance in a way similar to the transmission measurement as shown in Fig. 3a. The sample was fabricated using overlay electron beam lithography, followed by evaporation and lift-off of Si and Ge/Cr as well as dry etching to form the Si waveguide (see Methods). The pictures of the device before deposition of the SiO_2 cladding

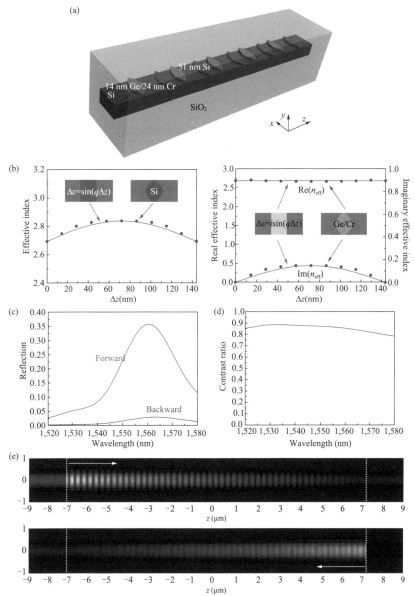

FIG. 2. Optical properties of the designed passive unidirectional reflectionless parity-time metamaterial. a, Periodically arranged 760-nm-wide sinusoidal shaped combo structures are applied on top of an 800-nm-wide Si waveguide embedded inside SiO$_2$ to mimic parity-time optical potentials, in which imaginary part modulation is implemented with 14nm Ge/24nm Cr structures and 51nm Si layers are for real part modulation. The designed parity-time metamaterial consists of 25 sets of top-modulated combo structures with a period of 575.5 nm and a width of 143.9 nm for each sinusoidal-shaped combo. b, Mode effective index with the real (left) and imaginary (right) parts of the modulated dielectric permittivity (red lines) as well as comparisons with their corresponding sinusoidal-shaped Si (left) and Ge/Cr (right) combos (blue dots). c, Simulated reflection spectra of the device in forward (red) and backward (blue) directions. d, Spectrum of contrast ratio of reflectivities, showing high contrast ratios over the studied wavelength range from 1,520 to 1,580 nm. e, Simulated electric field amplitude distribution of light in the device, where incidence is set at boundaries of the parity-time metamaterial (marked with white lines) along $+z/-z$ in the forward/backward direction.

are shown in Fig. 3b, c. In experiments, tapered fibres were used to couple light from fibres to waveguides and vice versa. The reflection spectra of the fabricated device have been measured for both forward and backward directions, as shown in Fig. 4a, respectively. Consistent with simulations, red-shifts of the resonance peaks are also observed in experimental measurements for both directions. The measured reflection spectra also show significantly distinguished characteristics in reflection: the reflectivity in the forward direction is about 7.5 dB stronger than that in the backward direction, indicating asymmetric optical properties owing to the parity-time symmetry phase transition. The corresponding high-contrast ratios over a broad band of telecom wavelengths in Fig. 4b further confirm the asymmetric reflection of the parity-time metamaterial associated with the exceptional point.

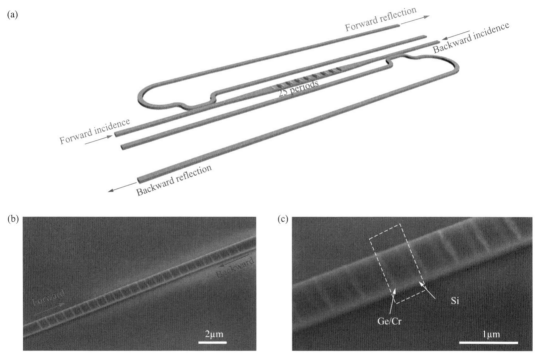

FIG.3. Experimental implementation of the passive unidirectional reflectionless parity-time metamaterial. a, Configuration of waveguides on the SOI platform to measure reflection from the device. A single-mode waveguide directional coupler with about 3 dB of coupling ratio is designed to maximize the detected signal (see Supplementary Information). Tapered waveguides are also introduced to transfer light from the single mode in the 400-nm-wide Si waveguides to the fundamental mode in the proposed waveguide device and vice versa. b, SEM picture of the whole device before deposition of SiO_2 cladding. The fabricated device consists of 25 periods of Ge/Cr and Si sinusoidal combo structures with a period of 575.5 nm on top of the Si waveguide. c, Zoom-in SEM picture of the device, where the boxed area indicates a unit cell. The remaining HSQ resist and two kinds of sinusoidal combos can be seen on top of the waveguide.

It is therefore evident that the presented parity-time metamaterial can successfully mimic the complicated quantum phenomena in classical on-chip optical waveguide systems. Similarly to previously investigated balanced gain/loss systems, proper engineering of the

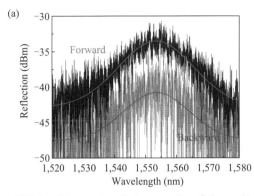
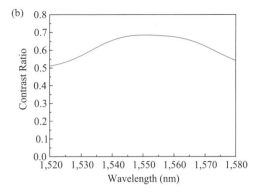

FIG.4. Measured optical properties of the parity-time metamaterial. a, Measured reflection spectra of the device through the waveguide coupler for both directions over a broad band of telecom wavelengths from 1,520 to 1,580 nm. Red and blue curves are Gaussian fits of raw data in forward (black) and backward (green) directions, respectively. b, Spectrum of contrast ratio of reflectivities obtained from the fitting data in a.

complex dielectric permittivity in a passive system also makes the exceptional point as well as its associated phase transition of parity-time symmetry become measurable quantities. The chip-scale parity-time metamaterial fabricated by conventional complementary metaloxide-semiconductor fabrication procedures clearly confirms the expected unidirectional and asymmetric characteristics in experiments. Moreover, the demonstrated unidirectional reflection can be converted to unidirectional invisibility by adding an additional linear amplifier to the presented parity-time metamaterial (see Supplementary Information). Further investigation of engineering the complex refractive index (linear and nonlinear parts) as well as geometric arrangement of these optical potentials is expected to create novel parity-time metamaterials with even more counterintuitive physical responses that once belonged only to the field of quantum theory, thus paving an approach to a new generation of photonic devices. For example, a four-port broadband unidirectional photonic device may construct backbones of chip-scale optical network analysers as its circuit counterpart in state-of-the-art microwave network analysers. Furthermore, the general design principle in our device in the optical domain can be extended to other frequencies and classical wave systems, such as unidirectional microwave invisibility for military applications and ultrasonic equipment for marine exploration[31].

Methods

Derivation of attenuation and coupling coefficients in coupled mode equations. The attenuation coefficient α indicates the system absorption due to the imaginary part of the parity-time potentials and the coupling coefficient κ is the mode overlap between forward and backward propagating light. Both of them can be derived at the exceptional point, $\delta = 1$, where the calculation is the most simplified. At the exceptional point, the modulated mode effective index can be approximately written as $n_{\text{eff}} \approx 2.69 + 0.15\cos(qz) - \text{i}0.15\sin$

(qz). Therefore the attenuation constant is calculated as $\alpha = n_{\text{imag}} k_0 = 0.15 k_0 = 0.61 \ \mu\text{m}^{-1}$, while the coupling coefficient is $\kappa = (k_0^2/2k_1)(\int E^* \Delta\varepsilon E \text{d}s / \int |E|^2 \text{d}s) = 0.48 \ \mu\text{m}^{-1}$. Both α and κ are validated using FDTD simulations. At the exceptional point, the coupled mode equations can be written as

$$\begin{cases} \dfrac{\text{d}A(z)}{\text{d}z} = -\dfrac{\alpha}{2\pi} A(z) \\ \dfrac{\text{d}B(z)}{\text{d}z} = -\text{i} \dfrac{\kappa}{4} A(z) + \dfrac{\alpha}{2\pi} B(z) \end{cases}$$

The corresponding transmission and reflection coefficients are $T = \exp(-\alpha L/\pi)$, $R_f = (\pi^2 \kappa^2 / 4\alpha^2) \sinh^2(\alpha L / 2\pi) \exp(-\alpha L/\pi)$, and $R_b = 0$, which can be used to fit the obtained reflection spectra from FDTD. α and κ in the best fitting results are about $0.61 \ \mu\text{m}^{-1}$ and $0.49 \ \mu\text{m}^{-1}$ (see Supplementary Information), consistent with the above analytical derivations.

Sample fabrication. The fabrication starts with an SOI wafer. Two layers of periodically arranged sinusoidal-shaped combo structures are first patterned with accurate alignment in polymethyl methacrylate by electron beam lithography, followed by electron beam evaporation of Ge/Cr and Si and lift-off in acetone, respectively, in two steps. Then the Si waveguide is defined by aligned electron beam lithography using hydrogen silsesquioxane (HSQ), followed by dry etching with mixed gases of SF_6 and C_4F_8. Because the developed HSQ becomes porous SiO_2 and its refractive index is similar to SiO_2, it remains on top of the waveguide and therefore the topology of the sinusoidal combos on top of the waveguide became obscure in scanning electron microscope (SEM) images (Fig. 3b, c). Finally, plasma enhanced chemical vapour deposition with mixed gases of SiH_4 and N_2O is used to deposit the cladding of SiO_2 on the entire wafer to increase the light coupling efficiency from tapered fibres to waveguides.

References and Notes

[1] Leonhardt, U. Optical conformal mapping. *Science* **312**, 1777 – 1780 (2006).

[2] Pendry, J. B., Schurig, D. & Smith, D. R. Controlling electromagnetic fields. *Science* **312**, 1780 – 1782 (2006).

[3] Bender, C. M. & Böttcher, S. Real spectra in non-Hermitian Hamiltonians having PT symmetry. *Phys. Rev. Lett.* **80**, 5243 – 5246 (1998).

[4] Bender, C. M. Making sense of non-Hermitian Hamiltonians. *Rep. Prog. Phys.* **70**, 947 – 1018 (2007).

[5] Bender, C. M., Brody, D. C. & Jones, H. F. Extension of PT-symmetric quantum mechanics to quantum field theory with cubic interaction. *Phys. Rev. D* **70**, 025001 (2004).

[6] Longhi, S. & Della Valle, G. Photonic realization of PT-symmetric quantum field theories. *Phys. Rev. A* **85**, 012112(2012).

[7] Makris, K. G., El-Ganainy, R., Christodoulides, D. N. & Musslimani, Z. H. Beam dynamics in PT-symmetric optical lattices. *Phys. Rev. Lett.* **100**, 103904 (2008).

[8] Musslimani, Z. H., El-Ganainy, R., Makris, K. G. & Christodoulides, D. N. Optical solitons in PT periodic potentials. *Phys. Rev. Lett.* **100**, 030402 (2008).

[9] Klaiman, S., Guenther, U. & Moiseyev, N. Visualization of branch points in PT symmetric waveguides. *Phys. Rev. Lett.* **101**, 080402 (2008).

[10] Longhi, S. Bloch oscillations in complex crystals with PT symmetry. *Phys. Rev. Lett.* **103**, 123601 (2009).

[11] Zheng, M. C., Christodoulides, D. N., Fleischmann, R. & Kottos, T. PT optical lattices and universality in beam dynamics. *Phys. Rev. A* **82**, 010103(R) (2010).

[12] Graefe, E. M. & Jones, H. F. PT-symmetric sinusoidal optical lattices at the symmetry-breaking threshold. *Phys. Rev. A* **84**, 013818 (2011).

[13] Miroshnichenko, A. E., Malomed, B. A. & Kivshar, Y. S. Nonlinearly PT-symmetric systems: Spontaneous symmetry breaking and transmission resonances. *Phys. Rev. A* **84**, 012123 (2011).

[14] Longhi, S. PT-symmetric laser absorber. *Phys. Rev. A* **82**, 031801 (2010).

[15] Chong, Y. D., Ge, L. & Stone, A. D. PT-symmetry breaking and laser-absorber modes in optical scattering systems. *Phys. Rev. Lett.* **106**, 093902 (2011).

[16] Ge, L. *et al*. Unconventional modes in lasers with spatially varying gain and loss. *Phys. Rev. A* **84**, 023820 (2011).

[17] Rüter, C. E. *et al*. Observation of parity-time symmetry in optics. *Nature Phys.* **6**, 192–195 (2010).

[18] Guo, A. *et al*. Observation of PT-symmetry breaking in complex optical potentials. *Phys. Rev. Lett.* **103**, 093902 (2009).

[19] Schindler, J., Li, A., Zheng, M. C., Ellis, F. M. & Kottos, T. Experimental study of active LRC circuits with PT symmetries. *Phys. Rev. A* **84**, 040101(R) (2011).

[20] Kulishov, M., Laniel, J., Bélanger, N., Azaña, J. & Plant, D. Nonreciprocal waveguide Bragg gratings. *Opt. Express* **13**, 3068–3078 (2005).

[21] Lin, Z. *et al*. Unidirectional invisibility induced by PT-symmetric periodic structures. *Phys. Rev. Lett.* **106**, 213901 (2011).

[22] Longhi, S. Invisibility in PT-symmetric complex crystals. *J. Phys. A* **44**, 485302 (2011).

[23] Jones, H. F. Analytic results for a PT-symmetric optical structure. *J. Phys. A* **45**, 135306 (2012).

[24] Mostafazadeh, A. Spectral singularities of complex scattering potentials and infinite reflection and transmission coefficients at real energies. *Phys. Rev. Lett.* **102**, 220402 (2009).

[25] Muga, J. G., Palaob, J. P., Navarroa, B. & Egusquizac, I. L. Complex absorbing potentials. *Phys. Rep.* **395**, 357–426 (2004).

[26] Cannata, F., Dedonder, J-P. & Ventura, A. Scattering in PT-symmetric quantum mechanics. *Ann. Phys.* **322**, 397–433 (2007).

[27] Regensburger, A. *et al*. Parity-time synthetic photonic lattices. *Nature* **488**, 167–171 (2012).

[28] Razzari, L. & Morandotti, R. Gain and loss mixed in the same cauldron. *Nature* **488**, 163–164 (2012).

[29] Kottos, T. Optical physics: Broken symmetry makes light work. *Nature Phys.* **6**, 166–167 (2010).

[30] Berry, M. V. Optical lattices with PT symmetry are not transparent. *J. Phys. A* **41**, 244007 (2008).

[31] Zhang, S., Xia, C. & Fang, N. Broadband acoustic cloak for ultrasound waves. *Phys. Rev. Lett.* **106**, 024301 (2011).

[32] We acknowledge critical support and infrastructure provided for this work by the Kavli Nanoscience Institute at Caltech. This work was supported by the NSF ERC Center for Integrated Access Networks

(no. EEC-0812072), the National Basic Research of China (no. 2012CB921503 and no. 2013CB632702), the National Nature Science Foundation of China (no. 11134006), the Nature Science Foundation of Jiangsu Province (no. BK2009007), the Priority Academic Program Development of Jiangsu Higher Education, and CAPES and CNPQ—Brazilian Foundations. M-H.L. also acknowledges the support of FANEDD of China.

[33] L.F. and M-H.L. conceived the idea. L.F., Y-L.X. and M-H.L. designed the device. Y-L.X., L.F. and M-H.L. performed the theoretical analysis of parity-time symmetry. W.S.F. and L.F. designed the chip and carried out fabrications and measurements. All the authors contributed to discussion of the project. Y-F.C. and A.S. guided the project. L.F. wrote the manuscript with revisions from other authors. L.F., Y-L. X. and W.S.F. contributed equally to this work.

[34] Supplementary information is available in the online version of the paper.

Plasmonic Airy Beam Generated by In-Plane Diffraction[*]

L. Li, T. Li, S. M. Wang, C. Zhang, and S. N. Zhu

National Laboratory of Solid State Microstructures, College of Physics,
College of Engineering and Applied Sciences, Nanjing University, Nanjing 210093, China

We report an experimental realization of a plasmonic Airy beam, which is generated thoroughly on a silver surface. With a carefully designed nanoarray structure, such Airy beams come into being from an in-plane propagating surface plasmon polariton wave, exhibiting nonspreading, self-bending, and self-healing properties. Besides, a new phase-tuning method based on nonperfectly matched diffraction processes is proposed to generate and modulate the beam almost at will. This unique plasmonic Airy beam as well as the generation method would significantly promote the evolutions in in-plane surface plasmon polariton manipulations and indicate potential applications in lab-on-chip photonic integrations.

The Airy wave packet is the only nontrivial one-dimensional (1D) solution for a wave propagation maintaining the nonspreading property, which was deduced from the Schrödinger equation in quantum mechanics for a free particle[1]. Since its recent observation in optics[2], intensive studies have been carried out, such as self-accelerating[3], ballistic dynamics[4], and self-healing[5,6], as well as the recent nonlinear generation[7] and possible applications[8—10]. To date, most of these optical Airy beams were generated in 3D free space. An intuitive extension of this unique wave packet to two dimensions would possibly forecast more fascinating physics and applications, especially as it is accommodated in the subwavelength plasmonics, which is another active field nowadays[11—15]. As has been envisioned theoretically[16], a plasmonic Airy beam would provide an effective means to route energy over a metal surface between plasmonic devices.

According the nature of the Airy wave packet[1], a 3/2-power phase modulation along the lateral dimension of beam is required[17,18]. It is commonly modulated to the 3-power phase type by a mask on an incident Gaussian beam with a followed Fourier transformation in generations of free space Airy beams[2—10]. This phase requirement is inherited for a surface plasmon polariton (SPP) Airy beam with a modified field form[16]. However, the conventional method tends to be hardly adopted in the SPP regime due to the complex transformation process and large spatial expense. Although an alternative approach was conceived by coupling a free-space-generated Airy beam into a planar plasmonic one[16], it

---

[*] Phys. Rev. Lett., 2011, 107(12): 126804

will inevitably bring other severe obstacles and remains a great inconvenience in the full in-plane manipulations.

Here, we report a new experimental realization of plasmonic Airy beams on a silver surface at a visible wavelength, which is accomplished by particular diffraction processes with a carefully designed nanocave array on a metal surface. This achieved plasmonic Airy beam directly reveals the nonspreading, self-bending, and self-healing properties and demonstrates the capacity of a transversely self-confined SPP beam in a planar dimension with lower propagation loss. The proposed new diffraction approach by designable nanostructures has exhibited its flexibility in beam tailoring, which may significantly stimulate further manipulations of SPPs in a planar dimension.

Using a periodic array on metal surfaces to manipulate SPP propagations has achieved great success in recent years[19-22]. These approaches used to change the SPP propagations can also be interpreted as a phase modulation, which is in coincidence with the descriptions in diffraction optics. In principle, it is quite possible to use an inhomogeneous array to change a linear phase of an incident SPP into a nonlinear one. Therefore, the Airy beam requiring 3/2 phase modulation is highly expected by diffractions in a nonperiodic array system. The scheme of our design is shown in Fig. 1. On the surface of a silver film (with SiO_2 as the substrate), an in-plane propagating SPP wave, generated by a grating coupling of a He-Ne laser ($\lambda_0 = 632.8$ nm), directly incidents into a nonperiodically arranged nanocave array. The diffracted SPP waves from nanocaves will interfere and ultimately build two SPP Airy beams on both sides.

In experiments, the nanocave array sample was fabricated by focused ion beam (FEI Strata FIB 201, 30 keV, 11 pA) milling on a 60-nm-thickness silver film, which has been deposited on a 0.2-mm-thickness SiO_2 substrate. The analysis of SPP wave propagation was performed by a home-built leakage radiation microscope (LRM) system[23,24] (for details, see [25]). The inset image in the lower right of Fig. 1 is a typical experimental result, which intuitively demonstrates the generation of SPP Airy beams that are very analogous to the schematic illustration, manifesting the self-bending, nonspreading, and multiple lobes. Here, the sample of the nanocave array is designed graded in the x direction and periodic in the z direction. Figure 2(a) depicts the top view of the graded nanocave array together with a grating coupler, and the detected SPP beam in the right branch is specifically shown in Fig. 2(b), where a set of Airy-like wave profiles is clearly manifested. Subsequently, we performed a theoretical calculation based on the Huygens-Fresnel principle[18] that all nanocaves in the array are considered as subsources radiating cylindrical surface waves with designed initial phases. The calculated beam trajectories [shown in Fig. 2(c)] are in good agreement with the experiment ones, although they are both imperfect due to limited diffraction elements and nonideal modulations.

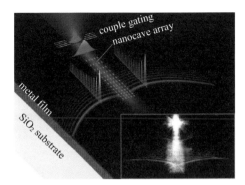 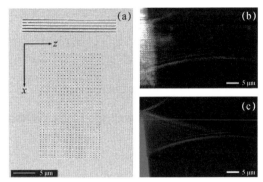

FIG. 1. (color online). Schematic of the generation of the SPP Airy beam. A laser beam is coupled into an in-plane propagating SPP wave by grating and incidents into a nonperiodically arranged nanocave array. Two SPP Airy beams are formed on both sides of the array by diffraction processes. The inset is a typical experimental result of a SPP Airy beam examined by the LRM system.

FIG. 2. (color online). (a) Top view of the graded nanocave array sample fabricated by a focused ion beam, where the lattice parameter is graded in the x dimension (a_x from 420 nm to 780 nm, grads $\Delta = 10$ nm), and the period in the z dimension is $p_z = 620$ nm. (b) Experimentally achieved SPP beam trajectories and (c) the calculated one.

To explain how the SPP Airy-like beam comes into being, a nonlinear phase modulation by diffractions from a nonperiodic array is introduced. As is well known, an incident SPP wave will be diffracted into a well-defined direction by a designed array governed by the Bragg condition, which can be clearly schemed out in the reciprocal space[26] [see Fig.3(a)]. However, if this condition is not perfectly satisfied, a preference diffraction will still occur with some sacrifice in intensity (as long as the deviation is not too large), owing to the elongated reciprocal lattice of the finite-scale array [see Fig. 3(b)] that is similar to the X-ray diffraction cases[27]. It is also proved by our experiments[28]. Therefore, a different lattice parameter is able to determine the preference diffraction direction, which can be regarded equivalently to yield an extra phase change of 2π from every local lattice (in the x direction). When a graded array is employed, the incident beam will diffract to different directions at different positions according to the local lattice parameters (for this small gradient case, $\Delta = 10$ nm). Thus, we can obtain the corresponding phase evolution from every lattice point in the incident SPP propagation (x direction) as $\phi(x) = \phi_0 + k_{spp}x - 2m\pi$. This phase evolution in turn manifests the gradually changed diffraction directions by the graded lattice.

From Fig. 2(a), a beaming angle ($\theta \sim 20°$) with respect to the z axis is found for the main lobe. This means that the lattice boundary (line $z=0$) is not the start line of this SPP Airy beam. We can deduce the phase information at a virtual starting line in the ξ axis that is perpendicular to the tangent of the beaming trajectory of the main lobe according to the principle of geometric optics as [see the scheme in Fig. 3(c)]

$$\phi(\xi) = \phi(x) - k_{spp}b = 2m\pi + k_{spp}x - k_{spp}b, \tag{1}$$

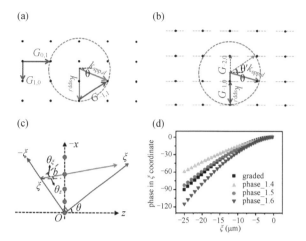

FIG. 3. (color online). Ewald construction for SPP diffraction direction with the Bragg condition is (a) satisfied and (b) not perfectly satisfied with limit diffraction elements in the z axis (elongated reciprocal lattices in the z axis are indicated). $k_{spp,i}$ and $k_{spp,d}$ are the incident and diffracted SPP wave vectors, respectively. $G_{0,1}$ and $G_{1,0}$ are two basic vectors of the reciprocal lattice. (c) Scheme of the phase transformation from the x axis to a virtual ξ axis, which can be designed with respect to the beaming angle θ for the main lobe of the SPP Airy beam. (d) Deduced phase distributions in the starting ξ axis together with the 1.4-, 1.5-, and 1.6-power phase modulations.

where

$$b = -\frac{x\tan(\theta_0)}{\cos(\theta_x)+\tan(\theta_0)\sin(\theta_x)},$$
$$\xi = \frac{x}{\cos(\theta_0)[1+\tan(\theta_x)\tan(\theta_0)]},$$
(2)

and $\sin(\theta_0)=\frac{\lambda_{spp}-a_0}{a_0}$, $\sin(\theta_x)=\frac{\lambda_{spp}-a_x}{a_x}$, and a_x is defined as the local lattice determined by the mean value of two distances before and after the lattice of x. According to the experimental result of the position of the original point of O ($a_0 \sim 450$ nm) and initial angle ($\theta \sim 20°$), we calculated the transformed phase $\phi(\xi)$ shown in Fig. 3(d) together with the results of 1.4-, 1.5-, and 1.6-power phase modulations. It is clearly seen that the deduced data from the graded array match the 3/2-power relation well, explaining the outcome of the Airy-like SPP beam.

Based on the above phase modulation method, a SPP Airy beam with a defined beaming direction can be generated by a proper nonperiodic array. With a given angle of θ, the corresponding phase along the x axis can be retrieved as

$$\psi(x) = -\frac{2}{3}\left(-\frac{\xi}{\xi_0}\right)^{3/2} - \frac{\pi}{4} - k\frac{\xi\sin\theta}{\cos(\theta-\theta_\xi)},$$
(3)

where $x=\xi\cos(\theta)+\xi\sin(\theta)\tan(\theta-\theta_\xi)$, $\theta_\xi = \arcsin[\partial_\xi\phi(\xi)]$, $\phi(\xi)$ is the phase satisfying the Airy function, and ξ_0 is a constant that determines the acceleration of Airy beam. According to the equivalent phase by diffraction $\phi_m(x) = kx + 2m\pi$, we can deduce the location of the mth diffraction unit by solving $\phi_m(x) = \psi(x)$ and ultimately retrieve the arrangement of the nanocave array. Figure 4(a) is the calculated SPP beam trajectories constructed by the diffraction processes with the designed array data shown in the inset image ($p_z = 640$ nm), corresponding to a horizontal beaming of $\theta = 0°$ with $\xi_0 = 1.08$. Figure 4(b) is the experimental result recorded by the LRM system, which well reproduces the calculated one, indicating the outcome of the SPP Airy beam. The inset image is the distribution of field intensity in the x axis picked up from the line of $z = 15$ μm (15 μm away from the start line), in which a definite Airy-like profile and small FWHM∼1.3 μm of the main lobe are clearly manifested. Similar to the previous result [Fig. 2(b)], this main lobe exhibits a weaker attenuation and preserves its narrow beam width over a long distance[28].

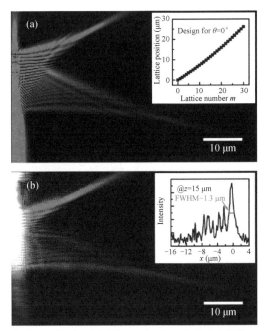

FIG. 4. (color online). (a) Theoretically calculated SPP propagation trajectories corresponding to a horizontal beaming ($\theta = 0°$) with a designed nonperiodic lattice in the x dimension (inset image) and $p_z = 640$ nm ($\xi_0 = 1.08$). (b) Experimentally recorded SPP beam trajectories by the LRM system with a beam profile at the propagation distance of $z = 15$ μm (inset image), for example.

In the following, we would like to make a detailed analysis on the propagation property of this achieved SPP Airy beam. Figure 5(a) depicts the normalized field intensity of the main lobe as a function of propagation distance with a comparison to that of a conventional SPP Gaussian beam, where the inset shows the directly coupled SPP Gaussian beam recorded by LRM (with an initial FWHM ∼3 μm). It is well demonstrated that the

Airy beam has a much longer propagation length (~50 μm) than the Gaussian beam(~15 μm). Considering the narrow beam width, the main lobe of the SPP Airy beam behaves like a self-confined in-plane waveguide with lower loss and suggests possible applications in guiding SPP waves. To further confirm the characteristics of the Airy beam, we plot the trajectory of the main lobe, which shows considerable coincidence with the analytical parabolic curve derived from our design, see Fig. 5(b). In addition, we artificially introduced two blocks in the beam paths to test the self-healing property, as shown in Figs. 5(c) and 5(d) for blocks of small (1.5×0.6 μm^2) and large (2.2×0.6 μm^2) rectangular holes, respectively. It is evident that the SPP Airy beams indeed heal up by themselves for both cases. Therefore, it is not doubted that a well-developed SPP Airy beam is accomplished in a designable way.

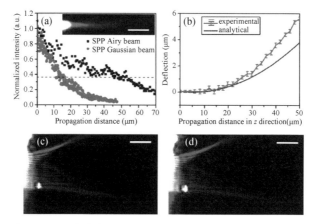

FIG. 5. (color online) (a) Field intensities of the main lobe of the SPP Airy beam and a conventional Gaussian beam as functions of propagation distance. The inset image is the LRM-detected Gaussian beam (initial FWHM ~3 μm). (b) Experimental trajectory of the main lobe compared with the analytical parabolic curve for the case of horizontal beaming. Experimental results of self-healing with respect to (c) a small rectangular block (1.5×0.6 μm^2) and (d) a larger one (2.2×0.6 μm^2). All scale bars equal 10 μm.

For further discussions, one point should be addressed that the beam width of main lobe(~1.3 μm) is near to the SPP wavelength(λ_{SPP}=610 nm) and the constant ξ_0=1.08 in Eq.(3) is not large enough considering the paraxial condition[29]. But it is still valid since such a SPP Airy beam still has a long dispersive distance (larger than 70 μm for the main lobe)[28]. From our additional simulations, if ξ_0 is further decreased, the Airy beam will disperse in a shorter distance and tend to totally collapse. Besides, in our approach the exact intensity modulation of the initial Airy function (the −1/4-power relation[1]) is not completely fulfilled. Fortunately, the graded system is able to build a localized wave packet at its propagation end (around the matched condition)[30], which usually has an asymmetric profile that is considerably analogous to the intensity envelope required by the Airy function. As for a more precise intensity modulation, we believe the beam would be improved by carefully tuning the diffraction elements. The phase modulation is the critical factor to achieve the Airy beam[17,18], and it is well proved to be fulfilled by such in-plane diffraction processes.

In conclusion, we have experimentally demonstrated the SPP Airy beam on a silver surface by particularly designed in-plane diffraction processes. The revealed SPP Airy beam exhibits unique features as expected, and the nonspreading property with lateral confinement for the main lobe over a long distance has implications in SPP manipulation and other related fields (e.g., arranging nanoparticles in nanoscale). Notably, the newly proposed method based on the nonperfectly matched diffraction effect allows for flexible modulations on the established plasmonic Airy beam almost at will, which may have more general applications in the wave-front tailoring as well as developing new kinds of photonic or plasmonic devices.

References and Notes

[1] M.V. Berry and N.L. Balazs, Am. J. Phys. **47**, 264 (1979).
[2] G. A. Siviloglou, J. Broky, A. Dogariu, and D. N. Christodoulides, Phys. Rev. Lett. **99**, 213901 (2007).
[3] G. A. Siviloglou, and D. N. Christodoulides, Opt. Lett. **32**, 979 (2007).
[4] G. A. Siviloglou, J. Broky, A. Dogariu, and D. N. Christodoulides, Opt. Lett. **33**, 207 (2008).
[5] J. Broky, G.A. Siviloglou, A. Dogariu, and D.N. Christodoulides, Opt. Express **16**, 12880 (2008).
[6] L. Carretero et al., Opt. Express **17**, 22432 (2009).
[7] T. Ellenbogen, N. Voloch-Bloch, A. Ganany-Padowicz, and A. Arie, Nat. Photon. **3**, 395 (2009).
[8] J. Baumgartl, M. Mazilu, and K. Dholakia, Nat. Photon. **2**, 675 (2008).
[9] P. Polynkin, M.Kolesik, J.V. Moloney, G. A. Siviloglou, and D. N. Christodoulides, Science **324**, 229 (2009).
[10] Y. L. Gu and G. Gbur, Opt. Lett. **35**, 3456 (2010).
[11] E. Ozbay, Science **311**, 189 (2006).
[12] D. K. Gramotnev and S. I. Bozhevolnyi, Nat. Photon. **4**, 83 (2010).
[13] W.L. Barnes, A. Dereux, and T.W. Ebbesen, Nature (London) **424**, 824 (2003).
[14] H. Ditlbacher, J.R. Krenn, G. Schider, A. Leitner, and F. R. Aussenegg, Appl. Phys. Lett. **81**, 1762 (2002).
[15] I.P. Radko et al., Laser Photon. Rev. **3**, 575 (2009).
[16] A. Salandrino and D. N. Christodoulides, Opt. Lett. **35**, 2082 (2010).
[17] D. M. Cottrell, J. A. Davis, and T. M. Hazard, Opt. Lett. **34**, 2634 (2009).
[18] Y. Kaganovsky and E. Heyman, Opt. Express **18**, 8440 (2010).
[19] L. L. Yin et al., Nano Lett. **5**, 1399 (2005).
[20] M.U. González et al., Phys. Rev. B **73**, 155416 (2006).
[21] A. B. Evlyukhin, S. I. Bozhevolnyi, A. L. Stepanov, and J. R. Krenn, Appl. Phys. B **84**, 29 (2006).
[22] A. Drezet et al., Nano Lett. **7**, 1697 (2007).
[23] A. Drezet et al., Mater. Sci. Eng. B **149**, 220 (2008).
[24] A. Drezet et al., Appl. Phys. Lett. **89**, 091117 (2006).

[25] A microscope objective (50×, NA=0.55) is used to focus the incident He-Ne laser, and another oil immersion objective (160×, NA=1.40) is used to collect the leakage radiation.

[26] A.-L. Baudrion *et al.*, Phys. Rev. B **74**, 125406 (2006).

[27] A. Guinier, *X-Ray Diffraction* (Freeman, London, 1963).

[28] See Supplemental Material at http://link.aps.org/supplemental/10.1103/PhysRevLett.107.126804 for preference diffraction, SPP Airy beam width, and paraxiality.

[29] A.V. Novistsky and D.V. Novitsky, Opt. Lett. **34**, 3430 (2009).

[30] S. M. Wang *et al.*, Appl. Phys. Lett. **93**, 233102 (2008).

[31] This work is supported by the State Key Program for Basic Research of China (No. 2012CB921501, No. 2009CB930501, and No. 2011CBA00200) and the National Natural Science Foundation of China (No. 11174136, No. 10974090, No. 60990320, and No. 11021403).

Collimated Plasmon Beam: Nondiffracting versus Linearly Focused*

L. Li, T. Li, S. M. Wang, and S. N. Zhu

National Laboratory of Solid State Microstructures, School of Physics, College of Engineering and Applied Sciences, Nanjing University, Nanjing 210093, China

We worked out a new group of collimated plasmon beams by the means of in-plane diffraction with symmetric phase modulation. As the phase type changes from 1.8 to 1.0, the beam undergoes an interesting evolution from focusing to a straight line. Upon this, an intuitive diagram was proposed to elucidate the beam nature and answer the question of whether they are nondiffracting or linear focusing. Based on this diagram, we further achieved a highly designable scheme to modulate the beam intensity (e.g., "lossless" plasmon). Our finding holds remarkable generality and flexibility in beam engineering and would inspire more intriguing photonic designs.

Surface plasmon polariton (SPP) is a combined excitation of an electron and photon, which provides a possible solution to integrate the light at a subwavelength scale and suggests many charming applications[1]. Among various research, a nondiffracting SPP beam is both of fundamental interest and useful functionality (e.g., guiding surface waves). As the counterpart of optical Airy beam in free space[2,3], the realization of Airy plasmon provides a good solution to achieve a nondiffracting surface wave without any nonlinearity[4-8]. Actually, beam engineering has attracted increasing interest nowadays (e.g., arbitrary convex trajectories[9,10], in the nonparaxial region[11-13]). Another aspect of high interest is the beam focusing that reflects the capability of people to concentrate the optical field, which has already received extensive studies in the plasmonic regime[14-17]. In some sense, the nondiffracting beam can be regarded as a particular kind of focusing with a preserved focal spot as the beam propagates, and it is worthwhile to engineer elaborately. However, the well demonstrated Airy plasmon would not always be convenient in on-chip integrations for its bending trajectory. To generate a straight SPP beam with strong field confinement and controllable intensity is of great importance.

In principle, a straight nondiffracting plasmon beam can be formed by the interference of two intersecting plane waves with a linear wave-front phase (just like the generation of the Bessel beam in free space[18]), which was indeed realized by Lin et al. very recently[19]. However, this linear phase is quite different from the 1.5-power phase type of the nondiffracting Airy beam. By carefully investigating them together with the previous

---

* Phys.Rev.Lett.,2013,110(4):046807

focusing case[17], we would find it should attribute to a distinct feature of the Airy beam—asymmetric as the 1.5-power phase in $x<0$ while there is exponential decay in $x>0$[20]. Thereafter, a straightforward question arises about using a symmetric 1.5-power phase to generate a straight non-diffracting SPP beam. If yes, it would possibly open up a new avenue to engineer nondiffracting beams. If not, what is the contribution of different phases and what kind of beam will be constructed?

In this letter, we utilize the nonperfectly matched (NPM) Bragg diffraction method[7] to establish arbitrary SPP beams with symmetric phase modulation as its type from 1.8 to 1.0. Self-collimated SPP beams are well characterized by leakage radiation microscopy (LRM), which is a well developed technique using an oil-immersed objective to extract the SPPs of large k vectors[21,22]. In some cases, these straight SPP beams appear to be less dispersive till collapse at certain distances. Meanwhile, they also look like the elongated focal spots. So, we raise a paradox of the beams: nondiffracting or focusing to a line? An intuitive diagram is proposed to illustrate the physical insight into this kind of beams with respect to different phase modulations. Based on this, a highly designable scheme is developed in a general sense with the capability of achieving a beam with a required intensity profile, by which an interesting "lossless" SPP beam is realized as an example. The new insight and full understanding of the beams are highlighted, and possible applications are discussed.

The strategy to modulate the SPP beam phase by NPM diffractions of a nonperiodic nanocave array has been illustrated in detail[7]. Briefly, when an SPP beam passes through a nanoarray, it will be diffracted to various directions by local units for constructive interference, resulting in an extra 2π phase change between every neighboring row. By careful design, a well diffracted SPP beam with desired lateral phase modulation is accessible. Figure 1(a) schematically shows the SPP diffraction by nonperiodic array, where different optimum diffractions correspond to different units with different angles θ for constructive interference. According to the phase distribution in the x axis, $\phi_m(x) = \phi_0 + k_{spp}x - 2m\pi$ (ϕ_0 is a reference phase of the incident wave and m is the sequence number of the lattice point), we are able to achieve any required phase type in a general form as $\psi(x) = -cx^n$.

First, we start from a symmetric $n=1.5$ phase modulation with a nanoarray covering a wider range of the local lattice in the x direction (a_x) from 826 nm to 477 nm. The detailed parameters were calculated by solving $\phi_m(x) = \psi(x)$ (with $c=0.6$) as depicted in Fig. 1(c). The retrieved phase distribution along the x axis shown as symbols in Fig. 1(d) is consistent with the designed one (solid curve). Having these data, we fabricated the sample by focus ion beam (Strata FIB 201, FEI Company) milling on a silver film with ~60 nm thickness on a quartz substrate. Figure 1(b) shows the scanning electron microscopy (SEM) image of the sample, where the unit is a rectangular nanohole with a size of 240×120 nm^2, depth of 20 nm, and local lattice of a_x in Fig. 1(c) and P_z of 610

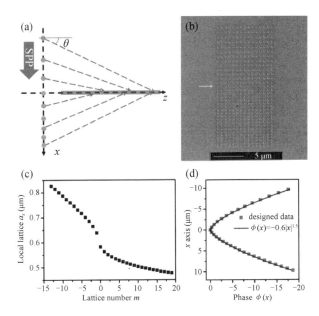

FIG. 1. (color online) (a) Scheme of the generating collimated SPP beam by in-plane diffractions. (b) SEM image of the non-periodic nanohole array, where an arrow is marked to indicate the symmetric center of the phase modulation. (c) Designed local lattice data a_x. (d) Retrieved phase distribution in lateral dimension of the beam (square symbol) compared with the designed phase type (solid curve).

nm. A white arrow in the figure indicates the symmetric position ($x=0$), where the local lattice matches the Bragg condition ($a_x \sim 610$ nm), and the lengths of the array in $\pm x$ directions are almost the same (both $\sim 9.7 \ \mu$m).

The experiment was carried out with an illumination of the He-Ne laser ($\lambda_0 = 632.8$ nm), and the SPP wave was launched by a grating with a period of 610 nm according to the SPP wavelength. Figure 2(a) shows the LRM recorded bidirectional SPP beams, revealing apparent straight collimation of the center lobes. To give more detailed information, beam intensities of cross sections within the selected region [in Fig. 2(a)] are plotted in Fig. 2(b), where a nearly nondispersive peak is well demonstrated with the FWHM kept $\sim 1.5 \ \mu$m within $\sim 30 \ \mu$m distance till it vanishes. Interestingly, the peak intensity experiences a slight increase to an abrupt drop as shown in the inset image. Theoretical calculation was subsequently performed on an SPP beaming with a well designed $n=1.5$ phase with an SPP attenuation length of $\sim 15 \ \mu$m[7], as the result shown in Fig. 3(a). Both the beam trajectory and beam profile (inset image) reproduce well the experimental results. Since the designed phase is symmetric, it is ready to accept that constructive interference always occurs in the center line. However, this interesting beam intensity profile would likely imply the underlying physics.

As has been illustrated in NPM Bragg diffraction[7], a different local lattice will lead

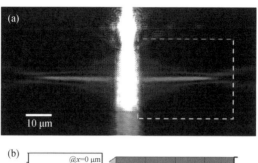
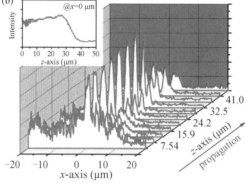

FIG.2. (color online) (a) LRM recorded SPP beam propagations as diffracted by the well designed nanoarray. (b) Detailed beam cross sections within the selected region in (a), inside which an inset image depicts the beaming profile of the center main lobe.

to an optimum diffraction angle θ [see Fig. 1(c)]. This angle can be derived from the differential of the phase function[23], as

$$\sin\theta = -\frac{1}{k}\frac{\partial\psi(x)}{\partial x} = \frac{cn}{k}x^{n-1} \tag{1}$$

by adopting $\psi(x) = -cx^n$. Thereafter, we can easily get the relation of a beam region in the z axis and the corresponding sources region (diffraction units) in the x axis as

$$z = x\cot\theta = \sqrt{\left(\frac{k}{cn}x^{2-n}\right)^2 - x^2}. \tag{2}$$

Calculating Eq.(2) with $n=1.5$ and $c=0.6$, we obtained a diagram of the correspondence curve of the beam in the z axis and the source in the x axis, as shown in Fig. 3(b). An apparent increasing slope of the curve indicates more sources contribute as the beam propagates, suggesting an enhanced beam in propagation. On the other hand, since the designed array is finite(-9.7 μm$<x<9.7$ μm), there is no source out of this range [indicated by dashed lines in Fig. 3(b)] to contribute to the beam, which thus leads to an abrupt drop in beam intensity at $z \sim 34$ μm.

For a straightforward generalization, we explored other cases of the phase type changing from $n=1.8$ to 1.0. Figures 4(a)–4(d) show the results of samples with fixed $c=0.6$ and $n=1.8, 1.6, 1.4,$ and 1.2, respectively. A clear evolution from a strong focusing

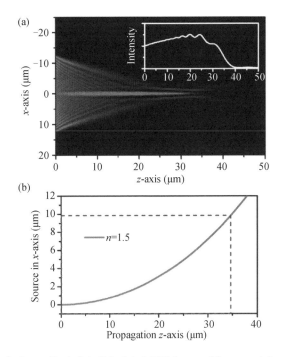

FIG. 3. (color online) (a) Calculated SPP beam with symmetric phase type $n=1.5$, where the inset image shows the intensity profile of the main lobe. (b) Source-to-beam correspondence curve in the x-z diagram, in which the dashed blue line indicates the region of the finite sources.

to a nearly nondiffracting beaming is demonstrated. Notably, the case of $n=1.0$ with $c=1.5$ is particularly shown in Fig. 4(e) (c is changed for a better visualized image), where a nondiffracting beam is clearly observed with the preserved beam shapes, shown in Fig. 4(f). Correspondingly, we calculated their source-to-beam diagrams in order to get an in-depth recognition of these beams, as the results shown in Fig. 4(g). Except for the linear case of $n=1.0$, others all have an increased tendency in their x-z curves, and the larger n, the larger the increment. The same as the previous $n=1.5$ case, there is a cutoff in the main lobe due to the finite array region, as indicated by a dashed line of $x=9.7$ μm. It does explain the experimental results. However, for the case of $n=1.8$, we will find that no matter how large the source provided is, the beam range is limited, and most sources only contribute to a narrow region around $z \sim 10$ μm. In this regard, it appears to be a focusing process. However, does it mean that the smaller n cases truly correspond to infinite beaming as long as the sources are infinite?

Revisiting Eq. (2), we will find it can be renormalized for some particular cases. For example, when $n=1$ it degenerates to a line with a slope of $\sqrt{\left(\frac{k}{cn}\right)^2 - 1}$; when $n=1.5$ it becomes a circle equation with a radius of $r = \frac{1}{2}\left(\frac{k}{1.5c}\right)^2$ and a center of $(r, 0)$; when $n=2$

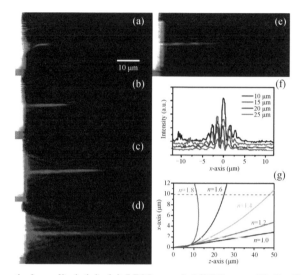

FIG. 4. (color online) (a)-(e) LRM recorded SPP beam with the diffracted phase modulation of $n=1.8$, 1.6, 1.4, 1.2, and 1.0, respectively ($c=1.5$ for $n=1.0$ and $c=0.6$ for others, and all in the same scale). (f) Recorded beam shapes ($n=1.0$) at different distances (10 μm, 15 μm, 20 μm, 25 μm, from top to bottom). (g) Source-to-beam correspondence curves for all cases, in which the finite source region is indicated by the blue dashed line.

it is still a circle with $r=k/2c$ and a center of $(0, 0)$ (see the Supplemental Material[24]). It is clear that the $x-z$ curve for $n=1.5$ in Fig. 3(b) is just a part of the circle arc, which implies the main lobe of beam will not go farther than the radius ($z=r=65.4$ μm) even if the source is infinite. As for other cases, the calculated correspondence $x-z$ curves reveal that their curvatures are determined by the coefficient c and phase factor n. It is evident that these curves only cover a finite region in the z dimension except for the $n=1.0$ case (see the Supplemental Material). In fact, the $n=1.0$ phase type truly corresponds to the case of two crossing SPP plane waves with the result of a nondiffracting cosine-Gauss beam[19]. Again in Fig. 4(g), the straight $x-z$ line of the $n=1.0$ case does indicate an equal contribution of sources to the beam everywhere, which is also the nature of the nondiffracting beam. However, the experimental result definitely shows us a decay beam, which is rightly because of the plasmonic loss as well as in Ref. [19] Now, it is well recognized that beams of the phase type $n \neq 1$ are not nondiffraction since they are all limited within a certain range. However, the small n cases have revealed the nearly nondiffracting characteristic and weaker side lobes if they are carefully designed. In this regard, these self-collimated beams are able to work as the nondiffracting one to some extent. It does provide a robust method to generate a straight SPP beam with a preserved narrow beamwidth, which holds more generality for further manipulations.

Furthermore, a highly designable scheme is developed to retrieve an arbitrary phase to achieve a desired beam intensity profile. As we know, the SPP wave always suffers from the large propagation loss even for the nondiffracting cosine-Gauss beam[19]. Our design has already manifested the capability to compensate the loss within a certain range by phase design (e.g., the $n=1.5$ case in Fig. 2). Indeed, the phase can be retrieved by strictly deducing a proper source contribution to balance the propagation loss and thus achieve an exact intensity-preserved "lossless" SPP beam. According to earlier discussions, we introduce a source-to-beam density as $\rho \propto \dfrac{\mathrm{d}x}{\mathrm{d}z}$. Assuming an SPP with an attenuation length of l and a fixed intensity of the unit source, the beam intensity in the z axis contributed by a unit source can be expressed as

$$I_i \propto \exp\left(-\frac{r}{l}\right) = \exp\left(-\frac{x}{l\sin\theta}\right). \tag{3}$$

Thereafter, we can artificially tune the source-to-beam density to get a constant beam intensity as $I(z) = \rho I_i = c$. Then, we have the equation:

$$\frac{\mathrm{d}x}{\mathrm{d}z}\exp\left(-\frac{x}{l\sin\theta}\right) = c. \tag{4}$$

Solving Eqs. (1), (2), and (4), we can obtain the phase distribution along the x axis for a loss compensated SPP beam. With the given $l=15$ μm[7], a particular phase distribution is numerically retrieved to realize a straight lossless SPP beam as the calculation result shown in Fig. 5(a). Figure 5(c) depicts the retrieved phase distribution together with the previous $n=1.5$ case for a comparison, which shows the lossless one has less curvature agreeing with less enhancement that is needed to balance the loss. Subsequently, we carried out the experiment by in-plane diffraction and successfully achieved this lossless beam as shown in Fig. 5(b), which almost reproduces the theoretical one. More clearly, Figs. 5(d) and 5(e) further depict the well achieved lossless profiles of the main lobe with a preserved intensity within a distance of $z < 35$ μm for both calculated and experimental results, respectively, which are in very good coincidence. Thus, an intensity-preserved and nearly nondiffracting SPP beam with strict design is well demonstrated.

It is reasonably believed that our design is capable of achieving almost any kind of intensity profile by introducing an arbitrary function of $F(z)$ as $I(z) = \rho I_u = cF(z)$ [e.g., an exponential increasing beam with $F(z) = \exp\left(\dfrac{z-z_0}{l'}\right)$; see the Supplemental Material[24]], which would possibly imply particular applications (e.g., in optical trapping[25]). Notably, this designable scheme is valid not only for the lossy SPP but also for other beam systems as the item of $\exp(-r/l)$ in Eq.(3) is replaced by a more general function. In addition to these novelties, the achieved SPP beams retain other unique properties, e.g., the self-healing property, which was also experimentally observed (see the Supplemental Material[24]).

In summary, we have achieved a group of self-collimated SPP beams by symmetric phase modulation. A paradox between a nondiffracting versus linear focusing was proposed

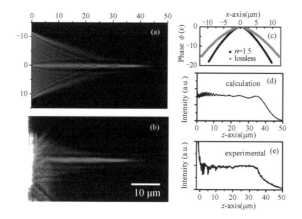

FIG. 5. (color online) Well achieved lossless SPP beams by (a) calculation and (b) experiment with a well designed phase distribution. (c) Retrieved phase distribution for the lossless beam compared with the previous $n=1.5$ case. (d) Calculated and (e) experimental intensity profiles of the main lobes with respect to the beam of (a) and (b), respectively, revealing very good preserved intensity within a distance of ～35 μm.

and insensitively studied. Utilizing an intuitive source-to-beam diagram, we successfully explained the nature of these beams. It is concluded that although some of these beams have a nondiffracting appearance, they are not the real ones except the particular case of $n=1.0$. As a result of constructive diffractions, these SPP beams can be regarded as a kind of the focusing with respect to a line but not a point. This newly discovered source-to-beam relationship offers us a powerful and flexible method to design an intensity controllable beam, and an interesting lossless SPP beam is designed and realized as an example. Our study gives a unique insight into the collimated SPP beam formation and is expected to inspire more intriguing phenomena and potential applications in beam engineering and nanophotonic manipulations.

References and Notes

[1] J.A. Schuller, E.S. Barnard, W.S. Cai, Y.C. Jun, J.S. White, and M. L. Brongersma, Nat. Mater. **9**, 193 (2010).

[2] G. A. Siviloglou, J. Broky, A. Dogariu, and D. N. Christodoulides, Phys. Rev. Lett. **99**, 213901 (2007).

[3] G. A. Siviloglou, J. Broky, A. Dogariu, and D. N. Christodoulides, Opt. Lett. **33**, 207 (2008).

[4] A. Salandrino and D. N. Christodoulides, Opt. Lett. **35**, 2082 (2010).

[5] P. Zhang, S. Wang, Y. M. Liu, X. B. Yin, C. G. Lu, Z. G. Chen, and X. Zhang, Opt. Lett. **36**, 3191 (2011).

[6] A. Minovich, A. Klein, N. Janunts, T. Pertsch, D. Neshev, and Y. Kivshar, Phys. Rev. Lett. **107**, 116802 (2011).

[7] L. Li, T. Li, S. Wang, C. Zhang, and S. Zhu, Phys. Rev. Lett. **107**, 126804 (2011).

[8] A. E. Klein, A. Minovich, M. Steinert, N. Janunts, A. Tünnermann, D. N. Neshev, Y. S. Kivshar, and T. Pertsch, Opt. Lett. **37**, 3402 (2012).

[9] E. Greenfield, M. Segev, W. Walasik, and O. Raz, Phys. Rev. Lett. **106**, 213902 (2011).

[10] L. Froehly, F. Courvoisier, A. Mathis, M. Jacquot, L. Furfaro, R. Giust, P. A. Lacourt, and J. M. Dudley, Opt. Express **19**, 16455 (2011).

[11] I. Kaminer, R. Bekenstein, J. Nemirovsky, and M. Segev, Phys. Rev. Lett. **108**, 163901 (2012).

[12] P. Zhang, Y. Hu, D. Cannan, A. Salandrino, T. Li, R. Morandotti, X. Zhang, and Z. Chen, Opt. Lett. **37**, 2820 (2012).

[13] P. Zhang, Y. Hu, T. Li, D. Cannan, X. Yin, R. Morandotti, Z. Chen, and X. Zhang, Phys. Rev. Lett. **109**, 193901 (2012).

[14] C. Zhao and J. S. Zhang, Opt. Lett. **34**, 2417 (2009).

[15] C. Zhao and J. S. Zhang, ACS Nano **4**, 6433 (2010).

[16] T. Tanemura, K. C. Balram, D. Ly-Gagnon, P. Wahl, J. S. White, M. L. Brongersma, and D. A. B. Miller, Nano Lett. **11**, 2693 (2011).

[17] L. Li, T. Li, S. M. Wang, S. N. Zhu, and X. Zhang, Nano Lett. **11**, 4357 (2011).

[18] J. Durnin, J. J. Micel, Jr., and J. H. Eberly, Phys. Rev. Lett. **58**, 1499 (1987).

[19] J. Lin, J. Dellinger, P. Genevet, B. Cluzel, F. de Fornel, and F. Capasso, Phys. Rev. Lett. **109**, 093904 (2012).

[20] M. V. Berry and N. L. Balazs, Am. J. Phys. **47**, 264 (1979).

[21] A. Drezet, A. Hohenau, D. Koller, A. Stepanov, H. Ditlbacher, B. Steinberger, F. R. Aussenegg, A. Leitner, and J. R. Krenn, Mater. Sci. Eng. B **149**, 220 (2008).

[22] A. Drezet, A. Hohenau, A. L. Stepanov, H. Ditlbacher, B. Steinberger, N. Galler, F. R. Aussenegg, A. Leitner, and J. R. Krenn, Appl. Phys. Lett. **89**, 091117 (2006).

[23] Y. Kaganovsky and E. Heyman, Opt. Express **18**, 8440 (2010).

[24] See Supplemental Material at http://link.aps.org/supplemental/10.1103/PhysRevLett.110.046807 for a diagram of the source-to-beam correspondence curves, a well-designed exponential increasing beam, and self-healing property.

[25] D. B. Ruffner and D. G. Grier, Phys. Rev. Lett. **109**, 163903 (2012).

[26] The authors would thank Q. J. Wang and X. N. Zhao for their help and support in the sample fabrications. This work is supported by the State Key Program for Basic Research of China (Grants No. 2012CB921501, No. 2010CB630703, No. 2009CB930501, and No. 2011CBA00200), the National Natural Science Foundation of China (Grants No. 11174136, No. 10974090, No. 11021403, and No. 60990320), and PAPD of Jiangsu Higher Education Institutions.

The Anomalous Infrared Transmission of Gold Films on Two-Dimensional Colloidal Crystals[*]

P. Zhan, Z. L. Wang, H. Dong, J. Sun, J. Wu,
Prof. H.-T. Wang, Prof. S. N. Zhu, Prof. N. B. Ming
National Laboratory of Solid State Microstructures
Nanjing University
Nanjing 210093 (P.R. China)
Prof. J. Zi
Surface Physics Laboratory, Fudan University
Shanghai 200433 (P.R. China)

Tailoring optical response using periodic nanostructures is one of the key issues in the current research on functional composite materials.[1-5] The anomalous light transmission through metallic films that have a regular array of submicro-meter holes[3-6] has stimulated much interest. This interest stems from both the underlying physics and also the perceived potential for applications in nanophotonics,[7] quantum-information processing,[8] nanolithography,[9] and surface-enhanced Raman scattering.[10]

Extraordinary transmission of light through an optically opaque metal film perforated with a 2D array of subwavelength holes was first reported by Ebbesen et al.[5] This unusual phenomenon can be understood as a result of diffractive coupling to evanescent surface plasmon polaritons (SPPs) that leads to a strong concentration of light at the metal surface, which then weakly tunnels through the holes in the film, reradiating by the inverse process on the exit side.[4,11-13] In order to explore the SPP properties of microstructured metal films, extensive efforts have been made to study their spectral response and dependence on geometrical parameters, such as the type of lattice symmetry, metal film thickness, and adjacent dielectric media.[14] Recent studies show that the hole shape has a significant effect on the optical transmission.[15-19] Nearly all the metallic films studied have been on a flat substrate and the hole arrays were made using focused ion-beam milling,[5,15,17,19] and electron-beam lithography[8] or interferometric lithography combined with reactive ion etching.[16,18]

Here we use nanosphere lithography[20] as the sample production technique. This approach has several advantages over the conventional lithographic and machining techniques, including the relative ease of casting large, high-quality, ordered

---

[*] Adv. Mater., 2006, 18(12):1612

nanomaterials and the low cost of implementation. Ordered arrays of gold half shells and nanocaps have been constructed by controlled gold vapor deposition with thicknesses less than 20 nm by using a 2D colloidal crystal (CC) as a substrate.[21,22] Baumberg's group has fabricated metallic nanocavity arrays by electrodeposition within the pores of CC templates and observed the excitation of the SPPs in metallic cavities that led to rich features in reflectivity spectra.[23] Very recently, Landström et al. have shown that the transmission spectra through a metal film formed on a 2D CC substrate are quite similar to those observed through subwavelength hole arrays in metal films.[24]

In this communication, we report a study on the infrared transmission properties of gold films patterned on 2D CCs. The fabricated metallodielectric structures have a strong surface corrugation as well as a 2D periodic pore array. We show that the SPPs on these curved surfaces display unusual dispersion properties, compared to those of metal films on flat substrates studied before. The dielectric property of the template spheres is also found to have a substantial effect on the transmission. More importantly, the transmission features vary dramatically as the gold film thickness is increased, with an apparent transition from the excitation of localized SPP resonance to extended SPP propagation at a critical metal film thickness. Our results will be useful for designing and fabricating new optical devices based on SPP excitation and this will stimulate further studies on the optical properties of metallic microstructures deposited on 2D CCs.

The ordered metallic microstructures were prepared by sputtering a thin gold layer onto a monolayer of dielectric microspheres self-assembled onto a quartz chip.[21] The 2D sphere arrays were crystallized by controlled evaporation from a colloidal solution within a channel formed using two quartz chips.[25] The microbeads were hemispherically covered with metal and the resulting gold film consists of a hexagonally close-packed (HCP) array of gold half-shells with a size dictated by the template spheres. The diameters of these hemispherical shells can be conveniently controlled from 200 nm to several micrometers by choosing colloidal beads with different sizes.

Figure 1 shows the scanning electron microscopy (SEM) image of a typical sample with a thin gold layer on a 2D silica CC substrate. In the center of the image, there is a vacancy through which the monolayer CC substrate can be clearly identified. Since the silica microspheres were densely packed, the gold half-shells on adjacent silica spheres were interconnected after a certain amount of deposition, thus forming a conducting network on the CC surface. In addition, due to the locally curved surface of the CC template, a lateral variation of the metal thickness was created on the spheres, with the thinnest layer at the equator of each sphere. The fabricated metallic microstructure reflects the order of the 2D CC template, whose assembly can be controlled using a variety of self-organization methods.[26]

The optical response of such metallic microstructures was measured by zero-order Fourier-transform infrared (FTIR) spectroscopy. In all transmission measurements, the

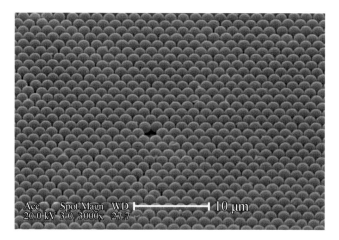

FIG.1. SEM image (tilted view) of a highly corrugated gold film deposited on a 2D CC assembled from 1.58 μm diameter silica spheres on a planar quartz chip. The thickness of the gold shells on top of the spheres is about 30 nm.

optical spot size on the samples was about 0.8 mm. All transmission spectra were normalized to the transmittance of a pure quartz substrate. We first present the transmission spectra of the modulated metallic networks under normal incidence of linearly polarized light.

Figure 2(a) shows typical experimental results for two corrugated gold films deposited on 2D CC substrates with different dielectric constants. Both films have a nominal thickness $t=30$ nm.[27] The CCs are assembled from silica spheres (1.58 μm in diameter) and polystyrene (PS) spheres (1.59 μm in diameter) on a quartz substrate. For the metal film patterned on a silica template [solid line in Fig. 2(a)], up to four transmission resonances are clearly seen. In particular, a strong, extraordinary transmission peak at 1935 nm is observed, a phenomenon similar to that observed in planar metal films perforated with a regular hole array.[5-10,13-19] This is compared to the negligible transmission observed for the homogeneous gold film deposited directly on a planar quartz substrate [dotted line in Fig. 2(a)]. The main resonance has a transmittance of about 23% which is about 2.5 times the projected area (about 9%) of the pores between the particles in the 2D plane of the structure.[28] Although the prepared samples have a good local ordering (Fig. 1), the crystal grains are typically tens of micrometers. Since the optical spot size is much larger than the single domain, the measured area is basically a region of multidomains, and as a consequence, the measured spectra are insensitive to the in-plane orientation of the patterned gold films[23] (results not shown).

The optical response of ordered assemblies of dielectric spheres partially coated with gold is a superposition of scattering diffraction and light reradiation via the excitations of SPPs on the metal films. It is noted that due to the incomplete shell morphology, the present microstructures allow strong coupling of the SPPs with the CC substrate. Thus,

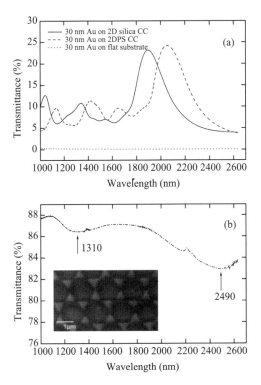

FIG. 2. (a) Transmission spectra of two textured gold films on different monolayer CCs assembled on a quartz chip. The silica and polystyrene (PS) spheres have nearly identical diameters (d_{silica} = 1.58 μm and d_{PS} = 1.59 μm). The dotted curve is for an unpatterned gold film on a flat quartz chip. (b) Transmission spectrum of a gold triangular nanoparticle array obtained by removing the silica spheres (d_{silica} = 1.58 μm) after gold deposition. Inset: SEM image of the gold triangular nanoparticles. All gold layers were deposited under the same conditions with the same thickness t = 30 nm.

reradiation of SPPs into photons should be affected by the dielectric properties of the colloidal spheres. This is demonstrated in Figure 2(a), in which red-shifts of the resonances were observed when a PS CC substrate with a higher refractive index (n_{PS} = 1.59, n_{silica} = 1.45) was used. Since both kinds of spheres chosen have nearly identical sizes (diameters, d_{silica} = 1.58 μm and d_{PS} = 1.59 μm), the wavelength location of the transmission peaks is expected to be in proportion to the refractive index of the template. This is in good agreement with the observations in Figure 2(a).

Note that during the sputtering process, gold is inevitably deposited onto the quartz chip through the interstices of the 2D CC, leading to the formation of ordered arrays of discrete triangular gold nanoislands [Fig. 2(b), inset]. These sharply pointed triangular islands also have strong SPP resonances.[20,29] In order to reveal their possible contribution to the anomalous transmission resonance, separate measurements of these gold islands have been performed by removing the gold-coated CCs after metal deposition. Figure 2(b) shows the transmittance spectrum for normal-incident light of an array of triangular gold islands; it was fabricated by removing the silica spheres in the sample shown in Figure 1 via tape stripping. For these uniform gold nanoparticles, the transmittance shows two weak minima at ca. 1310 nm and ca. 2490 nm. The broad band at 2490 nm is due to a dipole resonance of the gold nanoprisms. Such a localized SPP resonance has been observed in silver nanoprisms and have been shown to red-shift linearly with an increase in edge length.[20,29] The origin of the band at the shorter wavelength is not yet clear, but it is possibly because of a collective light scattering of the particle array. Nevertheless, these bands make a negligible positive contribution to the observed anomalous transmission of the metal film. Indeed, our assumption could be compromised if there exists a strong coupling between the array of the triangular islands and the array of the caps. However, it is understood that a strong near-field coupling happens only when the involved systems are closely located both spectrally and in spatially. Here, such a coupling should be weak as the resonances of the array of triangular islands are located far away from the main resonance of the whole structure [compare Fig. 2(a) and (b)], although the spatial separation between the arrays is on a submicrometer level.

To obtain more information about the SPP resonances of the microstructured gold film, we further measured zero-order transmission spectra for off-normal incidence. Angle-resolved measurements were performed using linearly polarized light whose electric field was perpendicular (s polarization) and parallel (p polarization) to the plane of incidence, respectively.

In Figure 3(a) and (b), the acquired transmission spectra are displayed separately for two polarizations. The angle of incidence was varied from $\theta = 0°$ to 20° with an interval of 2°. It is seen that the transmission resonances show quite different dispersion behaviors for different resonant modes, which depends on the polarization. For p-polarized incident light, the main SPP resonance initially shows little dispersion for $\theta < 8°$ in conjunction with a steady decrease in its intensity. Further increase in θ leads to the appearance of a small blue-shift ($\Delta\lambda = 30$ nm), although this is actually due to red-shifting of the resonance at ca. 1720 nm across the main resonance, followed by a clear red-shift. The two weak resonances at the blue edge of the main peak also show a decrease in intensity upon increasing the incident angle. For the two resonances with the highest energies in the spectrum at ca. 1050 nm and ca. 1330 nm, quite different dispersion properties are seen. The shorter-wavelength peak displays a red-shift, whereas the longer-wavelength one exhibits a blue-shift for $\theta < 14°$, followed by a red-shift on further increasing θ. It is noted that optically inactive surface modes under normal incidence could be excited under off-normal incidence, leading to new resonances in the spectra. This is confirmed in Figure 3(a) where new resonances develop as the angle of incidence is changed; the obvious resonance at 2250 nm shows a red-shift with increasing angle.

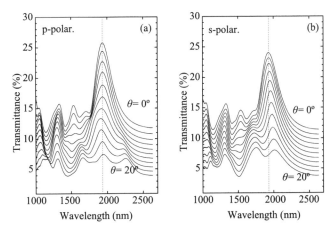

FIG.3. Transmission spectra of a textured gold film ($t=30$ nm) on a 2D silica CC ($d_{silica}=1.58$ μm) as a function of incident angle (θ). Spectra were taken for the structure in steps of 2° from 0° to 20° for (a) p polarization and (b) s polarization. The individual spectra are offset vertically by 3% from one another for clarity. The dotted vertical line indicates the wavelength $\lambda=1935$ nm of the main SPP resonance at normal incidence.

A different dispersion behavior, however, is observed for the main resonance at 1935 nm when the polarization of the incident light is changed to the s state [Fig. 3(b)]. The main resonance shows a noticeable dispersion even for s-polarized incident light, that is, it shifts steadily towards longer wavelength with decreasing intensity. In addition, in contrast to the case of p-polarization, the mode at ca. 1720 nm at the blue edge of the main resonance becomes more intense as θ increases.

Previous studies have demonstrated that for planar metal films drilled with regular hole arrays, the SPPs exhibit quite different dispersion features for incident light of different polarizations.[4,8,30,31] For example, the SPPs have been observed to exhibit a direction-independent dispersion in these metal films for s-polarized incident light.[5,8,30] This was explained as being due to the fact that the SPPs at the interfaces of a planar dielectric/metal interface are mainly longitudinal waves and, as a consequence, the coupling between the SPP and s-polarized light is independent of θ because, in this case, the electric field remained parallel to the planar interface.[11,12] In contrast, the main SPP resonance in our system has a noticeable dependence on incident angle under s polarization [Fig. 3(b)].

We have tried to assign the transmission peaks to the reciprocal lattice of the periodic structure. However, there was a large disagreement between the observed peak positions and the calculated values using the formula for labeling the transmittance peaks of planar metal films.[4,13] Currently, an exact calculation of the transmission and SPP dispersion of the interconnected half-shell arrays is not yet possible. However, a qualitative explanation can be made by considering the unique feature of the metal film [Fig. 4(a)]. Due to the quasi-3D property of the gold film, linearly polarized incident light always has an electric-field component that is somewhere perpendicular to the gold film surface, regardless of its polarization [Fig. 4(b)]. It is expected that the strong corrugation in the metal film could lead to a dramatic modification of the properties of the SPP eigenmodes from those in a flat metal film for both s - and p-polarized incident light. We suggest that sophisticated numerical calculations be implemented to simulate the surface plasmonic properties of this new kind of metal microstructure, by taking into account the 3D distribution of metal on the highly curved CC substrate.

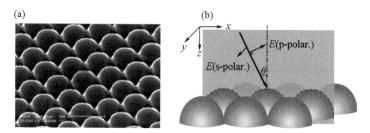

FIG.4. (a) SEM image of a textured gold film after removal of the silica cores, showing that the metal film is composed of interconnected metal half shells. (b) Representations of p - and s-polarized radiation incident upon the textured gold film at an angle θ.

An important issue in the study of plasmonics is the transition from a localized SPP to an extended SPP when the metallic nanoparticles merge to form a periodic hole array.[32] Here, the conducting metal network on a 2D CC is formed via gold nanobridges at the touching points of neighboring spheres, which can be established only after a certain amount of gold deposition.[21] In this case, a geometrical transition from isolated hemispherical metal shells to an interconnected periodic metal network occurs. This provides us with a convenient and novel route to study the transition from local SPPs on isolated metal shells and triangular islands to extended SPPs propagating along the metal network. Figure 5 shows the transmission spectra under normal incidence of a series of samples with different gold thicknesses, which were prepared by increasing the amount of gold deposit on a 2D silica CC supported on a quartz substrate. The thickness of the gold layer was varied from 0 (curve A) to 48 nm (curve N).

Curve A in Figure 5 corresponds to a 2D CC of bare silica spheres. The minimum transmittance at 1810 nm is due to the excitation of a surface eigenmode of the 2D dielectric CC.[33] For a sufficiently low thin metal layer (curve B with $t=5$ nm and curve C with $t=8$ nm), the sputtered gold forms isolated aggregates on the microspheres and in the triangular gaps. In such cases, light scattering by the 2D dielectric periodic structure dominates the transmission spectra, which are similar to that of the bare silica template (curve A).

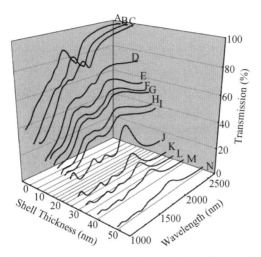

FIG.5. In situ transmission spectra of a series of gold-deposited 2D silica CCs. These samples were prepared by sequentially increasing the gold thickness on a silica CC ($d_{silica} = 1.58$ μm). The gold-shell thickness was increased from 0 (curve A) up to 48 nm (curve N). The dotted lines in the bottom plane are the corresponding projections of the transmission spectra, whose intersections with the shell-thickness axis show the thickness of the corresponding gold layer.

When the gold deposit is increased from 8 nm (curve C) to 11 nm (curve D), a large decrease in the overall transmittance is observed over the whole spectrum with the greatest decrease occurring in the long-wavelength region. Electrical-resistance measurements show that the template surface is still non-conducting, which indicates that a metal network has not yet formed. However, a continuous layer (nanocaps) may already be formed on each silica sphere[21] as well as in the triangular gaps between the spheres. In such cases, localized SPPs can be excited[32,34] and, as a consequence, the formation of these separated gold particles leads to the first strong redistribution of the photonic modes of the 2D dielectric photonic crystal (see curve E with $t = 14$ nm).

When the gold thickness is further increased up to about 25 nm, a conducting metallic network is created on the 2D CC surface. In this case, an abrupt decrease in transmittance in the long wavelength limit is observed and a sharp transmission peak with a high intensity is established simultaneously (compare spectra J and I in Fig. 5). We suggest that, upon the formation of gold nanobridges between touching silica spheres, a transition from localized SPPs to extended SPPs occurs. Large optical electric fields are then established via excitation of SPPs that assist light tunneling through the nanopore array, giving rise to very high transmittance at specific wavelengths.

In summary, we demonstrate a new avenue for tuning SPP properties by depositing metal films on a monolayer CC substrate. Extraordinary transmission resonances through the metal film are demonstrated with unique dispersion properties that depend on both s-

and p-polarizations of the incident light. Furthermore, a clear transition from localized SPPs to extended SPPs is observed that leads to a sudden attenuation of transmittance in the low-energy limit. These findings will stimulate further theoretical and experimental efforts in engineering the optical properties of these 2D ordered metallodielectric microstructures and exploit their potential applications in the near-infrared spectral region. The relative ease of growing high-quality CCs with a high ordering and large single domain, and the low cost of fabricating such plasmonic crystals with submicrometer periodicity promise applicability in areas ranging from biotechnology to optoelectronics.

Experimental

The monodisperse silica microspheres (size dispersion 1.9%) and PS microspheres (size dispersion 1%) used in the work were purchased from Duke Scientific Corps. The 2D silica and PS CCs were prepared by injecting an aqueous solution of colloidal dispersion with a suitable concentration into a channel that was formed from two parallel quartz slides separated by a U-shaped spacer. The quartz slides had been pretreated to render their surface hydrophilic by soaking in a solution of 30% hydrogen peroxide at 80℃ for 30 min. After drying in air, highly ordered CCs were grown within the channel under capillary force. The prepared 2D CCs acted as a topographic pattern and gold was then sputtered on the top of the microspheres in a vacuum of 5×10^{-6} Torr (1 Torr$\approx$133 Pa) at a rate of 1.2Ås^{-1} to the desired thickness by an ion-beam coater (IBC Model 682, Gatan Corp.). Through this process, ordered 2D arrays of metallodielectric beads with a hemispherical metal coverage were fabricated. The dielectric spheres could be removed, if needed, via suitable chemical etching by using toluene to remove the PS and HF to remove the silica spheres.

Sample structures were characterized by scanning electron microscopy (FEI Philips XL-30). Transmission spectra were obtained using a far-field FTIR spectrometer (Nicolet 5700). The optical spot size of the incident beam on the samples was about 0.8 mm. The numerical aperture set for the transmission was estimated to be less than 0.01.

References and Notes

[1] a) E. Yablonovitch, *Phys. Rev. Lett.* **1987**, *58*, 2059. b) S. John, *Phys. Rev. Lett.* **1987**, *58*, 2486.
[2] N. B. Ming, *Adv. Mater.* **1999**, *11*, 1079.
[3] J. B. Pendry, *Phys. Rev. Lett.* **2000**, *85*, 3966.
[4] W. L. Barnes, A. Dereux, T. W. Ebbesen, *Nature* **2003**, *424*, 824.
[5] T. W. Ebbesen, H. J. Lezec, H. F. Ghaemi, T. Thio, P. A. Wolff, *Nature* **1998**, *391*, 667.
[6] D. E. Grupp, H. J. Lezec, T. Thio, T. W. Ebbesen, *Adv. Mater.* **1999**, *11*, 860.
[7] H. J. Lezec, A. Degiron, E. Devaux, R. A. Linke, L. Martin-Moreno, F. J. Garcia-Vidal, T. W. Ebbesen, *Science* **2002**, *297*, 820.
[8] E. Altewischer, M. P. van Exter, J. P. Woerdman, *Nature* **2002**, *418*, 304.

[9] a) W. Srituravanich, N. Fang, C. Sun, Q. Luo, X. Zhang, *Nano Lett.* **2004**, *4*, 1085. b) Z. W. Liu, Q. H. Wei, X. Zhang, *Nano Lett.* **2005**, *5*, 957.

[10] a) A. G. Brolo, R. Gordon, B. Leathem, K. L. Kavanagh, *Langmuir* **2004**, *20*, 4813. b) A. G. Brolo, E. Arctander, R. Gordon, B. Leathem, K. L. Kavanagh, *Nano Lett.* **2004**, *4*, 2015.

[11] H. Raether, *Surface Plasmons on Smooth and Rough Surfaces and on Gratings*, Springer, Berlin **1988**.

[12] J. R. Sambles, G. W. Bradbery, F. Yang, *Contemp. Phys.* **1991**, *32*, 173.

[13] H. F. Ghaemi, T. Thio, D. E. Grupp, T. W. Ebbesen, H. J. Lezec, *Phys. Rev. B* **1998**, *58*, 6779.

[14] a) A. Degiron, H. J. Lezec, W. L. Barnes, T. W. Ebbesen, *Appl. Phys. Lett.* **2002**, *81*, 4327. b) T. J. Kim, T. Thio, T. W. Ebbesen, D. E. Grupp, H. J. Lezec, *Opt. Lett.* **1999**, *24*, 256.

[15] K. J. Klein Koerkamp, S. Enoch, F. B. Segerink, N. F. van Hulst, L. Kuipers, *Phys. Rev. Lett.* **2004**, *92*, 183 901.

[16] W. Fan, S. Zhang, B. Minhas, K. J. Malloy, S. R. J. Brueck, *Phys. Rev. Lett.* **2005**, *94*, 033 902.

[17] J. A. Matteo, D. P. Fromm, Y. Yuen, P. J. Schuck, W. E. Moerner, L. Hesselink, *Appl. Phys. Lett.* **2004**, *85*, 648.

[18] Y. H. Ye, D. Y. Jeong, Q. M. Zhang, *Appl. Phys. Lett.* **2004**, *85*, 654.

[19] R. Gordon, A. G. Brolo, A. McKinnon, A. Rajora, B. Leathem, K. L. Kavanagh, *Phys. Rev. Lett.* **2004**, *92*, 037 401.

[20] C. L. Haynes, R. P. van Duyne, *J. Phys. Chem. B* **2001**, *105*, 5599.

[21] J. C. Love, B. D. Gate, D. B. Wolfe, K. E. Paul, G. M. Whitesides, *Nano Lett.* **2002**, *2*, 891.

[22] J. Liu, A. I. Maaroof, L. Wieczorek, M. B. Cortie, *Adv. Mater.* **2005**, *17*, 1276.

[23] a) S. Coyle, M. C. Netti, J. J. Baumberg, M. A. Ghanem, P. R. Birkin, P. N. Bartlett, D. M. Whittaker, *Phys. Rev. Lett.* **2001**, *87*, 176 801. b) M. C. Netti, S. Coyle, J. J. Baumberg, M. A. Ghanem, P. R. Birkin, P. N. Bartlett, D. M. Whittaker, *Adv. Mater.* **2001**, *13*, 1368. c) T. A. Kelf, Y. Sugawara, J. J. Baumberg, M. Abdelsalam, P. N. Bartlett, *Phys. Rev. Lett.* **2005**, *95*, 116 802.

[24] L. Landstr.m, D. Brodoceanu, K. Piglmayer, G. Langer, D. Bäuerle, *Appl. Phys. A* **2005**, *81*, 15.

[25] a) Z. Chen, P. Zhan, Z. L. Wang, J. H. Zhang, W. Y. Zhang, N. B. Ming, C. T. Chan, P. Sheng, *Adv. Mater.* **2004**, *16*, 417. b) P. Zhan, J. B. Liu, W. Dong, H. Dong, Z. Chen, Z. L. Wang, Y. Zhang, S. N. Zhu, N. B. Ming, *Appl. Phys. Lett.* **2005**, *86*, 051 108.

[26] a) P. Jiang, J. F. Bertone, K. S. Hwang, V. L. Colvin, *Chem. Mater.* **1999**, *11*, 2132. b) S. O. Lumsdon, E. W. Kaler, J. P. Williams, O. D. Velev, *Appl. Phys. Lett.* **2003**, *82*, 949. c) N. D. Denkov, O. D. Velev, P. A. Kralchevsky, L. B. Ivanov, H. Yoshimura, K. Nagayama *Langmuir* **1992**, *8*, 3183. d) R. Micheletto, H. Fukuda, M. Ohtsu, *Langmuir* **1996**, *12*, 333. e) A. S. Dimitrov, K. Nagayama, *Langmuir* **1996**, *12*, 1303. f) Q.-H. Wei, D. M. Cupid, X. L. Wu, *Appl. Phys. Lett.* **2000**, *77*, 1641.

[27] Here t is defined as the thickness of the metal film on the top area of the template spheres.

[28] For the 2D CC assembled from $d = 1.58$ μm silica spheres, the equilateral triangular gaps between spheres is estimated to have an edge size of 420 nm.

[29] a) R. C. Jin, Y. C. Cao, E. C. Hao, G. S. Metraux, G. C. Schatz, C. A. Mirkin, *Nature* **2003**, *425*, 487. b) K. L. Kelly, E. Coronado, L. L. Zhao, G. C. Schatz, *J. Phys. Chem. B* **2003**, *107*, 668.

[30] W. L. Barnes, W. A. Murray, J. Dintinger, E. Devaux, T. W. Ebbesen, *Phys. Rev. Lett.* **2004**, *92*, 107 401.

[31] K. L. van der Molen, K. J. Klein Koerkamp, S. Enoch, F. B. Segerink, N. F. van Hulst, L. Kuipers,

Phys. Rev. B **2005**, *72*, 045 421.

[32] W. A. Murray, S. Astilean, W. L. Barnes, *Phys. Rev. B* **2004**, *69*, 165 407.

[33] H. Miyazaki, H. Miyazaki, K. Ohtaka, T. Sato, *J. Appl. Phys.* **2000**, *87*, 7152.

[34] C. Charnay, A. Lee, S. Man, C. E. Moran, C. Radloff, R. K. Bradley, N. J. Halas, *J. Phys. Chem. B* **2003**, *107*, 7327.

[35] We thank J. R. Sambles for a critical reading of the manuscript. We also thank C. T. Chan, Z. Y. Li, and Y. Y. Zhu for helpful discussions. This work was supported by a grant for the State Key Program for Basic Research of China and by the NSFC under Grant Nos. 10425415, 90501006, and 10534020. Z. L. Wang is grateful to the Distinguished Youth Foundation of the NSFC.

Localized and Delocalized Surface-plasmon-mediated Light Tunneling Through Monolayer Hexagonal-close-packed Metallic Nanoshells[*]

Chaojun Tang, Zhenlin Wang, Weiyi Zhang, Shining Zhu, and Naiben Ming

Department of Physics, National Laboratory of Solid State Microstructure,
Nanjing University, Nanjing 210093, China

Gang Sun

Beijing National Laboratory for Condensed Matter Physics, Institute of Physics,
Chinese Academy of Sciences, Beijing 100080, China

Ping Sheng

Department of Physics, Hong Kong University of Science and Technology,
Clear Water Bay, Kowloon, Hong Kong, China

We studied theoretically light transmission through a monolayer of hexagonal-close-packed nanoparticles consisting of a metallic shell and a dielectric core. We found that light can transmit through the dense particle assemblies via excitation of a variety of surface-plasmons (SPs). Localized SPs confined within metal nanoshells can mediate a narrow-band dispersionless transmission resonance (TR). Wide-band TRs were also observed as a result of strong near-field interparticle SP couplings, forming hybrid modes that are either localized at the nanogaps between adjacent particles or confined in the lattice pores, or distributed across the structure, each with distinct dispersion characteristics. Optical tuning strategies of these TRs are also elucidated that can allow for observation of SP anticrossing effects.

In the past decade, metal films perforated with a periodic array of subwavelength holes or slits attract much interest after the report of enhanced optical transmittance (EOT) by Ebbesen *et al*.[1] The EOT phenomenon is generally attributed to propagating surface-plasmon polaritons (SPPs),[2] which are excited by the incident light on the input side of the metal film, then evanescently tunnel to the exit side through the building up of strong electromagnetic fields above the apertures, and are finally re-emitted into optical far field. The SPP waves are Bragg scattered by the periodic apertures and thus are referred to as Bragg-type SPP modes. Although other theoretical models are also proposed,[3] many efforts have been made to study whether localized resonance modes in apertures can also bring EOT which should be independent of periodicity.[4]

Individual metallic nanoparticles can support localized surface-plasmons (SPs), whose resonance frequency depends on the particle size, shape, and composition, and also on the

---

[*] Phys.Rev.B, 2009, 80(16): 165401

surrounding medium.[5] At localized SP resonances, the huge electric field enhancement established on particle surface leads to a wealth of optical properties useful in surface enhanced Raman scattering,[6] optical antennas,[7] and optical tweezers.[8] For regularly spaced metal nanoparticles, the interactions of particle SP resonances can result in interesting collective optical properties. Lamprecht et al.[9] and Hicks et al.[10] have shown that the lineshape of these localized SPs can be controlled via coherent dipole far-field interactions. Recently extremely narrow plasmon resonances in one – and two-dimensional (2D) arrays of metal nanoparticles have been predicted[11–13] and observed experimentally,[14,15] which are revealed to be due to a diffraction coupling of localized SPs with a diffracted grazing wave at a Wood anomaly.[14] With these great efforts, a clear physical picture has been established about light interaction with two-dimensional (2D) metal nanoparticle arrays, where the excitation of dipole plasmon mode and its far-field interaction are dominant. However, less is known about the optical properties of dense metal nanoparticle arrays,[16] where higher-order SP modes and their near-field interactions could become extremely important for controlling the collective optical properties.[17] Recently, near-field interparticle localized SPs coupling in chains of closely spaced metal nanoparticles has been explored to guide electromagnetic energy with a lateral confinement below light diffraction limit.[18]

Some metallic nanostructures can support both localized and delocalized SPs and therefore exhibit rich optical phenomena related to their couplings when they are excited simultaneously. For example, Coyle et al.[19] and Kelf et al.[20] successfully identified the presence of both localized Mie and delocalized Bragg plasmons on nanostructured metal surfaces comprised of periodically arranged truncated spherical voids, through measurements of angle – and orientation-resolved optical reflectivity, and demonstrated their strong coupling thus producing bonding and antibonding mixed SPs by carefully controlling the sample thickness.[20] Sun and Chan[21] predicted frequency-selective enhanced absorption of light for a periodic lattice of dielectric spheres buried just beneath the surface of a metal substrate, by tuning localized Mie and delocalized Bragg plasmons into a strong coupling region.

In this paper, we study theoretically multiple light scatterings when light propagates through a 2D array of dense noble metal nanoshells with a dielectric core, arranged in a hexagonal-close-packed (HCP) lattice. We will show that light can transmit through the metallodielectric (MD) composite nanostructures with a high transmittance via a variety of SP excitations. We found that localized void plasmons, as an embodiment of single-particle Mie resonance, can mediate a narrow-band transmission resonance due to a high Q quality factor of noble metallic nanoshells. The strong interparticle interactions of SPs can form hybrid modes that can propagate across the MD nanostructures with Bragg-like dispersions, or modes that are highly localized near the gaps between adjacent nanoshells. In addition, the MD structures can support localized SPs that are mainly confined within the pores of the 2D lattice, which create a flat-dispersion transmission resonance under off-normal incidence. Furthermore, the anticrossing of the different SPs can lead to suppression of transmission. The distinct roles of these SPs and their interactions played in transmittance have been demonstrated clearly through the evaluation of transmission spectra as well as field distributions.

The transmission spectra and the field distributions are calculated with the electromagnetic wave layer-multiple-scattering theory formalism of Stefanou – Yannopapas – Modinos,[22] which has been shown to be highly efficient and accurate in calculations of electromagnetic properties of spherical objects. In this approach,[22] the electromagnetic field is first expanded into spherical harmonics about each spherical scatterer, and then the scattering properties of a periodical plane of scatterers are obtained by summation of all scattering events. Finally, the scattered wave field is transformed into a plane-wave representation. For pure dielectric structures, excellent numerical convergence can be achieved by including spherical waves with angular momentum index up to $L_{max}=7$ in the local expansion of electromagnetic waves.[23,24] Such a numerical convergence has also been tested for MD photonic band-gap crystals consisting of three-dimensional arrays of perfect metallic spheres in different lattices and excellent agreement between the band structure and the transmission code is obtained using $L_{max}=7$ in the angular momentum expansions and 37 2D reciprocal lattice vectors (whose cut-off length $R_{max}=22$, in units of the reciprocal of lattice period) in the plane-wave expansions.[25] For the closely spaced metallic nanoshells discussed here, however, a value of $L_{max}=7$ is not sufficient to obtain stable numerical results. In order to calculate precisely the scattering properties associated with multipole resonances of single MD spheres[26] and the effect of strong near-field coupling within the dense array, we took the angular momentum cut-off $L_{max}=19$ and used 109 2D reciprocal lattice vectors ($R_{max}=40$) to ensure a relative accuracy of 10^{-4} for spherical metallic nanoparticles with separations of at least 1 nm.

Figure 1(a) shows schematically the nanostructures to be studied. The coordinate is chosen such that the metallic nanoshells lie on the xy plane, with its origin located at the center of one shell. The incident light direction is defined by two angles, incidence angle θ (the angle between z axis and the wave-vector k) and azimuth angle φ (the angle between x axis and the wave-vector in-plane component k_{xy}). The composite spheres have a dielectric core of dispersionless dielectric constant ε_1 and a metallic coating nanoshell with the inner and outer radii denoted by R_1 and R_2, respectively. The MD structure has a lattice period of a. The relative permittivity of the metal is described by a Drude model: $\varepsilon_2 = 1 - \omega_p^2/[\omega(\omega+i\tau^{-1})]$, where ω_p is the plasma frequency and τ is the relaxation time related to energy loss. The parameter is taken to be $\hbar\omega_p = 9.2$ eV, which corresponds to the plasma energy of bulk silver.

In order to focus on the essence of the underlying physics, we begin our discussion by considering an ideal model in which the metal is lossless. The effect of absorption will be considered in the last part of the paper. Figure 1(b) presents the transmission spectrum of

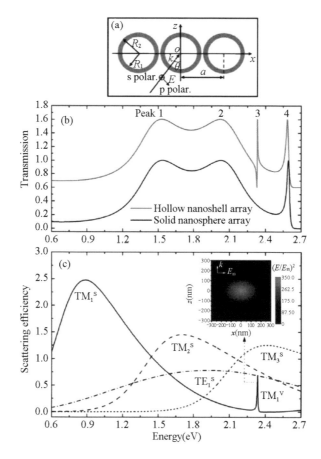

FIG. 1. (color online) (a) Schematic of a 2D HCP array of metallic hollow nanoshells in air. (b) Normal-incidence transmission spectrum for a 2D HCP array consisting of hollow Ag nanoshells with $R_1 = 200$ nm, $R_2 = 250$ nm, and $\varepsilon_1 = 1.0$, compared with that of solid Ag nanospheres with a diameter the same as the outer diameter of the nanoshells. The lattice period for both structures is $a = 520$ nm. The spectrum of the nanoshell array is vertically offset by 0.6 for clarity. (c) Partial scattering efficiencies of a single hollow Ag nanoshell for several dominant transverse magnetic ($TM_l^{V,S}$) and transverse electric ($TE_l^{V,S}$) modes, where the subscript l denotes the angular momentum index, and the superscripts V and S stand for voidlike and spherelike modes, respectively. The inset gives the normalized electric field intensity of a TM_1^V mode.

a 2D HCP array of hollow metallic nanoshells under normal illumination of light. The structural parameters are $R_1 = 200$ nm, $R_2 = 250$ nm, $a = 520$ nm, and $\varepsilon_1 = 1.0$. In Fig. 1(b), four conspicuous peaks with perfect transmission are observed which are located at 1.53 eV, 2.02 eV, 2.33 eV, and 2.59 eV, corresponding to wavelengths 810 nm (λ_1), 614 nm (λ_2), 532 nm (λ_3), and 479 nm (λ_4), respectively. The first three transmission resonances have a wavelength larger than the lattice period a. While the wavelength of the highest-energy resonance is slightly smaller than a, it is still much larger than the thinnest waist size of horn-shaped pores among adjacent shells in the lattice. It is noted that only a zero-order diffraction transmittance channel is open for the above MD structure in the spectral range discussed here. Thus, a metallic nanoshell array allows for light tunneling on a sub-wavelength scale, similar to 2D sub-wavelength nanohole arrays studied before.[1,2]

It is noted that the multiple transmission resonances are mediated by the excitation of SPs which will be shown to have different origins. We first discuss the SP mode that leads to the sharp transmission resonance at peak 3. For comparison, Fig. 1(b) also presents the transmission spectrum of a similar structure but composed of solid metallic spheres with an identical size as the nanoshell outer radius. We see that the solid-nanosphere assembly has almost the same spectral features as the nanoshell aggregate, except for the absence of peak 3. This suggests that this narrow-band resonance is due to the excitation of a void mode of individual metallic nanoshells. To confirm this, partial scattering efficiencies of a single metallic nanoshell were calculated for the dominant transverse magnetic ($TM_l^{V,S}$) and transverse electric ($TE_l^{V,S}$) modes, by using the Mie theory[27] and are plotted in Fig. 1 (c). Here, the subscript l denotes the angular momentum index, and the superscripts V and S, respectively, stand for voidlike and spherelike modes of single metallic nanoshells as these modes are either confined within the inner space (for voidlike modes) or localized on the shell surface (for spherelike modes). The existence of these two types of SP modes makes spherical core/shell nanostructures especially attractive due to a unique structural tunability of the particle plasmons.[28]

It is clearly seen that the TM_1^V mode is responsible for peak 3 as they have nearly the same energy. A similar mechanism for light tunneling through thin metal films embedded with dielectric spheres on the surface has been proposed recently.[29] In fact, the TM_1^V mode is a dipole plasmon resonance excited within the void of the nanoshells. The inset of Fig. 1 (c) gives the electric field intensity distribution at such a resonance, from which we see that the fields are exclusively confined within the dielectric core. Such a strong localization makes the resonance frequency dependent only on the dielectric permittivity or the size of the core, but essentially independent of the structure periodicity. Since this mode is trapped and thus long lived when metal absorption is neglected, peak 3 has a very narrow bandwidth.

In order to reveal the physical origins for the other three transmission peaks in Fig. 1(b) which are basically the same for both MD nanostructures composed of either solid spheres or nanoshells of the same size, the electric field distributions around the nanoshell array were calculated for resonances at λ_1, λ_2, and λ_4. The corresponding numerical results are plotted in Fig. 2. At resonance λ_1, the fields are seen to be concentrated in the nanogaps between adjacent spheres along the x direction (the polarization direction of incident light), known as hot spot.[30] The resonance at λ_2, with its field distributions shown in Figs. 2(c) and 2(d), has a similar field pattern as the resonance at λ_1, except that the four spots on the sphere surface near the hot spot become distinguishable in intensity. Based on the field distributions and on the evolution of transmission spectra obtained by increasing the angular momentum cutoff L_{max} from 1 to higher values (not shown here), we conclude that peaks 1 and 2 are mainly relevant to excitation of the TM_2^S and TM_3^S modes, which are quadrupole and hexapole SP resonances on the outer surface of single metallic nanoshells, respectively. Nevertheless, due to SP interactions among spheres the two peaks have obvious energy shift and field distribution deformation with respect to the individual sphere SP modes.

The resonance at λ_4 has a character different from the transmission resonances at λ_1 and λ_2 discussed above. For example, the enhanced field is no longer confined within the nanogaps between spheres but extends into the surrounding medium (i.e., air) from the particle surface with multiple nodes and a longer decay length. The unique field distribution associated with a huge field enhancement (with a maximum of ~ 300) suggests that a kind of propagating surface mode may be excited at this resonance. Since the metal nanospheres are densely packed, SP interactions on adjacent spheres are strong enough for the SP to hop from sphere to sphere to form a propagating surface wave.[18] As we will show later, this mode corresponding to peak 4 follows approximately the dispersion characteristics of a Bragg-type SPP.[1,2]

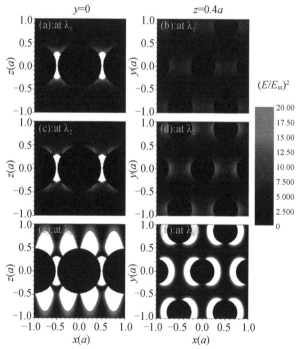

FIG. 2. (color online) Normalized electric field intensity $(E/E_{in})^2$ on the plane of $y = 0$ (left column) and $z = 0.4a$ (right column) for peak 1 (top panels), peak 2 (middle panels), and peak 4 (bottom panels). The maximum in the color scale is 20 and values in white regions are greater than 20. Incident field E_{in} is polarized alone the x axis.

Note that peak 4 will become narrow in bandwidth upon increasing sphere separations, and will eventually disappear when the separation is too large because in this situation, plasmon hopping between spheres via near-field coupling is suppressed. To demonstrate this, Fig. 3 shows the normal-incidence transmission spectra of a series of hollow metallic nanoshell arrays with increasing separation from 20 nm to 50 nm in steps of 10 nm. It is evident that peak 4 gets progressively narrower as the distance between metallic spheres increases, and finally vanishes when a is increased to 550 nm. In contrast, there is almost no change in peak 3 upon increasing a, due to the highly localized nature of the TM_1^V mode. Peaks 1 and 2 show only a slight shift in position for this small change in a as it could lead to a slight modification in the intersphere coupling of the TM_2^S and TM_3^S plasmon modes.

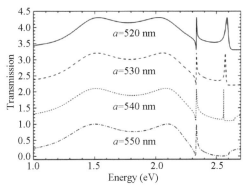

FIG. 3. Normal-incidence transmission spectra of 2D HCP arrays of hollow metallic nanoshells for different array periods $a=520$ nm, 530 nm, 540 nm, and 550 nm. Other structural parameters are the same as those used in Fig.1. Individual spectra are vertically offset by 1.1 from one another for clarity.

Now we turn to study the dispersion properties of the different resonant modes, by calculating the transmittance under off-normal incidence of light. For this purpose, Fig. 4 plots the transmission coefficients of the HCP hollow nanoshell array as a function of incidence angle θ and photon energy at two typical azimuth angles $\varphi=0°$ and $30°$ for both p and s polarizations. It is seen that the TM_1^V voidlike mode has an energy which is independent of the incident direction and polarization of light because of its highly localized nature and the high symmetry of the spherical cavities. It appears as a nondispersive bright horizontal line located at 2.33 eV in all four intensity maps in Fig. 4.

As is clearly seen in Fig. 4, peak 4 splits in energy into several modes at oblique incidence whose positions are all angle-dependent, suggesting a delocalization nature of these modes. It is interesting to see whether some of spectral features can be identified with Bragg-scattered SPP modes. To do this, yellow dashed lines are overlaid in Fig. 4 that describe the dispersion relations calculated by matching the momentum of a SPP on a smooth metal-dielectric interface with the reciprocal vector G_{mn} of a 2D hexagonal lattice: $E_{SPP} = \hbar c [(\varepsilon_2 + 1)/\varepsilon_2]^{1/2} |G_{mn} + k_{xy}|$, where c is light speed in vacuum, and m and n are integers. The analytical expression predicts well the position of peak 4 at normal incidence of light. Interestingly, some of the peaks and dips do follow the dispersion lines to a good approximation after splitting [see, for example, the lines labeled as G_{-10} G_{-11} in Fig. 4(a) and G_{01} G_{0-1} in Fig. 4(c)], further supporting the picture that this transmission peak is a propagating SPPs-mediated resonance. However, many dispersive features have no correspondence to the planar interface SPP dispersion folded by Bragg scattering (yellow lines). This is not surprising as the plasmonic bands formed by the hopping of SPs on disjointed spheres should be different from SPPs of planar metal films interrupted by subwavelength scattering objects.[1,2]

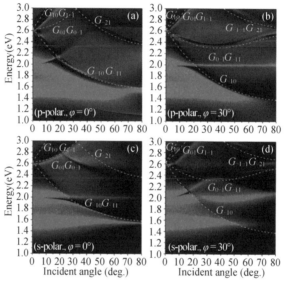

FIG. 4. (color online) Dispersion maps for a 2D HCP array of metallic nanoshells at two azimuth angles $\varphi=0°$ and $30°$ for both p and s polarizations of incident light. Structural parameters are the same as those used in Fig. 1. The linear color scale for transmittance is set from black (minimum $T=0$) to red (maximum $T=1$). Overlaid yellow dashed lines represent the dispersion relations of SPPs modes associated with reciprocal vectors G_{mn} of the 2D lattice.

For the spherelike modes contributing to two broad transmission bands in the low-energy region at normal incidence, the mode at $\omega_1=1.53$ eV becomes narrower in bandwidth as θ is increased (especially for s polarization), since the mode is in essence localized at the nanogaps between adjacent metallic nanoshells. The resonance centered at $\omega_2=2.02$ eV gradually merges into the lowest dispersion line of the Bragg-type mode after $\theta\sim10°$, due to its interaction with SPP mode [see Fig. 4(a)-4(c)] but sustains as a wide-band peak with small dispersion when the coupling becomes weak [see Fig. 4(d)].

In the dispersion maps of Fig. 4 we found a highly localized transmission resonance. This resonant mode shows up as a horizontal, relatively narrow transmission band centered at about 2.0 eV in Figs. 4(b) and 4(c). To get more insight into this narrow-band mode, we plot in Fig. 5 the corresponding electric field distributions in the six planes located at different distances from the xy plane ($z=0$) when the MD structure is illuminated with a p-polarized light with an energy $\omega=2.0$ eV under $\theta=30°$ and $\varphi=30°$. Clearly, the excited-field pattern is quite different from that of the modes excited at normal incidence of light (Fig. 2, right column). As is shown in the four planes $z=\pm 0.3a$ and $z=\pm 0.4a$ which are symmetrically around $z=0$, the fields are mainly concentrated in the pores of the lattice. This field localization in the pore region can still be resolved even at distances $z=\pm 0.5a$, although the maximum intensity is much reduced. Based on the field distribution characteristics, we could refer to this resonant mode as a porelike mode. Strong localization of this mode also occurs in Fig. 4(c) at exactly the same energy, but becomes weak in Fig. 4(d) and eventually vanishes in Fig. 4(a). The reason for this is unclear yet, but may be related to the anisotropy of the pore shape, which leads to different coupling strengths with the far field. Since the porelike mode is a hybridization of the plasmons on individual nanoshells, it contains an admixture of nanoshell plasmons of different angular momenta.

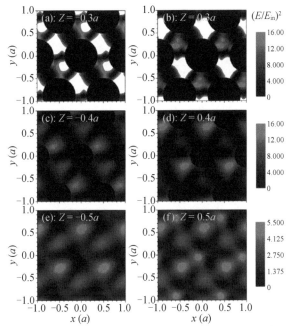

FIG. 5. (color online) Normalized electric field intensity $(E/E_{in})^2$ in the six planes of $z=\pm 0.3a$ (top row), $z=\pm 0.4a$ (middle row), and $z=\pm 0.5a$ (bottom row) for a p-polarized light with energy $\omega=2.0$ eV incident at $\theta=30°$ and $\varphi=30°$.

The resonance position of the porelike mode could be tuned by varying the lattice period a and thus the pore size for arrays of metallic nanoshells with fixed structural parameters. Figure 6 shows the same dispersion maps as in Fig. 4 for the same metallic nanoshells arranged in a HCP lattice, but with a period of 501 nm thus with an intersphere separation of only 1 nm. In this nearly touching case, the two spherelike modes get further overlapped and the porelike mode almost moves out of the bandwidth of the spherelike modes at small angles of incidence, as shown in Figs. 6(b) and 6(c). Such a decrease in period a also results in a redshift of the porelike mode, possibly because part of its field is squeezed out of the pores, leading to an increase in the mode volume. Note that the porelike mode exhibits a stronger localization in Figs. 6(a) and 6(d) than in Figs. 4(a) and 4(d), appearing as a flat narrow transmission band centered at 1.8 eV from about $\theta=10°$ to 35°. In addition, the yellow dispersion lines of Bragg-scattered SPPs move upward as a whole as compared with those in Fig. 4, since the SPPs energies will increase when a is decreased. Although these dispersion lines have many deviations from the practical transmission resonance positions, they still show an agreement with the spectral features for some resonances revealed numerically in Fig. 6 to a good approximation. We mention that the energy of the voidlike mode TM_1^V is identical to that in Fig. 4 because of its highly localized nature.

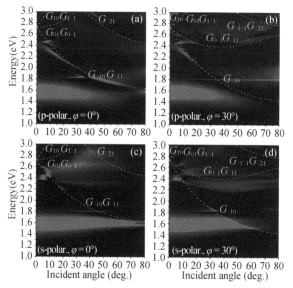

FIG.6. (color online) The same as Fig. 4 but for the lattice period $a=501$ nm.

Next we discuss the effects of interactions between different resonant modes on the transmittance. We first study the coupling between the porelike and spherelike modes. In Fig. 4 a noticeable gap around 2.0 eV is observed from about $\theta=5°$ to 15° in the wide transmission band of spherelike modes for all polarization and azimuth angles. It is clear that this transmission gap arises from the anticrossing of the two modes and this effect makes the spherelike mode centered at $\omega_2=2.02$ eV be cut into two parts. Such a gap also occurs in Fig. 6 at about 1.8 eV within the region of smaller angles of incidence for a 1 nm intersphere separation. Note that we do not observe the gaps at normal incidence in Figs. 4 and 6. This could be possibly because the porelike mode is excitable only at off-normal incidence of light. Very recently Tserkezis *et al.* have pointed that some hybrid plasmon modes in 2D square arrays of metallic nanoshells are inactive at normal incidence, and only could be excited at oblique incidence.[31]

To observe the effect as a result of coupling between wide-band spherelike and narrow-band voidlike modes, the relative permittivity ε_1 of the dielectric core is increased from unity to tune the voidlike modes into the lower-energy regime where the spherelike modes are located. Figure 7 shows the calculated transmission spectra of the MD nanostructures at normal incidence when ε_1 is varied from 1.0 to 3.0 in steps of 0.5. It is seen that although the increase in the dielectric permittivity within the metallic nanoshells does not bring about a large variation in the overall transmission, a sharp zero-transmittance dip is always created as the TM_1^V voidlike mode is tuned across the broad band of the spherelike modes. This is the result of anticrossing (or spectral interference) between localized modes within and outside thin metallic nanoshells. Analogous interference phenomenon is commonly referred to as the Fano resonance in atomic physics.[32] When the TM_1^V mode is tuned below the lower edge of the spherelike modes, a narrow-band transmission resonance is observed again for $\varepsilon_1 = 3$. Upon increasing ε_1, it is noted that the voidlike modes at higher energies will be simultaneously redshifted into the spectral range of interest, which can enable similar anticrossing effects, as is seen in Fig. 7.

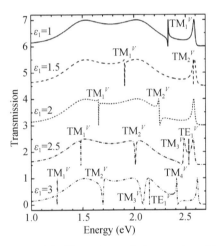

FIG. 7. Normal-incidence transmission spectra of 2D HCP arrays of metallic nanoshells. The dielectric core within the perfect metallic nanoshells takes different values of relative permittivity $\varepsilon_1 = 1.0, 1.5, 2.0, 2.5$, and 3.0. Other structural parameters are the same as those used in Fig. 1. Individual spectra are vertically offset by 1.5 from one another for clarity.

In addition to the coupling between localized SP modes, by optimizing the structural parameters, our MD nanostructures can also allow interesting avoided crossing between propagating SPPs and various localized modes. Such a phenomenon has been discussed in other plasmonic structures.[20,29,33] For example, narrow zero-transmittance dips due to anticrossing between voidlike TM_2^V and TE_1^V modes and a SPP mode can be observed in Fig. 7, when ε_1 takes the values of 1.5 and 2.5, respectively. Similar low-transmission dips as a result of anticrossing between localized cavity modes and delocalized SPPs have been predicted in thin metal films buried with a monolayer dielectric sphere array by de Abajo et al.[29] Similarly, the avoided crossing between highly dispersive SPPs and localized spherelike modes can also be seen in Figs. 4 and 6, which produces small gaps with a negligible transmittance.

We now consider the influence of light absorption by the shell metal. Figures 8(a)–8(c) shows, respectively, the normal-incidence transmission, reflection, and absorption spectra of a 2D HCP array of metallic nanoshells with different values of the loss parameter in the Drude model: $\hbar\tau^{-1}=0$ eV, 0.05 eV, 0.1 eV, and 0.2 eV. In Fig. 8(a), it is seen that the salient features remain robust in the presence of absorption even when $\hbar\tau^{-1}$ is increased to 0.2 eV. The two relatively narrow resonances, i.e., peaks 3 and 4 previously revealed in Fig. 1(b), however, are more affected by losses, which are smoothed out progressively as τ^{-1} increases, implying that their lifetime becomes more and more short. The reason is partly due to the larger Ohm loss in metal, as is clearly seen in Fig. 8(c), owing to the larger electric-field enhancements on the inner and outer surfaces of metallic nanoshells when voidlike and SPPs modes are excited, respectively.

On the other hand, for the two broad-band spherelike modes, the reflection at resonances is nearly equal to zero [less than 0.01, see Fig. 8(b)] for all values of τ^{-1}, which indicates that the reduction in transmittance mainly arises from the Ohm loss in metal. For peaks 3 and 4, the larger the value of τ^{-1} is, the stronger is the reflection at the narrowband resonances. This means the radiative loss (reflection) will also contribute to the decrease in the transmittance.

Note that for the sharp resonance of the TM_1^V mode, the absorption peak decreases with increasing τ^{-1} [see Fig. 8(c)], unlike the behavior observed for the other three resonances. At first glance, this seems to be somewhat counterintuitive. But, this is not hard to be understood when realizing that light must penetrate into the metal shell in order to excite the voidlike Mie mode. Certainly, the increase in τ^{-1} will reduce the coupling strength of this mode, and thus the field enhancement on the inner surface of the metal shell. Consequently, the absorption peak amplitude will decrease when τ^{-1} is increased. When the metal is very lossy (τ^{-1} is large), the cavity mode cannot be excited and the corresponding absorption will vanish (not shown here).

Finally, it should be mentioned that the transmission peaks correspond to the reflectivity minima and absorbance maxima, a feature consistent with the involvement of SPs.[34] The fields at the metal surface are enhanced when the incident light couples to a SP mode, thus aiding transmission. However, the SP mode is also damped by absorption in the metal so that enhanced absorption has the same conditions that give an enhanced transmission.

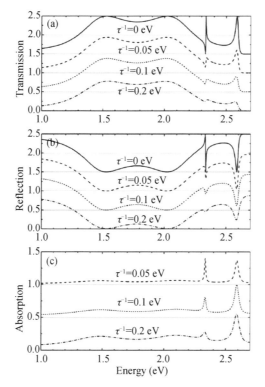

FIG. 8. Normal-incidence transmission (a), reflection (b), and absorption (c) spectra of a 2D HCP array of hollow metallic nanoshells for the shell metal to have different values of loss in the Drude model: $\hbar\tau^{-1}=0$ eV(perfect), 0.05 eV, 0.1 eV, and 0.2 eV. Other structural parameters are the same as those used in Fig. 1. Individual spectra are vertically offset by 0.5 from one another for clarity. Horizontal light-gray lines are guides to the eye.

In conclusion, we have shown that light can tunnel through 2D dense HCP arrays of metal nanoshells via excitations of voidlike, spherelike Mie modes, Bragg-scattered SPPs, and porelike localized resonance modes. Voidlike Mie modes can be controlled by the size and dielectric constant of the dielectric core. The other three types of modes can be tuned by increasing the permittivity of surrounding medium or by scaling the array size. Moreover, transmission suppression phenomena due to anticrossing between different plasmon modes are revealed. Our study will be helpful to understand the physical mechanisms of anomalous light transmittance through a variety of metallic nanostructures. On the other hand, huge enhancement of electric fields at plasmon resonances suggests that the MD system can serve as a substrate for surface-enhanced spectroscopies.[30]

References and Notes

[1] T. W. Ebbesen, H. J. Lezec, H. F. Ghaemi, T. Thio, and P. A. Wolff, Nature (London) **391**, 667 1998; H. F. Ghaemi, T. Thio, D. E. Grupp, T. W. Ebbesen, and H. J. Lezec, Phys. Rev. B **58**, 6779

(1998).

[2] J. A. Porto, F. J. García-Vidal, and J. B. Pendry, Phys. Rev. Lett. **83**, 2845 (1999); L. Salomon, F. Grillot, A. V. Zayats, and F. de Fornel, *ibid.* **86**, 1110 (2001); L. Martín-Moreno, F. J. García-Vidal, H. J. Lezec, K. M. Pellerin, T. Thio, J. B. Pendry, and T. W. Ebbesen, *ibid.* **86**, 1114 (2001).

[3] M. M. J. Treacy, Phys. Rev. B **66**, 195105 (2002); H. J. Lezec and T. Thio, Opt. Express **12**, 3629 (2004).

[4] F. Yang and J. R. Sambles, Phys. Rev. Lett. **89**, 063901 (2002); K. J. Klein Koerkamp, S. Enoch, F. B. Segerink, N. F. van Hulst, and L. Kuipers, *ibid.* **92**, 183901 (2004); Z. Ruan and M. Qiu, *ibid.* **96**, 233901 (2006); J. W. Lee, M. A. Seo, D. H. Kang, K. S. Khim, S. C. Jeoung, and D. S. Kim, *ibid.* **99**, 137401 (2007).

[5] K. L. Kelly, E. Coronado, L. L. Zhao, and G. C. Schatz, J. Phys. Chem. B **107**, 668 (2003).

[6] S. Nie and S. R. Emory, Science **275**, 1102 (1997); H. Xu, E. J. Bjerneld, M. Käll, and L. Börjesson, Phys. Rev. Lett. **83**, 4357 (1999).

[7] P. Mühlschlegel, H. J. Eisler, O. J. F. Martin, B. Hecht, and D. W. Pohl, Science **308**, 1607 (2005).

[8] M. Righini, G. Volpe, C. Girard, D. Petrov, and R. Quidant, Phys. Rev. Lett. **100**, 186804 (2008); A. N. Grigorenko, N. W. Roberts, M. R. Dickinson, and Y. Zhang, Nat. Photonics **2**, 365 (2008).

[9] B. Lamprecht, G. Schider, R. T. Lechner, H. Ditlbacher, J. R. Krenn, A. Leitner, and F. R. Aussenegg, Phys. Rev. Lett. **84**, 4721 (2000).

[10] E. M. Hicks, S. Zou, G. C. Schatz, K. G. Spears, R. P. Van Duyne, L. Gunnarsson, T. Rindzevicius, B. Kasemo, and M. Käll, Nano Lett. **5**, 1065 (2005).

[11] K. T. Carron, W. Fluhr, M. Meier, A. Wokaun, and H. W. Lehmann, J. Opt. Soc. Am. B **3**, 430 (1986).

[12] S. Zou, N. Janel, and G. C. Schatz, J. Chem. Phys. **120**, 10871 (2004).

[13] V. A. Markel, J. Phys. B **38**, L115 (2005).

[14] V. G. Kravets, F. Schedin, and A. N. Grigorenko, Phys. Rev. Lett. **101**, 087403 (2008).

[15] B. Auguié and W. L. Barnes, Phys. Rev. Lett. **101**, 143902 (2008).

[16] Z. Chen, P. Zhan, Z. L. Wang, J. H. Zhang, W. Y. Zhang, N. B. Ming, C. T. Chan, and P. Sheng, Adv. Mater. **16**, 417 (2004).

[17] B. N. Khlebtsov, V. A. Khanadeyev, J. Ye, D. W. Mackowski, G. Borghs, and N. G. Khlebtsov, Phys. Rev. B **77**, 035440 (2008); F. Le, D. W. Brandl, Y. A. Urzhumov, H. Wang, J. Kundu, N. J. Halas, J. Aizpurua, and P. Nordlander, ACS Nano **2**, 707 (2008).

[18] S. A. Maier, P. G. Kik, H. A. Atwater, S. Meltzer, E. Harel, B. E. Koel, and Ari A. G. Requicha, Nature Mater. **2**, 229 (2003).

[19] S. Coyle, M. C. Netti, J. J. Baumberg, M. A. Ghanem, P. R. Birkin, P. N. Bartlett, and D. M. Whittaker, Phys. Rev. Lett. **87**, 176801 (2001).

[20] T. A. Kelf, Y. Sugawara, J. J. Baumberg, M. Abdelsalam, and P. N. Bartlett, Phys. Rev. Lett. **95**, 116802 (2005); T. A. Kelf, Y. Sugawara, R. M. Cole, J. J. Baumberg, M. E. Abdelsalam, S. Cintra, S. Mahajan, A. E. Russell, and P. N. Bartlett, Phys. Rev. B **74**, 245415 (2006).

[21] G. Sun and C. T. Chan, Phys. Rev. E **73**, 036613 (2006).

[22] N. Stefanou, V. Yannopapas, and A. Modinos, Comput. Phys. Commun. **113**, 49 (1998); **132**, 189 (2000).

[23] V. Yannopapas, N. Stefanou, and A. Modinos, Phys. Rev. Lett. **86**, 4811 (2001).

[24] Z. L. Wang, C. T. Chan, W. Y. Zhang, Z. Chen, N. B. Ming, and P. Sheng, Phys. Rev. E **67**, 016612

(2003).

[25] W. Y. Zhang, X. Y. Lei, Z. L. Wang, D. G. Zheng, W. Y. Tam, C. T. Chan, and P. Sheng, Phys. Rev. Lett. **84**, 2853 (2000); Z. L. Wang, C. T. Chan, W. Y. Zhang, N. B. Ming, and P. Sheng, Phys. Rev. B **64**, 113108 (2001).

[26] M. I. Tribelsky and B. S. Luk'yanchuk, Phys. Rev. Lett. **97**, 263902 (2006).

[27] C. F. Bohren and D. R. Huffman, *Absorption and Scattering of Light by Small Particles* (Wiley, New York, 1983).

[28] E. Prodan, C. Radloff, N. J. Halas, and P. Nordlander, Science **302**, 419 (2003).

[29] F. J. García de Abajo, G. Gómez-Santos, L. A. Blanco, A. G. Borisov, and S. V. Shabanov, Phys. Rev. Lett. **95**, 067403 (2005).

[30] F. J. García-Vidal and J. B. Pendry, Phys. Rev. Lett. **77**, 1163 (1996).

[31] C. Tserkezis, G. Gantzounis, and N. Stefanou, J. Phys.: Condens. Matter **20**, 075232 (2008).

[32] U. Fano, Phys. Rev. **124**, 1866 (1961).

[33] Y. J. Bao, R. W. Peng, D. J. Shu, M. Wang, X. Lu, J. Shao, W. Lu, and N. B. Ming, Phys. Rev. Lett. **101**, 087401 (2008).

[34] W. L. Barnes, W. A. Murray, J. Dintinger, E. Devaux, and T. W. Ebbesen, Phys. Rev. Lett. **92**, 107401 (2004).

[35] We thank C. T. Chan, L. Zhou, and J. Zi for helpful discussions and C. T. Chan for a critical reading of the manuscript. Acknowledgments for financial support are for C. T., Z. W. from the State Key Program for Basic Research of China and NSFC under grants No. 10425415, No. 10734010, No. 50771054, and No.10804044, for G.S. from NSFC under grant No. 10674157, and for S.Z. from NSFC under grant No. 10534020.

Experimental Observation of Sharp Cavity Plasmon Resonances in Dielectric-metal Core-shell Resonators*

Ping Gu,[1] Mingjie Wan,[1] Qi Shen,[1] Xiaodan He,[1] Zhuo Chen,[1,2]
Peng Zhan,[1,2] and Zhenlin Wang[1,2]

[1] *School of Physics and National Laboratory of Solid State Microstructures,*
Nanjing University, Nanjing 210093, China
[2] *Collaborative Innovation Center of Advanced Microstructures, Nanjing 210093, China*

We report on the experimental realization of dielectric-metal core-shell resonators with a nearly perfect metal shell layer by physically depositing metal onto the self-supporting dielectric colloids. Sharp electric and magnetic-based cavity plasmon resonances are experimentally observed, whereas increasing the metal shell thickness increases their Q-factors while narrowing their linewidths. In particular, a high Q-factor up to ~ 100 with a correspondingly narrow linewidth down to ~ 12 nm is experimentally obtained at a dipolar magnetic cavity plasmon resonance. Simulations and analytical Mie calculations show excellent agreements with the experimental results and demonstrate strong optical field confinement of such three-dimensional resonators.

Metallic nanostructures supporting collective electron excitations, known as surface plasmons (SPs), have the ability to concentrate light into subwavelength volumes and produce highly localized fields, which enables their use in a wide range of nanophotonics technologies and devices.[1,2] The low quality factors (Qs) of plasmon resonances, however, severely hamper their applications such as biosensors and plasmon rulers,[2-4] where sharp spectral responses with high quality factors are much desired. Recently, the Fano interference between subradiant and superradiant plasmon modes has been suggested as a promising technique to improve the Q (narrow the linewidth) of plasmon resonances.[2,5,6] However, even with such metallic Fano-resonant systems, the experimentally achievable Q is limited to the order of 10.[2,4] Although periodically patterned metal nanoparticles have been reported to show sharp Fano resonances with linewidths down to 5 nm, they are based on diffractive coupling mechanism, which requires a near-perfect periodic arrangement of the nanoparticles and is extremely sensitive to the incident angles.[7,8] More recently, hybrid plasmon-photon resonances with high Q values in the thousands have been theoretically predicted in metal-coated whispering gallery mode (WGM) microcavities such as hollow silver tubes and silver-coated microtoroids.[9,10]

---

* Appl.Phys.Lett.,2015,107(14):141908

Although there have even been experimental demonstrations based on silver-coated microdisks or microbottle resonators,[11,12] but their dimensions corresponding to those of the WGM dielectric microcavities are inevitably much larger than the wavelength of the incident light.[13]

Dielectric-metal core-shell resonators (DMCSRs) are versatile structures in nanophotonics because they have been predicted to support both sphere and cavity plasmons.[14] Previous experimental investigations of DMCSRs mainly focus on the spectrally broad sphere plasmons within the quasistatic limit.[15—18] Although the hybridization between the sphere and cavity plasmons has been exploited to explain the effect of the core-shell aspect ratio or the symmetry breaking on the sphere plasmons,[19,20] the cavity plasmon itself is hardly observed in these experiments because it only couples weakly with the incident light under the quasistatic limit and may suffer a further damping from the interband transitions in metals.[16] When the core size is increased beyond the quasistatic regime, it has been theoretically predicted that cavity plasmons with narrow linewidths can interact strongly with the incident optical field because of the phase retardation effect[21] and can give rises to Fano resonances to maintain the high quality factors of the cavity plasmons.[22—24] However, due to the particular challenge in the wet-chemistry based preparation of core-shell structures with a relatively large (comparable to the resonance wavelength) dielectric core and a perfect (dense, smooth, and complete) metal shell layer,[19,25—27] the sharp cavity plasmons and its induced Fano resonances in DMCSRs have not been experimentally demonstrated.[14,21—24]

In this letter, we report on the realization of spherical DMCSRs with a nearly perfect metal shell by physically depositing silver films onto the both sides of self-supporting polystyrene (PS) colloids. The prepared DMCSRs are experimentally demonstrated to be capable of supporting high-Q magnetic and electric-based cavity plasmons at free-space wavelengths longer than their physical dimensions. As the silver shell thickness goes beyond its skin depth, the Q-factors are improved to ~ 30 for the observed electric cavity plasmons, and as high as ~ 100 for a dipolar magnetic cavity plasmon resonance with a corresponding linewidth ~ 12 nm, which is almost an order of magnitude higher than the previously reported Q-factors of ~ 10 in Fano-like resonances supported by the individual plasmonic nanostructures.[2] Our experimental findings are supported by excellent agreements with simulations and analytical calculations, which also reveal a strong 3D confinement of resonant optical fields within the dielectric core. These DMCSRs could open a pathway for realizing ultrasensitive bio-sensors, lasing, and nonlinear optical devices with a high-performance.

To prepare the nearly perfect DMCSRs, a simple two-step approach is developed and illustrated schematically in Fig. 1(a). In the first step, a monolayer of monodisperse PS spheres with a diameter of $D \approx 1.0$ μm is first self-assembled on the water surface using a modified Langmuir-Blodgett method,[28] and then transferred onto a substrate with tens of micrometer-sized through-holes to form self-supporting PS colloids by exploiting the strong interparticle van der Walls interactions and the interactions between the particles and the grid substrate as well [Fig. 1(b)]. Note that the existence of the through-holes ensures the accessibility of both the upper and lower half surfaces of the PS colloids during the subsequent deposition process. In the second step, thin metal films with an identical thickness (t) are successively plasma sputtered onto the upper and lower half-surfaces of self-supporting PS colloids to wrap a nearly perfect metal shell layer around each colloid. Figure 1(c) shows a top-view scanning electron microscopy (SEM) image of the resultant DMCSRs with a shell thickness of $t \sim 30$ nm. The coated silver shell is nearly complete, except that nano-windows with an opening angle of $\sim 20°$ are present in its equator region where the original PS colloids touch each other [the inset of Fig. 1(c)]. The nano-windows can be used to easily introduce specimen or integrate functional materials (e.g., gain materials) into the resonators for future sensing or nanolasing applications. It is also worth noting that our method can be used to create core-shell structures with various core sizes consisting of virtually any metal and dielectric materials.

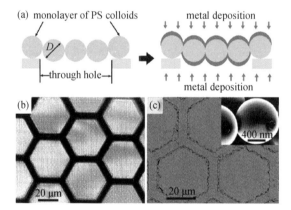

FIG. 1. (a) Schematic illustration of the DMCSRs prepared by successively depositing thin metal films onto the both upper and lower half-surfaces of the self-supporting PS colloids. (b) Optical microscopy image of a monolayer of PS colloids ($D \approx 1.0$ μm) transferred onto a substrate with through-holes. (c) SEM image of the prepared DMCSRs with a shell thickness of $t \sim 30$ nm. Inset shows the side-view SEM image of the DMCSRs, where the PS cores are dissolved away to more clearly reveal the shell morphology.

The optical properties of the DMCSRs are characterized using a Fourier-transform infrared (FTIR) spectrometer, in which a quasi-collimated white light beam passes through a linear polarizer and is slightly focused onto the samples with an off-axis parabolic mirror [Fig. 2(a)]. The zeroth-order extinction (1 − transmission) spectra of the prepared

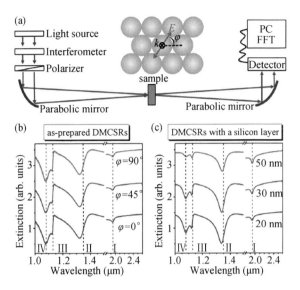

FIG. 2. (a) Schematic of the FTIR transmission measurement setup. (b) Measured extinction spectra of the as-prepared DMCSRs at normal incidence for three different electric field orientation angles. (c) The extinction spectra measured at normal incidence for the DMCSRs coated with extra silicon layers of different thicknesses.

DMCSRs with a shell thickness $t=30$ nm are first measured at normal incidence for three different electric field orientation angles $\varphi=0°, 45°$, and $90°$. Figure 2(b) presents that the results are almost exactly the same regardless of the incident electric field orientations. The most remarkable spectral features are the four distinct resonances marked as Ⅰ, Ⅱ, Ⅲ, and Ⅳ, locating at the wavelengths of $\lambda \approx 1970$ nm, 1350 nm, 1130 nm, and 1080 nm, respectively. Our previous theoretical studies have demonstrated that a periodic array of DMCSRs can support cavity, sphere, and pore-like plasmon modes in the zeroth-order wavelength region.[22] The sphere and pore-like plasmon modes are sensitive to the dielectric permittivity of the surrounding medium, because their surface charges are predominantly located on the outer surface of the shell.[22,29] On the contrary, the cavity plasmon modes are associated with the surface charges on the inner surface of the shell and are therefore sensitive to the dielectric constant of the dielectric core.[22,29] Based on such different dependencies, it is easy to distinguish the cavity plasmon modes from the sphere and pore-like counterparts. For this purpose, an extra layer of silicon with different thicknesses is deposited on the outer surface of the as-prepared DMCSRs. Figure 2(c) shows that the extinction spectra measured for the DMCSRs with such silicon layer. The four resonances are still clearly visible and maintain their original resonance wavelengths, which is consistent with predictions that the cavity plasmons should be insensitive to the surrounding medium,[22,29] and therefore provides a strong evidence that the observed resonances are due to the cavity plasmon modes. In addition, the coated silicon film (and thus the increased dielectric constant of the surrounding medium) can result in the expected red-shifts of the sphere plasmon modes[22,29] and consequently alter coupling between cavity and sphere plasmon modes,[22—24] which make the resonance Ⅱ narrower and the resonances Ⅲ and Ⅳ shallower [Fig. 2(b)] compared with the original resonances in the as-prepared DMCSRs [Fig. 2(a)].

In order to identify the observed resonances, numerical simulations are conducted using the three-dimensional finite-element-method (FEM) software COMSOL Multiphysics. Figure 3(a) shows the extinction spectra calculated at normal incidence for different electric field orientation angles corresponding to the experimental conditions applied in Fig. 2(b). The simulated domain is a cuboid containing one complete and four separate one-quarter of the DMCSRs with the geometrical parameters estimated from Fig. 1(c), and its four sides (xz- and yz-planes) are applied with periodic boundary conditions [inset in Fig. 3(a)]. The refractive index of PS is assumed to be 1.59, and the permittivity of silver is taken from the experimental data by Johnson and Christy.[30] The simulated results show an excellent agreement with our measurements, faithfully reproducing the main features of the measured spectra. Furthermore, our simulations also yield the spatial distributions of the field intensity (normalized by the incident electric field intensity) for

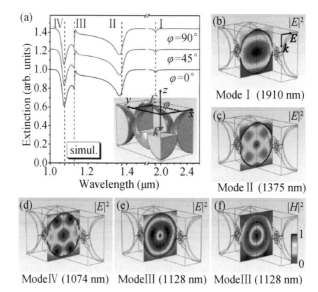

FIG. 3. (a) Extinction spectra calculated at normal incidence for three different polarization angles. Inset shows the cuboid simulation domain with a portion of structure being removed for clarity, and the incident light direction (k) and the electric field orientation angle (φ) are also shown. (b)-(d) Normalized electric field intensity distributions of the mode Ⅰ, Ⅱ, and Ⅳ, respectively. (e) and (f) Normalized electric and magnetic field intensity distributions of the mode Ⅲ, respectively.

each resonance wavelength at $\varphi = 0°$, as summarized in Figs. 3(b)–3(f). The single-[Fig. 3(b)], four-[Fig. 3(c)], and six-lobed [Fig. 3(d)] electric field intensity distributions within the dielectric core reveal that the resonances Ⅰ, Ⅱ, and Ⅳ correspond to the excitations of the electric dipolar, quadrupolar, and octupolar cavity plasmons in the DMCSRs, respectively. Interestingly, the electric field for the resonance Ⅲ is found to be circulated inside the dielectric core [Fig. 3(e)], and the magnetic field is highly concentrated at the center of the core [Fig. 3(f)]. Such electric and magnetic field distributions imply that the resonance Ⅲ is related to the excitation of magnetic-based dipolar cavity plasmon mode. Furthermore, optical fields at each cavity plasmon resonance are mostly confined within the dielectric core, confirming strong light confinement capabilities of the DMCSRs.

The observed cavity plasmons can be further analyzed using Mie theory,[31] due to their highly localized nature [Figs. 3(b)–3(f)]. Figure 4(a) shows the analytically calculated total extinction efficiency of an isolated DMCSR, and again the measured extinction spectrum for comparison. In the calculations, the DMCSR is simplified to a 980-nm-diameter PS sphere concentrically surrounded by a uniform 21-nm-thick silver shell, while all the dielectric constants remain the same as in the simulations. Figure 4(a) shows that there is a very good one-to-one correspondence between experimental (red line) and theoretically predicted resonances (blue line). One of the main advantages of the analytical Mie solution is its ability to decompose the obtained spectra into separate multipolar contributions, characterized by electric (a_l) and magnetic (b_l) scattering coefficients, where l is the index of angular momentum.[31] Figure 4(b) shows the results of this analysis on the individual DMCSR (solid lines) as well as on a solid silver sphere with the same outer radius (dashed lines). In the displayed wavelength range, only the first three electric terms (a_1, a_2, a_3) and the first magnetic term (b_1) of Mie expansion have dominant contributions. It is clearly seen from Fig. 4(b) that spectrally broad peaks marked with TM_1^S, TM_2^S, TM_3^S, and TE_1^S in the spectra of the individual DMCSR (solid lines) also appear in the solid silver sphere (dashed lines), revealing that they correspond to the excitations of the sphere Mie plasmons. However, narrow spectral features (marked by TM_1^V, TM_2^V, TM_3^V, and TE_1^V) that appear in the spectra of the DMCSR are completely absent in the solid silver sphere, providing an evidence that they are directly related to the excitations of cavity plasmons. Therefore, as indicated by vertical stripes with the same colour in Figs. 4(a) and 4(b), the resonances Ⅰ, Ⅱ, Ⅳ, and Ⅲ could be attributed to the excitations of the cavity Mie plasmon modes TM_1^V, TM_2^V, TM_3^V, and TE_1^V, respectively.

In the following, we demonstrate that increasing the silver shell thickness is an efficient way to reduce the linewidths, and correspondingly increase the Q-factors of cavity plasmons. Figure 5(a) shows the measured extinction spectra for the DMCSRs with

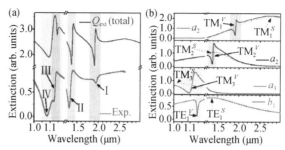

FIG. 4. (a) Calculated total extinction efficiency of an individual DMCSR (blue line) and measured extinction for as-prepared DMCSRs (red line). (b) Decomposed extinction spectra of a DMCSR (solid lines) and a solid silver sphere (dashed lines). Vertical stripes with the same colour in (a) and (b) indicate a one-to-one correspondence.

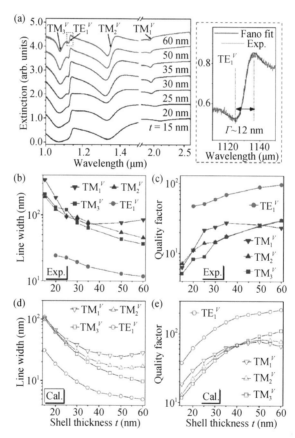

FIG. 5. (a) Measured extinction spectra of DMCSRs with different shell thicknesses. Inset in (a) shows the enlarged extinction spectrum for $t=60$ nm and its Fano fitting in the vicinity of the magnetic-based TE_1^V Fano resonance. (b) and (c) The resonance linewidths and quality factors extracted from the best fit to the experimentally measured extinction spectra, respectively. (d) and (e) The resonance linewidths and quality factors extracted from the best fit to the analytically calculated extinction spectra, respectively.

different shell thicknesses. The resonances present a progressive decrease in the linewidth and a gradual blue-shift (more pronounced for the resonance TM_1^V) upon increasing the shell thickness, which is attributed to the stronger confinement of the cavity plasmon modes at larger shell thickness.[21] To precisely describe this resonance narrowing effect, the measured extinction spectrum is fitted in the vicinity of each resonance using a Fano formula $F(\varepsilon) = \sigma_{bg} + \sigma_0 (\varepsilon + q)^2/(1 + \varepsilon^2)$, where σ_0 and σ_{bg} are the normalized and background extinction, q is the asymmetry parameter, and $\varepsilon = 2(\lambda - \lambda_{res})/\Gamma$ with λ_{res} and Γ being the position and linewidth of cavity plasmon resonance, respectively.[23] The inset in Fig. 5(a) shows an example of such a Fano fitting for the resonance TE_1^V in the DMCSRs with a shell thickness of $t=60$ nm. The fitted curve (olive line) is in good agreement with the experimental spectrum (red line). With the extracted spectral positions (λ_{res}) and linewidths (Γ), the Q-factor is calculated as $Q = \lambda_{res}/\Gamma$. Figures 5(b)–5(e) summarize the linewidths and Q-factors of both experimentally observed and theoretically predicted cavity plasmon resonances for the DMCSRs with silver shells having different thicknesses. Although experimentally achieved linewidths [Fig. 5(b)] or Q-factors [Fig. 5(c)] are broader or lower than the predicted values [Figs. 5(d) and 5(e)], which are most likely due to the broadening mechanism introduced by electron scattering in the thin metal shell, the PS core size distribution, and the roughness of the deposited silver shell layer, both the observed and predicted resonances follow the same trends of linewidth narrowing and improvement in Q-factors with increasing the silver shell thickness. Figures 5(b)–5(e) also show that as the shell thickness increases beyond the optical skin depth of ~30 nm for silver the observed linewidth narrowing and Q-factor improvement begin to converge as the additional can no longer increase the optical field confinement.[21] Furthermore, compared with the electric cavity plasmons, the magnetic cavity plasmon resonance is found to have a much narrower linewidth and a higher Q-factor, because it produces relatively smaller field enhancements near the inner surface of the shell [Figs. 3(b)–3(f)], and thus there are less power penetration into the silver walls. For example, when the silver shell thickness is increased to $t=60$ nm, the experimentally achieved Q-factor is improved to 20~30 (the linewidth is narrowed to 36 nm~84 nm) for the electric cavity plasmons TM_1^V, TM_2^V, and TM_3^V, while a particularly high value of ~100 (a narrow linewidth of ~12 nm) for the magnetic cavity plasmon resonance TE_1^V is observed [Figs. 5(b) and 5(c)].

In summary, high-quality DMCSRs with nearly perfect metal shells are realized by physically depositing metal onto a self-supporting monolayer of dielectric colloids. Sharp Mie cavity plasmons with narrow linewidths down to ~12 nm and high Q-factors up to ~100 are experimentally observed. Numerical simulations and analytical calculations reproduce the experimental results very well and reveal that the cavity plasmons supported by the DMCSRs have an ability to provide strong three-dimensional confinement of optical fields. Such desirable characteristics coupled with the ease of fabrication make the DMCSRs suitable for applications such as ultrasensitive bio-sensors,[20] low-threshold lasing,[32] slow-light, and nonlinear optical devices.[4]

References and Notes

[1] K. L. Kelly, E. Coronado, L. L. Zhao, and G. C. Schatz, J. Phys. Chem. B **107**, 668 (2003).

[2] B. Luk'yanchuk, N. L. Zheludev, S. A. Maier, N. J. Halas, P. Nordlander, H. Giessen, and C. T. Chong, Nat. Mater. **9**, 707 (2010).

[3] F. Hao, P. Nordlander, Y. Sonnefraud, P. V. Dorpe, and S. A. Maier, ACS Nano **3**, 643 (2009).

[4] N. Liu, M. Hentschel, T. Weiss, A. P. Alivisatos, and H. Giessen, Science **332**, 1407 (2011).

[5] U. Fano, Phys. Rev. **124**, 1866 (1961).

[6] A. E. Miroshnichenko, S. Flach, and Y. S. Kivshar, Rev. Mod. Phys. **82**, 2257 (2010).

[7] W. Zhou and T. W. Odom, Nat. Nanotechnol. **6**, 423 (2011).

[8] V. G. Kravets, F. Schedin, and A. N. Grigorenko, Phys. Rev. Lett. **101**, 087403 (2008).

[9] A. Rottler, M. Bröll, S. Schwaiger, D. Heitmann, and S. Mendach, Opt. Lett. **36**, 1240 (2011).

[10] Y. F. Xiao, C. L. Zou, B. B. Li, C. H. Dong, Z. F. Han, and Q. H. Gong, Phys. Rev. Lett. **105**, 153902 (2010).

[11] B. Min, E. Ostby, V. Sorger, E. U. Avila, L. Yang, X. Zhang, and K. Vahala, Nature **457**, 455 (2009).

[12] A. Rottler, M. Harland, M. Bröll, M. Klingbeil, J. Ehlermann, and S. Mendach, Phys. Rev. Lett. **111**, 253901 (2013).

[13] K. J. Vahala, Nature **424**, 839 (2003).

[14] A. L. Aden and M. Kerker, J. Appl. Phys. **22**, 1242 (1951).

[15] S. J. Oldenburg, R. D. Averitt, S. L. Westcott, and N. J. Halas, Chem. Phys. Lett. **288**, 243 (1998).

[16] H. Wang, Y. P. Wu, B. Lassiter, C. L. Nehl, J. H. Hafner, P. Nordlander, and N. J. Halas, Proc. Natl. Acad. Sci. U.S.A. **103**, 10856 (2006).

[17] S. Mukherjee, H. Sobhani, J. B. Lassiter, R. Bardhan, P. Nordlander, and N. J. Halas, Nano Lett. **10**, 2694 (2010).

[18] J. A. Fan, C. Wu, K. Bao, J. Bao, R. Bardhan, N. J. Halas, V. H. Manoharan, P. Nordlander, G. Shvets, and F. Capasso, Science **328**, 1135 (2010).

[19] E. Prodan, C. Radloff, N. J. Halas, and P. Nordlander, Science **302**, 419 (2003).

[20] F. Hao, Y. Sonnefraud, P. V. Dorpe, S. A. Maier, N. J. Halas, and P. Nordlander, Nano Lett. **8**, 3983 (2008).

[21] J. J. Penninkhof, L. A. Sweatlock, A. Moroz, H. A. Atwater, A. Blaaderen, and A. Polman, J. Appl. Phys. **103**, 123105 (2008).

[22] C. J. Tang, Z. L. Wang, W. Y. Zhang, S. N. Zhu, N. B. Ming, G. Sun, and P. Sheng, Phys. Rev. B **80**, 165401 (2009).

[23] A. E. Miroshnichenko, Phys. Rev. A **81**, 053818 (2010).

[24] J. S. Parramon and D. Jelovina, Nanoscale **6**, 13555 (2014).

[25] S. J. Oldenburg, J. B. Jackson, S. L. Westcott, and N. J. Halas, Appl. Phys. Lett. **75**, 2897 (1999).

[26] S. L. Westcott, J. B. Jackson, C. Radloff, and N. J. Halas, Phys. Rev. B **66**, 155431 (2002).

[27] Z. Chen, P. Zhan, Z. L. Wang, J. H. Zhang, W. Y. Zhang, N. B. Ming, C. T. Chan, and P. Sheng,

Adv. Mater. **16**, 417 (2004).

[28] C. C. Ho, P. Y. Chen, K. H. Lin, W. T. Juan, and W. L. Lee, Appl. Mater. Interfaces **3**, 204 (2011).

[29] N. J. Halas, S. Lal, S. Link, W. S. Chang, D. Natelson, J. H. Hafner, and P. Nordlander, Adv. Mater. **24**, 4842 (2012).

[30] P. B. Johnson and R. W. Christy, Phys. Rev. B **6**, 4370 (1972).

[31] C. F. Bohren and D. R. Huffman, *Absorption and Scattering of Light by Small Particles* (Wiley, New York, 1983).

[32] J. Pan, Z. Chen, J. Chen, P. Zhan, C. J. Tang, and Z. L. Wang, Opt. Lett. **37**, 1181 (2012).

[33] The authors thank the State Key Program for Basic Research of China (SKPBRC) under Grant Nos. 2012CB921501 and 2013CB632703, and the National Nature Science Foundation of China (NSFC) under Grant Nos. 11174137, 91221206, 11274160, 11321063, and 51271092 for their support.

Magnetic Field Enhancement at Optical Frequencies Through Diffraction Coupling of Magnetic Plasmon Resonances in Metamaterials[*]

C. J. Tang, P. Zhan, Z. S. Cao, J. Pan, Z. Chen, and Z. L. Wang

Department of Physics and National Laboratory of Solid State Microstructure,
Nanjing University, Nanjing 210093, China

We studied theoretically the diffraction coupling of magnetic resonances in metamaterials consisting of a planar rectangular array of metal rod pairs with a dielectric spacer. A narrow-band mixed mode was observed due to a strong interaction between magnetic resonances in individual pairs of metal rods and an in-plane propagating collective surface mode arising from the array periodicity. Upon the excitation of this mixed mode, a five-fold enhancement of magnetic field in the dielectric spacer was achieved as compared with the purely magnetic resonance. It was also found that only a collective surface mode with its magnetic field parallel to the array plane could mediate the excitation of such a mixed mode.

Metallic nanoparticles exhibit rich optical properties and have many applications because they support surface plasmon(SP) resonances associated with a huge enhancement of electric fields in their vicinity.[1] Due to radiative or nonradiative (absorptive) damping, however, the particle SP resonances usually have a broad bandwidth or short lifetime, which will limit their applications in some cases. Now recent studies have shown that the SP bandwidth could be tuned through interparticle near-field interaction[2,3] or far-field diffraction coupling[4—8] in periodic arrays of metallic nanoparticles. In the later case, a narrow-band plasmon mode near particle SPs was predicted[4,5] and observed experimentally,[6—8] when the array period is close to the particle resonance wavelength. It is well known that the arrangement of metallic nanoparticles into a periodic lattice can give rise to an in-plane propagating collective surface mode (also referred to as lattice resonance), which appears near the Wood anomaly.[9] The narrow-band mode steps from the interaction between a particle SP mode and this type of collective surface mode.[4—9] Because of a narrow bandwidth, such a hybridized plasmon mode is anticipated to find potential applications. For example, it has been used to enhance and direct fluorescent emission in periodic plasmonic nanoantenna arrays.[8]

In metamaterials, specially shaped metallic nanoparticles, like split-ring resonators (SRRs)[10] and paired rods[11,12] or nanodisks,[13] can induce a magnetic moment counteracting the external magnetic field and thus produce diamagnetic responses, termed magnetic SP resonances to differentiate them from electric SP resonances observed in usual

---

[*] Phys.Rev.B, 2011, 83(4): 041402

metallic nanoparticles. These metallic nanostructures have been widely employed as artificial magnetic atoms to fabricate negative-permeability or negative-refractive-index metamaterials[14] with peculiar electromagnetic properties.[15] Although the interactions of electric SP resonances in metallic nanostructures have been extensively studied and well understood, less is known about the interactions of magnetic SP resonances which could result in interesting physical phenomena.[16] Very recently, a classical analog for electromagnetic induced transparency (EIT) observed in a three-level atomic system[17] has been demonstrated in metamaterials made of coupled SRRs.[18]

On the other hand, achieving magnetic field enhancement at optical frequencies is now drawing increasing attention,[19] due to its potential applications such as magnetic nonlinearity[20] and magnetic sensors. However, in light-matter interactions, the magnetic component of light generally plays a negligible role since it is very weak.[21] Therefore, seeking new mechanisms to enhance the magnetic field becomes quite important. It is well known that a metallic nanogap can provide huge electric field enhancement at electric SP resonances.[22] Very recently, its complementary structure, namely, the metallic nanowire, was proposed to achieve extraordinary enhancement of magnetic field at terahertz (THz) frequencies based on Babinet's princinple.[23] But, it is still a challenge to achieve huge magnetic field enhancement in the region of optical frequencies.

In this Rapid Communication, we propose an approach to enhancing magnetic fields at optical frequencies via the diffraction coupling of magnetic SP resonances in metamaterials consisting of two-dimensional (2D) periodic arrays of paired metal rods. We found a narrow-band mixed mode due to the strong interaction between magnetic SP resonances and an in-plane propagating collective surface mode of the array. At the optical resonance of this mixed mode, the maximum of magnetic field intensity is about 450 times of the incident field. The magnetic fields in the metal rod pairs are enhanced to nearly 5 times larger than those at the magnetic resonance of individual pairs of metal rods. More interestingly, we found that only a collective surface mode with its electric (magnetic) field perpendicular (parallel) to the array plane could mediate such a coupling.

Figure 1(a) schematically shows the metamaterials to be studied, which are 2D arrays of two parallel Ag rods and a glass spacer between them. The length, width, and height of the Ag rods are, respectively, $l=150$ nm, $w=100$ nm, and $h=50$ nm. The dielectric spacers have identical dimensions and a refractive index $n=1.45$. The relative permittivity of Ag is described by a Drude model: $\varepsilon=1-\omega_p^2/[\omega(\omega+i\tau^{-1})]$, where ω_p is the plasma frequency and τ is the relaxation time related to energy loss. The parameters are taken to be $\hbar\omega_p=9.2$ eV and $\hbar\tau^{-1}=0.02$ eV.[24] The coordinates are chosen such that the Ag rod pairs lie on the xy plane, with its origin located at the center of one of the rod pairs. The array periods along the x and y axes are p_x and p_y, respectively. The electric field E_{in}, magnetic field H_{in}, and wave vector K_{in} of the incident light are along the x, y, and z axes, respectively.

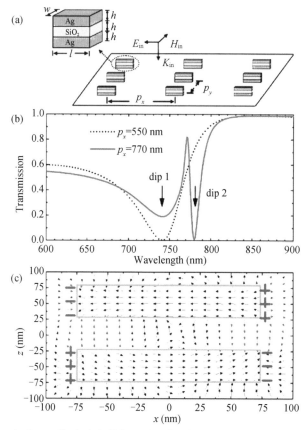

FIG. 1. (color online) (a) Schematic of a 2D rectangular array of pairs of parallel Ag rods with length $l = 150$ nm, width $w = 100$ nm, and height $h = 50$ nm, separated by a glass layer with a refractive index $n = 1.45$. (b) Normal-incidence transmission spectra of two 2D rectangular arrays of Ag rod pairs for different periods along the x axis, $p_x = 550$ nm (black dotted line) and 770 nm (red solid line), but the same period along the y axis, $p_y = 200$ nm. (c) Electric field vectors mapped on the xOz plane across the center of one rod pair in the array for dip 1 in (b). Arrows represent field direction and colors show field strength with red larger and black smaller. Two green rectangles outline the regions of two metal rods. Blue signs "+" and "−" stand for positive and negative charges, respectively.

Figure 1(b) presents the normal-incidence transmission spectra of two typical 2D rectangular arrays of Ag rod pairs, calculated with the commercial software package COMSOL MULTIPHYSICS. The two arrays have the same period in the y direction ($p_y = 200$ nm) but different periods in the x direction ($p_x = 550$ nm and 770 nm). A broad transmission dip (labeled as dip 1) centered at $\lambda_1 = 740$ nm is observed for both arrays. However, for the array with $p_x = 770$ nm, as shown by the red solid line in Fig. 1(b), a relatively narrow transmission dip (labeled as dip 2) appears at $\lambda_2 = 780$ nm.

The position of the broad dip is almost independent of the period p_x, implying that this transmission dip arises from the excitation of resonance in individual Ag rod pairs. In fact, such a resonance state is a magnetic SP mode. To show this, Fig. 1(c) maps the electric field vectors on the xoz plane across the center of one rod pair for dip 1, at a time when the incident field reaches its maximum on the xoy plane. It is evident that at this individual particle resonance the electric fields in the upper and lower metal rods oscillate out of phase and produce antiparallel currents, accompanied by an antisymmetric charge distribution on the pair rod ends. The antiparallel ohmic currents, together with the displacement currents in the dielectric layer, form a current loop that induces a magnetic moment.[12]

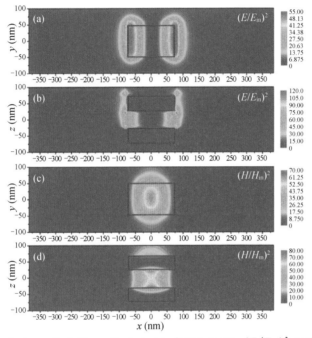

FIG. 2. (color online) Normalized electric field intensity $(E/E_{in})^2$ on the plane of $z=0$ (a) and on the plane $y=0$ (b) for dip 1. (c) and (d) The same as (a) and (b), respectively, but for normalized magnetic field intensity $(H/H_{in})^2$. Gray rectangles outline the regions of glass layers in (a) and (c), and those of Ag rods in (b) and (d).

In order to better understand the properties of magnetic SP resonance, in Fig. 2 we plot the electric and magnetic field distributions at the resonance $\lambda_1 = 740$ nm. It is clear that the electric fields are mainly concentrated near the end points of metal rods [see Figs. 2(a) and 2(b)] and the magnetic fields are highly confined between the metal rods [see Figs. 2(c) and 2(d)], which are characteristics of a magnetic SP resonance of individual pairs of metal rods.[12] The formation of such a magnetic resonance can be well understood through plasmon hybridization.[13]

In Fig. 3, we plot the corresponding electromagnetic field distributions for the dip 2 resonance located at $\lambda_2 = 780$ nm. Although the field patterns are almost the same as the cases shown in Fig. 2, the magnetic fields in the region between the Ag rods and the electric fields near the rod end points become much stronger, with a nearly 5 times enhancement. In particular, the maximum magnetic field is enhanced to be about 450 times of the incident field [please see Fig. 3(d)].

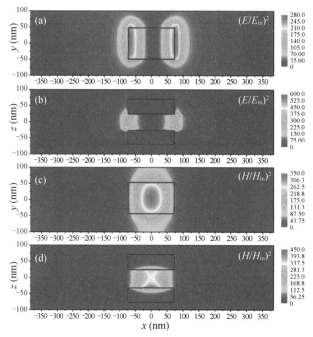

FIG. 3. (color online) The same as Fig. 2 but for dip 2 marked in Fig. 1(b).

In the following, we will show that such a huge enhancement of magnetic fields at optical frequency results from the strong coupling between the magnetic SP resonances in metal rod pairs and an in-plane propagating collective surface mode signaled by the Wood anomaly.[4-9] This strong coupling leads to the formation of two mixed modes, i.e., the high- and low-energy states, whose energies can be calculated with a coupled oscillator model[25]: $E_{+,-} = (E_{SP} + E_{Wood})/2 \pm \sqrt{\Delta/2 + (E_{SP} - E_{Wood})^2/4}$. Here, E_{SP} and E_{Wood} are the energies of the magnetic SP and Wood anomalies, respectively; and Δ stands for the coupling strength. The wavelength of the Wood anomaly can be calculated by matching the wave vector K_{in} of incident light in the surrounding medium (air) with the reciprocal vector $G_{m,n}$ of a 2D rectangular lattice under normal incidence[26]: $\lambda_{Wood}^{m,n} = 1/\sqrt{(m/p_x)^2 + (n/p_y)^2}$, where m and n are integers related to different diffraction orders. E_{SP} is determined by the geometrical and material parameters of individual pairs of metal rods, but independent of the array periods. Through numerical simulations, we found that for individual pairs of metal rods E_{SP} is about 1.647 eV, corresponding a wavelength of 753 nm. Figure 4(a) shows the transmission spectra of a series of 2D rectangular arrays of Ag rod pairs, with

p_x being varied from 650 nm to 850 nm in steps of 20 nm. The period along the y axis is kept constant $p_y = 200$ nm. In each transmission spectrum, two transmission dips are observed: one is narrow and the other is broad. The open black circles in Fig. 4(b) summarize the dependence of the positions of these transmission dips on p_x. By taking the coupling strength to be $\Delta = 310$ meV in the above coupled oscillator model, we can predict the positions of transmission dips for different periods p_x. The two branches of red lines in Fig. 4(b) give the predicted results. Clearly, they are in a good agreement with the locations dictated from the transmission spectra. The black line and the horizontal green line in Fig. 4(b) show the positions of the Wood anomaly and magnetic SPs, respectively. At the crossing of these two lines, the positions of transmission dips present an obvious anticrossing, which is a characteristic of the strong coupling between the collective surface mode and magnetic SP. Away from this strong coupling regime, the transmission dip positions follow approximately one of these two lines.

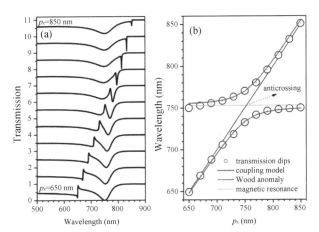

FIG. 4. (color online) (a) Normal-incidence transmission spectra of a series of 2D rectangular arrays of Ag rod pairs, with p_x varied from 650 nm (top line) to 850 nm (bottom line) in steps of 20 nm but with p_y kept constant at 200 nm. Individual spectra are vertically offset from one another by 1.0 for clarity. Other structural parameters are the same as those used in Fig. 1(b). (b) The p_x dependences of the positions (open black circles) of transmission dips in (a), and the predicted positions (two red curved lines) using a coupling model of the Wood anomaly and magnetic SP resonance. The Wood wavelength $\lambda_{\text{Wood}}^{1,0} = p_x$ (black diagonal line) and the position (horizontal green line) of magnetic SP resonance are also shown.

In a series of arrays investigated in Fig. 4(a), the period p_y is set to be 200 nm, a value much smaller than the spectral range of interest from 500 nm to 900 nm. As a result, the diffraction channel in the y direction is closed, and only the diffraction channel in the x direction is opened. In this situation, the above formula for the wavelength of Wood anomaly reduces to a simple form: $\lambda_{\text{Wood}}^{m,0} = p_x/m$. A first-order diffraction corresponding to

$\lambda_{\text{Wood}}^{1,0} = p_x$ ($m=1$) is predicted, as shown by black line in Fig. 4(b). The other higher order ($m=2,3,\ldots$) diffraction wavelengths are all smaller than 500 nm. Therefore, our arrays support only a first-order collective surface mode propagating along the x direction. By differentiating the x, y, and z components of the total electromagnetic fields at resonance (not shown here), we found that the electric field of this surface mode is along the z axis, and its magnetic field is along the y axis. Because the surface mode has a magnetic field of the same direction as the induced magnetic moment in the Ag rod pairs, it can strongly interact with the magnetic SP resonance in each pair of metal rods when grazing the metamaterial surface. Such a strong interaction can suppress radiative damping since the electromagnetic fields of the surface mode are trapped in the lattice,[9,27] thus forming a narrow-band mixed mode at a wavelength slightly larger than the Wood wavelength $\lambda_{\text{Wood}}^{1,0}$.

By varying the array periods while keeping the other conditions unchanged, we have also studied other situations in which the diffraction channel along the x direction is closed but the one along the y direction is opened (not shown here). A narrow-band mode is also observed when the period p_y approaches the magnetic SP resonance wavelength. Nevertheless, in these cases the collective surface mode propagating in the y direction has an electric (magnetic) field parallel to the x (z) axis. Strictly speaking, the narrow-band mode does not arise from the diffraction coupling of magnetic SP resonances, but is more likely a result of the diffraction coupling of electric SP resonances, since the magnetic field of the surface mode is orthogonal to the induced magnetic moment (along the y direction) in the Ag rod pairs, that is, the surface mode cannot interact directly with magnetic SP resonances of individual pairs of Ag rods. In fact, at the narrow-band mode resonance, the magnetic fields in the metal rod pairs become weaker, rather than getting enhanced, as compared at the pure magnetic SP resonance.

In conclusion, we have shown that in metamaterials consisting of 2D periodic arrays of pairs of metal rods, the interaction between magnetic resonances of the metal rod pairs and an in-plane propagating collective surface mode signaled by the Wood anomaly can form a narrow-band hybrid mode. It is found that only a collective surface mode with its magnetic field parallel to the induced magnetic moment in the rod pairs could mediate such an interaction. The huge magnetic field enhancement at the hybrid mode resonance suggests the metamaterials have a potential to explore optical phenomena based on enhanced magnetic fields, such as second-harmonic generation from magnetic nonlinearity.[20]

References and Notes

[1] U. Kreibig and M. Vollmer, *Optical Properties of Metal Clusters* (Springer, Berlin, 1995).

[2] S. Linden, J. Kuhl, and H. Giessen, Phys. Rev. Lett. **86**, 4688 (2001); G. Gantzounis, N. Stefanou, and N. Papanikolaou, Phys. Rev. B **77**, 035101 (2008).

[3] C. L. Haynes *et al.*, J. Phys. Chem. B **107**, 7337 (2003); A. Bouhelier *et al.*, *ibid*. **109**, 3195 (2005).

[4] K. T. Carron *et al.*, J. Opt. Soc. Am. B **3**, 430 (1986); V. A. Markel, J. Mod. Opt. **40**, 2281 (1993);

J. Phys. B **38**, L115 (2005); S. Zou, N. Janel, and G. C. Schatz, J. Chem. Phys. **120**, 10871 (2004); S. Zou and G. C. Schatz, *ibid*. **121**, 12606 (2004).

[5] X. M. Bendaña and F. J. García de Abajo, Opt. Express **17**, 18826 (2009); B. Auguié, X. M. Bendaña, W. L. Barnes, and F. J. García de Abajo, Phys. Rev. B **82**, 155447 (2010).

[6] B. Lamprecht *et al.*, Phys. Rev. Lett. **84**, 4721 (2000); N. Félidj *et al.*, J. Chem. Phys. **123**, 221103 (2005); E. M. Hicks *et al.*, Nano Lett. **5**, 1065 (2005).

[7] V. G. Kravets, F. Schedin, and A. N. Grigorenko, Phys. Rev. Lett. **101**, 087403 (2008); B. Auguié and W. L. Barnes, *ibid*. **101**, 143902 (2008); Y. Chu *et al.*, Appl. Phys. Lett. **93**, 181108 (2008); A. Bitzer *et al.*, Opt. Express **17**, 22108 (2009).

[8] G. Vecchi, V. Giannini, and J. Gómez Rivas, Phys. Rev. Lett. **102**, 146807 (2009).

[9] F. J. García de Abajo, Rev. Mod. Phys. **79**, 1267 (2007).

[10] C. Enkrich, M. Wegener, S. Linden, S. Burger, L. Zschiedrich, F. Schmidt, J. F. Zhou, Th. Koschny, and C. M. Soukoulis, Phys. Rev. Lett. **95**, 203901 (2005).

[11] S. Zhang, W. Fan, N. C. Panoiu, K. J. Malloy, R. M. Osgood, and S. R. J. Brueck, Phys. Rev. Lett. **95**, 137404 (2005).

[12] G. Dolling *et al.*, Opt. Lett. **30**, 3198 (2005); V. M. Shalaev *et al.*, *ibid*. **30**, 3356 (2005); J. F. Zhou *et al.*, *ibid*. **31**, 3620 (2006).

[13] T. Pakizeh *et al.*, Opt. Express **14**, 8240 (2006); C. Tserkezis, N. Papanikolaou, G. Gantzounis, and N. Stefanou, Phys. Rev. B **78**, 165114 (2008).

[14] C. M. Soukoulis, S. Linden, and M. Wegener, Science **315**, 47 (2007); V. M. Shalaev, Nature Photon. **1**, 41 (2007).

[15] V. G. Veselago, Sov. Phys. Usp. **10**, 509 (1968); J. B. Pendry, Phys. Rev. Lett. **85**, 3966 (2000); R. A. Shelby, D. R. Smith, and S. Schultz, Science **292**, 77 (2001).

[16] S. Linden, M. Decker, and M. Wegener, Phys. Rev. Lett. **97**, 083902 (2006); H. Liu, D. A. Genov, D. M. Wu, Y. M. Liu, J. M. Steele, C. Sun, S. N. Zhu, and X. Zhang, *ibid*. **97**, 243902 (2006); N. I. Zheludev *et al.*, Nature Photon. **2**, 351 (2008).

[17] S. E. Harris, Phys. Today **50**, 36 (1997).

[18] N. Liu, S. Kaiser, and H. Giessen, Adv. Mater. **20**, 4521 (2008); P. Tassin, L. Zhang, Th. Koschny, E. N. Economou, and C. M. Soukoulis, Phys. Rev. Lett. **102**, 053901 (2009); R. Singh, C. Rockstuhl, F. Lederer, and W. Zhang, Phys. Rev. B **79**, 085111 (2009).

[19] M. Burresi *et al.*, Science **326**, 550 (2009).

[20] M. W. Klein *et al.*, Science **313**, 502 (2006).

[21] L. D. Landau and E. M. Lifshitz, *Electrodynamics of Continuous Media* (Wiley, New York, 1984).

[22] H. Xu, E. J. Bjerneld, M. Käll, and L. Borjesson, Phys. Rev. Lett. **83**, 4357 (1999).

[23] S. Koo, M. S. Kumar, J. Shin, D. S. Kim, and N. Park, Phys. Rev. Lett. **103**, 263901 (2009).

[24] V. Yannopapas, A. Modinos, and N. Stefanou, Phys. Rev. B **60**, 5359 (1999).

[25] C. Symonds *et al.*, New J. Phys. **10**, 065017 (2008).

[26] R. W. Wood, Philos. Mag. **4**, 396 (1902); Phys. Rev. **48**, 928 (1935).

[27] G. Vecchi, V. Giannini, and J. Gómez Rivas, Phys. Rev. B **80**, 201401(R) (2009).

[28] We greatly acknowledge financial support from the State Key Program for Basic Research of China and NSFC under grant Nos. 10425415, 10734010, 50771054, 10804044, and 11021403.

第四章 声学超晶格和声子晶体
Chapter 4　Acoustic Superlattices and Sonic Crystals

4.1　Acoustic Superlattice of LiNbO$_3$ Crystals and Its Applications to Bulk-wave Transducers for Ultrasonic Generation and Detection up to 800 MHz

4.2　High-frequency Resonance in Acoustic Superlattice of LiNbO$_3$ Crystals

4.3　Ultrasonic Spectrum in Fibonacci Acoustic Superlattices

4.4　Ultrasonic Excitation and Propagation in an Acoustic Superlattice

4.5　High-frequency Resonance in Acoustic Superlattice of Periodically Poled LiTaO$_3$

4.6　Bulk Acoustic Wave Delay Line in Acoustic Superlattice

4.7　Negative Refraction of Acoustic Waves in Two-dimensional Sonic Crystals

4.8　Acoustic Backward-Wave Negative Refractions in the Second Band of a Sonic Crystal

4.9　Negative Birefraction of Acoustic Waves in a Sonic Crystal

4.10　Extraordinary Acoustic Transmission through a 1D Grating with Very Narrow Apertures

4.11　Acoustic Surface Evanescent Wave and its Dominant Contribution to Extraordinary Acoustic Transmission and Collimation of Sound

4.12　Tunable Unidirectional Sound Propagation through a Sonic-Crystal-Based Acoustic Diode

4.13　Acoustic Asymmetric Transmission based on Time-dependent Dynamical Scattering

4.14　Acoustic Cloaking by a Near-zero-index Phononic Crystal

4.15　Acoustic Phase-reconstruction near the Dirac Point of a Triangular Phononic Crystal

4.16　Topologically Protected One-way Edge Mode in Networks of Acoustic Resonators with Circulating Air Flow

第四章 声学超晶格和声子晶体

陈延峰

介电体的基本物理性质包括弹性、压电、介电、弹光、声光、电光和非线性光学性能等。介电体超晶格材料的思想是：利用这类材料在一维、二维或三维空间调制周期或准周期的超晶格，实现对相应的物理参数进行调制，从而设计和研究这类人工微结构材料所具备的新效应和新功能。这一章讨论对压电系数周期性调制的介电体超晶格——声学超晶格，以及对弹性常数调制的介电体超晶格——声子晶体。两者实现的功能是对声波的激发、耦合和传播的调控。在本章中，选择了 16 篇论文，介绍了这两类介电体超晶格对声波振幅、频率、波矢等进行调控的理论和实验研究。

我们提出了声学超晶格的概念，理论表明声学超晶格超高频超声波的激发频率与超晶格的周期成反比，而与厚度无关。如果用通常均质材料来制备工作频率为几百兆到几千兆的体波超声器件，其晶片总厚度要减到几十微米到几微米，对当前超声工程中的工艺系统是有困难的，而制备周期为数微米的声学超晶格则比较方便。物理上，超晶格中的畴界都可看为 δ 声源，当这些声源产生的超声波相长干涉时，即产生增强超声激发。理论和实验表明，声学超晶格具有超声波频率与超晶格周期成反比，超声波强度与超晶格的周期数的平方成正比的特点，而作为超声换能器，其阻抗匹配可以由周期数、超晶格面积确定。实验上利用生长层技术制备了周期在 10 微米左右的声学超晶格，获得了 500 MHz 左右的超高频超声激发，研制了超高频声谐振器和换能器。理论和实验研究了准周期声学超晶格的自相似超声激发谱。4.1 描述了声学超晶格作为超高频换能器的结果，4.2 演示了 500 MHz～800 MHz 的谐振器，4.3 是对声学超晶格准周期结构激发谱的研究，4.4 发展了声学超晶格激发和传播的理论，分析了声学超晶格作为声学器件的优点，4.5 在实验上实现了室温极化 $LiTaO_3$ 声学超晶格在不同激发条件下的声波模式的调控，而 4.6 描述了利用周期渐变的宽带声学体波延迟线的实验。

继 Yablonovitch 和 John 提出光子晶体之后[Phys. Rev. Lett. 58, 2059 (1987), Phys. Rev. Lett. 58, 2486 (1987)]，声子晶体的概念也被提出[Phys. Rev. Lett. 71, 2022 (1993)]。其出发点是：声波在周期结构中的传播形成布洛赫波。与电子在半导体中的传播以及光波在光子晶体中的传播类似，由于周期性的散射，声波在布里渊区边界出现能隙，而在布里渊附近，声子出现强色散。这与声波在均匀介质中的线性色散完全不同。声子晶体的提出，为声波的调控提供了一个全新的原理。

声子晶体的制备可以利用多种材料，例如无机非金属材料、金属材料和高分子材料，甚至流体等，只要两种材料具有不同的弹性模量和声速。当然，实际上，声子晶体的应用，还要兼顾到其他的性能，如力学强度等。声子晶体的制备技术也很多，如机械加工、3D 打印、自组装、微加工方法等。经过二十多年，人们发展了声子晶体的计算、模拟和设计方法，包括平面波方法、转移矩阵方法、多重散射方法、时域有限差分(FDTD)和有限元方法等,这些方法都有各自的优点。具体的体系,都有最适合的方法。最近出现的遗传算法可以根据具体性能的要求,逆向设计最佳的声子晶体结构,达到优化的性能。

我们在声子晶体中首先实现的效应是声波的负折射。负折射效应是 1968 年由苏联科学家 Veselago 提出[Sov. Phys. Usp. 10, 509(1968)],所针对的是电磁波。材料对于电磁波的响应由介电常数 ε 和磁导率 μ 描述,当两者均为负时,电磁波(光波)能够在材料中传播,但其折射率为负,同时,其中电场-磁场-波矢三者满足左手定则,因此,这类材料也被称为左手材料,或者双负材料。自然界中一直没有发现这类材料,直到 2001 年 Smith 与 Pendry 合作,利用金属双劈裂环结构实现了负 μ,进一步制备了左手材料,从而实现了微波段电磁波的负折射[Science 292, 77(2001)]。几乎同时,MIT 的 Joannopoulos 小组报道了在光子晶体中实现光波负折射的理论[Phys. Rev. B 65, 201104(R) (2002)]。与前者不同,光子晶体中负折射的机制是光子带隙带边的强色散,它使得光子晶体等频线强烈弯曲,从凸变到凹,从而使得光波的群速度从正折射变成负折射。我们的问题是:在声子晶体中是否能够实现负折射?通过分析声子晶体的能带色散、等频线和 FDTD 模拟,我们先后在实验上实现了第一能带的负折射(4.7)、第二能带的负折射和可调谐的正、负折射等现象(4.8)。虽然在流体中声波是纯纵波,但在由固体与流体构成的声子晶体中,出现负折射,甚至双负折射(4.9)。这些工作揭示了声子晶体中新颖的折射规律。

1998 年 Ebbesen 报道的光波穿过小孔金属薄膜发生异常透射的现象[Nature391, 667 (1998)]引起了人们特别的关注,其发生的机制引起了一系列争论。我们试图考察声波是否有类似的现象。通过实验发现:声波在穿过一维声子晶体时,同样将出现声波的异常透射增强。为了理解这个现象,我们提出了基于衍射动力学的理论,很好地计算了透射的增强因子和对应的波长(4.10)。接着,研究了声波透过一块仅有一条狭缝同时具有周期凹槽结构的固体板的传播规律(4.11)。实验发现:尽管只有一个狭缝,波长远大于狭缝宽度的声波也可以有显著的增强透射,同时,伴随着自准直效应。为了理解这一现象,我们发展了基于结构散射和远场耦合的理论,成功地解释了这一实验。这个工作证实了声波同样具有异常透射增强效应,同时发现,在具有周期结构的固体表面,存在沿表面传播的表面波,其物理机制与光波在金属表面形成表面等离激元类似,即在具有周期结构的固体表面具有负的有效弹性模量或负的有效密度,因而形成了一种新的表面波,它具有亚波长特征,这种由于结构而产生的声表面波不同于人们熟知的瑞利波和板波。

二极管是最基本的电子器件,其基本功能是控制电子的单向流动,而半导体的电子二极管基于半导体 p-n 结,人们对此非常熟悉。同样都是波,光波和声波的单向传播现象并不普遍。原因是控制光波和声波的波动方程中,时间反演和宇称都是对称的,另外的原因是,与电子波相比,光波和声波与物质相互作用弱,因此,通常很难控制光波和声波只沿着单向传播。在光学中,实现光波单向传播的基本途径有三个:一是通过外加磁场,使得磁光晶体打破时间反演对称;二是利用材料的非线性效应;三是利用时间变化调制材料的物理性质。我们的问题是,在声学中,如何实现声波的单向传播?我们提出,利用声子晶体的各向异性能带结构和方向带隙,构造了宇称破缺的声子晶体:一侧为平界面,另一侧是具有衍射结构的界面,该结构提供了模式跃迁的波矢。从平界面入射的声波因为位于带隙中不能传播,而从另一侧入射的声波因为被散射到通带而得以传播。这个原理被称为声单向传播的"波矢跃迁"机制,其优点是:不需要强声场,具有大带宽和高透射率。4.12 描述了这个工作。该机制还能拓展到其他形式的波动系统中,如光波、水表面波等。声单向传播的实现为声波调控提供了新思路,为实现声信息处理的声二极管元件研究开辟了新途径。我们进一步研究了声子结构单元随时间变化的声子晶体(4.13),发现了声波的类拉曼效应,利用这个效应,结合中心频率为入射波频率的滤波器,实验演示了时变系统的单向声波二极管。

4.14 报道了利用二维蜂窝结构,声子晶体(声石墨烯)具有零折射率,因而具有声波隐身的功能。4.15 描述了在声石墨烯中,在 Dirac 点频率附近声波传播时的散射和波前重构的有趣现象。

最近,关于拓扑性的问题引起了人们的高度关注,4.16 是关于拓扑声子晶体的理论研究。以旋转气流环为单元,构造了一个二维声子晶体,由于气流旋转,在其中模拟了矢量势,从而打破了晶体的时间反演对称性。通过紧束缚模型,计算了这个声子晶体的能带,并计算了整个布里渊区中不同能带的陈数,证明了这个声子晶体是拓扑非平庸的。通过研究边界态的传播特性,揭示了由拓扑性导致的奇特边界态的主要特征:其体态为绝缘态而边界是传播态、边界态是类自旋与波矢方向绑定的手性传播态、边界态对杂质和缺陷不敏感的鲁棒性。有关声子晶体的拓扑性研究方兴未艾。

声波广泛地影响着人类的生活。工业、国防和医疗健康等各个领域,都离不开声波。有目的地设计和应用声学材料,将极大地改变人类的生活,因此,发展新型声学材料,具有十分重要的意义。我们对介电体超晶格材料中的声学超晶格和声子晶体方面的研究,从一个侧面展示了人工微结构在调控材料的声学性质,改变材料的功能,实现新颖的效应方面的巨大潜力。这个方面的研究和应用,大有可为。

Chapter 4　Acoustic Superlattices and Sonic Crystals

Yanfeng Chen

The physical properties of dielectrics can be characterized in many aspects, including their elastic, piezoelectric, dielectric, elasto-optic, acousto-optic, electro-optic, nonlinear optical effect, and so on. The corresponding coefficients for these effects can be spatially modulated either periodically or quasi-periodically in one-dimension (1D), 2D or even 3D fashion to create dielectric superlattices, new kind of artificial microstructured materials, which exhibit exotic physical effects and new functionalities never shown by their bulk counterparts. This chapter main concerns acoustic superlattices (ASL) formed by the periodic modulation of piezoelectric coefficient and sonic crystals (also called phononic crystals or acoustic crystals) created by the periodic modulation of elastic coefficient. These two artificial microstructures are catalogued in one in terms of their ability to control the excitation, coupling, and propagation of acoustic waves.

The concept of ASL was firstly proposed by Ming et al. Our theoretical analysis suggested that the excitation frequency of super-high-frequency ultrasonic waves is independent of the sample thickness, but rather inversely proportional to the period of the superlattice in the sample. If a conventional homogenous crystal was used to make a bulk wave ultrasonic transducer with an operation frequency ranging from hundreds of MHz to a few GHz, the crystal thickness would have to be thinned down to tens of micrometers or even a few micrometers, which poses a great fabrication challenge in the field of ultrasonic engineering. However, fabricating ASL with a spatial period on the order of micrometers is instead very straightforward based on the readily available micro-fabrication technology. In an ASL, each domain boundary between two different modulation sections can be considered as a δ-source for acoustic wave excitation. An enhanced ultrasonic wave can be coherently generated only if all the sources from the spatially separated domain boundaries in the ASL are in phase. In this regard, it is shown in both theory and experiment that the frequency of the generated ultrasonic wave is inversely proportional to the period of the ASL and its intensity is proportional to the square of the number of the periods of the ASL. For device application, the impedance of ultrasonic transducers made with the ASL is determined by the number of periods as well as the area of the ASL. We have experimentally fabricated ASL with various periodically modulated dopant concentrations in the crystal growth of $LiNbO_3$ using the high temperature Czochralski method. For example, one of the as-grown ASLs has a period of 10 μm, giving rise to the excitation of super-high-frequency ultrasonic waves with a frequency ∼500 MHz. Building upon this fabrication technology, we further developed super-high-frequency acoustic oscillators and transducers. Besides, this technology enabled us to fabricate quasi-periodic ASL and study its self-similarity related novel phenomena, e.g., self-similar excitation process. In this chapter, 4.1 presents a transducers operating at high frequency made of ASL, 4.2 demonstrates 500 MHz∼800 MHz resonators made of ASLs. 4.3 describes the excitation spectrum of quasi-periodic ASL revealing self-similarity. 4.4 presents a systemic theory and experiments of ASL based acoustic devices. 4.5 reports the relationship between the excitation scheme vs acoustic wave mode for ASL made by pulse electric field poling technique, and 4.6 does a wide-band transducer and its delay line prepared by modulation length gradually varying ASL.

The concept of sonic crystal was proposed in 1993 [Phys. Rev. Lett. 71, 2022(1993)] after the invention of photonic crystal by Yablonovitch and John in 1987 [Phys. Rev. Lett. 58, 2059 (1987), Phys. Rev. Lett. 58, 2486 (1987)]. Analogous to electrons in crystals and optical waves in photonic crystals, acoustic waves, when propagating in sonic crystals, also behave as Bloch waves due to periodic scatterings, which in turn results in band-gap at the boundaries of Brillouin Zones. Near these boundaries, acoustic wave-vectors show very strong frequency dependence, a key feature distinct from that of a homogenous medium. This special feature in sonic crystals has enabled a completely different route to manipulate acoustic waves.

Since its proposal in 1993, sonic crystals have undergone rapid development. The material choice for fabricating sonic crystals has been greatly extended to include inorganic non-metallic materials, metals, polymers, and even fluids. In fact, any two materials with different elastic constants and/or acoustic velocities can be utilized to construct sonic crystals, but in practical applications, other restraints, e.g., structural mechanical strength, need also to be taken into account. In addition to the abundant choice of materials, a variety of fabrication techniques for sonic crystals have also been developed, which could include machining, 3D printing, self-assembling, micro/nano-fabrication, etc.

Besides the remarkable experimental progress in the field of sonic crystals, a comprehensive theoretical framework has also been established over the past two decades. Nowadays, many theoretic tools, including plane wave expansion method, transfer matrix method, multi-scattering method, time domain finite difference method, finite element method and so on, have been developed for designing sonic crystals and simulating acoustic wave propagation in them. It should be noted that none of these tools is meant to replace other ones. Each of these tools has its own advantages as well as shortcomings when applied to study sonic crystal related topics. In recent years, genetic algorithms has been employed to design sonic crystals in an inverse or object-oriented fashion, where the target response is first defined and then the whole parameter space is searched evolutionarily to find a sonic structure that has the desired response.

The first exotic physical phenomenon we have realized in sonic crystals was the negative refraction of acoustic waves. Negative refraction was first proposed as a novel and counter-intuitive optical effect in the seminal work by Soviet Union physicist Veselago in 1968 [Sov. Phys. Usp. 10, 509 (1968)]. The constitution equations characterizing material's electromagnetic response contain two macroscopic material parameters, permittivity ε and permeability μ. Veselago's work indicated that when both ε and μ were simultaneously negative, an electromagnetic wave (optical wave) could still propagate in this exotic material but with a negative index of refraction, and the wave's electric field E, magnetic field H, and wave-vector k now satisfy left-handed rule instead of the well-known right-handed rule in a common medium. Because of this, such materials are also called left-handed materials (LHM), or double-negative materials. LHM has never been discovered so far in nature. The first LHM was realized in 2001 by Smith and Pendry [Science 292, 77(2001)] in a 2D array of an artificial unit cell combining a double-split ring with a metallic strip. Almost at the same time, Joannopoulos' group from MIT reported an alternative way to realize negative refraction in photonic crystals (PC) [Phys. Rev. B 65, 201104(R) (2002)]. However, the physical mechanism responsible for negative refraction in PC is completely different from that in LHM. The strong dispersion near the PC's bandgap deforms its initial convex equi-frequency contour (EFC), resulting in a concave EFC. Consequently, the group velocity, determined by the gradient of the EFC, experiences negative refraction rather than the commonly observed positive refraction. The question was obviously raised whether negative refraction can also occur for acoustic waves in sonic crystals. By studying the dispersion relation and the EFC of sonic crystals, we demonstrated in simulation using finite-difference-time-domain(FDTD) method and later in experiment negative refraction in the first (4.7) and the second band (4.8), and tunable negative and positive refraction. In addition, we shown that negative refraction and more importantly negative bi-refraction could occur in the sonic crystals composed by solids and fluids, although only the longitudinal acoustic modes are supported in fluids (4.9). All of these results reveal that the law of refraction has to be revised to include the structural influence from sonic crystals.

In 1998, Ebbesen reported extraordinary optical transmission (EOT) through a thin metal film with perforated holes on a transparent plate [Nature391, 667(1998)]. His work soon attracted much attention and later raised an intense debate on whether the surface plasmonic polariton (SPP) plays a key role in EOT. We shown in experiment that a similar extraordinary acoustic transmission (EAT) phenomenon could also occur for acoustic waves transmitting through a 1D sonic crystal. In order to understand this phenomenon, we developed a theory based on diffraction dynamics, with which the resonance frequency and corresponding enhancement factor of EAT could be predicted (4.10). Next, we studied the general behavior of acoustic transmission through a perforated slit surrounded by periodic grooves on a plate. Surprisingly, this configuration, although with only one slit, can also support enhanced transmission for acoustic wavelengths much larger than the slit width, and the transmitted acoustic waves exhibits a clear self-collimation feature (4.11). To analyze this result, we developed an analytic model taking into account structural scattering from the grooved interface and its coupling with the far-field radiation from the slit. Using this model, we successfully explained the EAT effect observed in experiment. Our further theoretical investigation indicated that the EAT is aided by a surface acoustic mode at the surface of the grooved plate. The formation of such a unique acoustic mode is very similar to the surface plasmon polariton (SPP), where a metal with negative real part of index of refraction is crucial. In our acoustic system, negative material parameters including effective density or Young's modulus can be realized by using a periodic feature on the surface of solids. Similar to the SPP, the surface acoustic wave has a wavelength scale much smaller than its bulk counterpart and thus it exhibits many different properties compared with the well-known Rayleigh wave and plate wave.

Diode is a fundamental electronic device, which restricts the flow of electrons in only one direction. Although electrons, optical waves and acoustic waves exhibit very similar wave behaviors, optical and acoustic diode with one-way propagation feature are rarely observed. This is because the fundamental physical description for optical and acoustic waves (Maxwell and Newton equations) obeys time-reversal symmetry (TRS) and space inversion symmetry/parity symmetry in common matters and because optical and acoustic waves have relative weak interaction with matters. For optical waves, there are three basic approaches to realize one-way propagation, 1) breaking the TRS with a magnetic field applied to magneto-optic crystals, 2) using nonlinear optical effects, and 3) exploiting time-varying material properties. Our goal is to seek for novel approaches to achieve acoustic one-way propagation. We proposed a scheme based on the anisotropic band structure and directional band-gap of sonic crystals. The space inversion symmetry/parity symmetry is broken by using an asymmetric acoustic structure composed of two different sections, a sonic crystal and a diffractive structure. When the acoustic incidence is from the side of the sonic crystal, its directional band-gap prohibits acoustic wave from transmitting into the crystal. However, when the acoustic incident is from the side of the diffractive structure, the acoustic wave experiences multiple orders of diffraction acquiring wave-vectors outside of the band-gap and thus propagates through the sonic crystal, leading to one-way propagation. This momentum/wave-vector-jumping-based one-way propagation has a number of key advantages, including threshold-less operation, broad bandwidth, and high transmission efficiency. Such a physical principle could be easily extended to other wave systems, e.g., optical waves and fluid surface waves. 4.12 presents this work. Our work greatly expands the application scope of artificial acoustic microstructures, providing a new scheme to manipulate the propagation of acoustic waves and opening a new door to novel acoustic information processing technologies. Recently, we have studied non-static sonic crystals including time-varying elements by rotating an elliptical-shaped blade (4.13). We demonstrated Raman-like acoustic effect, wherein the incident acoustic waves are scattered by the time-varying elements generating acoustic "Stokes" and "Anti-Stokes" components. By combining the rotating rod with narrow-band filters operating at the central frequency, we were able to construct an acoustic diode with unidirectional propagation.

In addition, we designed zero-index sonic crystals, in which acoustic cloaking was realized experimentally (4.14). 4.15 describes interesting experiment revealing scattering and reconstructing process of acoustic waves that pass through Dirac point of 2D sonic crystals with honeycomb lattice.

Recently, topology related condensed matter has attracted much attention in worldwide. 4.16 presents our work to search some possible new phenomena of topology related sonic crystals. By introducing air-flow-circulated fluid in acoustic resonator, a mimicking vector potential could be established as an acoustic counterpart of magnetic field for electron. By using the rings to construct 2D honeycomb lattice, we construct a type of topological sonic crystal. The mimicking vector potential breaks TRS, which results in topological nature for sonic crystals. We calculated the Chern number of acoustic energy bands based on tight-binding model, and found some of topological nontrivial states. We further revealed exotic edge states with topological characteristics: forbidden bulk states but with permitting edge states, pseudo-spin locking momentum chiral propagation of the edge states, and robust character immune disorder and defect along the edges. In fact, acoustic topological phase is an emerging research field, which is currently under rapid development.

There is no doubt that acoustic wave impacts almost every aspects of human life. Its related technology has been widely used in industry, defense, and health-care. Hence, any innovation on acoustic materials would have great influence on human life. The studies we have conducted over the past twenty years on the acoustic functional materials in the frame of the dielectric superlattices have uncovered the remarkable power of artificial microstructures in manipulating the material properties and the propagation behavior of acoustic waves. We believe that the full potential of acoustic functional materials has yet to be unleashed and this research field will have a promising future.

Acoustic Superlattice of LiNbO$_3$ Crystals and Its Applications to Bulk-wave Transducers for Ultrasonic Generation and Detection up to 800 MHz[*]

Yong-yuan Zhu and Nai-ben Ming

Laboratory of Solid State Microstructures, Nanjing University, People's Republic of China

Wen-hua Jiang and Yong-an Shui

Institute of Acoustics, Nanjing University, People's Republic of China

In this letter theoretical calculations and experimental results of ultrasonic generation and detection up to 800 MHz using an acoustic superlattice of LiNbO$_3$ crystals are reported. The experiment is in good agreement with theory. Transducers with an insertion loss of nearly 0 dB at 555 MHz and a 5.8% 3 dB bandwidth have been made.

Ultrasonic transducers operating at frequencies above 100 MHz are made either by thin-film deposition techniques or by bonding thin piezoelectric plates to the substrate and lapping them to the required thickness.[1-3] Here we report a novel method which uses an acoustic superlattice (ASL) of LiNbO$_3$ crystal, a single crystal with periodic laminar ferroelectric domain structures, to generate and detect ultrasonic waves with frequencies in the range 200 MHz – 800 MHz.

The ASL of LiNbO$_3$ crystal was prepared by the Czochralski method, and the LiNbO$_3$ melt was doped with 0.3wt.% of yttrium. The ASL was induced by the concentration gradient of the dopant in the rotational growth striations.[4,5] In practice, the periodicity of ASL of the order of microns can be obtained.

In the ASL, the piezoelectric tensor is not a constant, but a periodic function of the spatial coordinate. For simplicity, we assume that the ASL of the LiNbO$_3$ crystal is arranged in such a way that the Z axis is perpendicular to the lamellae. The thicknesses of the positive and negative domains are a and b, respectively; the number of periods is N; i.e., the periodic length of the ASL is $a+b$ and its total thickness is $N(a+b)$, as shown in Figs. 1(a) and 1(b). We also assume that the electrode faces of transducers made of the ASL are the {0001} faces which are parallel to the lamellae of the ASL and that one face is free and the other is fully matched to a transmission medium which is a Z-cut single domain LiNbO$_3$ crystal. Under these conditions, if the effect of electrodes is neglected, a longitudinal planar wave propagating along the Z axis is excited and the electrical

[*] Appl.Phys.Lett.,1988,53(15):1381

impedance Z_i of transducers can be derived by use of the Green's function method,[6] which is

$$Z_i = R_i - jX_i,$$
$$R_i = (4h_{33}^2 v/\omega^2 C_{33}^D A_{12}) \{\sin[kN(a+b)/2]/\sin[k(a+b)/2]\}^2 \{\langle\cos[k(a-b)/2] - \cos[k(a+b)/2]\rangle \times \cos[kN(a+b)/2] - \sin[k(a-b)/2]\sin[kN(a+b)/2]\}^2,$$
$$X_i = (4h_{33}^2 v/\omega^2 C_{33}^D A_{12}) \{\sin[kN(a+b)/2]/\sin[k(a+b)/2]\}^2 \{\langle\cos[k(a-b)/2] - \cos[k(a+b)/2]\rangle \times \cos[kN(a+b)/2] - \sin[k(a-b)/2]\sin[kN(a+b)/2]\} \times \{\langle\cos[k(a-b)/2] - \cos[k(a+b)/2]\rangle \sin[kN(a+b)/2] + \sin[k(a-b)/2]\cos[kN(a+b)/2]\} + (h_{33}^2 v/\omega^2 C_{33}^D A_{12}) \{4N\cot[k(a+b)/2] - 4N\cos[k(a-b)/2]/\sin[k(a+b)/2] - \langle 3 - 2\cos(ka/2) - 2\cos(kb/2) + \cos[k(a+b)/2]\rangle \times \sin[kN(a+b)]/2\sin^2[k(a+b)/2]\} + N(a+b)/\omega A_{12}\varepsilon_{33}^s, \quad (1)$$

where ω is the angular frequency, v the sound velocity, k the wave vector, A_{12} the area of electrode, h_{33}, C_{33}^D, ε_{33}^s are the piezoelectric, elastic, and dielectric constants, respectively.

From Eq. (1) it can be seen that the resonance frequencies can be obtained by setting $\sin[k(a+b)/2] = 0$. Thus we have

$$f_n = nv/(a+b). \quad n = 1, 2, 3, \ldots \quad (2)$$

Obviously the resonance frequencies are determined by the periodic length of the ASL rather than its total thickness. The thinner the laminar domains, the higher the resonance frequencies. In practice, a thickness of laminar domains of several microns can be achieved,[4,5] which corresponds to the resonance frequencies of hundreds megahertz to several gigahertz. Besides, since Z_i is a function of N and A_{12}, R_i and X_i can be regulated by an adequate choice of the value of N and A_{12}. For transducers made of Z-cut single domain LiNbO$_3$ crystals with an electromechanical coupling constant $K = 0.17$, under high-frequency operation the static capacitance is the main part of the impedance at resonance frequency. As a result, the insertion loss of the transducer is very high.[7] In our case, however, R_i can be equal to or even larger than X_i by choosing N and A_{12} suitably. For example, in the special case of $a = b$, at the fundamental resonance frequency $f_1 = v/2a$, Eq. (1) can be simplified to

$$Z_i = R_i - jX_i = (8NK^2/\pi - j)/\omega_0 C_0. \quad (3)$$

Here $C_0 = \varepsilon_{33}^s A_{12}/2Na$ is the static capacitance. Let $A_{12} = 2.5 \times 2.5$ mm^2, $N = 15$, $a = 6.6$ μm, then $R_i = 50$ Ω and $X_i = 45$ Ω. The transducers thus fabricated will have an insertion loss near 0 dB in a 50 Ω measurement system. As for the case of unequal a and b, numerical calculations have shown that the real part of the impedance will decrease with the difference between a and b. In the limiting case, i.e., when $N = 1$ and $b = 0$ or $a = 0$, Eq. (1) gives the result which is identical to that for transducers made of single domain LiNbO$_3$ crystals.[7]

If the normal of the laminar domain boundaries does not coincide with the Z axis, but lies in Y - Z plane, and the electrodes of the transducer are still made parallel to the lamellae of the ASL, then one quasi-longitudinal wave (QLW) and one quasi-shear wave

(QSW) will be generated by the transducer.[8] In such a case, theoretical analysis has shown that the electrical impedance Z_i consists of two parts, one for QLW and the other for QSW. Both are very similar to Eq. (1). The expression for the resonance frequency, Eq. (2), still holds except for the substitution of v_L (or v_S) for v; i.e.,

$$f_n^{L(S)} = n v_{L(S)}/(a+b), n = 1,2,3,\ldots \quad (4)$$

here $v_L(v_S)$ is the sound velocity of QLW (QSW).

In order to verify the predictions given above, a set of transducers with their resonance frequencies in the range of 200–800 MHz has been made of the ASL of LiNbO$_3$ crystals as shown in Fig. 1(c). The normals of the domain boundaries of the ASL are not parallel to the Z axis, but in the Y–Z plane. The angles between normals and the Z axis are shown in Table 1. The two electrode faces are both parallel to the lamellae. Each transducer was indium bonded to the {0001} face of a single domain LiNbO$_3$ crystal sample separately. The sample, as a transmission medium, had its faces polished flat to about one-tenth of a wavelength and parallel to each other within several minutes. The length of the samples was 3.0 mm and the dimensions of the electrode area were 2.5×2.5 mm^2.

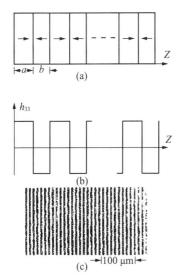
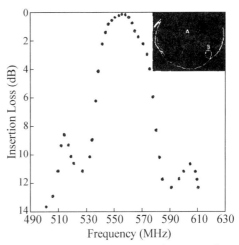

FIG. 1. Acoustic superlattice of a single LiNbO$_3$ crystal. (a) Schematic diagram of ASL (the arrows indicate the directions of spontaneous polarization). (b) Corresponding piezoelectric coefficient as a periodic function of Z. (c) Optical photomicrograph of ASL of LiNbO$_3$ crystal revealed by etching.

FIG. 2. Insertion loss vs frequency for transducer No. 1. Its Smith chart is shown on the upper right corner with scan width $\Delta f = $ 1300 MHz. A—QLW, B—QSW.

In order to evaluate the performance of the transducers, both frequency domain and time domain measurements are made.

In frequency domain measurements, a Hp8505A network analyzer has been used. We measured the reflection coefficients of the transducers. A typical result is shown on the upper right corner in Fig. 2. Two resonance frequencies were observed. One is for QLW and the other for QSW with frequencies of 555 MHz and 279 MHz, respectively. Theoretical values, calculated from Eq. (4) with $a+b=13.2$ μm, $v_L=7300$ m/s, and $v_S=3600$ m/s,[9] are 553 MHz for QLW and 273 MHz for QSW. The resonance frequencies measured and calculated are listed in Table 1. It may be noted that the experimental results are in good agreement with the theoretical ones.

TABLE 1. Comparison between theoretical predictions and experimental results.

Transducer	Periodicity of superlattice $a+b$ μm	Number of periods N	Angle between normal of domain boundaries and Z axis (deg)	Resonance frequency of the QLW (MHz)		Resonance frequency of the QSW (MHz)	
				Cal.[a]	Meas.	Cal.[a]	Meas.
No.1	13.2	15	~5	553	555	273	279
No.2	10.8	55	~50[b]	648	641	383	380
No.3	9.6	62	~30[b]	750	764	415	418

[a] Reference [9].
[b] Pulling direction not to be [001].

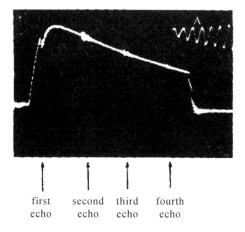

FIG. 3. Oscilloscope photograph of multiple echoes produced by transducer No.1. The magnified view of the second echo is shown on the upper right corner labeled A. Its vertical scale is 10 mV/div. and the horizontal scale is 10 ns/div.

According to Sittig, if the dissipation from dielectric, sound absorption and other losses is neglected, the insertion loss (IL) of a transducer can be expressed as[10]

$$IL = -20\log(1-R^2), \quad (5)$$

here R is the modulus of the reflection coefficient. Figure 2 gives the insertion loss versus frequency for transducer No. 1. An insertion loss of nearly 0 dB at 555 MHz was achieved, which is in good agreement with the theory. The 3 dB bandwidth of transducer No. 1 is 5.8%. The curve of insertion loss shown in Fig. 2 is very similar to that of surface-wave interdigital transducers.[6] Both have a periodic structure.

In time domain measurements, the pulse-echo technique was used. The transducers were excited by a δ pulse and echoes were observed. For lack of a sufficiently sensitive oscilloscope only the echoes of QSW for transducer No. 1 were detected. They are shown in Fig.3. The frequency of the reflected signal was determined to be 273 MHz which is consistent both with the theoretical one and with the result obtained from the frequency domain measurement given above.

In conclusion, the above-mentioned results suggest that the ASL of $LiNbO_3$ crystals can be used to generate and detect ultrasonic waves of very high frequencies. Hence the technique will have potential applications in acoustic devices.

References and Notes

[1] N.F.Foster, G. A. Coquin, G. A. Rozgonyi, and F. A. Vannatta, IEEE Trans. Sonics Ultrason. **SU-15**, 28 (1968).

[2] E. K. Sittig and H. D. Cook, Proc. **IEEE 56**, 1375 (1968).

[3] E. A. Gerber, T. Lukaszek, and A. Ballato, IEEE Trans. Microwave Theory Tech. **MTT-34**, 1002 (1986).

[4] N. B. Ming, J. F. Hong, and D. Feng, J. Mater. Sci. **17**, 1663(1982).

[5] N. B. Ming, J. F. Hong, and D. Feng, Acta. Phys. Sin. **31**, 104(1982).

[6] R. F. Mitchell, Philips Res. Repts. Suppl. 3, 1 (1972).

[7] A. H. Meitzler and E. K. Sittig, J. Appl. Phys. **40**, 4341 (1969).

[8] B. A. Auld, *Acoustic Fields and Waves in Solids* (Wiley-Interscience, New York, 1937), Vol. 1, p. 272.

[9] J. DeKlerk, in *Proceedings of the International School of Physics "Enrico Fermi" Vol. 63, A Physical Approach to Elastic Surface Waves*, edited by D. Sette (North-Holland, Amsterdam, 1976), p. 437.

[10] E. K. Sittig, in *Progress in Optics Vol. 10, Elastooptic Light Modulation and Deflection*, edited by E. Wolf (North-Holland, Amsterdam, 1972), p. 231.

[11] We would like to thank Professor D. Feng and J. F. Hong for valuable discussions. We gratefully acknowledge Zh. N. Lu, Y. S. Yang, and Y. Z. Li for technical assistance. This research was supported by National Natural Science Foundation of China.

High-frequency Resonance in Acoustic Superlattice of LiNbO₃ Crystals[*]

Yong-yuan Zhu and Nai-ben Ming
Laboratory of Solid State Microstructures,
Nanjing University, People's Republic of China

Wen-hua Jiang and Yong-an Shui
Institute of Acoustics, Nanjing University, People's Republic of China

The relationship between the resonance frequency and the periodicity of the acoustic superlattice of LiNbO₃ crystals has been obtained by analysis of the excitation and propagation of elastic waves in this kind of material with which high-frequency resonance in the range of 500 – 800 MHz has been realized experimentally. The experimental results are in good agreement with the theoretical ones. The applications of acoustic devices operating at frequencies high above 500 MHz are to be expected.

The establishment of a one-to-one correspondence between periodic growth striations and ferroelectric domain structures[1-4] has made it possible to prepare a single crystal, such as LiNbO₃,[5,6] LiTaO₃,[7] and triglycine sulphate,[8] with periodic laminar ferroelectric domain structures (PLFDS's). Some of the interesting new physical properties of such materials have already been discovered.[9-11] This present letter is concerned with the high-frequency acoustic resonance in such kind of materials. LiNbO₃ single crystals with PLFDS's are used as a representative example in the analysis and experiments.

In what follows, we will call PLFDS the acoustic superlattice (ASL). In such an ASL, the piezoelectric tensor is no longer a constant, but a periodic function of spatial coordinates. Assuming that the ASL is arranged along the z axis perpendicular to the lamellae, a and b are the thicknesses of the positive and negative domains, respectively, N is the number of periods, then the periodicity of the ASL is $a+b$ and its total thickness is $N(a+b)$, as shown in Figs. 1(a) and 1(b). If the {0001} planes are the electrode faces that are parallel to the lamellae of the ASL, a longitudinal planar wave propagating along the z axis will be excited by the action of an alternating external electric field. The electric impedance of the resonators can be derived by using the Green's function method[12] to solve the wave equations, which is

$$z = (2h_{33}^2/j\omega k A_{12} C_{33}^D) \times \{-4N\sin(ka/2)\sin(kb/2)/\sin[k(a+b)/2] - \langle \sin^2[k(a-b)/2]/\sin^2[k(a+b)/2]\rangle \times \tan[Nk(a+b)/2]\} + N(a+b)/j\omega A_{12}\varepsilon_{33}^s, \qquad (1)$$

---

[*] Appl. Phys. Lett., 1988, 53(23): 2278

where ω is the angular frequency, k the wave vector, A_{12} the area of the electrodes, h_{33}, C_{33}^D, and ε_{33}^s are the piezoelectric, the elastic, and the dielectric coefficients of LiNbO$_3$, respectively. The first term in Eq. (1) related to the wave vector is the motional impedance and the second term is the static impedance, i.e., the capacitive reactance. Note that on setting $b=0$ or $a=0$ in Eq. (1), the result will be simplified to the one for the ordinary LiNbO$_3$ crystals with single domain.[13] It can be seen from Eq. (1) that the relationship between the antiresonance frequency and the periodicity of the ASL can be obtained if $z \to \infty$ in which there are two interesting cases. The first one comes from the condition $\sin[k(a+b)/2]=0$ and we have

$$f_{main}^n = nv/(a+b), n=1,2,3,\ldots \qquad (2)$$

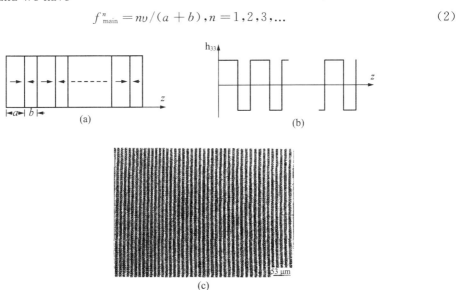

FIG. 1. Acoustic superlattice of LiNbO$_3$: (a) schematic diagram of acoustic superlattice (the arrows indicate the directions of the spontaneous polarization), (b) corresponding piezoelectric coefficient as a period function of z, (c) optical photomicrograph of acoustic superlattice of LiNbO$_3$ revealed by etching, lamellae perpendicular to the z axis.

where v is the velocity of the longitudinal wave propagating along the z axis. In this case we may call f_{main}^n the main antiresonance frequency and f_{main}^1 is the fundamental frequency of the main antiresonances. It is worth noting that the main antiresonances are determined only by the periodicity of the ASL, $a+b$, not by the total thickness of the ASL. As we know, the thickness of a resonator working at several hundred megacycles per second with ordinary materials, such as the single domain LiNbO$_3$ crystals, is too thin to be fabricated by ordinary processing techniques. However, it is easy to grow the ASL of LiNbO$_3$ crystals with the thickness of each lamellae of several microns.[5,6,11] Therefore, it is possible to fabricate the acoustic devices operating at frequencies high above 500 MHz by using the ASL. The second condition for resonance can be obtained when $\tan[Nk(a+b)/2] \to \infty$. If we take f_{main}^1 for the reference frequency, through some algebraic operations, we can get

$$f_s^m = mv/2N(a+b), m = \pm 1, \pm 3, \pm 5, \ldots \qquad (3)$$

We may call the set of f_s^m the satellite-like antiresonance frequencies which locate on both sides of the main one and are related to the total thickness of the ASL, $N(a+b)$ only. The frequency difference between any two adjacent satellite-like antiresonances is

$$\Delta f_s = f_s^{m+2} - f_s^m = f_{\text{main}}^1/N. \qquad (4)$$

If a equals b, Eq. (1) can be simplified to

$$z = (2Na/j\omega A_{12}\varepsilon_{33}^s) \cdot$$
$$\times \{1 - (h_{33}^2 \varepsilon_{33}^s / C_{33}^D)[\tan(ka/2)/(ka/2)]\}. \qquad (5)$$

In this case there exist only main antiresonances, no satellite-like ones. It is clear that the satellite-like anti-resonances result from the inequality of the thicknesses of the positive and the negative domain lamellae. This phenomenon is similar to that in x-ray or electron diffraction. They all stem from the appearance of the longer period apart from the short one. Here, as we know, in solving the wave equation by using the Green's function method, the image sources must be introduced in order to satisfy the free boundary conditions at both electrode faces.[12] Thus a longer periodicity, $2N(a+b)$, is superimposed on the original ASL. Nevertheless, when $a=b$, the longer periodicity, $N(a+b)$, does not appear.

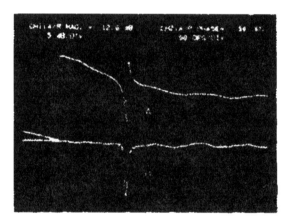

FIG.2. Reflection coefficient of resonator No.3 in rectangular coordinates. The horizontal scale is frequency centered at 650 MHz with a scan width $f=1300$ MHz, curve A for phase and curve B for magnitude.

In order to verify the predictions given above, a set of resonators with working frequencies in the range of 500–800 MHz has been made of the ASL of LiNbO$_3$ crystals, as shown in Fig. 1 (c), which were grown from a Czochralski-growth system by utilizing the periodic growth striation technique.[1,5] The periodicity of the ASL inside the resonators is along the z axis perpendicular to the lamellae and the two electrode faces are {0001} planes parallel to the lamellae. Under the action of an external periodic electric field, an elastic wave is excited inside the resonator through the piezoelectric effect and it forms a stationary wave when Eq. (2) or (3) is satisfied.

The reflection coefficients of the resonators have been measured by means of a Hp8505 network analyzer and a typical result is shown in Fig. 2. It can be seen that there is one main antiresonance in the range of 5 – 1300 MHz. The resonance frequency is 552 MHz, close to the theoretical value, 554 MHz, which is calculated from Eq. (2) with $a+b=$ 13.2 μm and $v=7320$ m/s (Ref. [14]) for resonator No. 3. The measured and calculated resonance frequencies are listed in Table 1. It may be noted that the experimental values are very close to the theoretical ones. The deviation here is due to several reasons such as the irregularity of the periodicity of the ASL, the lamellae not being exactly perpendicular to the z axis, and so on.

TABLE 1. Relationship between the antiresonance frequency f_{main}^1 and the periodicity of acoustic superlattice $a+b$.

Resonator	Periodicity of superlattice $a+b$ (μm)	Frequency of resonance (MHz)		Error (%)
		f_{main}^1 cal.	mes.	
No. 1	10.0	732	710	3
No. 2	11.0	665	686	3
No. 3	13.2	554	552	0.4
No. 4	13.3	550	553	0.5

Figure 3 has manifested that there does exist a sequence of satellite-like resonances located on both sides of the main resonance. Table 2 shows that the measured frequency difference between any two adjacent satellite-like resonances, $\overline{\Delta f_s}$ is in good agreement with the theoretical one calculated from Eq. (4).

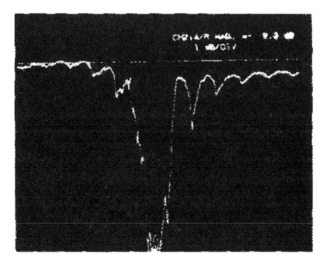

FIG. 3. Magnitude of the reflection coefficient of resonator No. 4 in rectangular coordinates. The horizontal scale is frequency centered at 553 MHz with a scan width $f=200$ MHz.

TABLE 2. Frequency difference between any two adjacent satellite-like antiresonances, $\Delta f_s = f_s^{m+2} - f_s^m$.

	m	1	3	5	7	9
Δf_s (MHz) (measured values)	f_s^m	560.3	577.9	595.1	612.4	629.6
	Δf_s		17.6	17.2	17.3	17.2
	$\overline{\Delta f_s}$			17.3		

$\Delta f_s = f_{\text{main}}^1 / N$ (theoretical values) (MHz)	N	32
	$a+b$	13.3 μm
	v	7320 m/s
	f_{main}^1	550
	Δf_s	17.2

The theoretical and the experimental results given above have demonstrated that it is possible to use this kind of material to fabricate acoustic devices working at frequencies above 500 MHz. And different exciting schemes are currently being investigated for determining optimum characteristics for device applications.

References and Notes

[1] N. B. Ming, J. F. Hong, and D. Feng, J. Mater. Sci. **27**, 1663 (1982).
[2] J. F. Hong, Z. M. Sun, Y. S. Yang, and N. B. Ming, Physics **9**, 5 (1980).
[3] N. B. Ming, J. F. Hong, Z. M. Sun, and Y. S. Yang, Acta Phys. Sin. **30**, 1672 (1981).
[4] Z. M. Sun, N. B. Ming, and D. Feng, Acta Inorganic Mater. **1**, 207 (1986).
[5] N. B. Ming, J. F. Hong, and D. Feng, Acta Phys. Sin. **31**, 104 (1982).
[6] J. F. Hong and Y. S. Yang, Acta Opt. Sin. **4**, 821 (1986).
[7] W. Wang, Q. Zou, Z. Geng, and D. Feng, J. Cryst. Growth **79**, 706 (1986).
[8] W. Wang and M. Qi, J. Cryst. Growth **79**, 758 (1986).
[9] D. Feng, N. B. Ming, J. F. Hong, Y. S. Yang, J. S. Zhu, and Y. N. Wang, Appl. Phys. Lett. **37**, 607 (1980).
[10] Y. H. Xue, N. B. Ming, J. S. Zhu, and D. Feng, Acta Phys. Sin. **32**, 1515 (1983); Chin. Phys. **4**, 554 (1984).
[11] A. Feisst and P. Koidl, Appl. Phys. Lett. **47**, 1125 (1985).
[12] R. F. Mitchell, Philips Res. Repts. Suppl. **3**, 1 (1972).
[13] D. A. Berlincourt, D. R. Curran, and H. Jaffe, in *Physical Acoustics*, edited by W. P. Mason (Academic, New York, 1964), Vol. 1A, p.169.
[14] A. W. Warner, M. Onoe, and G. A. Coquin, J. Acoust. Soc. Am. **42**, 1223 (1966).
[15] We are grateful to Professor D. Feng and J. F. Hong for valuable discussions and Y. S. Yang and Y. Z. Li for technical assistance. This research was supported by the National Natural Science Foundation of China.

Ultrasonic Spectrum in Fibonacci Acoustic Superlattices[*]

Yong-yuan Zhu and Nai-ben Ming

Laboratory of Solid State Microstructures, Nanjing University, People's Republic of China

Wen-hua Jiang

Institute of Acoustics, Nanjing University, People's Republic of China

A new type of superlattice, a Fibonacci acoustic superlattice, is presented. Its ultrasonic spectrum, which is excited by the piezoelectric effect, has been studied theoretically and experimentally. The results are in good agreement with each other.

The discovery of quasicrystalline phases in metallic alloys[1] has opened a new field in solid-state physics. Since then, a lot of work has been done on the vibrational properties of the one-dimensional quasiperiodic superlattice, a heterostructure with quasiperiodic ordering of multilayers.[2—6] The spectra observed were those of thermally excited phonons and the superlattices are made from materials such as GaAs-AlAs, Nb-Cu, etc. with their modulated wavelength of several nanometers comparable with the wavelength of a de Broglie wave. However, there exists another type of superlattice. It is made from a piezoelectric single crystal and its modulated wavelength is in the range of several micrometers to tens of micrometers, which is comparable with the wavelengths of ultrasonic waves. We are interested in the ultrasonic spectrum excited by the piezoelectric effect which have extensive applications in acoustic devices. In our previous work,[7,8] we investigated the ultrasonic excitation by the piezoelectric effect in a periodic superlattice of a single crystal of $LiNbO_3$ with periodic laminar ferroelectric-domain structures. On that basis, we have studied theoretically, in this paper, the ultrasonic spectrum excited by a Fibonacci superlattice made from $LiNbO_3$ crystals. We have also verified quantitatively the self-similar structures of the ultrasonic spectrum experimentally. A good agreement with the theory is obtained.

The piezoelectric acoustic superlattice can be viewed as a structure composed of a series of sound δ sources arranged periodically or quasiperiodically. These δ sources are generated by the discontinuity of the piezoelectric stress which is the product of a piezoelectric coefficient and electric field. So, there are two ways to prepare a periodic or

---

[*] Phys.Rev.B, 1989,40(12):8536

quasiperiodic acoustic superlattice. The first one is to use piezoelectric crystals such as a LiNbO$_3$ single crystal with periodic or quasiperiodic laminar ferroelectric-domain structures induced by the growth striations (Figs. 1 and 2).[9—12] The sign of a piezoelectric coefficient of the layered crystal changes alternately. In this case, the piezoelectric coefficient is discontinuous at the ferroelectric-domain boundaries which become sound sources under the action of an alternating external electric field and the excited acoustic waves are bulk waves. The second one is to fabricate an interdigital surface-wave transducer with electrode intervals varying periodically or quasiperiodically (Figs. 3 and 4). The transducers are easily fabricated by a standard photolithographic technique.[13,14] Aluminum electrodes are deposited on a piezoelectric substrate such as a single-domain crystal of LiNbO$_3$. When an alternating external voltage is applied onto the electrodes, an electric field is built beneath the surface of the substrate as shown in Figs. 3(b) and 4(b), which can be simplified to Figs. 3(c) and 4(c).[15] The electric field generated beneath the surface of the substrate alternates its sign periodically or quasiperiodically. In this case, the electric field is discontinuous on the bisect line of each electrode beneath the surface of the substrate which is also a sound source and the excited acoustic waves are surface waves.

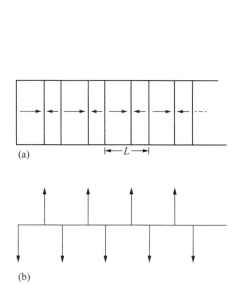

FIG. 1. Acoustic superlattice of LiNbO$_3$ crystals with periodic ferroelectric-domain structures. (a) Schematic diagram of periodic acoustic superlattice with the periodicity of L (the arrows indicate the directions of the spontaneous polarization). (b) Corresponding sound δ sources.

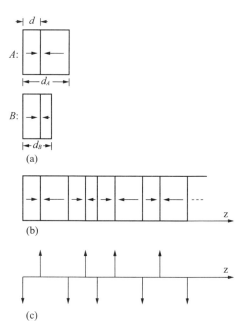

FIG. 2. Fibonacci acoustic superlattice of LiNbO$_3$ crystals. (a) The two blocks of a Fibonacci acoustic superlattice, each composed of one positive and one negative ferroelectric domain. (b) Schematic diagram of a Fibonacci acoustic superlattice (the arrows indicate the directions of the spontaneous polarization). (c) Corresponding sound δ sources.

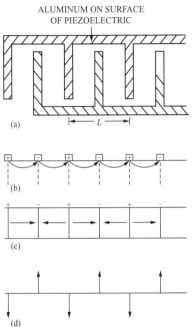

FIG. 3. (a) Periodic interdigital transducer schematic diagram with the periodicity of L. (b) and (c) are side views of the interdigital transducer, showing electric field patterns inside the thin skin of the substrate. (b) Actual electric field pattern. (c) Simplified electric field pattern. (d) Corresponding sound δ sources.

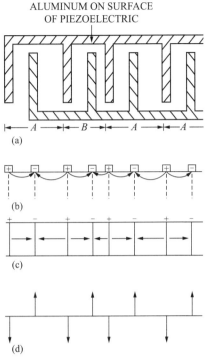

FIG. 4. (a) Quasiperiodic interdigital transducer schematic diagram. (b) and (c) are side views of the interdigital transducer, showing electric field patterns inside the thin skin of the substrate. (b) Actual electric field pattern. (c) Simplified electric field pattern. (d) Corresponding sound δ sources.

In both cases, the physical mechanism of ultrasonic excitation is obvious. According to Bömmel and Dransfeld,[16] under the action of an external electric field, the discontinuity in piezoelectric stress at the ferroelectric-domain boundaries (or on the bisect line of each electrode beneath the surface of the substrate) must be balanced by a strain $S(u_m)$. Here u_m represents the position where the discontinuity takes place. This strain will propagate as a wave:

$$S(u) = S(u_m)\cos(\omega t - ku). \tag{1}$$

So each domain boundary (or the bisect line of each electrode where the discontinuity of electric field takes place) can be viewed as a sound δ source [see Figs. 1(b), 2(c), 3(d), and 4(d)].

Because of this similarity, in what follows we will limit ourselves to $LiNbO_3$ crystals with quasiperiodic laminar ferroelectric-domain structures. The results obtained will suit each other.

The Fibonacci acoustic superlattice of a $LiNbO_3$ crystal consists of two fundamental blocks of A and B. The widths of these two blocks are different, each composed of one positive and one negative ferroelectric domain as shown in Fig. 2(a). The quasiperiodicity can be realized with a Fibonacci sequence of blocks A and B fulfilling relations:[17]

$$S(1)=|A|, \ S(2)=|AB|, \ S(3)=|AB:A|,$$
$$S(4)=|ABA:AB|, \ S(n)=|S(n-1):S(n-2)|, \tag{2}$$

and the ratio of the thicknesses of A, B blocks just equals the golden mean, i.e.,

$$d_A/d_B = \tau = (1+\sqrt{5})/2. \tag{3}$$

The Fibonacci acoustic superlattice thus obtained is shown in Fig. 2(b).

We assume that the lateral dimensions of the Fibonacci acoustic superlattice are much larger than the width of the ferroelectric domains and that the normal of the domain boundaries coincides with the z axis; then the one-dimensional model is applicable. Under these conditions, when an alternating voltage applied on the $\{0001\}$ faces of the sample, a longitudinal planar wave propagating along the z axis will be excited inside the superlattice. It obeys the wave equation:

$$\partial^2 u_3/\partial z^2 - (1/v^2)\partial^2 u_3/\partial t^2 = (2h_{33}D_3/C_{33}^D)\sum_m (-1)^m \delta(z-z_m), \tag{4}$$

where u_3 represents the particle displacement in the z direction, v is the sound velocity, h_{33}, C_{33}^D, and D_3 are the piezoelectric, elastic coefficients, and electric displacement, respectively. $\{z_m\}$ are the positions of the ferroelectric-domain boundaries.

Equation (4) is a fundamental wave equation governing the excitation and propagation of sound waves in a piezoelectric superlattice. The term on the right-hand side stands for the exciting source of the sound wave. As above mentioned, the ultrasonic waves in such a superlattice are excited by a series of δ sources at the domain boundaries, which is also shown in Eq. (4). Therefore, we can make use of the impulse response model[13,14] to obtain the ultrasonic spectrum excited by the Fibonacci acoustic superlattice of $LiNbO_3$ crystals.

As can be seen in Fig. 2(c), the δ sources can be divided into two parts, one for positive δ sources, the other for negative δ sources. Each forms a Fibonacci sequence. The negative is displaced a distance d relative to the positive. Performing Fourier transforms on the positive terms and the negative terms on the right-hand side of Eq. (4) separately, and adding them together, we can get the ultrasonic spectrum for an infinite array, which is

$$H(k) \propto \sin(kd/2)\sum_{m,n}(\sin X_{m,n}/X_{m,n}) \times \delta(k-2\pi(m+n\tau)/D), \tag{5}$$

where $X_{m,n} = \pi\tau(m\tau-n)/(1+\tau^2)$, $D = \tau d_A + d_B$, and k is the wave vector. In deriving Eq. (5), the projection method has been used.[18,19] The appearance of the term $\sin(kd/2)$ is due to the relative displacement between the positions of positive δ sources and that of negative δ sources. The resonant peaks in the ultrasonic spectrum can be obtained from the δ functions in Eq. (5), which is $k=2\pi(m+n\tau)/D$, or

$$f_{m,n} = v(m+n\tau)/D, \tag{6}$$

where m, n are integers. The most significant resonant peaks in ultrasonic spectrum occur at $f = f_{m,n}$ for which $X_{m,n}$ is small. This means that n/m must be close to τ. It is well known that the best rational approximants to τ occur when m and n are successive Fibonacci integers,[20] $(m, n) = (F_{p-1}, F_p)$, where $F_{p+1} = F_p + F_{p-1}$ and $(F_0, F_1) = (0, 1)$. For these values of (m, n), $f_{m,n}$ takes the form of (integer) $v\tau^p/D$ and these resonant peaks are labeled τ^p.

The results are much similar to that of acoustic-phonon transmission[5] except for the substitution of $f_{m,n} = v(m + n\tau)/D$ for $f_{m,n} = v(m + n\tau)/2D$ and of peaks for dips. These can be explained as follows. In the case of excitation of ultrasonic waves, if the piezoelectric medium is a superlattice one, the ultrasonic waves will be excited on the ferroelectric-domain boundaries. The waves coming from successive boundaries will interfere with each other. Those satisfying the constructive interference will appear as resonant peaks in the ultrasonic spectrum. Whereas in the case of acoustic-phonon transmission when the phonons, reflected from successive interfaces satisfy the constructive interference, they will appear as dips in transmission spectrum. Obviously, the peaks in the former case just correspond to the dips in the latter case. The occurrence of the factor $\frac{1}{2}$ in the expression for dip frequencies is due to the fact that the path difference between phonons reflected from the adjacent interfaces is twice as large as the path difference between the ultrasonic waves excited on the adjacent ferroelectric-domain boundaries.

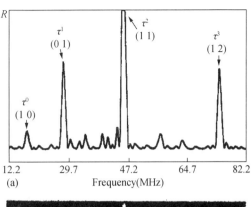

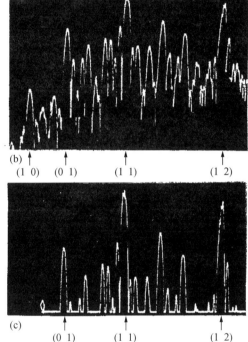

FIG. 5. Ultrasonic spectrum for an eighth-generation Fibonacci acoustic superlattice. (a) Ultrasonic spectrum calculated for a LiNbO$_3$ crystal. R, relative value of radiation resistance. (b) Ultrasonic spectrum measured. (c) The same as (b) with the amplitude magnified. In both (b) and (c) the amplitude is in logarithmic scale.

We have calculated the ultrasonic spectrum excited by the Fibonacci acoustic superlattice numerically. In order to be compared with the experiments only the ultrasonic spectrum of the superlattice of the eighth Fibonacci generation has been shown in Fig. 5 (a), with the parameters selected to be in line with the experiments. It is worth noting that the resonant peaks obey Eq. (6). In this figure, several frequencies predicted by the expression $f = f_{m,n} = (m,n)$ are indicated by arrows. As expected, the main peaks are observed at $f_{m,n}$ for which m and n are neighboring Fibonacci numbers, as indicated by τ^p.

Detailed calculations have shown that the resonant peaks densely fill reciprocal space in a self-similar manner. It is just expected. It is well known that the self-similarity exists in the spectra of phonon,[2-6] electron energy,[4] polariton,[21] and in X-ray diffraction,[17] etc. Here again, in the ultrasonic spectrum excited by the Fibonacci acoustic superlattice of LiNbO$_3$ crystals, the selfsimilarity also exists. It may be said that the selfsimilarity is a common feature of the Fibonacci superlattices and is a reflection of the self-similar structure of Fibonacci superlattices in reciprocal space.

Experimentally, we have fabricated several quasiperiodic surface-wave interdigital transducers of the eighth Fibonacci generation. Aluminum electrodes are deposited on the surface of a single-domain crystal of LiNbO$_3$. The intervals between the electrodes varied successively according to the Fibonacci sequence. In this case, the surface-wave velocity $v = 3944$ m/sec. The other parameters are $d = 39.0$ μm, $d_B = 62.3$ μm, and $d_A = 100.8$ μm which were measured by an optical microscope. The ultrasonic spectra are measured, which is the ratio of output voltage to input voltage. In the measurement, the frequency scan range of the frequency spectrum analyzer is selected to be from 12.2 MHz to 82.2 MHz. The output voltage signals which satisfy the resonant conditions, i.e., $f_{m,n} = v(m + n\tau)/D$, are strengthened. The others are diminished. Figures 5(b) and 5(c) show one of these results. In the ultrasonic spectrum many sharp peaks appear at nonequal intervals. It is clear that there exists a one to one correspondence between the resonant peaks in Figs. 5(a) and 5(b) or 5(c). Table 1 shows the results calculated and measured. The agreement is quite satisfactory.

TABLE 1. Comparison between resonance frequencies calculated and measured for a quasiperiodic surface-wave interdigital transducer.

Vibrational mode	Resonance frequency	f (MHz)
(m,n)	calc.	meas.
(1,0)	17.5	17.5
(0,1)	28.3	28.4
(1,1)	45.8	45.8
(1,2)	74.1	74.5

In order to compare the ultrasonic spectrum of the Fibonacci acoustic superlattice with that of the periodic acoustic superlattice, Figs. 6(a) and 6(b) show the theoretical and the experimental ultrasonic spectra obtained from a periodic acoustic superlattice of a LiNbO$_3$ crystal with periodic laminar ferroelectric-domain structures. The analysis and the experiment are analogous to our previous work.[7,8] The resonant frequency measured is 555 MHz, close to the theoretical one, 553 MHz, which is calculated from $f=v/L$ with $L=13.2$ μm and $v = 7300$ m/sec. Here L is the periodicity of the periodic acoustic superlattice. Obviously Figs. 5 and 6 bear no resemblance to each other.

In conclusion, we have studied and obtained some of the features of the ultrasonic spectrum excited by a Fibonacci acoustic superlattice. Two ways to fabricate such a superlattice are presented.

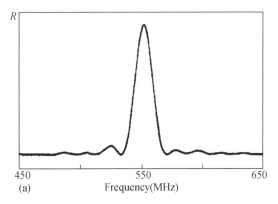

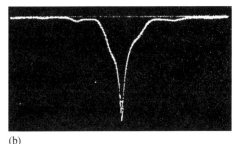

FIG. 6. Ultrasonic spectrum for a periodic acoustic superlattice of a LiNbO$_3$ crystal. (a) Ultrasonic spectrum calculated. R, relative value of radiation resistance. (b) Measured magnitude of the reflection coefficient. The horizontal scale is frequency centered at 550 MHz with a scan width $f=200$ MHz.

The superlattice is different from the traditional one in that it is a piezoelectric single crystal and the modulated wavelength is much larger. The ultrasonic spectrum is excited by the piezoelectric effect. The theoretical results are quantitatively verified by the experiments.

References and Notes

[1] D. Schechtman, I. Blech, D. Gratias, and J. W. Cahn, Phys. Rev. Lett. **53**, 1951 (1984).
[2] J. P. Lu, T. Odagaki, and J. L. Birman, Phys. Rev. B **33**, 4809 (1986).
[3] M. Kohmoto and J. R. Banavar, Phys. Rev. B **34**, 563 (1986).
[4] F. Nori and J. P. Rodriguez, Phys. Rev. B **34**, 2207 (1986).
[5] S. Tamura and J. P. Wolfe, Phys. Rev. B **36**, 3491 (1987).
[6] X. K. Zhang, H. Xia, G. X. Cheng, A. Hu, and D. Feng, Phys. Lett. A **136**, 312 (1989).
[7] Y. Y. Zhu, N. B. Ming, W. H. Jiang, and Y. A. Shui, Appl. Phys. Lett. **53**, 1381 (1988).
[8] Y. Y. Zhu, N. B. Ming, W. H. Jiang, and Y. A. Shui, Appl. Phys. Lett. **53**, 2278 (1988).
[9] N. B. Ming, J. F. Hong, and D. Feng, J. Mater. Sci. **27**, 1663 (1982).
[10] N. B. Ming, J. F. Hong, Z. M. Sung, and Y. S. Yang, Acta Phys. Sin. (in Chinese) 30, 1672 (1981).
[11] J. F. Hong and Y. S. Yang, Acta Opt. Sin. (in Chinese) **4**, 821 (1986).

[12] A. Feisst and P. Koidl, Appl. Phys. Lett. **47**, 1125 (1985).

[13] R. H. Tancrel and M. G. Holland, Proc. IEEE **59**, 395 (1971).

[14] C. S. Hartmann, D. T. Bell, Jr., and R. C. Rosenfeld, IEEE Trans. Sonics Ultrason. **20**, 80 (1973).

[15] W. R. Smith, H. M. Gerard, J. H. Collins, T. M. Reeder, and H. J. Shaw, IEEE Trans. Microwave Theory Tech. **17**, 856 (1969).

[16] H. E. Bömmel and K. Dransfeld, Phys. Rev. **117**, 1245 (1960).

[17] R. Merlin, K. Bajema, R. Clarke, F. Y. Juang, and P. K. Bhattacharya, Phys. Rev. Lett. **55**, 1768 (1985).

[18] V. Elser, Phys. Rev. B **32**, 4892 (1985).

[19] R. K. P. Zia and W. J. Dallas, J. Phys. A **18**, L341 (1985).

[20] V. Hoggatt, *Fibonacci and Locas Numbers* (Mifflin, Boston, 1969).

[21] H. R. Ma and C. H. Tsai, Phys. Rev. B **35**, 9295 (1987).

[22] This work was supported by the National Natural Science Foundation of China.

Ultrasonic Excitation and Propagation in an Acoustic Superlattice*

Yong-yuan Zhu

CCASTA, P.O. Box 8730, Beijing 100080, People's Republic of China and National Laboratory of Solid State Microstructures, Nanjing University, Nanjing 210008, People's Republic of China

Nai-ben Ming

Center for Condensed Matter Physics and Radiation Physics, China Center of Advanced Science and Technology (World Laboratory), P.O. Box 8730, Beijing 100080, People's Republic of China

The ultrasonic excitation and propagation in an acoustic superlattice made of a LiNbO$_3$ crystal are studied. Some interesting phenomena are predicted theoretically. The resonance is determined by the periodicity of the acoustic superlattice (ASL). The acoustic power emitted by the ASL into the transmission medium is directly proportional to N^2 (N is the number of domains of the ASL). In view of these two features, the ASL may have potential applications in acoustic devices operating at frequencies of several hundred megahertz to several gigahertz. Another interesting phenomenon is the existence of a satellitelike resonance, whose frequency is related to the total thickness of the ASL. Using the ASL, resonators and transducers have been fabricated with operating frequencies in the range of 500 – 1000 MHz. Satellitelike resonances have been observed. Transducers with a very low insertion loss at 555 MHz and a 5.8% 3 dB bandwidth have been fabricated. The experiments basically confirm the theoretical predictions.

1. Introduction

In 1969 Esaki and Tsu proposed the idea of a semiconductor superlattice with periodicity on the order of a nanometer, comparable with the wavelength of a de Broglie wave.[1] In the subsequent 20 years we have witnessed tremendous research activity on semiconductor superlattices, both experimentally and theoretically, all over the world.[2]

With further investigation, it was found that the constituents of a superlattice are not restricted to semiconductors; other materials, such as metals, can also be used.[3] Recently we have proposed a new type of superlattice that is made of a ferroelectric single crystal and whose periodicity is in the range of several micrometers—comparable with the wavelengths of light and ultrasonic waves. We have systematically studied its growth and the mechanism of its formation.[4–6] Meanwhile, we have studied its physical properties,

---

* J.Appl.Phys.,1992,72(3):904.

including the nonlinear optical effect[7—10] and the piezoelectric effect.[11—13] Some unparalleled features have been found that are of potential application in device fabrication.

The importance of investigation on such classical systems lies not only in their practicality but also in their similarity in certain respects to quantum-mechanical systems. They can be used to imitate the salient features of quantum-mechanical systems owing to the fact that classical waves (mechanical or electromagnetic) in classical systems, and electrons (considered as "waves") in a quantum-mechanical system, are both governed by a wave equation. With classical systems, eigenvalues, eigenfunctions,[14] and properties of quasicrystals[13,15] were measured directly. In a random classical system, even the phenomenon of Anderson localization[16] was observed. These are difficult if not impossible to obtain in quantum-mechanical systems.

In this paper we will report our theoretical analysis and experimental results obtained by using this kind of superlattice, i.e., a single LiNbO$_3$ crystal with periodic laminar ferroelectric domain structures, to generate and detect ultrasonic waves with frequencies up to 1000 MHz. The structure of the paper is as follows: Section 2 is devoted to the theoretical analysis of acoustic excitations using the Green's function method.[17] We obtain a resonance frequency, which is inversely proportional to the periodicity of the superlattice, i.e., the thickness of the domains. On that basis, we successfully fabricate a set of resonators and transducers operating at 100 MHz, which is described in Sec. 3. Section 4 includes discussions, while Sec. 5 has a brief summary and conclusions.

2. Theory

In what follows, we will take as an example a LiNbO$_3$ crystal with periodic laminar ferroelectric domain structures and will call it an acoustic superlattice (ASL). In such an ASL, the positive and negative domains arrange periodically in one direction (say the z axis, Fig. 1). The usually accepted rectangular coordinate systems[18] for these two types of domains are shown in Fig. 1 (b). The two are interrelated by a 180° rotation around the x axis. In this case, all the odd-rank tensors will change signs from one domain to the next. Thus the piezoelectric tensor, being of third rank, is no longer a constant through the ASL, but a periodic function of the spatial coordinate z [see Fig. 1 (c)]. Under the action of an alternating external electric field, the domain walls, where the piezoelectric coefficient shows discontinuity, can be viewed as sound δ sources[19] as shown in Fig. 1 (d). The ultrasonic waves excited by these sound δ sources will interfere with each other. Those satisfying the condition for constructive interference will appear as resonance peaks in the ultrasonic spectrum. This is the physical basis for the ultrasonic excitation with the ASL.

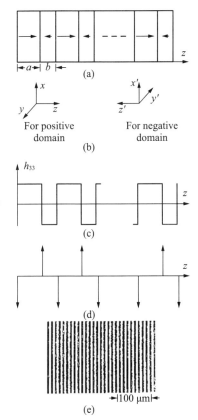

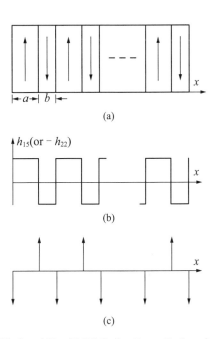

FIG. 1. ASL of LiNbO$_3$ for the excitation of one pure longitudinal wave: (a) schematic diagram of the ASL (the arrows indicate the directions of the spontaneous polarization); (b) the usually accepted rectangular coordinates associated with positive and negative domains, respectively; (c) corresponding piezoelectric coefficient as a periodic function of z; (d) corresponding sound δ sources; (e) optical photomicrograph of the ASL of LiNbO$_3$ revealed by etching, lamellae perpendicular to the z axis.

FIG. 2. ASL of LiNbO$_3$ for the excitation of one quasilongitudinal and one quasishear wave: (a) schematic diagram of the ASL (the arrows indicate the directions of the spontaneous polarization); (b) corresponding piezoelectric coefficient as a periodic function of x; (c) corresponding sound δ sources.

This excitation scheme (Fig. 1), determined by the periodic arrangement of ferroelectric domains of the ASL, will produce a pure longitudinal wave propagating along the z axis. There are many other excitation schemes. For example, Fig. 2 shows one that will produce one quasilongitudinal wave and one quasishear wave, both propagating along the x axis. Without the loss of generality, the discussion below will be restricted to the excitation of a pure longitudinal wave.

Here we assume that the ASL arrange along the z axis and that the domain walls lie in the xy plane. The thicknesses of the positive and negative domains are a and b, respectively, with the number of periods equal to N, i.e., the periodicity of the ASL is $a+b$ and its total thickness is $N(a+b)$, as shown in Fig. 1(a). Also we assume that the electrode face is parallel to the xy plane and that the transverse dimensions are very large compared with an acoustic wavelength so that a one-dimensional model is applicable. Under these conditions, a longitudinal planar wave propagating along the z axis will be excited by the action of an alternating external electric field. With $\partial/\partial x = \partial/\partial y = 0$, the piezoelectric equations[17] pertaining to this case are

$$T_3 = C_{33}^D S_3 + (-1)^n h_{33} D_3, \tag{1}$$

$$E_3 = (-1)^n h_{33} S_3 + (\varepsilon_{33}^s)^{-1} D_3, \tag{2}$$

$n = 1, 2, 3, \ldots, 2N$, where T_3, S_3, E_3, and D_3 are the stress, strain, electric field, and electric displacement, respectively. C_{33}^D, h_{33}, and ε_{33}^s are the elastic, piezoelectric, and dielectric coefficients, respectively. Newton's equation of motion is

$$\frac{\partial T_3}{\partial z} = \rho \frac{\partial^2 U_3}{\partial t^2}, \tag{3}$$

where ρ represents the mass density and U_3 the particle displacement from equilibrium position in the z direction.

Differentiating Eq. (1) twice with respect to time and Eq. (3) once with respect to z we obtain

$$\frac{\partial^2 T_3}{\partial t^2} = C_{33}^D \frac{\partial^3 U_3}{\partial t^2 \partial z} + (-1)^n h_{33} \frac{\partial^2 D_3}{\partial t^2}, \tag{4}$$

$$\rho \frac{\partial^3 U_3}{\partial z \partial t^2} = \frac{\partial^2 T_3}{\partial z^2}. \tag{5}$$

By reversing the order of differentiation we can substitute Eq. (4) into Eq. (5) and obtain

$$\frac{\partial^2 T_3}{\partial t^2} - \left(\frac{C_{33}^D}{\rho}\right) \frac{\partial^2 T_3}{\partial z^2} = (-1)^n h_{33} \frac{\partial^2 D_3}{\partial t^2}. \tag{6}$$

This is the fundamental wave equation governing the excitation and propagation of stress in a piezoelectric ASL. The presence of the source term on the right-hand side makes it an inhomogeneous equation. This term describes the excitation of the waves, while the homogeneous lefthand side describes their propagation.

If we assume that all variables have the same time dependence $\exp(j\omega t)$, then Eq. (6) becomes

$$\frac{\partial^2 T_3}{\partial z^2} + k^2 T_3 = (-1)^n h_{33} k^2 D_3, \tag{7}$$

where $k = \omega/v$ is the wave vector and $v = \sqrt{C_{33}^D/\rho}$ represents the sound velocity.

We will use the Green's function method[17] to solve the wave equation. In order to satisfy the free-boundary condition, a particular technique must be used that is the so-called "method of images". The method consists of noting that homogeneous boundary conditions can be satisfied along a plane boundary by introducing one fictitious "image source" for every real source. This image source is of opposite polarity if zero T_3 is required on the boundary. It is placed on the opposite side of the boundary plane and the same distance from the boundary as the real source (Fig. 3). Physically, this is equivalent to treating the waves reflected by the boundary as the waves emitted by the "image sources." The boundary may now be forgotten and the solution obtained merely by analyzing the effects of the real and image sources. The solution is valid only on the "real" side of the boundary, of course.

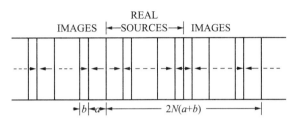

FIG. 3.　ASL with its images. A longer periodicity, $2N(a+b)$, appears.

The method of images makes heavy use of the Green's function for boundaries at infinity. In our case, the Green's function for one dimension is

$$G(z',z) = (-j/2k)\exp(-jk|z'-z|), \tag{8}$$

where z' and z are the positions of the sound source and the observation point, respectively.

According to the Green's function method, the stress at a point within the ASL is the sum of the stress waves generated from sound source points, including images, on both sides. It can be expressed as[17]

$$T_3(z) = -\left(\frac{1}{2}\right)jk\left(\exp(-jkz)F(h_{33}D_3) - 2j\int_z^\infty h_{33}(z')D_3(z')\sin[k(z'-z)]dz'\right). \tag{9}$$

where h_{33} and D_3 include both real-source and image-source distributions, and $F(h_{33}D_3)$ is

$$F(h_{33}D_3) = h_{33}D_3\int_{-\infty}^\infty f(z)e^{jkz}dz, \tag{10}$$

where $f(z) = +1$, if z is in the positive domains, and -1, if z is in the negative domains.

In order to understand the electrical properties of the ASL we need to know its impedance, which expresses the relationship between applied voltage V and electric current I. Here

$$-V = \int E_3 dz, \tag{11}$$

$$I = j\omega A_{12}D_3, \tag{12}$$

where A_{12} is the electrode area and ω is the angular frequency.

In the following, we will discuss two different boundary conditions: one for resonators and the other for transducers.

2.1　Resonators

The boundary conditions are

$$T_3(z=0) = T_3[z=N(a+b)] = 0. \tag{13}$$

Under these conditions, the images must be introduced repeatedly in order to satisfy the stress-free boundary conditions on both electrode faces. In doing so, a longer periodicity, $2N(a+b)$, is superimposed on the original ASL as shown in Fig. 3. Using the Green's function method, we obtain the stress in the positive domains,

$$T_3^p = h_{33}D_3 + T, \tag{14}$$

and the stress in the negative domains,

$$T_3^n = h_{33}D_3\{-1 + 2\cos(k[z-(n+1)a-nb])\} + T, \tag{15}$$

with

$$T = h_{33}D_3 \left\{ -4\sin\left(\frac{1}{2}kb\right) \frac{\sin\frac{1}{2}kn(a+b)}{\sin\frac{1}{2}k(a+b)} \times \sin\left[k\left(z-\frac{1}{2}(n+1)a-\frac{1}{2}nb\right)\right] - \frac{\cos\left\{\frac{1}{2}k[(N-1)a+(N+1)b]\right\}}{\cos\left[\frac{1}{2}kN(a+b)\right]\sin\left[\frac{1}{2}k(a+b)\right]} \sin(kz) + \frac{\sin k\left[z-\frac{1}{2}(a+b)\right]}{\sin\left[\frac{1}{2}k(a+b)\right]} \right\}, \tag{16}$$

where for positive domains, $n(a+b) < z < (n+1)a + nb$ and for negative domains, $(n+1)a + nb < z < (n+1)(a+b)$.

We also obtain the impedance of the ASL,

$$Z = \left(\frac{2h_{33}^2 v}{j\omega^2 A_{12} C_{33}^D}\right) \left[-4N \frac{\sin\left(\frac{1}{2}ka\right)\sin\left(\frac{1}{2}kb\right)}{\sin\frac{1}{2}k(a+b)} - \frac{\sin^2\left[\frac{1}{2}k(a-b)\right]}{\sin^2\left[\frac{1}{2}k(a+b)\right]} \tan\left(\frac{1}{2}kN(a+b)\right) \right]$$
$$+ \frac{N(a+b)}{j\omega A_{12}\varepsilon_{33}^s}. \tag{17}$$

From Eq. (17), two different kinds of resonances can be predicted according to the condition $z \to \infty$.

First, letting $\sin[k(a+b)/2] = 0$, we have

$$f_{\text{main}}^n = nv/(a+b), n = 1,2,3,\ldots \tag{18}$$

Here we call f_{main}^n the main resonance (MR) frequency and f_{main}^1 is the fundamental frequency of the MRs. The fact that the MRs are determined only by the periodicity of the ASL, $a+b$, and not by its total thickness, is of great practical significance. As we know, the thickness of a resonator working at hundreds of megahertz with ordinary materials, such as the single-domain $LiNbO_3$ crystal, is several or tens of micrometers, which is too thin to be fabricated by ordinary processing techniques. However, according to Eq. (18), we can see that the thinner the laminar domains, the higher the resonance frequencies. In practice, a thickness of laminar domains of several micrometers can be achieved, which corresponds to resonance frequencies of hundreds of megahertz to several gigahertz.

Therefore, it is possible to fabricate acoustic devices operating at very high frequencies by using the ASL.

Second, let $\tan[kN(a+b)/2]\to\infty$ and take f_{main}^1 for the reference frequency; then we have
$$f_s^m = mv/[2N(a+b)] = mf_{\text{main}}^1/2N. m = \pm 1, \pm 3, \pm 5,\ldots \qquad (19)$$
We call f_s^m the satellitelite resonance (SLR) frequency that is located on either side of the main one and is related to the total thickness of the ASL, $N(a+b)$, only.

The frequency difference between any two adjacent SLRs is given by
$$\Delta f_s = f_s^{m+2} - f_s^m = f_{\text{main}}^1/N, \qquad (20)$$
which is just one-Nth of the f_{main}^1.

If the thicknesses of positive domains and negative domains are equal, i.e., $a=b$, Eq. (17) can be simplified to
$$Z = \frac{2Na}{j\omega A_{12}\varepsilon_{33}^s}\left[1 - \frac{h_{33}^2\varepsilon_{33}^s}{C_{33}^D}\frac{\tan\left(\frac{1}{2}ka\right)}{\frac{1}{2}ka}\right]. \qquad (21)$$

In this case there exist only main resonances, no satellitelike ones. In solving the wave equation by using the Green's function method, the image sources were introduced in order to satisfy the free-boundary conditions at both electrodes. Thus a longer periodicity, $2N(a+b)$, is superimposed on the original ASL. Nevertheless, when $a=b$, the longer periodicity does not appear. This phenomenon is similar to that in X-ray or electron diffraction. This common feature originates from the common nature of the three interaction systems. First, the acoustic wave, X-ray, and electron can all be treated as waves. Second, the media involved in the interaction processes all possess superstructures.

If $a=0$ (or $b=0$), Eq. (17) can be reduced to
$$Z = \frac{L}{j\omega A_{12}\varepsilon_{33}^s}\left[1 - \frac{h_{33}^2\varepsilon_{33}^s}{C_{33}^D}\frac{\tan\left(\frac{1}{2}kL\right)}{\frac{1}{2}kL}\right]. \qquad (22)$$

It is the same as that of a resonator made of a single-domain $LiNbO_3$ crystal with thickness equal to L ($L=Nb$ or Na).[20]

The comparison of Eq. (21) with Eq. (22) shows that Eq. (21) is an equation equivalent to that for a system of $2N$ identical piezoelectric resonators of thickness a connected in series. To explain this, we must resort to Eqs. (14) and (15). We find that, under stress-free boundary conditions, the stresses on the domain walls are also equal to zero. That is to say, the ASL can be considered to be composed of $2N$ individual crystal plates each oscillating independently. However, this simplification is not shared by the case with $a\neq b$.

2.2 Transducers

The transducer consists of an ASL, coated with two electrodes, and a transmission medium. They are bonded together. The boundary conditions are assumed to be

$$T_3^{\text{ASL}} = \begin{cases} 0, & \text{at } z=0 \\ T_3^{\text{TM}}, & \text{at } z=N(a+b) \end{cases} \qquad (23)$$

where TM is the transmission medium. That is, one face of the ASL is free and the other face is fully matched to a transmission medium, which we consider to be semi-infinite. Using the same analysis we have obtained $Z = R - jX$,

$$R = \frac{4h_{33}^2 v}{\omega^2 C_{33}^D A_{12}} \left[\frac{\sin\left[\frac{1}{2}kN(a+b)\right]}{\sin\left[\frac{1}{2}k(a+b)\right]} \right]^2 \left\{ \left[\cos\left(\frac{1}{2}k(a-b)\right) - \cos\left(\frac{1}{2}k(a+b)\right) \right] \times \right.$$

$$\left. \cos\left(\frac{1}{2}kN(a+b)\right) - \sin\left(\frac{1}{2}k(a-b)\right) \sin\left(\frac{1}{2}kN(a+b)\right) \right\}^2,$$

$$X = \frac{4h_{33}^2 v}{\omega^2 C_{33}^D A_{12}} \left[\frac{\sin\left[\frac{1}{2}kN(a+b)\right]}{\sin\left[\frac{1}{2}k(a+b)\right]} \right]^2 \left\{ \left[\cos\left(\frac{1}{2}k(a-b)\right) - \cos\left(\frac{1}{2}k(a+b)\right) \right] \times \right.$$

$$\left. \cos\left(\frac{1}{2}kN(a+b)\right) - \sin\left(\frac{1}{2}k(a-b)\right) \sin\left(\frac{1}{2}kN(a+b)\right) \right\} \left\{ \left[\cos\left(\frac{1}{2}k(a-b)\right) - \cos\left(\frac{1}{2}k(a+b)\right) \right] \sin\left(\frac{1}{2}kN(a+b)\right) + \sin\left(\frac{1}{2}k(a-b)\right) \cos\left(\frac{1}{2}kN(a+b)\right) \right\} +$$

$$\frac{h_{33}^2 v}{\omega^2 C_{33}^D A_{12}} \left[4N\cot\left(\frac{1}{2}k(a+b)\right) - 4N \frac{\cos\left[\frac{1}{2}k(a-b)\right]}{\sin\left[\frac{1}{2}k(a+b)\right]} - \{3 - 2\cos(ka) - 2\cos(kb) + \cos[k(a+b)]\} \frac{\sin[kN(a+b)]}{2\sin^2\left[\frac{1}{2}k(a+b)\right]} + \frac{N(a+b)}{\omega A_{12} \varepsilon_{33}^s} \right]. \qquad (24)$$

The resonance frequency, i.e., the operating frequency of the transducer we are interested in, can be obtained by setting $\sin[k(a+b)/2]=0$. The result is the same as that of the resonator. The difference here is that no SLR exists. This is due to our assumption that in transducers the ASL is fully matched to a semi-infinite transmission medium. No longer periodicity has been introduced by the use of the image method; consequently, no satellitelike resonance exists.

If $a=b$, Eq. (24) can be reduced to

$$R = \frac{h_{33}^2 v}{\omega^2 C_{33}^D A_{12}} \tan^2\left(\frac{1}{2}ka\right) \sin^2 2Nka,$$

$$X = \frac{h_{33}^2 v}{\omega^2 C_{33}^D A_{12}} \left[4N + \frac{1}{2}\tan\left(\frac{1}{2}ka\right)\sin(4Nka)\right] \times \tan\left(\frac{1}{2}ka\right) + \frac{N(a+b)}{\omega A_{12}\varepsilon_{33}^s}. \quad (25)$$

This expression is similar to that derived for interdigital transducers from an equivalent circuit.[21] The difference is due to the fact that in our case, one face of the ASL ($z=0$) is stress free, whereas the interdigital transducer is fully matched to the medium on both sides.

From Eqs. (24) and (25) we can see that Z is a function of N and A_{12}, hence R and X can be controlled by an adequate choice of the values of N and A_{12}. For transducers made of z-cut single-domain LiNbO$_3$ crystals (electromechanical coupling constant $K=0.17$), under high-frequency operation, the static capacitance dominates the impedance at resonance. As a result, the insertion loss of the transducer is very high.[22] In our case, however, R can be equal to or even larger than X by choosing N and A_{12} suitably. For example, in the special case of $a=b$, at the fundamental resonance frequency $f_{\text{main}}^1=v/2a$, Eq. (25) can be simplified to

$$Z = R - jX = \frac{1}{2\pi f_{\text{main}}^1 C_0}\left(\frac{8NK^2}{\pi} - j\right), \quad (26)$$

where $C_0 = \varepsilon_{33}^s A_{12}/2Na$ is the clamped capacitance. If $A_{12}=5.0$ mm^2, $N=15$, and $a=6.6$ μm, then $R=50$ Ω and $X=45$ Ω at $f_{\text{main}}^1=554$ MHz. The transducers thus fabricated will have a very low insertion loss in a 50 Ω measurement system. As for the case of $a\neq b$, from Eq. (24) we can show that the real part of the impedance will decrease with the difference between a and b. For example, we consider the situation when the transducer is at its fundamental frequency, i.e., $k(a+b)/2=\pi$. Then Eq. (24) reduces to

$$R = \frac{4h_{33}^2 N^2 v}{\omega^2 C_{33}^D A_{12}}\left[\cos\left(\frac{1}{2}k(a-b)\right)+1\right]^2. \quad (27)$$

If $a=b$, we have

$$R = \frac{16h_{33}^2 N^2 v}{\omega^2 C_{33}^D A_{12}}, \quad (28)$$

but if $b=2a$, we have $\left(\frac{1}{2}ka=\pi/3\right)$

$$R = \frac{9h_{33}^2 N^2 v}{\omega^2 C_{33}^D A_{12}}. \quad (29)$$

Thus we can see that inequalities in the thicknesses of the positive and negative domains greatly decrease the resistance.

In the limiting case, i.e., when $N=1$ and $b=0$ (or $a=0$), Eq. (24) reduces to the normal formula for transducers made of single-domain LiNbO$_3$ crystals.[20]

If the normal of the laminar domain walls does not coincide with the z axis, but lies in the yz plane, and the electrodes of the transducer are still made parallel to the lamellae of the ASL, then one quasilongitudinal wave (QLW) and one quasishear wave (QSW) will be generated by the transducer.[23] In such a case, theoretical analysis has shown (see Appendix) that the electrical impedance Z consists of two parts, one for the QLW and the other for the QSW. The expression for the resonance frequency, Eq. (18), still holds except for the substitution of v_L (or v_S) for v, i.e.,

$$f_n^{L(S)} = nv_{L(S)}/(a+b), \quad n = 1, 2, 3, \ldots \tag{30}$$

where $v_{L(S)}$ is the sound velocity of the QLW (QSW).

Under a more general condition, i.e., the normal of the walls is along an arbitrary direction, there will be three waves excited, one QLW and two QSWs. This situation will not be treated here.

3. Experimental Results

3.1 Preparation of the ASL

The ASL crystal can be fabricated from a congruent melt of $LiNbO_3$ doped with about 0.3 wt % yttrium, either by means of the so-called "current-modulated method"[24] or by means of rotational striations.[4,5,25] For the latter method, we have purposely displaced the rotational axis of the growing crystal from the symmetry axis of the temperature field. In doing so, periodic fluctuations of temperature, growth rate, effective distribution coefficient, and solute concentrations within the crystal are produced during Czochralski growth. The interior rotational striations are compositional variations induced by the growth-rate fluctuations. The solute in the crystal is generally ionized but not completely shielded, especially when the temperature is decreasing and passing through the ferroelectric phase transition. Thus, a nonuniform solute distribution is equivalent to a nonuniform space-charge distribution in the crystal and a nonuniform internal field is produced in it, which can induce the ions of lithium and niobium within the lattice to displace preferentially at a temperature close to the Curie point, thus forming the ASL crystal.[4,6] Figure 1 (e) shows an optical photomicrograph of the ASL revealed by etching.

3.2 Electrical Measurements for Resonators

The ASL crystals were grown along the z axis. A set of resonators were made ranging from 500 MHz to 1000 MHz. The domain walls within the resonators are perpendicular to the z axis and the electrodes of the resonators are the xy planes parallel to the domain wall. Under the action of an external periodic electric field an acoustic wave is excited inside the resonator through the piezoelectric effect and it will form a stationary wave when Eq. (18) or Eq. (19) is satisfied.

With a HP 8505 network analyzer we have measured the reflection coefficients of the resonators. Its principle is as follows. If the impedance of a load is not equal to that of the electrical measurement system, the electrical energy will be reflected by the load. When a resonator is in oscillation, its impedance will vary with the frequency. In the vicinity of resonance, the impedance of a resonator changes greatly. This will be reflected in the reflection coefficient. Figure 4 shows the result of resonator no. 5. In Fig. 4 curve A represents the frequency dependence of the phase, curve B the frequency dependence of the magnitude. Near the resonance frequency of 552 MHz, both the phase and the magnitude of the reflection coefficient vary markedly. It can be seen that there is only one MR in the range of 0 – 1300 MHz, in agreement with the theory. The resonance frequency is 552 MHz, close to the theoretical value of 554 MHz calculated from Eq. (18) with $a+b=13.2$ μm and $v=7320$ m/s for resonator no. 5. The measured and calculated resonance frequencies are listed in Table 1. It may be noted that the experimental values are very close to the theoretical ones.

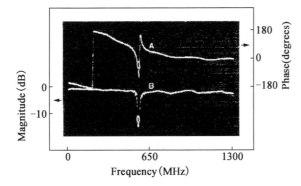

FIG. 4. Reflection coefficient of resonator no. 5 in rectangular coordinates. The horizontal scale is frequency centered at 650 MHz with the scan width $f=1300$ MHz; curve A for phase and curve B for magnitude.

Figure 5 shows a series of satellitelike resonances located on both sides of the main resonance. To our knowledge, this phenomenon has not been previously observed in resonators. Note that the thicknesses of the positive and negative domains are not equal in the sample. Table 2 shows that the measured frequency difference between any two adjacent SLRs, and $\overline{\Delta f_s}$ is in good agreement with the theoretical value calculated from Eq. (20).

3.3 Electrical Measurements for Transducers

Transducers were made from the ASL. Here the growth direction of the crystal deviates from the z axis and lies in the yz plane; therefore the normals of the laminar domain walls of the ASL are not parallel to the z axis. The angles between the normals and the z axis are shown in Table 3. The two electrodes are both parallel to the lamellae. Each

TABLE 1. Relationship between the MR frequency f^1_{main} and the periodicity of ASL resonators.

Resonator	Thickness of positive domain a (μm)	Thickness of negative domain b (μm)	Periodicity of ASL $a+b$ (μm)	Frequency of MAR f^1_{main} (MHz)	
				Calc.	Meas.
No. 1	2.8	4.6	7.4	989	975
No. 2	2.0	6.2	8.2	892	882
No. 3	4.0	6.0	10.0	732	710
No. 4	5.4	5.6	11.0	665	686
No. 5	6.2	7.0	13.2	554	552
No. 6	5.7	7.6	13.3	550	553

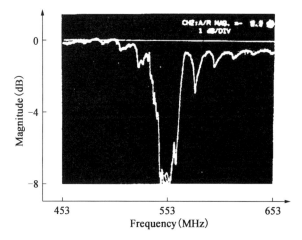

FIG. 5. Magnitude of the reflection coefficient of resonator no. 6 in rectangular coordinates. The horizontal scale is frequency centered at 553 MHz with the scan width $f=200$ MHz.

TABLE 2. Frequency difference between any two adjacent SLRs for ASL resonator no. 6, $\Delta f_s = f_s^{m+2} - f_s^m$.

	m	1	3	5	7	9
Δf_s (MHz) (measured values)	f_s^m	560.3	577.9	595.1	612.4	629.6
	Δf_s		17.6	17.2	17.3	17.2
	$\overline{\Delta f_s}$			17.3		
$\Delta f_s = f^1_{main}/N$ (theoretical values) (MHz)	N			32		
	$a+b$			13.3 μm		
	v			7320 m/s		
	f^1_{main}			550		
	Δf_s			17.2		

ASL is indium bonded to the xy plane of a single-domain LiNbO$_3$ plate separately. The plate is used as a transmission medium. Its faces were polished parallel to each other within several minutes and flat to about one-tenth of the wavelength of an optical wave used for examining the plate. The thickness of the plate is 3.0 mm and the dimensions of the rectangular electrodes are about 5.0 mm^2.

In order to evaluate the performance of the transducers, both frequency-domain and time-domain measurements were made.

In frequency-domain measurements, a HP 8505 network analyzer was used to measure the reflection coefficient of the transducers. A typical curve is shown on the upper right-hand-side corner in Fig. 6. Two resonance frequencies are observed—one for the QLW and the other for the QSW with frequencies of 555 MHz and 279 MHz, respectively. Theoretical values, calculated from Eq. (30) with $a+b=13.2$ μm, $v_L=7300$ m/s, and $v_S=3600$ m/s,[26] are 553 MHz for the QLW and 273 MHz for the QSW. Table 3 lists the measured and calculated resonance frequencies. The theoretical and experimental results show satisfactory agreement.

TABLE 3. Comparison between theoretical predictions and experimental results for ASL transducers.

Transducer	Periodicity of super lattice $a+b$ (μm)	Number of periods N	Angle between normal of domain boudaries and Z axis (degrees)	Resonance frequency of QLW (MHz)		Resonance frequency of QSW (MHz)	
				Calc.	Meas.	Calc.	Meas.
No. 1	13.2	15	5	553	555	273	279
No. 2	10.8	55	50	648	641	383	380
No. 3	9.6	62	30	750	764	415	418

As far as the transducer is concerned, the insertion loss is a very important parameter. According to Sittig,[27] the insertion loss (IL) of a transducer can be expressed as

$$IL = 2(ML + DL), \quad (31)$$

where ML is the "matching loss" and DL the "dissipation loss" that measures the acoustic power absorbed in the transducer due to internal dissipation from dielectric, sound absorption, and other losses. In transducers made of single-domain LiNbO$_3$ crystals, experiments showed that DL≪ML,[27] so that the transducer loss and its dependence on frequency is mainly determined by ML. In order to estimate the insertion loss, we neglect the dissipation loss, obtaining[27]

$$IL = -20\log(1-r^2), \quad (32)$$

where r is the magnitude of the reflection coefficient. Using Eq. (32), we have calculated the insertion loss versus frequency for transducer no. 1 from the measured value of r. The results are shown in Fig. 6. The insertion loss at 555 MHz is very low, in good agreement with the prediction. The 3 dB fractional bandwidth of transducer no. 1 is about 5.8%. The insertion loss curve shown in Fig. 6 is very similar to that of surface-wave interdigital transducers.[17] Both devices have a periodic structure.

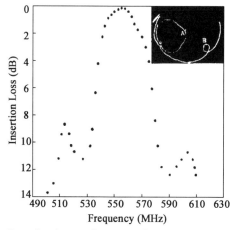

FIG. 6. Insertion loss vs frequency for transducer no. 1. Shown on the upper right-hand-side corner is its reflection coefficient in polar coordinates with the scan width $f = 1300$ MHz; A: QLW; B: QSW.

In time-domain measurements, the pulse-echo technique was used. The transducers were excited by a δ pulse and echoes were observed. For a lack of a sufficiently sensitive oscilloscope, only the echoes of the QSW for transducer no. 1 were detected. These are shown in Fig. 7. A detail of the first echo is shown on the upper right-hand-side corner in Fig. 7, labeled A. From its oscillation period $t = 3.66$ ns, the frequency of the excited ultrasonic wave was determined to be 273 MHz. This is consistent both with the theoretical resonance frequency and the frequency-domain measurements. The measured delay time is 1.7 μs, consistent with the theoretical value calculated from the relation:

$$\tau = 2d/v_s, \tag{33}$$

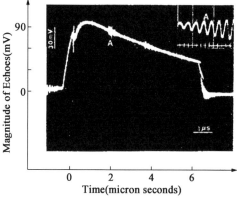

FIG. 7. Oscilloscope photograph of multiple echoes produced by transducer no. 1. The magnified view of the first echo is shown on the upper right-hand-side corner labeled A. Its vertical scale is 10 mV/div and the horizontal scale is 10 ns/div.

where d is the acoustic transit distance. For transducer no. 1, $d = 3.1$ mm. If the transmission medium is an unknown material, then we can use Eq. (33) to evaluate the sound velocity of that material.

4. Discussion

4.1 The effect of Echo Signals on the Reflection Coefficient

In the case of transducers, each ASL is indium bonded to the xy plane of a transmission medium made of a singledomain $LiNbO_3$ crystal separately. The generated acoustic wave will transmit into the transmission medium and propagate along the direction normal to the surface of the transmission medium. When it hits the free surface of the transmission medium (parallel to the bonded surface), the wave is reflected back and reverses its propagation direction. The acoustic impedance of the indium bond material is not matched to both the ASL and the transmission medium. Thus, the signal will propagate back and forth inside the transmission medium. Each time the signal hits the bond surface, part of it will be transmitted back to the ASL and be detected by the ASL. This signal will interfere with the original one and cause ripples in the curve of the reflection coefficient (see Fig. 8). The phase difference between the two is $2\omega d/v$, where d is approximately the thickness of the medium plus half of the thickness of the ASL and v is the sound velocity of $LiNbO_3$. At certain frequencies, when the condition:

$$2\omega d/v = 2n\pi \tag{34}$$

is satisfied, the ripple will be at its maximum. The difference between any two adjacent maxima is

$$\Delta f_1 = v/2d. \tag{35}$$

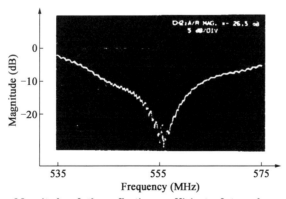

FIG. 8. Magnitude of the reflection coefficient of transducer no. 1 in rectangular coordinates. The horizontal scale is frequency centered at 555 MHz with the scan width $f = 40$ MHz. The frequency difference between any two adjacent ripples is 0.58 MHz.

In some cases, if the transmitted signal is strong enough, not only the first echo but also the second echo may interfere with the original one. By the same procedure, we obtain
$$\Delta f_2 = v/4d. \tag{36}$$

Figure 8 exhibits our experimental results from the transducer. The measured frequency difference Δf_{meas} is
$$\Delta f_{\text{meas}} = 0.58 \text{ MHz}. \tag{37}$$

In our case, we take $v = 7300$ m/s and $d = 3.1$ mm in Eq. (36), and the results agree well with the experiments.

From Fig. 8 we can see that the ripples can be divided into two parts: a strong one alternating with a weaker one. Here we believe two interfering echo signals have been detected.

4.2 The Periodicity of the ASL

As mentioned before, the ASL can be viewed as composed of a series of sound δ sources. When an electrical impulse is applied, each individual sound source will launch a sound δ pulse. These sound pulses will be successively transmitted into the transmission medium. The time interval between any two sound pulses is determined by the spatial separation between sound sources. The time interval and the spatial separation differ from each other by a proportionality factor v, the sound velocity. Thus from $v_S = 3600$ m/s and $t = 3.66$ ns, determined in Sec. 3, the periodicity of the ASL can be estimated, which is 13.2 μm, the same as the value measured by optical microscopy.

4.3 Power Flow Density in the Transmission Medium

The transducer discussed here can be used in many ways, e.g., acoustic delay lines, acousto-optic deflectors, acousto-optic modulators, etc. In applications of acousto-optic devices, the intensity of the diffracted light is directly proportional to the acoustic power. Therefore it is worth-while discussing the power flow density emitted by the ASL.

First we must derive the acoustic Poynting vector that, in our case, is
$$\boldsymbol{P} = -\boldsymbol{v} \cdot \boldsymbol{T} = -v_3 T_3 \hat{k}, \tag{38}$$
where v_3 is the velocity of the particle in vibration, $\hat{k}$ is a unit vector along the z axis, and
$$T_3 = -\frac{1}{2} jk e^{j(\omega t - kz)} \int h_{33}(z') D_3(z') e^{jkz'} dz' = -2h_{33} D_3 \frac{\sin\left[\frac{1}{2} kN(a+b)\right]}{\sin\left[\frac{1}{2} k(a+b)\right]} \times$$
$$\left[\left(\cos\frac{1}{2}k(a-b) - \cos\frac{1}{2}k(a+b)\right) \cos\frac{1}{2}kN(a+b) - \sin\frac{1}{2}k(a-b)\sin\frac{1}{2}kN(a+b)\right] e^{j(\omega t - kz)}. \tag{39}$$

In obtaining Eq. (39), both real sources and image sources are included.

v_3 can be obtained from

$$\frac{\partial T_3}{\partial z} = \rho \frac{\partial^2 U_3}{\partial t^2} = \rho \frac{\partial v_3}{\partial t}, \qquad (40)$$

which is

$$v_3 = -(v/C_{33}^D) T_3. \qquad (41)$$

Neglecting the exponential term, we have

$$\boldsymbol{p} = \hat{k} \frac{4h_{33}^2 D_3^2 v}{C_{33}^D} \left| \frac{\sin\frac{1}{2}kN(a+b)}{\sin\frac{1}{2}k(a+b)} \right|^2 \times \left[\left(\cos\frac{1}{2}k(a-b) - \cos\frac{1}{2}k(a+b) \right) \times \right.$$

$$\left. \cos\frac{1}{2}kN(a+b) - \sin\frac{1}{2}k(a-b)\sin\frac{1}{2}kN(a+b) \right]^2. \qquad (42)$$

The power flow density is then expressed as

$$P = \boldsymbol{p} \cdot \hat{k}. \qquad (43)$$

In order to gain an insight into the merit of the ASL, let us consider a special case of $a = b$. Equation (43) becomes

$$P = \frac{4h_{33}^2 D_3^2 v}{C_{33}^D} \left| \frac{\sin(kNa)}{\cos\frac{1}{2}ka} \right|^2 \sin^2\left(\frac{1}{2}ka\right) \cos^2 kNa. \qquad (44)$$

At resonance, i.e., $\frac{1}{2}ka = \pi/2$,

$$P = 4(2N)^2 (h_{33}^2 D_3^2 v/C_{33}^D). \qquad (45)$$

If the ASL degenerates into only one domain, that is, $N=1$ and $b=0$, then we have

$$P = 4(h_{33}^2 D_3^2 v/C_{33}^D). \qquad (46)$$

Comparing Eq. (46) with Eq. (45), we find that the power flow density of the ASL is proportional to N^2 (N is the number of domains of the ASL). This shows that with an ASL transducer, the power flow density in the transmission medium can be greatly increased. This is very useful in practical applications.

5. Summary

We have analyzed theoretically the ultrasonic excitation and propagation in the ASL by the piezoelectric effect and deduced the expressions for electrical impedances of both resonators and transducers. The resonance conditions show that there are two kinds of resonances existing in the ASL. One is the main resonance, which is determined by the period of the ASL. This is very useful for acoustic devices working at hundreds of megahertz to several gigahertz. The other one, the satellitelike resonance frequency, is related to the total thickness of the ASL. This phenomenon is similar to the satellite in x-ray or electron diffraction due to the existence of a superstructure. By an adequate choice of the electrode area and the number of domains, the radiation resistance or the real part of the impedance of the ASL transducer can be made equal to or even larger than its reactance. The transducers thus fabricated will be very efficient.

Experimentally, we have prepared a set of resonators and transducers. Electrical measurements verify our predictions. Resonance frequencies up to 1000 MHz have been detected in the ASL with a period of several micrometers. The satellitelike resonance phenomenon has been observed for the first time in acoustics. This phenomenon is not shared by transducers because of a lack of a longer periodicity. Transducers with an insertion loss close to 0 dB at 555 MHz were made.

Discussions have been made about some interesting problems. The acoustic power emitted by the ASL into the transmission medium is directly proportional to the square of the number of domains, which is favorable in many practical applications. Owing to the fact that the two faces of the transmission medium are parallel, the reflected signals interfere with the main signal causing reflection ripple; therefore, the electrical characteristics of the transducer are affected. In practice, this should be avoided by either tilting one face relative to the other or coating some sound-absorbing material on the free surface of the medium. In short, there remains much to be done in the future.

6. Appendix

Consider an acoustic wave propagating along a direction lying in the yz plane of a LiNbO$_3$ crystal. This problem may be approached by transforming the constitutive matrices to the rotated coordinate system. We assume that the propagating direction is $\hat{z}'$, which is a unit vector along the normal of domain walls of the ASL, and that the angle between the normal and the z axis is θ. Then the piezoelectric equations to be used here are

$$T'_3(\theta) = C'_{33}(\theta)S'_3 + C'_{34}(\theta)S'_4 - h'_{33}(\theta)D'_3, \tag{A1}$$

$$T'_4(\theta) = C'_{34}(\theta)S'_3 + C'_{44}(\theta)S'_4 - h'_{34}(\theta)D'_3, \tag{A2}$$

$$E'_3(\theta) = -h'_{33}(\theta)S'_3 - h'_{34}(\theta)S'_4 + [\varepsilon'_{33}(\theta)]^{-1}D'_3, \tag{A3}$$

where

$$C'_{33}(\theta) = n^4 C^D_{11} + 2m^2 n^2 (C^D_{13} + 2C^D_{44}) + 4mn^3 C^D_{14} + m^4 C^D_{33},$$
$$C'_{34}(\theta) = -mn^3 C^D_{11} - mn(m^2 - n^2)(C^D_{13} + 2C^D_{44}) - n^2(3m^2 - n^2)C^D_{14} + m^3 n C^D_{33},$$
$$C'_{44}(\theta) = m^2 n^2 (C^D_{11} - 2C^D_{13} + C^D_{33}) + 2mn(m^2 - n^2)C^D_{14} + (m^2 - n^2)^2 C^D_{44},$$
$$h'_{33}(\theta) = mn^2(h_{31} + 2h_{15}) - n^3 h_{22} + m^3 h_{33},$$
$$h'_{34}(\theta) = m^2 n(h_{33} - h_{31}) - n(m^2 - n^2)h_{15} + mn^2 h_{22},$$
$$[\varepsilon'_{33}(\theta)]^{-1} = n^2 (\varepsilon^s_{11})^{-1} + m^2 (\varepsilon^s_{33})^{-1},$$
$$m = \cos\theta, \quad n = \sin\theta.$$

For simplicity, we derive the electrical impedance for an ASL with either of its faces fully matched to a semi-infinite transmission medium separately and $a = b$. Using the Green's function method we obtain

$$Z = R - jX,$$

$$R = \frac{2M_1 v_1}{\omega^2 A'_{12}} \tan^2\left(\frac{1}{2}k_1 a\right) \sin^2 k_1 Na + \frac{2M_2 v_2}{\omega^2 A'_{12}} \tan^2\left(\frac{1}{2}k_2 a\right) \sin^2 k_2 Na,$$

$$X = -\left[\frac{4NM_1 v_1}{\omega^2 A'_{12}} \tan\frac{1}{2}k_1 a + \frac{M_1 v_1}{\omega^2 A'_{12}} \tan^2\left(\frac{1}{2}k_1 a\right) \sin(2k_1 Na) + \frac{4NM_2 v_2}{\omega^2 A'_{12}} \tan\frac{1}{2}k_2 a + \right.$$

$$\left. \frac{M_2 v_2}{\omega^2 A'_{12}} \tan^2\left(\frac{1}{2}k_2 a\right) \sin(2k_2 Na)\right] + \frac{1}{\omega A'_{12}} \Bigg(M_1 d + M_2 d - \frac{C'_{44}(\theta) h'_{33}(\theta) - C'_{34}(\theta) h'_{34}(\theta)}{C'_{33}(\theta) C'_{44}(\theta) - C'^2_{34}(\theta)} \times$$

$$h'_{33}(\theta) d - \frac{C'_{33}(\theta) h'_{34}(\theta) - C'_{34}(\theta) h'_{33}(\theta)}{C'_{33}(\theta) C'_{44}(\theta) - C'^2_{34}(\theta)} h'_{34}(\theta) d \Bigg) + \frac{d}{\omega \varepsilon'_{33}(\theta) A'_{12}}, \qquad (A4)$$

where $d = 2Na$, A'_{12} is the electrode area,

$$v_1^2 = \{C'_{33}(\theta) + C'_{44}(\theta) - \sqrt{[C'_{33}(\theta) - C'_{44}(\theta)]^2 + 4C'^2_{34}(\theta)}\}/2\rho,$$

$$v_2^2 = \{C'_{33}(\theta) + C'_{44}(\theta) + \sqrt{[C'_{33}(\theta) - C'_{44}(\theta)]^2 + 4C'^2_{34}(\theta)}\}/2\rho,$$

$$M_1 = \left(h'_{33}(\theta) - \frac{C'_{33}(\theta) - \rho v_1^2}{C'_{34}(\theta)} h'_{34}(\theta)\right) \times$$

$$\frac{[C'_{33}(\theta) C'_{44}(\theta) - C'^2_{34}(\theta)] h'_{33}(\theta) - [C'_{44}(\theta) h'_{33}(\theta) - C'_{34}(\theta) h'_{34}(\theta)] \rho v_2^2}{\rho (v_1^2 - v_2^2)[C'_{33}(\theta) C'_{44}(\theta) - C'^2_{34}(\theta)]},$$

$$M_2 = \left(h'_{33}(\theta) - \frac{C'_{33}(\theta) - \rho v_2^2}{C'_{34}(\theta)} h'_{34}(\theta)\right) \times$$

$$\frac{-[C'_{33}(\theta) C'_{44}(\theta) - C'^2_{34}(\theta)] h'_{33}(\theta) + [C'_{44}(\theta) h'_{33}(\theta) - C'_{34}(\theta) h'_{34}(\theta)] \rho v_1^2}{\rho (v_1^2 - v_2^2)[C'_{33}(\theta) C'_{44}(\theta) - C'^2_{34}(\theta)]}.$$

References and Notes

[1] L. Esaki and R. Tsu, IBM J. Res. Dev. **14**, 686 (1970).
[2] L. L. Chang, in *Synthetic Modulated Structures*, edited by B. C. Giessen (Academic, New York, 1985).
[3] I. K. Schuller, Phys. Rev. Lett. **44**, 1597 (1980).
[4] N. B. Ming, J. F. Hong, and D. Feng, J. Mater. Sci. **17**, 1663 (1982).
[5] N. B. Ming, J. F. Hong, and D. Feng, Acta Phys. Sin. **31**, 104 (1982).
[6] J. Chen, Q. Zhou, J. F. Hong, W. S. Wang, N. B. Ming, and D. Feng, J. Appl. Phys. **66**, 336 (1989).
[7] D. Feng, N. B. Ming, J. F. Hong, J. S. Zhu, Z. Yang, and Y. N. Wang, Appl. Phys. Lett. **37**, 607 (1980).
[8] Y. H. Xue, N. B. Ming, J. S. Zhu, and D. Feng, Chin. Phys. **4**, 554 (1984).
[9] J. Feng, Y. Y. Zhu, and N. B. Ming, Phys. Rev. B **41**, 5578 (1990).
[10] Y. Y. Zhu and N. B. Ming, Phys. Rev. B **42**, 3676 (1990).
[11] Y. Y. Zhu, N. B. Ming, W. H. Jiang, and Y. A. Shui, Appl. Phys. Lett. **53**, 1381 (1988).
[12] Y. Y. Zhu, N. B. Ming, W. H. Jiang, and Y. 4. Shui, Appl. Phys. Lett. 53, 2278 (1988).
[13] Y. Y. Zhu, N. B. Ming, and W. H. Jiang, Phys. Rev. B **40**, 8536 (1989).
[14] S. He and J. D. Maynard, Phys. Rev. Lett. **62**, 1888 (1989).
[15] J. P. Desideri, L. Macon, and D. Sornette, Phys. Rev. Lett. **63**, 390 (1989).
[16] S. He and J. D. Maynard, Phys. Rev. Lett. **57**, 3171 (1986).
[17] R. F. Mitchell, Philips Res. Suppl. **3**, 1 (1972).

[18] Batcher *et al.*, Proc. IRE **37**, 1378 (1949).

[19] H. E. Bommel and K. Dransfeld, Phys. Rev. **117**, 1245 (1960).

[20] D. A. Berlincourt, D. R. Curran, and H. Jaffe, in *Physical Acoustics*, edited by W. P. Mason (Academic, New York, 1964), Vol. 1A, p. 169.

[21] W. R. Smith, H. M. Gerard, J. H. Collins, T. M. Reeder, and H. J. Shaw, IEEE Trans. Microwave Theory Tech. **MTT-17**, 856 (1969).

[22] A. H. Meitzler and E. K. Sittig, J. Appl. Phys. **40**, 4341 (1969).

[23] B. A. Auld, *Acoustic Fields and Waves in Solids* (Wiley-Interscience, New York, 1973), Vol. 1, p. 272.

[24] J. F. Hong and Y. S. Yang, Acta Opt. Sin. (in Chinese) **4**, 821 (1984).

[25] J. F. Hong, Z. M. Sun, Y. S. Yang, and N. B. Ming, Physics (in Chinese) **9**, 5 (1980).

[26] J. DeKlerk, in *Proceedings of the International School of Physics "Enrico Fermi"*, *Vol. 63: A Physical Approach to Elastic Surface Waves*, edited by D. Sette (North-Holland, Amsterdam, 1976), p. 437.

[27] E. K. Sittig, in *Progress in Optics Vol. 10: Elasto-optic Light Modulation and Deflection*, edited by E. Wolf (North-Holland, Amsterdam, 1972), p. 231.

[28] We are grateful to professor Y. A. Shui and W. H. Jiang for valuable discussions and technical assistance. This research was supported by a grant for key research project from the State Science and Technology Commission of China.

High-frequency Resonance in Acoustic Superlattice of Periodically Poled LiTaO$_3$[*]

Yan-Feng Chen, Shi-Ning Zhu, Yong-Yuan Zhu, and Nai-Ben Ming

*National Laboratory of Solid-State Microstructures, Nanjing University,
Nanjing 210093, People's Republic of China*

Biao-Bing Jin and Ri-Xing Wu

Center of Superconductor Electronics, Nanjing University, Nanjing 210093, People's Republic of China

An electric poling method has been used to prepare microstructured LiTaO$_3$ crystals with periodically inverted-ferroelectric domains. By using these crystals as acoustic superlattices, both an "in-line" scheme and a "cross-field" scheme for acoustic excitation have been realized. The experimental results are in good agreement with the theoretical analysis. It is expected that these results may be applied to a bulk-acoustic device operating at a frequency high above 450 MHz.

Since Esaki and Tsu proposed the idea of a semiconductor superlattice[1] with periodicity on the order of a nanometer, comparable with the wavelength of a de Brogie wave, we have witnessed tremendous research activity on the semiconductor superlattice. The concept of the superlattice has been extended to other materials, such as metals[2] and dielectrics.[3]

Recently we have proposed a new type of superlattice[4] that is made of a ferroelectric single crystal with 180° domains arranged periodically with a periodicity in the range of either visible light or ultrasonic waves. We called it an acoustic superlattice (ASL) for the ultrasonic case[4] and an optical superlattice (OSL) for the visible light case.[5] We have studied systematically the excitation and propagation of acoustic waves in the ASL, and demonstrated resonators and transducers operating at frequencies high above 500 MHz, using periodically inverted-domain LiNbO$_3$ crystals grown by the Czochralski method.[6] Recently much progress has been made in the techniques used to prepare microstructured ferroelectric crystals. Among them, the electrically poling approach has attracted much attention due to its convenience and accuracy in manipulating domain formation.[7—10] In this way, the periodic ferroelectric domain can be "written" in the crystal by applying a pulsed electric field through a periodic electrode which is patterned on the surface of the crystal by the standard microelectronic process. Using microstructured crystals produced by this method, such as LiNbO$_3$ and LiTaO$_3$ (LT), a quasi-phase-matched second

---

[*] Appl.Phys.Lett.,1997,70(5):592

harmonic generation has been realized.[7–10] In this letter, we first used the periodically poled LiTaO$_3$ to build an ASL, and realized "in-line" and "cross-field" acoustic excitation schemes with resonant frequencies high above 450 MHz, revealing interesting properties for device applications.

LT is a well-known ferroelectric crystal which possesses superior piezoelectric properties: a high electromechanical coupling coefficient, a low transmission loss, and a high chemical stability, suitable for the microwave acoustic applications. We used a c-cut LT single domain slab, 0.5 mm thick, as the sample. Both c faces of the sample were polished, upon which 0.1-μm-thick Al films were deposited as electrodes. On the $+c$ face of the sample there was a plane electrode, and on the $-c$ face there was a periodic electrode fabricated by a standard photolithography and wet-etching technique. The width of the "grating strip" of the periodic electrode was about 0.3 – 0.35 of the period. The sample was connected to a high-voltage pulse generator. The pulse-field treatment was carried out at room temperature. Applying a field strength of 22 kV/mm and 1 : 3 duty cycle, we successfully fabricated a 0.5-mm-thick LT sample with a period of $\Lambda = 8.5$ μm. After being immersed in a mixture of one part HF and two parts HNO$_3$ (by volume) for 10 min at 100 ℃, the sample shows the reversed domain structures (Fig. 1). The details were published in a previous report.[8]

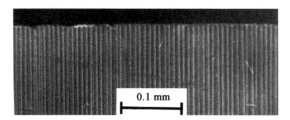

FIG. 1. The cross section (y face) of the periodic domain structured in a 0.5-mm-thick poled LT sample with an 8.5 μm period.

For a given ASL, acoustic waves can be excited through two different schemes. One is an in-line scheme with the acoustic propagation vector parallel to the applied electric field. The other is a cross-field scheme, which is characterized by an electric field perpendicular to the propagation vector. The former has been treated in detail both theoretically and experimentally, while the latter has recently been analyzed theoretically.[4] Figure 2 is a schematic diagram used in this study. In this case, the domain boundaries are parallel to the y-z plane, and the domains are arranged periodically along the x axis. The directions of their spontaneous polarization are parallel to the z axis; up arrows correspond to the positive domains and down arrows represent the negative domains. Table 1 lists the details of the samples we prepared. The acoustic properties of those superlattices were investigated with an Hp8205 network analyzer. Figures 3 and 4 show two typical experimental results, and Table 2 lists the measurement results of a set of resonators made of periodically poled LT.

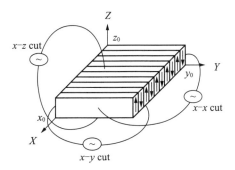

FIG. 2. A schematic diagram of the ultrasonic excitation in an acoustic superlattice made of a microstructured LT in which the domains are arranged periodically along the x axis, and the spontaneous polarization directions of this domain are parallel to the z axis. The excitation by electrodes coated on the x faces (x-x cut) corresponds to the in-line scheme, while electrodes coated on the y faces (x-y cut) and the z faces (x-z cut) excite acoustic waves by the cross-field scheme.

TABLE 1. The samples used in the experiment.

Sample	Area of electrode (mm^2)	Sample's geometry (mm)			The scheme of excitation
		x_0	y_0	z_0	
x-x cut	4×0.5	1	4	0.5	Cross field
x-y cut	4×0.5	4	0.4	0.5	In line
x-z cut	5×3	5	3	0.5	In line

According to the theory of ASL deduced by the Green function method,[4,11] the resonant frequency of an ASL is determined by the period of domain modulation instead of the total thickness, i.e., $f=v/(a+b)$, where v is the velocity of the acoustic wave, a, b is the thickness of the + domain and the − domain, respectively, and, in this study, $a=b$. Two shear waves (S1 and S2) are excited in the x-x cut ASL, and a longitudinal wave (L) is excited in the x-y cut or the x-z cut ASLs. In the above three cases, all the waves excited propagate along the x axis. By using the Equation in Ref. [4] and the data of LT in Refs. [12] and [13], theoretical resonant frequencies are calculated, as shown in Table 2. As expected, two resonances, 391 MHz(S2) and 466 MHz(S1), were detected in the x-x cut sample (Fig. 3). In order to reveal clearly the resonance at 391 MHz, marked by A, the differentiation of the reflection coefficient (S11) is shown in the inset in Fig. 3. The cross-field scheme was carried out in the x-y cut (Fig. 4) and x-z cut ASL. In Fig. 4, a resonance of 660 MHz was detected with some ripples appearing outside the bandpass. This result is similar to acoustic surface wave device characteristics,[14] indicating that these ripples may result from an acoustic wave propagating along the surface of the x-y cut ASL which will be discussed elsewhere. It is worth noting that the calculated value agrees very well with the measured results. In the x-z cut ASL, the measured reflection coefficients at the resonant frequencies are much smaller ($-0.3 \sim -0.7$ dB) than those in the x-x cut and the x-y cut (-2.5 dB). The reason is that the electromechanical coupling coefficient (k) of the x-z cut sample is almost an order smaller than that of the x-y cut and x-z cut (Table 2). The same reason can be used to explain the resonance (S2) in the x-x cut sample described above. The full width at half maximum (FWHM) of the sample is 14 MHz in the x-x cut, 8 MHz in the x-y cut, and 2 MHz in the x-z cut.

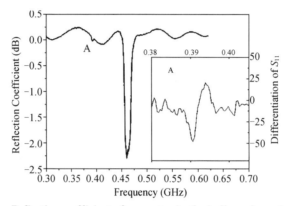

FIG. 3. Reflection coefficient of resonator in the in-line scheme in the x-x cut sample is in rectangular coordinates. The horizontal scale is frequency centered at 600 with the scan width of 300 MHz. The inset on the right (marked by A) is the differentiated $S11$, indicating an obvious resonance at 391 MHz.

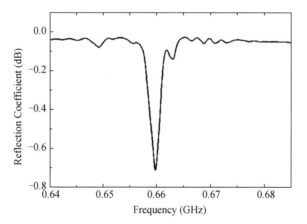

FIG. 4. In an x-y cut ASL, an acoustic wave operating at 660 MHz was excited by the cross-field scheme.

TABLE 2. Comparison between theoretical and experimental results for an ASL resonator.

Sample	Mode	Period of ASL(N)	Velocity (10^3 m/s)	Resonance cal. (MHz)	Frequency meas. (MHz)	k
x-x cut	S1	118	4.16	489	466	0.44
	S2	118	3.30	395	391	0.02
x-y cut	L	470	5.65	652	660	0.17
x-z cut	L	588	5.65	652	661	0.01

The special excitation schemes described here are of important interest to device applications. With the cross-field scheme (x-y cut and x-y cut), acoustic transducers can be attached to a transmission medium, which plays a detrimental role in the acousto-optic device, without an electrode inserted between them. This work also demonstrates that the electrically poling technique allows one to fabricate artificially different kinds of patterns, such as a chirped transducer and an apodized stagger-tuned comb (as a transducer or a filter).[15] These are important in radar and communication systems and in the past have only been realized in surface acoustic wave (SAW) devices. Furthermore, the advantage of the microstructured crystal is that it can be used at a higher frequency than SAW with the same modulation wave-length due to its higher acoustic velocity (LT: 5650 m/s in x-x cut microstructured LT, and 3310 m/s in SAW). It should be noted that an operating frequency high above 1 GHz can be easily realized in such a device.

In summary, we first use an electrically poled ASL LT to fabricate a set of acoustic resonators operating at a frequency high above 450 MHz. With the ASL, some special excitation schemes have been realized. These developments offer the promise of building some artificial microstructures in single crystal bulk wave ultrasonic devices and provide exciting prospects for a bulk wave ultrasonic device.

References and Notes

[1] L. Esaki and R. Tsu, IBM J. Res. Dev. **14**, 686 (1970).

[2] L. L. Chang, in *Synthetic Modulated Structures*, edited by B. C. Giessen (Academic, New York, 1985).

[3] N. B. Ming, Progress in natural science **4**, 554 (1994); E. Wiener-Avnear, Appl. Phys. Lett. **65**, 1784 (1994); H.-M Christd, L. A. Boatner, J. D. Budai, M. F. Chisholm, L. A. Gea, P. J. Marrero, and D. P. Norton, *ibid*. **68**, 1488 (1996).

[4] Y. Y. Zhu, N. B. Ming, J. Appl. Phys. **72**, 904 (1991); Y. Y. Zhu, S. N. Zhu, Y. Q. Qing, and N. B. Ming, *ibid*. **79**, 2221 (1996); Y. Y. Zhu, S. N. Zhu, and N. B. Ming, J. Phys. D **29**, 185 (1996).

[5] B. Xu and N. B. Ming, Phys. Rev. Lett. **71**, 1003 (1993); 71, 3959 (1993).

[6] N. B. Ming, J. F. Hong, and D. Feng, J. Mater. Sci. **17**, 1663 (1982); Y.Y. Zhu, N. B. Ming, W. H. Jiang, and Y. A. Shui, Appl. Phys. Lett. **53**, 2278 (1988); **53**, 1381 (1988).

[7] G. A. Magel, M. M. Fejer, and R. L. Byer, Appl. Phys. Lett. **56**, 108 (1990).

[8] S. N. Zhu, Y. Y. Zhu, Z. Y. Zhang, H. F. Wang, H. Shu, J. F. Hong, C. Z. Ge, and N. B. Ming, J. Appl. Phys. **77**, 5481 (1995); Appl. Phys. Lett. **67**, 320 (1995).

[9] H. Ito, C. Takyu, and H. Inaba, Electron. Lett. **27**, 1221 (1991).

[10] M. Yamada, N. Nada, M. Saitoh, and K. Watanabe, Appl. Phys. Lett. **62**, 435 (1993).

[11] R. F. Mitchell, Philips Res. Rep. Suppl. **3**,1 (1972).

[12] T. Yamada and N. Niizeki, Jpn. J. Appl. Phys. **6**, 151 (1967); **7**, 292 (1968); **8**, 1127 (1969).

[13] A. W. Warner, M. Onoe, and G. A. Coquin, J. Acoust. Soc. Am. **42**, 1223 (1967).

[14] M. G. Nolland and L. T. Claiborne, Proc. IEEE **62**, 582 (1974); *Piezoelectricity*, edited by C. Z. Rosen, B. V. Hiremath, and R. Newnham (AIP, New York, 1992), p. 469.

[15] R. F. Tancrell and R. C. Williamson, Appl. Phys. Lett. **19**, 456(1971).

[16] This work is supported by a grant for the Key Research Project in Climbing Program. The author is an acknowledged Ke-Li fellow.

Bulk Acoustic Wave Delay Line in Acoustic Superlattice[*]

Ruo-Cheng Yin, Si-Yuan Yu, Cheng He, Ming-Hui Lu, and Yan-Feng Chen

Department of Materials Science and Engineering and National Laboratory of Solid-State Microstructures, Nanjing University, Nanjing 210093, People's Republic of China

The bulk acoustic wave delay lines are demonstrated both theoretically and experimentally based on the acoustic superlattices. In one bulk acoustic wave delay line, two collinear acoustic superlattices are separately located, where the generation and detection of the bulk acoustic waves are executed, respectively. A basic bulk acoustic wave delay line working at 819 MHz with the delay time of 2.46 μs was realized. Furthermore, by employing two enantiomorphous aperiodic acoustic superlattices, a broadband dispersive bulk acoustic wave delay line with the center frequency at 295 MHz has also been fabricated and investigated.

Acoustic superlattice (ASL) consists of artificially patterned ferroelectric structures with different domain inversions, giving rise to exotic electromechanical coupling. It has attracted much attention in the aspects of acoustic excitation,[1—4] phononic polariton,[5—7] and other integrated acoustoelectronic devices and applications.[8,9] By using the electrical poling method,[10] the piezoelectric coefficient in the ASLs could be efficiently manipulated and modulated to realize tunable generation of bulk acoustic waves (BAWs). Among the past works, the feasibility and reliability of acoustic excitation in single ASLs, relying on the inverse piezoelectric effect, have been well demonstrated and developed.

On the other hand, the electric excitation and modulation induced by the generated acoustic waves in the ASL, via the piezoelectric effect, need to be further investigated. In this paper, we will focus on the modulation of electric signals in ASLs and demonstrate the feasibilities of ASLs to be efficient delay lines. Compared to the past studies of using BAWs to build up delay lines that consists of a thin single domain piezoelectric crystal layer deposited on an elastic medium,[11] sophisticated design of ASLs is able to provide more freedom to tailor the properties of the delay line by controlling the piezoelectric domain's arrangement and thus demonstrate more complex manipulation of electric signals.

Under a uniform external electric field, instead of interdigital transducers,[12] the piezoelectric domain walls are subjected to a regular strain due to the regular change in the piezoelectric coefficients, effectively acting as localized sound sources. The generated

---

[*] Appl.Phys.Lett.,2010,97(9):092905

acoustic wave is congruent and can be converted back to the electric signals with the piezoelectric effect by implementing a second ASL. Therefore, an electric delay line based on the BAWs is, in principle, achievable with integrating two separated collinear ASLs as follows: one is working as the launching transducer and the other is operating as the receiving transducer, respectively.

Figure 1 gives the schematic diagram of our design in which the ASL transducer could be either periodic [Fig. 1(b)] or varying gradually in domain periods [Fig. 1(c)] with electrically poled lithium tantalite ($LiTaO_3$). The ASLs are zx-cut, i.e., the ferroelectric domains have the antiparallel spontaneous polarizations arranging along the x axis, and the domain walls lie in the yz plane. A crossed time harmonic electric field E_2 is applied in the y direction to cover the entire launching ASL as an input electric signal. The BAWs excited by the launching ASL propagates along the x axis, as indicated by the arrow in Fig. 1. The corresponding piezoelectric equations are[6]

$$T_1 = C_{11}^E S_1 - e_{22}(x) E_2,$$
$$P_2 = -e_{22}(x) S_1 + \varepsilon_0 (\varepsilon_{11}^s - 1) E_2. \quad (1)$$

Here T_1, S_1, E_2, and P_2 are the stress, strain, electric field, and polarization, respectively. C_{11}^E, $e_{22}(x)$, and ε_{11}^s are the elastic, piezoelectric, and dielectric coefficients, respectively. $e_{22}(x)$ equals $+e_{22}$ in positive domains, whereas $-e_{22}$ in negative domains. The resonant frequency is given by $f = v/\lambda$, with $\lambda \equiv \Lambda$, where λ is the wavelength, Λ is the period of ASL, and v is the velocity of the BAW, respectively.

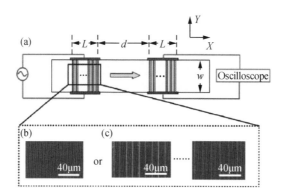

FIG. 1. (color online) (a) The schematic diagram of ASL-based delay lines utilizing propagating BAWs in ASLs. The ASLs could be either (b) periodically or (c) gradually varying periodically poled lithium tantalite. The arrow represents the propagation direction of acoustic wave.

Equation (1) can be applied to the receiving ASL as well but works inversely in contrast to the launching ASL, that is, when a longitudinal acoustic wave propagates through the receiving ASL, the positive domains' expansion combined with the negative ones' contraction causes the appearance of charges of the same sign (positive or negative) on the same side of the two different domains.[2-4] The entire receiving ASL is polarized synchronously as a consequence, the generated BAWs act as a delay line of the input electric signal in the whole process.

To demonstrate such a delay effect, we experimentally prepared the nondispersive delay lines using 500 μm thick zx-cut $LiTaO_3$ wafers (see Fig. 1). Two collinear periodic ASLs were fabricated on the wafer by the electrical poling method,[10] with the same length of $L=5$ mm, width of $w=3$ mm, and period of $\Lambda=7$ μm. Figure 1(b) shows the local topography of the etched ASL by scanning electronic microscopy (SEM). The smallest distance between two ASLs is $d=9.5$ mm. Another delay line with the similar parameters except $d=14$ mm was also fabricated to have a comparison.

We first measured the electrical response of a single ASL in the network analyzer. It is evident that a resonant frequency around 810 MHz was constructed as a strong reflection dip, as shown in Fig. 2(a), confirming the generation of BAWs. Aluminum was deposited onto the y facet over a length of 6 mm as electrodes, covering the two ASLs respectively [Fig. 1(a)]. On the launching ASL, two electrodes was ultrasonically welded to a 50 Ω impedance matching network, and a rf signal generator, respectively. While on the receiving ASL, one electrode was also ultrasonically welded to a 50 Ω impedance matching network and the other was accessed into the oscilloscope. The signal delay has been clearly detected on the receiving ASL as shown in Fig. 2(b) when a single pulse of sinusoidal signals at 819 MHz is applied on the launching ASL [see the curve on top in Fig. 2(b)]. The measured delays τ are 1.68 μs and 2.46 μs for two samples of $d=9.5$ mm and 14 mm, respectively, consistent with calculations of 1.68 μs and 2.47 μs with the sound velocity of 5650 m/s. Furthermore, since the propagation loss of BAWs in $LiTaO_3$ is negligible, the amplitude of the received signal barely changes from 160 mV to 158 mV with different lengths of delay lines.

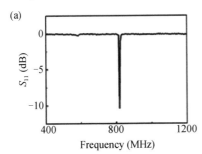
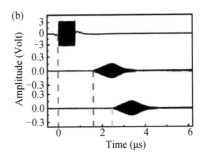

FIG. 2. (color online)(a)S_{11} vs f_r in a single ASL for the nondispersive delay line. (b) On top is the applied electric signal. Underneath are the measured delay times with different interval distances. The dashed lines indicate the different delay times.

The delay line above is limited by a narrow working regime in frequency since the perfect periodicity of ASLs. To implement a broadband operation, we constructed two aperiodic ASLs with Λ varying gradually, and the corresponding resonant frequency thus covers from v/Λ_1 to v/Λ_2. The concrete parameters are $\Lambda_1 = 17.5$ μm and $\Lambda_2 = 21$ μm, with $\Delta\Lambda = 0.01$ μm from the beginning to the end of ASLs. Figure 1(c) shows the SEM images of different regions in etched ASLs. Here $L = 9$ mm, $w = 3$ mm, and the launching and receiving ASLs are arranged enantiomorphously in the x direction with $d = 6$ mm. The measured reflection scattering parameter S_{11} exhibit a broadband excitation of BAWs [see Fig. 3(a)]. At a certain frequency, only the specific ferroelectric domains with the same periods interact, whereas other domains do not participate in the process. It is therefore expected to create a dispersive delay effect with shorter delays at higher frequencies and longer delays at lower frequencies.[13] As shown in Fig. 3(b), the horizontal distance between two discrete resonance curves is proportional to the delay time τ in frequency f, which can be expressed as

$$\tau = \left(\frac{d}{v} - \frac{\Lambda_1^2}{v \cdot \Delta\Lambda}\right) + \frac{v}{f^2 \cdot \Delta\Lambda}. \qquad (2)$$

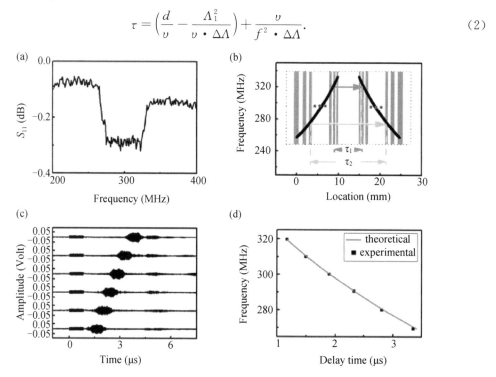

FIG. 3. (color online)(a) The dependence of S_{11} on frequencies. (b) Schematic diagram of the dispersive BAW delay line in ASL. The embedded frame (dotted gray lines) represents the ASLs, where the gray and white rectangles refer to the aperiodic arranged domains with opposite polarizations. The dotted black lines indicate the discrete relationships between the resonant frequencies and the locations. The arrows represent the propagation of BAWs with both ends pointing out the launching and receiving locations, and τ_1 (τ_2) is the corresponding delay, respectively. (c) The measured output signals from 270 MHz to 320 MHz with an interval of 10 MHz. (d) The relationship of frequencies with the measured and calculated delay times.

Equation (2) reveals the discrete relationship between the resonant frequencies and delay times. A similar setup like the nondispersive delay line is implemented to test the dispersive device. Figure 3(c) shows the measured output signals from 270 MHz to 320 MHz with an interval of 10 MHz. The weaker square pulses appearing at the beginning of the window are the radiation signals from the launching ASL. Extracted from Fig. 3(c), the duration of time delays in frequency is experimentally constructed in Fig. 3(d), showing excellent agreement with the theoretical calculations by Eq. (2). The compression ratio, approximately TB, is used to characterize the pulse compression, where T is the duration of delay time and B is the bandwidth. This aperiodic ASL-based delay line is centered at a frequency of 295 MHz with $B=50$ MHz and $T=2.2$ μs, so the TB product is 110, which means the ratio between the incident and output pulse length is about 110. At a certain resonant frequency, since the domains involved in resonant vibration is strongly limited by the local period of the aperiodic ASLs, the intensity of the output signal is weaker than that with uniform periodic ASLs.

Due to the dispersion relationship, the ASL-based delay line could be further applied to the pulse compression. The ideal case would be that we can utilize the entire frequency band to have the maximum pulse compression. However, we are limited by the characteristics of our function generator that the generated electric pulse only has the bandwidth of 8 MHz and the time duration of 0.4 μs with a center frequency of 277 MHz. As shown in Fig. 4, this broadband pulse was applied to the delay line and observed as a radiation square signal at the beginning, followed by the output compressed pulse with a very sharp waveform. The theoretical TB product is 3.2, and the measured result is approximately 3.0. It is evident that the input electrical pulse has been significantly compressed with much smaller pulse width. The rightmost sharp pulses result from multiple reflections of BAWs.

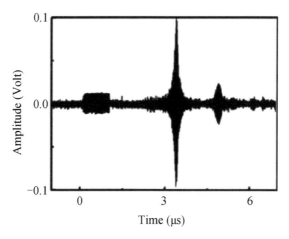

FIG. 4. The practical pulse compression with the center frequency of 277 MHz, the bandwidth of 8 MHz, and the time duration of 0.4 μs.

In conclusion, we have demonstrated the basic principle of ASL-based electric delay lines and fabricated and tested the delay lines on the zx-cut $LiTaO_3$ wafers in experiment. The sophisticated ASL engineering provides us great freedom to have the electric delay line operated at either a single resonant frequency (periodic ASLs) or a broad frequency band (aperiodic ASLs). Especially the aperiodic ASL-based delay line shows a nonlinear dispersion that has been further applied to pulse shaping. With the great freedom in the ASL design and the well-developed fabrication technique, more complicated and tunable ASL-based acoustoelectronic devices can be expected with more complex functionalities in pulse delay, pulse engineering, and signal processing.

References and Notes

[1] Y. Y. Zhu, N. B. Ming, W. H. Jiang, and Y. A. Shui, Appl. Phys. Lett. **53**, 1381 (1988).
[2] Y. Y. Zhu and N. B. Ming, J. Appl. Phys. **72**, 904 (1992).
[3] Y. Y. Zhu, S. N. Zhu, and Y. Q. Qin, J. Appl. Phys. **79**, 2221 (1996).
[4] Y. F. Chen, S. N. Zhu, Y. Y. Zhu, N. B. Ming, B. B. Jin, and R. X. Wu, Appl. Phys. Lett. **70**, 592 (1997).
[5] Y. Q. Lu, Y. Y. Zhu, Y. F. Chen, S. N. Zhu, N. B. Ming, and J. Feng, Science **284**, 1822 (1999).
[6] Y. Y. Zhu, X. J. Zhang, Y. Q. Lu, Y. F. Chen, S. N. Zhu, and N. B. Ming, Phys. Rev. Lett. **90**, 053903 (2003).
[7] C. P. Huang and Y. Y. Zhu, Phys. Rev. Lett. **94**, 117401 (2005).
[8] D. Yudistira, D. Janner, S. Benchabane, and V. Pruneri, Opt. Lett. **34**, 3205 (2009).
[9] H. Gnewuch, N. K. Zayer, C. N. Pannell, G. W. Ross, and P. G. R. Smith, Opt. Lett. **25**, 305 (2000).
[10] M. Yamada, N. Nada, and K. Watanabe, Appl. Phys. Lett. **62**, 435 (1993).
[11] D. Liufu and K. C. Kao, J. Vac. Sci. Technol. A **16**, 2360 (1998).
[12] R. M. White and F. W. Voltmer, Appl. Phys. Lett. **7**, 314 (1965).
[13] R. H. Tancrell, M. B. Schulz, H. H. Barrett, L. Davies, and M. G. Holland, Proc. IEEE **57**, 1211 (1969).
[14] The work was jointly supported by the National Basic Research Program of China (Grant No. 2007CB613202) and the National Nature Science Foundation of China (Grant No. 50632030). We also acknowledge the support the Nature Science Foundation of China (Grant No.10874080) and the Nature Science Foundation of Jiangsu Province (Grant No. BK2009007).

Negative Refraction of Acoustic Waves in Two-dimensional Sonic Crystals[*]

Liang Feng,[1] Xiao-Ping Liu,[1] Yan-Bin Chen,[1] Zhi-Peng Huang,[1] Yi-Wei Mao,[2]
Yan-Feng Chen,[1] Jian Zi,[3] and Yong-Yuan Zhu[1]

[1] *National Laboratory of Solid State Microstructures, Nanjing University,
Nanjing 210093, People's Republic of China*

[2] *Institute of Acoustics, Nanjing University, Nanjing 210093, People's Republic of China*

[3] *Surface Physics Laboratory, Fudan University, Shanghai 200433, People's Republic of China*

Negative refractions of acoustic waves in a two-dimensional (2D) sonic crystal were studied both experimentally and theoretically. By calculating the acoustic band structure and equifrequency surfaces, we theoretically analyzed the acoustic single-beam negative refraction in the first band operating in the ultrasonic regime. A 2D square sonic crystal was constructed with 2.0-mm-diam steel cylinders arranged as square arrays of 2.5 mm lattice constant. A scanning transmission measurement of spatial distribution was carried out to establish the acoustic refraction of a single Gaussian beam. It was demonstrated that the negative refraction is strongly dependent on both frequencies and incident angles, which shows great potential in acoustic devices.

Negative refractions of electromagnetic (EM) waves were theoretically predicted in the left-handed material (LHM) by Veselago in 1968.[1] Due to its unique physical properties and potential applications, the study of negative refractions has been a hot focus of interest recently. There are two approaches to realize EM negative refraction, plasmon resonances in metamaterials[2-7] and multiple scatterings in photonic crystals (PCs).[8-13]

In metamaterials, plasmon resonances of two-dimensional (2D) arrays consisting of split-ring resonators and wires can be used to fabricate LHM, which as a whole has both negative effective permittivity and negative permeability in a certain microwave range.[5-7] Based on the above prediction, a flat LHM slab with $\varepsilon=-1$ and $\mu=-1$ was proposed by Pendry to realize a so-called perfect lens[14] in which both propagating and evanescent waves contribute to the image, and some relevant phenomena have been experimentally demonstrated.[15]

In photonic crystals, the existence of negative refractions is due to intense multiple scatterings near the Brillouin-zone boundaries.[8-13] Similar to the perfect lens in LHM, negative refractions in PCs could also result in a flat lens, called superlens, whereby it is

---

[*] Phys.Rev.B, 2005, 72(3): 033108

possible to obtain the transmission amplitude for evanescent waves to produce a real image.[12,13,16] This phenomenon was observed experimentally in the microwave and infrared ranges,[17,18] showing great promise in photoelectronic applications.

Recently, negative refractions were also realized in other kinds of classical waves. Liquid surface waves in a periodic structure of copper cylinders were characterized by band structures resulted from multiple scatterings,[19] leading to negative refraction and a superlensing effect.[20] The same phenomena have been realized recently for acoustic waves.[21,22] In this paper, we studied, both experimentally and theoretically, the dependence of the negative refraction of acoustic waves on frequencies and incident angles in a 2D sonic crystal (SC).

In SCs, Lamé coefficients $\lambda(r)$, $\mu(r)$, and the (mass) density $\rho(r)$ are modulated periodically and acoustic waves can be described in terms of a band structure, as in the case of electrons and photons.[23-25] The wave equation can be written as[26]

$$\frac{\partial^2 u^i}{\partial^2 t} = \frac{1}{\rho}\left\{\frac{\partial}{\partial x_i}\left(\lambda \frac{\partial u^l}{\partial x_l}\right) + \frac{\partial}{\partial x_l}\left[\mu\left(\frac{\partial u^i}{\partial x_l} + \frac{\partial u^l}{\partial x_i}\right)\right]\right\}, \quad i,l=1,2,3 \tag{1}$$

where u^i ($i=1,2,3$) are the Cartesian components of the displacement vector $u(r)$, and x^l ($l=1,2,3$) are the Cartesian components of the position vector.

In the present experiment, in order to manufacture a 2D SC, an aluminum plate was drilled to be a square array of holes with the radius of 1.0 mm and the lattice constant of 2.5 mm. In the lateral direction of the plate, holes were arranged in the (1, 1) direction with 79 layers and in the perpendicular direction with 27 layers. The 2D SC was constructed by inserting 250-mm-long steel cylinder rods with the radius of 1.0 mm into the periodically drilled plate. Hence, our SC was a 2D square SC with steel cylinders in air background [$\rho_{steel} = 7800$ kg/m³, $\rho_{air} = 1.21$ kg/m³, $c_{steel} = 6100$ m/s, $c_{air} = 334.5$ m/s (sound velocity in air at 0℃)], in which the lattice constant is $a = 2.5$ mm and the radius of cylinders is $R = 1.0$ mm, resulting in a filling fraction $\pi R^2/a^2$ of approximately 50%.

For the SC consisting of steel and air, only the longitudinal waves are allowed. Then Eq. (1) could be simplified as follows:[27]

$$\frac{1}{\lambda}\frac{\partial^2 u}{\partial^2 t} = \nabla \cdot \left(\frac{\nabla u}{\rho}\right). \tag{2}$$

By using the plane-wave expansion (PWE) method[28] and applying the Bloch theorem, Eq. (2) could yield the eigenvalue equation as

$$\sum_{G'}[\omega^2 \lambda^{-1}_{G-G'} - \rho^{-1}_{G-G'}(K+G)\cdot(K+G')]u_{G'} = 0, \tag{3}$$

where K is restricted within the first Brillouin zone and G is the reciprocal vector. By using 289 plane waves, the band structure is calculated as shown in Fig. 1(a). With more plane waves, such as 625 plane waves, the calculated band structure in the first and second bands is the same too. Based on this band structure, the acoustic equifrequency surface (EFS) [Fig. 1(b)] is constructed for the acoustic waves propagating from air to the SC with the interface normal along the (1, 1) direction ($\Gamma-M$ in k space). It is well known, for the square lattice, that the lowest band has $k \cdot \partial\omega/\partial k \geqslant 0$ everywhere within the first Brillouin zone. In other words, the group velocity is never opposite to the phase velocity. From Fig. 1(b) we can see that for frequencies that correspond to all convex contours, the group velocity points to the M point. In this case, the incidence and refraction will stand on the same side of the interface normal, resulting in the acoustic negative refraction. Otherwise, when the EFS is concave around the M point, the group velocity $\partial\omega/\partial k$ points away from the M point, leading to the positive refraction.

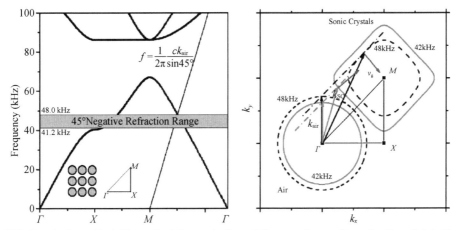

FIG. 1. (color online) Theoretical demonstration of the acoustic negative refraction. (a) is the band structure of the SC. The shading region denotes the negative refractive frequency range from 41.2 kHz to 48.0 kHz with the incident beam making 45° with $\Gamma-M$ direction. (b) is the EFS in k space of the air and SC at 42 kHz (solid) and 48 kHz (dashed), respectively; k_{SC} and v_g are the wave vector and group velocity in the SC, respectively.

To verify the analysis of the refraction of acoustic waves, the fabricated SC was arranged with 27 layers in the propagating direction and 79 layers in the lateral direction. A scanning transmission measurement of spatial distributions covering frequencies from 37 kHz to 44 kHz was carried out by various incident angles. In our experiment (Fig. 2), the SC with the interface of $\Gamma-M$ direction [(1,1) in real space] was placed between two transducers (Airmar AR41, USA), one as an emitter and the other as a receiver. A continuous acoustic incident Gaussian beam from the emitting transducer was generated by a function generator (Agilent 33120A, USA). To eliminate the angular divergence and obtain a good Gaussian shape, we carried out a 100-mm-long, 15-mm-thick sponge loop with an inner radius of 15 mm as an absorbing waveguide between the SC and the emitting transducer. So a good Gaussian beam could be obtained by absorbing acoustic waves with large angles. We measured the amplitudes with both the absorber and no absorber. The results (the inset of Fig. 2) showed that the width of the half peak decreased 1/2 times when waves propagated through the absorber, indicating a good Gaussian shape with angular divergence less than 7°. The receiving transducer was mounted on a goniometer that ran along the lateral direction parallel with the SC interface. The detector was positioned at 10 cm away from the refraction surface to eliminate near field effects. The transmission signal was acquired by a digital sampling oscilloscope with a temporal resolution of 2.5 ns. The refraction was considered negative (positive) if the emerging beam is detected at the same (different) side of the surface normal as the incident beam.

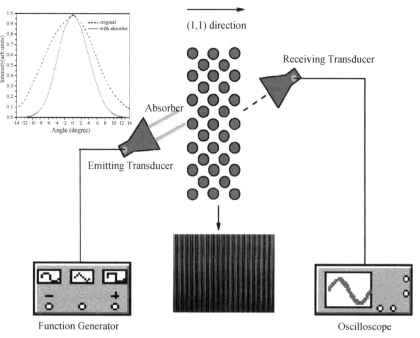

FIG. 2. (color online) Schematic of the experimental setup used to measure the transmission of ultrasonic wave in a SC, consisting of two transducers, a flat rectangular slab of steel cylinders, a function generator, and an oscilloscope. The SC, the rectangular slab of steel cylinders shown as the middle-bottom inset, is placed between two transducers and the left-top inset shows measurements of the amplitude comparisons with the absorber (solid) to without the absorber (dashed).

The dependence of refractions on frequencies is shown in Fig. 3 with a,b for the incident angle of 45° and c,d for the case of 30°, respectively. In the experiment, both the emitter and receiver were oriented to the same angle of 45° or 30° with respect to the surface normal. For a better contrast, the data are represented in terms of the average transmission intensity. For the frequencies below 41 kHz shown in Fig. 3(a), the center of the outgoing Gaussian beam is shifted to the right and the positive refraction is observed. Above 42 kHz, the beam center shifts to the left, corresponding to the negative refraction. Notice that the refractive angle is dependent on frequencies. By employing PWE methods with 289 plane waves, the refractive direction with the frequency was calculated as shown in Fig. 3(b). The negative refraction for the incident angle of 45° starts to appear at the frequency of 41.2 kHz, which is very close to the experimental measurements with the refractive angle of 1.7° at 41 kHz and −2.9° at 42 kHz, respectively. Above the negative-refraction-starting frequency of 41.2 kHz, the higher the frequency, the larger the negative refractive angle. In the range of scanning frequencies from 37 kHz to 44 kHz, the negative refractive angle reached its maximum of theoretical $-27.8°$ and experimental $-28.0°$ at 44 kHz. The experimental measurements and theoretical calculations for the incident angle of 30° are shown in Figs. 3(c) and 3(d). The negative-refraction-starting frequency was theoretically calculated at 41.9 kHz, which agrees very well with our experimental results by 2.1° at 41 kHz and $-0.2°$ at 42 kHz. Meanwhile, at the same frequency, the negative refractive angle for the incident angle of 45° is always greater than that with 30°.

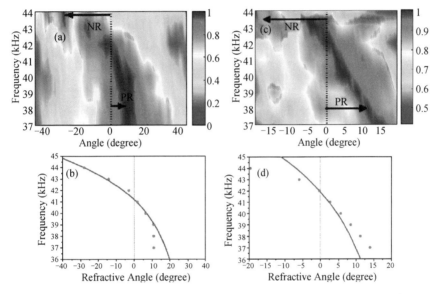

FIG. 3. (color online) The dependence of negative refractions on frequencies. (a) and (c) are the average acoustic transmission intensity vs frequencies (kHz) and angles for the SC with the incident angle of (a) 45° and (c) 30°. The refraction was established as the negative refraction (NR) or the positive refraction (PR) if the maximum transmission intensity is detected at the left or right side of the surface normal, respectively. The average intensity scale varies from 0 to 1. (b) and (d) are comparisons of experimental measurements (dots) and theoretical simulations (lines) of refractive angles vs frequencies with the incident angle of 45° and 30°.

In the experiment, results for 30° incidence are not as good as those for 45°, which is resulted from the angular divergence and the big size (5-cm-diam) of the receiving transducer. Both characters of the receiver result in a low resolution to the refraction beam. For a bigger incident angle, the refractive angle is greater so that the refraction could be easily resolved and the receiver's resolution is high enough to get accurate measurements. While for a smaller incident angle, the refractive angle is also smaller, for example, the maximum around 7° at 44 kHz for 30° incidence, so the low resolution of the receiver leads to the bigger aberrance. To reduce this inaccuracy, a perfect Gaussian beam is needed, however, unfortunately it is quite difficult to realize in the regime of several tens of kHz.

These observations could be well understood by examining the acoustic anisotropic EFS [Fig. 1(b)]. The corresponding acoustic EFS and its curvature determine the direction of acoustic refraction in SCs. With the increase of frequencies, the SC's EFSs are easier to convex around the M point so that negative refractions will occur at higher frequencies. The higher operating frequency results in the smaller EFS around the M point; and the curvature of the corresponding point at the EFS is greater, which leads to a bigger negative refractive angle. When the EFS is small enough, however, there exists a stopping frequency because the total internal reflection occurs at the directional band gap for a single acoustic beam, which could be calculated by using a similar method that determines the upper limit of all angle negative refraction (AANR).[12] The shading region in Fig. 1(a) denotes the negative refractive frequency range from 41.2 kHz to 48.0 kHz with the incident beam making 45° with $\Gamma - M$ direction. Differed from $f = ck_{air}/2\pi$ calculated in AANR, we calculate the dispersion line by $f = (1/2\pi)ck_{air}/\sin 45° = \sqrt{2}ck_{air}/2\pi$, whose intersection with the band structure indicates the frequency 48.0 kHz with the maximum negative refractive angle of $-90°$.

The negative refraction is also strongly dependent on the incident angles of the acoustic beam. The anisotropy of the SC's EFSs could also be used to describe relations between negative refractions and incident angles at a constant frequency. To demonstrate this effect, the angles of refraction are theoretically calculated by the PWE method with 289 plane waves and experimentally measured with a variety of incident angles at 43 kHz as depicted in Fig. 4. We got good agreements between theoretical calculations and experiment measurements, for example, the refractive angle theoretically $-15.6°$ and experimentally $-14.2°$ with the incident angle of 45°, and theoretically $-55.9°$ and experimentally $-55.0°$ with the incident angle of 60°, respectively. If the angle of incidences is bigger than 70°, there exists the directional band gap; then the acoustic wave will experience total internal reflection and no refraction exists. So the acoustic dispersion and anisotropy might result in the abundant nature of refraction in arbitrary directions and will lead to some additional interesting effects as PCs, for example, acoustic collimating effects.

In addition, AANR (Ref. 12) is an important character for negative refraction and its application, which in acoustic waves was discussed in a three-dimensional sonic crystal of carbide beads in water[22] and a 2D system of water cylinders in mercury.[21] In the SC consisting of steel and air, calculated with the method described above, there is no AANR existing in our current crystal of $R=0.4a$ and the steel-in-air structure of $R=0.36a$ in Ref. [21]. However, we found that AANR may appear at 41.2 kHz when R is greater than 1.05 mm ($R=0.42a$), and the frequency range of AANR is to be enlarged by increasing the filling fraction of steel cylinders. When $R=1.1$ mm ($R=0.44a$), it is from 38.1 kHz to 40.1 kHz that AANR could be realized. So with the high filling fraction, AANR may also appear in the steel-in-air SC, which is very convenient to be used in applications.

To summarize, the acoustic single-beam negative refraction in the lowest band of 2D

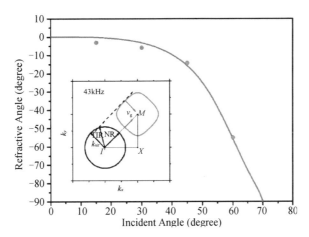

FIG. 4. (color online) Comparison of measured (dots) and calculated (lines) angles of refractions versus angles of incidences at 43 kHz. The inset is the EFSs at 43 kHz, in which NR and TIR indicate the incident angles with the negative refraction and total internal reflection, respectively.

square SCs was theoretically analyzed by calculating the acoustic band structure and equifrequency surfaces and experimentally established in a 2D SC in the ultrasonic regime by scanning transmission measurements of spatial distributions of a single Gaussian beam. In order to obtain the larger negative refractive angle, it is necessary to make the bigger incident angle at the higher frequency. The acoustic negative refraction could be tuned to the desired frequency range by adjusting the lattice constant and constructing materials of the SC, for example, with immersing our structure into liquids such as water. In addition, acoustic negative refractions in our SC provide the possibilities to construct an acoustic superlens in air to realize both acoustic propagating and evanescent waves imaging.

References and Notes

[1] V. G. Veselago, Sov. Phys. Usp. **10**, 509 (1968).
[2] J. B. Pendry *et al.*, Phys. Rev. Lett. **76**, 4773 (1996).
[3] J. B. Pendry *et al.*, J. Phys.: Condens. Matter **10**, 4785 (1998).
[4] J. B. Pendry *et al.*, IEEE Trans. Microwave Theory Tech. **47**, 2075 (1999).
[5] D. R. Smith *et al.*, 3Phys. Rev. Lett. **84**, 4184 (2000).
[6] R. A. Shelby *et al.*, Appl. Phys. Lett. **78**, 489 (2001).
[7] R. A. Shelby, D. R. Smith, and S. Schultz, Science **292**, 77 (2001).
[8] M. Notomi, Phys. Rev. B **62**, 10696 (2000).
[9] E. Cubukcu *et al.*, Nature (London) **423**, 604 (2003).
[10] B. Gralak, S. Enoch, and G. Tayeb, J. Opt. Soc. Am. A **17**, 1012 (2000).
[11] C. G. Parazzoli *et al.*, Phys. Rev. Lett. **90**, 107401 (2003).
[12] C. Luo *et al.*, Phys. Rev. B **65**, 201104(R)(2002).
[13] C. Luo, S. G. Johnson, and J. D. Joannopoulos, Appl. Phys. Lett. **81**, 2352 (2002).
[14] J. B. Pendry, Phys. Rev. Lett. **85**, 3966 (2000).

[15] A. A. Houck et al., Phys. Rev. Lett. **90**, 137401 (2003).

[16] C. Luo et al., Phys. Rev. B **68**, 045115 (2003).

[17] P. V. Parimi et al., Nature (London) **426**, 404 (2003).

[18] E. Cubukcu et al., Phys. Rev. Lett. **91**, 207401 (2003).

[19] X. Hu et al., Phys. Rev. E **68**, 037301 (2003).

[20] X. Hu et al., Phys. Rev. E **69**, 030201(R)(2004).

[21] X. Zhang and Z. Liu, Appl. Phys. Lett. **85**, 341(2004).

[22] S. Yang, J. H. Page, Z. Liu, M. L. Cowan, C. T. Chan, and P. Sheng, Phys. Rev. Lett. **93**, 024301 (2004).

[23] E. Yablonovitch, Phys. Rev. Lett. **58**, 2059(1987).

[24] S. John, Phys. Rev. Lett. **58**, 2486 (1987).

[25] J. D. Joannoupoulos, R. D. Meade, and J. N. Winn, *Photonic Crystals*(Princeton University Press, Princeton, NJ, 1995).

[26] E. N. Economou and M. Sigalas, J. Acoust. Soc. Am. **95**, 1734 (1994).

[27] F. Wu, Z. Liu, and Y. Liu, Phys. Rev. E **66**, 046628(2002).

[28] M. Plihal and A. A. Maradudin, Phys. Rev. B **44**, 8565(1991).

[29] The work was jointly supported by the National 863 High Technology Program, the State Key Program for Basic Research of China, and the National Nature Science Foundation of China (Grant No. 50225204).

Acoustic Backward-Wave Negative Refractions in the Second Band of a Sonic Crystal[*]

Liang Feng,[1] Xiao-Ping Liu,[1] Ming-Hui Lu,[1] Yan-Bin Chen,[2] Yan-Feng Chen,[1] Yi-Wei Mao,[3] Jian Zi,[4] Yong-Yuan Zhu,[1] Shi-Ning Zhu,[1] and Nai-Ben Ming[1]

[1] *National Laboratory of Solid State Microstructures, Nanjing University, Nanjing 210093, China*
[2] *Department of Materials Science and Engineering, University of Michigan, Ann Arbor, Michigan 48109, USA*
[3] *Institute of Acoustics, Nanjing University, Nanjing 210093, China*
[4] *National Laboratory of Surface Physics, Fudan University, Shanghai 200433, China*

Acoustic negative refractions with backward-wave (BW) effects were both theoretically and experimentally established in the second band of a two-dimensional (2D) triangular sonic crystal (SC). Intense Bragg scatterings result in the extreme deformation of the second band equifrequency surface (EFS) into two classes: one around the K point and the other around the Γ point of the reduced Brillouin zone. The two classes can lead to BW negative refractions (BWNRs) but with reverse negative refraction dependences on frequencies and incident angles. Not only BWNR but BW positive refraction can be present at EFSs around the K point, so it is possible to enhance the resolution of acoustic waves with a subdiffraction limit regardless of refractions, which is no analogy in both left-handed material and SCs' first band. These abundant characters make refractions in the second band distinguished.

A kind of unusual material with both permittivity and permeability simultaneously negative was first predicted by Veselago in 1968, named left-handed material (LHM) because when light propagates in it, the electric field, the magnetic field, and the wave vector form a left-handed set[1]. Accompanying the left-handedness, the wave vector is opposite to the energy flow, which is called the backwardwave (BW) effect (in anisotropic media such as photonic and sonic crystals, a general situation for the BW effect should be considered as following that, although the wave vector and energy flow are always not collinear, BW still exists because the energy flow is forward but the wave vector is backward). An array of split ring resonators and wires[2-4], a kind of LHM proposed by Pendry et al., has been established in experiments to realize BW negative refraction (BWNR) in microwave range by Smith et al.[5-7]. And through BWNRs, a superlens by LHM can be constructed to make use of both evanescent and propagating waves to produce a real subwavelength image beyond the diffraction limit due to phase compensations induced by the BW effect[8-11].

Recently, the negative refraction was also realized in photonic crystals (PCs)[12-18].

---

[*] Phys.Rev.Lett.,2006,96(1):014301

There are two cases of negative refractions in PCs. One is because of intense scatterings in the lowest band with negative refractions but without a negative index[12,13], and the other located in higher bands is the left-handed electromagnetism with a BW effect as BWNR in LHM[14−16], showing both the negative refraction and the negative refractive index. With the phase compensations by BW effects, it is also possible to obtain the transmission amplitude for evanescent waves to construct a PC based superlens[17], which was observed experimentally in the infrared region[18] and expected in the visible region.

Since last year the negative refraction has gone into acoustics, leading to an acoustic superlens based on sonic crystals (SCs) [19,20]. Although there is no corresponding left-handed set in SCs, it is still highly anticipated to obtain the acoustic BWNR with phase compensations to amplify acoustic evanescent waves. As the left-handed electromagnetism in PCs, BWNR can also exist in SCs' counterparts with an effective negative index[21,22]. Regarding the BWNR, the previous discussions[14−18,21,22] mainly defined the wave propagating along some high-symmetry directions (Γ-M or Γ-K), where the group velocity is antiparallel to the wave vector, so the refraction direction could be determined by the negative index. Except for the negative refractive index, however, in a general case with wave propagating along any directions, there are no detailed discussions about BWNR in both PC and SC, which is inevitable to stimulate us to consider more general BWNRs and find out their dependences on both frequencies and incident angles. By addressing this question in this Letter, BWNRs inside the Brillouin zone (BZ) with the wave propagating along any arbitrary low-symmetry direction are first carried out, showing more complicated and interesting characters than negative refractions in the lowest band.

In the present experiment, in order to propose a 2D triangular SC, an aluminum plate was designed to be drilled as a triangular array of holes with the lattice constant of 4.5 mm. Holes were arranged as the Γ-K direction with 99 layers in the lateral direction of the plate, and Γ-M with 20 layers in the perpendicular direction. By inserting 250-mm-long steel cylinders with the radius of 1.0 mm into the periodically drilled plate, a 2D triangular SC was constructed with steel cylinders in air background [ρ_{steel} = 7800 kg/m^3, ρ_{air} = 1.21 kg/m^3, c_{steel} = 6100 m/s, c_{air} = 344.5 m/s (sound velocity in air at 20℃)].

A scanning transmission measurement of spatial distributions covering frequencies from 45.5 kHz to 56 kHz was carried out with various incident angles. In our experiment (Fig. 1), the triangular SC with the normal as Γ-M direction was placed between two transducers (Airmar AT50, USA), one as an emitter and the other as a receiver. A 50-cycle-pulse acoustic Gaussian beam from the emitting transducer was generated by a function generator (Agilent 33120A, USA). To eliminate the angular divergence and obtain a good Gaussian shape, we used a 100-mm-long, 15-mm-thick sponge loop with an inner radius 15 mm as an absorbing waveguide between the SC and the emitting transducer. So a good Gaussian beam could be obtained by absorbing acoustic waves with large angles. The receiving transducer was mounted on a goniometer that ran along the lateral direction parallel with the SC interface (Γ-K direction). The detector was positioned at 10 cm away from the refraction surface to eliminate near field effects. In the experiment, both the emitter and the receiver were oriented to the same angle with the normal of the crystal interface. The transmission signal was acquired by a digital sampling oscilloscope with a temporal resolution of 2.5 ns. The position where the most intense signal was observed was regarded as the refraction of the acoustic wave. And the refraction was considered negative (positive) if the emerging beam is detected at the same (different) side of the surface normal as the incident beam.

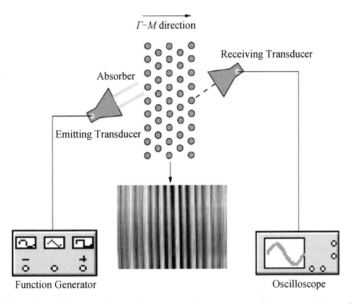

FIG. 1. Schematic of the experimental setup used to measure the transmission of ultrasonic wave in a 2D triangular SC, consisting of two transducers, an absorber, a function generator, and an oscilloscope. The SC, the rectangular slab of steel cylinders shown as the inset, is placed between two transducers with the normal as the Γ-M direction.

The BWNR in the SC could be well understood by examining the acoustic band structure and equifrequency surface (EFS). The band structure of the 2D triangular SC is calculated by both the plane wave expansions method and multiple scattering theory, which is shown in Fig. 2(a). In the second band, where the frequency of the Γ point is higher than those of other points, EFSs [Fig. 2(b)] move inwards, indicating that $v_g \cdot k < 0$. Since the group velocity is always forward (positive), the negative wave vector is expected, indicating a BW effect and producing an effective negative index $n_{eff} = k_{SC}/k_{air}$.

In studying the detailed relation of acoustic refractions, it is difficult to determine the refraction only by n_{eff} except in some particular high-symmetry directions (Γ-M or Γ-K)[14-16]. For the features inside the BZ, the propagating direction of acoustic waves must be marked in acoustic EFSs [Fig. 2(b)], in which we make a discussion on the refraction of a 30° incidence from air with the normal along the SC's Γ-M direction as we did in our experiment. With the BW effect, the right-down forward wave vector in air will find the corresponding point at the left panel of the SC's EFS because of the negative left-down wave vector in the SC, and lead to the right-up group velocity. Then in this case, the incidence and refraction stand on the same side of the interface normal (Γ-M direction), showing a BWNR.

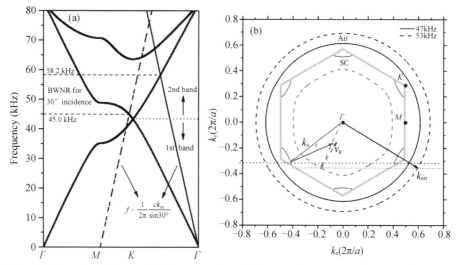

FIG. 2. (color online). Theoretical demonstrations of the acoustic negative refraction. (a) The band structure of the SC. The region between 45.0 kHz and 58.2 kHz denotes the negative refractive frequency range for the incident beam making 30° with the SC's Γ-M direction. (b) The equifrequency surfaces in k space of the air and SC at 47.0 kHz (solid line) and 53.0 kHz (dashed line), respectively; k_{SC} and v_g are the wave vector and group velocity in the SC, respectively.

In the second band, strong Bragg scatterings at the BZ boundaries result in the extreme deformation of EFSs of acoustic waves in the SC. The continuous EFS in the second band are dramatically distorted to several pieces, and these pieces could be presented in the reduced BZ scheme [Fig. 2(b)]. Dependent on the frequency of acoustic waves, there are different kinds of EFSs: one is the dramatically changed EFS around the K point, and the other is the gradually varying EFS around the Γ point. So there are two kinds of EFSs that lead to two different characteristic cases for BWNRs as well as their interesting relations with frequencies. When the frequency is below 48.8 kHz (M point), EFSs are around the K point and become larger from the K to M point. So for 30° incidence, k_{air} will find its first corresponding SC's EFS point at the K-M BZ boundary with a negative refractive angle of $-90°$. And when the frequency is higher, the corresponding point will be far from the BZ boundary and the EFS's curvature is less, making the refraction less negative. When the frequency is above 48.8 kHz, EFSs are around the Γ point. With the increase of frequencies, the curvature of the corresponding point at the EFS is greater, which results in a bigger negative refractive angle. And when the EFS is small enough, there exists a stopping frequency, with the negative refractive angle of $-90°$. To define a single beam BWNR frequency range, the similar method to determine the upper limit of all angle negative refraction could be applied here[11]. The region between two dashed lines in Fig. 2(a) denotes the BWNR frequency range from 45.0 to 58.2 kHz for 30° incidence. The dispersion line $f = \frac{1}{2\pi} \times \frac{ck_{air}}{\sin 30°}$ was calculated from both M and Γ points, whose intersections with the M-K and Γ-K directions correspond to the starting frequency 45.0 kHz and the ending frequency 58.2 kHz, respectively.

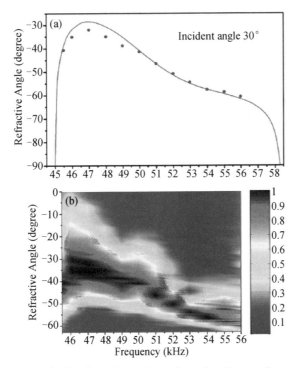

FIG. 3. (color online). The dependence of negative refractions on frequencies. (a) The comparison of experimental measurements (dots) and theoretical simulations (line) of refractive angles versus frequencies with the incident angle of 30°. (b) The average acoustic transmission intensity versus frequencies and angles with the 30° incidence. The average intensity scale varies from 0 to 1.

In the BWNR frequency range for 30° incidence from 45.0 kHz to 58.2 kHz, the dependence of refractions on frequencies is shown in Fig. 3 with Fig. 3(b) the measured data represented in terms of the average transmission intensity, which agreed well with the theoretical calculation in Fig. 3(a). The negative refractive angle reached its minimum of theoretical $-28.4°$ and experimental $-32.0°$ at 47.0 kHz. From 45.0 kHz to 47.0 kHz, the sharp change of the refraction is related to the dramatically changed EFSs around the K point due to intense Bragg scatterings. And from 47.0 kHz to 48.8 kHz, though EFSs are still around the K point, it is the transient region for EFSs changed from the K to Γ point so that the curvature becomes greater and the refraction changes more negatively. From 48.8 to 58.2 kHz, the Bragg scatterings are less, corresponding to the gradually varying

EFSs around the Γ point, and the refraction becomes more negative slowly. Hence, because of two different frequency dependent characters for EFSs, the relation of BWNRs with frequencies is not monotonically changed, different from the characters in the first band published before[12,18,19].

BWNRs are also strongly dependent on the incident angles of the acoustic beam. For both different BWNRs, however, the dependences of refractions on incident angles are reverse. The SC's EFSs (Fig. 2) could also be used to describe relations between negative refractions and incident angles at a constant frequency. For the EFS around the K point at 47.0 kHz, there are two directional band gaps, one as the incident angle from 0° to 21.4° and the other from 38.6° to 90°. Only when the incident angle is between 21.4° and 38.6° can k_{air} find its corresponding SC's EFS point, which also first takes place at the K-M BZ boundary, producing the negative refractive angle of $-90°$ with the incident angle of 21.4°. And with raising the incident angles, the refraction will become from negative to positive, and reach its positive maximum 30° with the incident angle of 38.6°. While for the EFS around the Γ point at 53.0 kHz, the EFS is smaller than air's so that the acoustic wave will experience total internal reflection and no refraction exists if the angle of incidences is greater than 43.6°. With enlarging the incident angle from 0° to 43.6°, the refractive angle will monotonically change to be more negative from 0° to $-90°$. To demonstrate this effect, the angles of refractions are theoretically calculated and experimentally measured with a variety of incident angles at 47.0 kHz and 53.0 kHz as depicted in Fig. 4, showing good agreements between theoretical calculations and experiment measurements. So both absolutely reverse refractions might result in the abundant nature of refractions in arbitrary directions and lead to some additional interesting effects, such as BW positive refraction (BWPR).

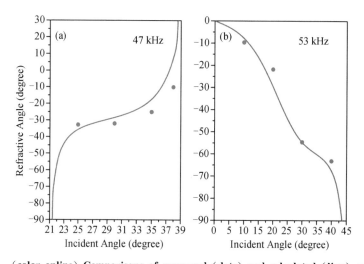

FIG. 4. (color online) Comparisons of measured (dots) and calculated (line) angles of refractions versus angles of incidences at 47.0 kHz (a) and 53.0 kHz (b).

BW effects taking place in the second band of SCs are quite different from that in LHM and SC's first band as shown in the illustration Fig. 5. In fact, BWNR inherent from EFSs around the Γ point in the second band is something like the negative refractions in LHM[2—7] where an effective negative index results in approximately isotropic EFSs. However, EFSs around the K point attribute to two kinds of BW effects including both BWNR and BWPR. Especially for the BWPR, there is no analogy in both LHM and SC's first band. These abundant characters are attributed to the extremely distorted EFSs due to intense Bragg scatterings near BZ boundaries. Hence in the second band, BWPR is the most important character distinguished from both LHM and the first band. Comparing to these effects in PC's second band, BWNR and BWPR in SC are analogous to left-handed negative and positive refractions in PC, respectively. With BW effect induced phase compensations, regardless of BWNR and BWPR, evanescent waves in the second band are always amplified to enhance the resolution of acoustic waves with a subdiffraction limit[11].

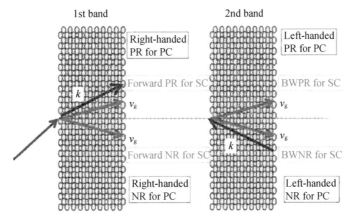

FIG. 5. (color online) Illustration of refractions in both the first and second bands of SC, as well as their counterparts in PC.

To summarize, the acoustic negative refraction with BW effects in the second band of a 2D triangular SC was experimentally established and theoretically analyzed inside the BZ. Resulted from the intense scatterings near the BZ boundary, two different EFSs around K and Γ points, respectively, in the second band attribute to two kinds of absolutely reverse BWNRs as well as their interesting dependences on both frequencies and incident angles. In the second band, both BWNR and BWPR exist, which is distinguished from that in the SC's first band and LHM. With phase compensations, BW effects in our SC provide the possibilities to obtain the transmission amplitude for evanescent waves with both negative and positive refractions, then enhance the resolution of acoustic waves and obtain an acoustic subwavelength image, which have great potential in ultrasonic biosensing and medical measurements. And this effect could also extend to PC and other wave propagations in the periodic structures, exhibiting a universal promise.

References and Notes

[1] V. G. Veselago, Sov. Phys. Usp. **10**, 509 (1968).

[2] J. B. Pendry, A. J. Holedn, W. J. Stewart, and I. Youngs, Phys. Rev. Lett. **76**, 4773 (1996).

[3] J. B. Pendry, A. J. Holedn, D. J. Robbins, and W. J. Stewart, J. Phys. Condens. Matter **10**, 4785 (1998).

[4] J. B. Pendry, A. J. Holedn, D. J. Robbins, and W. J. Stewart, IEEE Trans. Microw. Theory Tech. **47**, 2075 (1999).

[5] D. R. Smith, W. J. Padilla, D. C. Vier, S. C. Nemat-Nasser, and S. Schultz, Phys. Rev. Lett. **84**, 4184 (2000).

[6] R. A. Shelby, D. R. Smith, S. C. Nemat-Nasser, and S. Schultz, Appl. Phys. Lett. **78**, 489 (2001).

[7] R. A. Shelby, D. R. Smith, and S. Schultz, Science **292**, 77 (2001).

[8] J. B. Pendry, Phys. Rev. Lett. **85**, 3966 (2000).

[9] A. A. Houck, J. B. Brock, and I. L. Chuang, Phys. Rev. Lett. **90**, 137401 (2003).

[10] L. Feng, X. P. Liu, M. H. Lu, and Y. F. Chen, Phys. Lett. A **332**, 449 (2004).

[11] N. Fang, H. Lee, C. Sun, and X. Zhang, Science **308**, 534 (2005).

[12] C. Luo, S. G. Johnson, J. D. Joannopoulos, and J. B. Pendry, Phys. Rev. B **65**, 201104(R) (2002).

[13] E. Cubukcu, K. Aydin, E. Ozbay, S. Foteinopoulou, and C. M. Soukoulis, Nature (London) **423**, 604 (2003).

[14] M. Notomi, Phys. Rev. B **62**, 10 696 (2000).

[15] S. Foteinopoulou and C. M. Soukoulis, Phys. Rev. B **67**, 235107 (2003).

[16] P. V. Parimi, W. T. Lu, P. Vodo, J. Sokoloff, J. S. Derov, and S. Sridhar, Phys. Rev. Lett. **92**, 127401 (2004).

[17] K. Guven, K. Aydin, K. B. Alici, C. M. Soukoulis, and E. Ozbay, Phys. Rev. B **70**, 205125 (2004).

[18] A. Berrier, M. Mulot, M. Swillo, M. Qiu, L. Thylén, A. Talneau, and S. Anand, Phys. Rev. Lett. **93**, 073902 (2004).

[19] X. Zhang and Z. Liu, Appl. Phys. Lett. **85**, 341 (2004).

[20] S. Yang, J. H. Page, Z. Liu, M. L. Cowan, C. T. Chan, and P. Sheng, Phys. Rev. Lett. **93**, 024301 (2004).

[21] J. H. Page, A. Sukhovich, S. Yang, M. L. Cowan, F. Van Der Biest, A. Tourin, M. Fing, Z. Liu, C. T. Chan, and P. Sheng, Phys. Status Solidi B **241**, 3454 (2004).

[22] C. Qiu, X. Zhang, and Z. Liu, Phys. Rev. B **71**, 054302 (2005).

[23] The work was jointly supported by the National 863 High Technology Program, the State Key Program for Basic Research of China, and the National Nature Science Foundation of China (Grant No. 50225204).

Negative Birefraction of Acoustic Waves in a Sonic Crystal[*]

Ming-Hui Lu[1], Chao Zhang[1], Liang Feng[1], Jun Zhao[1],
Yan-Feng Chen[1], Yi-Wei Mao[2], Jian Zi[3], Yong-Yuan Zhu[1],
Shi-Ning Zhu[1] And Nai-Ben Ming[1]

[1] *National Laboratory of Solid State Microstructures and Department of Materials Science and Engineering, Nanjing University, Nanjing 210093, People's Republic of China*

[2] *Institute of Acoustics, Nanjing University, Nanjing 210093, People's Republic of China*

[3] *National Laboratory of Surface Physics, Fudan University, Shanghai 200433, People's Republic of China*

Optical birefringence and dichroism are classical and important effects originating from two independent polarizations of optical waves in anisotropic crystals[1]. Furthermore, the distinct dispersion relations of transverse electric and transverse magnetic polarized electromagnetic waves in photonic crystals can lead to birefringence more easily[2-6]. However, it is impossible for acoustic waves in the fluid to show such a birefringence because only the longitudinal mode exists. The emergence of an artificial sonic crystal (SC) has significantly broadened the range of acoustic materials in nature[7-18] that can give rise to acoustic bandgaps and be used to control the propagation of acoustic waves. Recently, negative refraction has attracted a lot of attention and has been demonstrated in both left-handed materials and photonic crystals[19-26]. Similar to left-handed materials and photonic crystals, negative refractions have also been found in SCs[14-18]. Here we report, for the first time, the acoustic negative-birefraction phenomenon in a two-dimensional SC, even with the same frequency and the same 'polarization' state. By means of this feature, double focusing images of a point source have been realized. This birefraction concept may be extended to other periodic systems corresponding to other forms of waves, showing great impacts on both fundamental physics and device applications.

The birefringence phenomenon due to different polarization states has attracted wide attention. However, in the higher-frequency bands of photonic crystals, birefraction could occur even for the same polarization state[5,6] because the same frequency might be located in different bands with different wavevectors owing to the overlap of some bands. Its acoustic counterpart should also exist in SCs. Herein, a two-dimensional triangular SC is fabricated with a periodic lattice constant of 4.5 mm and a cylinder radius of 1.0 mm. The propagation of acoustic waves in this SC within the second and the third bands is analysed by the plane-wave-expansion (PWE) and multiple-scattering (MS) methods and simulated using the finite-difference time-domain (FDTD) technique (see the Supplementary

[*] Nat. Mater., 2007, 6(10): 744

Information). We demonstrate that acoustic birefraction can occur in the SC for the same 'polarization' of acoustic waves, and a novel focusing phenomenon has been realized with the acoustic negative birefraction effect.

To understand well the negative refraction in the SC, the acoustic band structure and the equi-frequency surface (EFS) were calculated by the PWE and MS methods as shown in Fig. 1(a). In both the second and the third bands, the frequency of the Γ point is the highest, so the EFSs [Fig. 1(a), right] move inwards, indicating that $v_g \cdot k < 0$ (v_g and k are the group velocity and wavevector, respectively), which leads to the backward-wavevector (BWV) effect and produces an effective negative index $n_{eff} = k_{sc}/k_{air}$ (k_{sc} and k_{air} are the wavevectors in the sonic crystal and air respectively). In Fig. 1(a) (left), the most striking character is that the second and third frequency bands overlap above 63 kHz. In other words, waves at frequencies above 63 kHz could have two crystal momenta, thereby leading to two different phase velocities and two different group velocities in the SC. The propagating direction of acoustic waves could be analysed through the EFSs inside the Brillouin zone [Fig. 1(a) right]. Let us consider how a beam incident from air at 7° to the normal to the SC's Γ - M direction will be refracted at the frequency of 73 kHz. Owing to the BWV effect, the right-down forward wavevector in air will find the corresponding point at the left-down part of the SC's EFS, on the basis of conservation of momentum. At the corresponding point, the curvature normal orientates the right-up group velocity, so the refractive and the incident beams stand on the same side of the Γ - M direction, producing negative refraction. The right panel of Fig.1(a) shows two refractive directions respectively corresponding to the second and the third frequency bands, which is illustrated by the (FDTD) numerical simulation as shown in Fig.1(b), with one refractive angle of about −24° due to the third band, and the other of −50° due to the second band (the minus sign in this letter means negative refraction). The two refractive angles are also calculated (by PWE) and are shown in the inset of Fig.4(b). Experimentally, with a continuous gaussian beam with an incident angle of 7° to the SC's surface normal at 73 kHz, the spatial intensity and the phase distribution of the acoustic pressure field were mapped out as shown in Fig. 1(c). This obviously manifests that two refractive outgoing beams show birefraction of acoustic waves. The two refractive angles in the experiment were −27.5° and −50.1°, respectively, in good agreement with the simulation in Fig. 1(b). To demonstrate the acoustic birefraction more clearly, the phase distribution was measured simultaneously, shown at the top of Fig. 1(c). The equi-phase surface of two outgoing beams is parallel, indicating that both of them originate from the same incident beam. Moreover, they are out of phase spatially because the birefracted beams experience two different propagation lengths, which leads to the dark area separating the two beams. It should be noticed that there are some anomalous outgoing beams marked by black arrows in Fig. 1(b), which are also observed in the experiment and marked by blue dashed arrows in Fig. 1(c). The wavefronts of the beams are not parallel to those of the incident beam, so

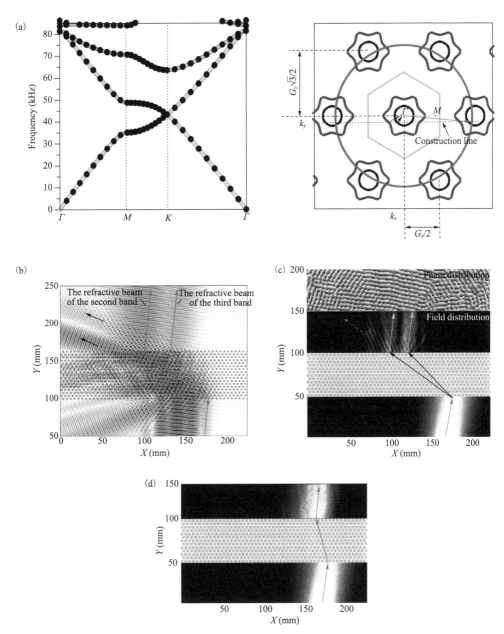

FIG.1. The theoretical analysis and experimental results of the birefraction of an acoustic wave in the SC with triangular lattices. (a), Theoretical demonstrations of the acoustic negative birefraction. In the band structure of the SC (left panel), red open circles are results calculated by PWE and black filled circles are those calculated by MS. In the extended presentation of the EFS of air and the SC at 73 kHz in k space (right) the blue, black and red lines are for the third band, for the second band and for air, respectively. The red arrow indicates k_{air}, and the blue and the black arrows mean the v_g of the third band and the second band. The green line is the boundary of the Brillouin zone. Herein, G_x and G_y are the basic reciprocal vectors of the sonic crystal in the x and y directions respectively. (b), The FDTD simulations of the spatial field distribution of the acoustic pressure field at 7° incidence and 73 kHz. (c),(d), The spatial distributions of the intensity and the phase of the acoustic pressure field at 7° incidence and 73 kHz (c) and the single negative refraction at 60 kHz (d), respectively. Colours changing from bright to dark indicate the value of the intensity or phase, changing from the highest to the lowest.

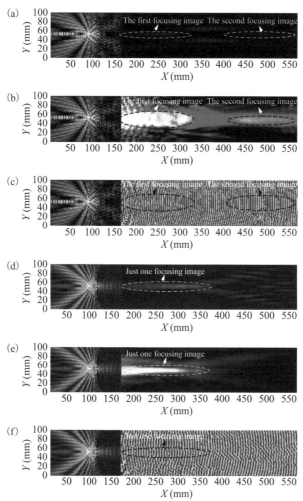

FIG.2. The simulation and experimental results of the point imaging with the SC at 73 kHz and 60 kHz. (a), (d), The FDTD simulation of the field distribution of a point source 15 mm in front of the SC and its focusing images through the SC sample with 16 layers at 73 kHz (a) and 60 kHz (d). (b), (c), (e), (f), The experimental spatial intensity and phase distribution of its focusing images are measured at 73 kHz [(b), (c)] and 60 kHz [(e), (f)] respectively. The dashed ellipses denote the regions of the focusing images. Colours changing from bright to dark indicate the value of the intensity or the phase, changing from the highest to the lowest.

we believe such beams are due to higher-order diffraction[27]. When the frequency is located in one band, for example 60 kHz in the second band, only one negative refraction beam could be observed with a refractive angle of $-9.7°$ [Fig. 1(d)], consistent with the calculation result of $-10°$ [Fig. 4(b) below]. Hence, it is confirmed that we have experimentally demonstrated the acoustic birefraction phenomenon.

On the basis of such negative birefractions as described above, what kind of image can we obtain? At 73 kHz, the imaging effect is simulated [Fig. 2(a)]. There are two coherent elongated focusing images. To confirm these effects, the imaging experiment at 73 kHz was carried out with a point source positioned 15 mm in front of the SC. Two elongated focusing images denoted by the black dashed line are illustrated in Fig. 2(b), which are consistent with the result simulated by FDTD. Figure 2(c) demonstrates the phase pattern of the double-focusing effects. Along the X-direction, the shape of the equi-phase surface changes gradually from a backward conelike pattern at the outgoing surface of the SC to forward at the position of 300 mm, showing the first focusing image, and then becomes nearly flat at the position of 400 mm, like a quasi-directive emission, indicating the second focusing image, which is also consistent with the simulated phase pattern [Fig. 2(a)]. Nevertheless, we can say the double focusing images are established using a plate acoustic 'superlens' based on negative birefractions. Compared with 73 kHz, only one focusing image could be expected at 60 kHz on the basis of negative refractions, and, indeed, only one elongated focusing image has been obtained, as shown in Fig. 2(d)-(f). Figure 2(f) shows the equi-phase surface evolves from a backward conelike phase pattern at the outgoing surface of the SC to a forward conelike phase pattern, showing the elongated imaging effect from 161 mm to 350 mm, which is consistent with the numerical simulation (FDTD). Obviously, the focusing images at the frequencies of 73 kHz and 60 kHz are not good point images, and eventually become quasi-directive emission. We also found that each elongated image shows some distortions and variations in intensity distribution. We think this might be attributable to the interference between the incident wave of the source and the reflective wave of the SC's surface resulting from the impedance mismatch between the air and the SC[14].

To obtain a good image, the imaging experiment was carried out at the frequency of 49 kHz, which is also located in the second band of the SC. The simulation and experimental results are shown in Fig. 3. The measured intensity distribution [Fig. 3(b)] of the pressure field distinctly shows a point image with high quality at 250 mm, well consistent with the simulation at 235 mm [Fig. 3(a)]. In Fig. 3(c), the equi-phase surfaces gradually change from a backward pointing conelike pattern at the position of 200 mm to a forward pointing conelike pattern at the position of 275 mm, clearly manifesting the point focusing effect. To interpret the diversity of the point image's quality at two frequencies of 60 kHz and 49 kHz, the EFSs and the corresponding relationships between refractive angles versus incident angles are calculated, as shown in Fig. 4. The EFS at 49 kHz is more circular than that at 60 kHz, indicating nearly isotropic refractive index and resulting in a good dependence of refractive angle on incident angles with an effective negative index of about -0.90 [Fig. 4(b)], which is closer to -1. However, at 60 kHz and 73 kHz, both of which are far away from the band edge, the EFSs are not circular and their anisotropies increase, showing smaller effective refractive index. Therefore, only the wave with small incident angle can pass through the SC. In other words, owing to the high contrast of the refractive index, only a few or even one predominant plane-wave component contributes in the Floquet—Bloch wave. Hence, the focusing imaging is elongated and both the forward and backward conelike phase patterns decrease, even becoming nearly flat, which denotes that a nearly directive emission is produced[14,27].

All these imaging experiments (Figs 2 and 3) confirm that the acoustic birefraction occurs even for waves with the same 'polarization' state when the acoustic wave's frequency is located simultaneously in two bands with two different Bloch wavevectors, corresponding to two different Bloch states (see Supplementary Information, Fig. S3). If the birefraction is double negative, it will result in double focusing images. In fact, the birefraction is not always double negative. For different SCs, there might be different cases. For example, in a square lattice SC of 2.0-mm-diameter steel cylinders embedded in air with a lattice constant of 2.5 mm, for the incident angle of 15° at 95 kHz, we could obtain two refractive beams: one is positive refraction with the BWV effect and the other is negative refraction with the BWV effect[17], as shown in Supplementary Information, Fig. S4. In other cases, the birefraction could be double positive as well.

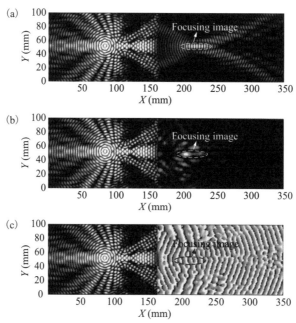

FIG.3. The simulation and experimental results of the point imaging with the SC at 49 kHz. (a), The simulations of the field distribution of a point source and its focusing image through 16 layers at 49 kHz. (b), The experimentally measured intensity distribution of the focusing image behind the SC sample. (c), The corresponding phase map for the focusing images. Colours changing from bright to dark indicate the value of the intensity or the phase, changing from the highest to the lowest.

Although this experiment studies sound propagation in a two-dimensional SC, reasonably, the birefraction phenomenon could be analogically extended to other periodical structures corresponding to other forms of waves such as photonic crystals for optical waves, semiconductors for electrons, plasmonic crystals for plasmons[28] and so on. In these structures, the band structures can be engineered and the corresponding birefractions originate from the overlap of different bands instead of the conventional optical birefringence inherent in the anisotropy of the real crystals. Especially for the case of an optical wave in a photonic crystal, the birefraction means two eigen-Bloch states, which are two indistinguishable paths in the photonic crystal, just like Young's double-slit experiment. According to the particle aspect of photons, a photon birefracted in the photonic crystal cannot be detected at two separated positions simultaneously; that is, the detection of a photon at one point eliminates the probability of detecting the photon at the other point. For a pair of highly correlated photons, quantum interference might be observed when both photons can reach the two foci simultaneously when the different paths

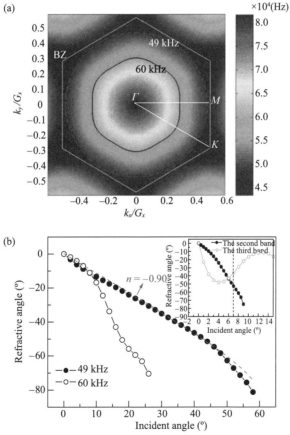

FIG.4. The EFS and the corresponding relationships between the refractive angle and the incident angle of 49, 60 and 73 kHz. (a), The EFSs of the SC at 60 and 49 kHz in the k space. Colours denote the values of the frequencies of the EFSs. The red dashed, the blue and the green lines denote EFSs at 49 and 60 kHz and the first Brillouin zone boundary, respectively. (b), The relationships between the refractive angles and the incident angles at 49 and 60 kHz. The red dashed line is the fitting line of the curve at 49 kHz with an effective index of -0.90. The inset is the dependence of refractive angles on incident angles at 73 kHz located in the second band and the third band.

are indistinguishable in the photonic crystal[29]. We believe the study of different Bloch states of waves in periodic structures will have great impact on basic physics in photo-electronics and quantum photonics, and will bring great potential applications.

In summary, the acoustic negative birefraction phenomenon has been established in a two-dimensional SC composed of an array of steel cylinders embedded in air. By this effect, a novel double-focusing-imaging phenomenon has been realized. These two refractive waves are coherent, which might be used for division of a wavefront in acoustic holography and acoustic communication. The artificial ultrasonic periodic composite material gives a means to control the propagation of acoustic waves and show many new effects and novel phenomena such as acoustic negative birefraction. We believe it will open a new horizon in the design of new acoustic devices.

References and Notes

[1] Born, M. & Wolf, E. *Principles of Optics* (Pergamon, Oxford, 1975).

[2] Netti, M. C. *et al*. Optical trirefringence in photonic crystal waveguides. *Phys. Rev. Lett.* **86**, 1526 – 1529 (2001).

[3] Elliott, J., Smolyaninov, I., Zheludev, N. & Zayats, A. Wavelength dependent birefringence of surface plasmon polaritonic crystals. *Phys. Rev. B* **70**, 233403 (2004).

[4] Zharov, A. A., Zharova, N. A., Noskov, R. E., Shadrivov, I. V. & Kivshar, Y. S. Birefingent left-handed metamaterials and perfect lenses for vectorial fields. *New J. Phys.* **7**, 220 – 228 (2005).

[5] Kosaka, H. *et al*. Superprism phenomena in photonic crystals. *Phys. Rev. B* **58**, R10096 – R10099 (1998).

[6] Gajić, R., Meisels, R., Kuchar, F. & Hingerl, K. Refraction and rightness in photonic crystals. *Opt. Express* **13**, 8596 – 8605 (2005).

[7] Kushwaha, M. S., Halevi, P., Dobrzynski, L. & Djafari-Rouhani, B. Acoustic band structure of periodic elastic composites. *Phys. Rev. Lett.* **71**, 2022 – 2025 (1993).

[8] Montero de Espinosa, F. R., Jimenez, E. & Torres, M. Ultrasonic band gap in a periodic two-dimensional composite. *Phys. Rev. Lett.* **80**, 1208 – 1211 (1998).

[9] Liu, Z. *et al*. Locally resonant sonic materials. *Science* **289**, 1734 – 1736 (2000).

[10] Sánchez-Pérez, J. V. *et al*. Sound attenuation by a two-dimensional array of rigid cylinders. *Phys. Rev. Lett.* **80**, 5325 – 5328 (1998).

[11] Gorishnyy, T., Ullal, C. K., Maldovan, M., Fytas, G. & Thomas, E. L. Hypersonic phononic crystals. *Phys. Rev. Lett.* **94**, 115501 (2005).

[12] Cheng, W., Wang, J. J., Jonas, U., Fytas, G. & Stefanou, N. Observation and tuning of hypersonic bandgaps in colloidal crystals. *Nature Mater.* **5**, 830 – 836 (2006).

[13] Cervera, F. *et al*. Refractive acoustic devices for airborne sound. *Phys. Rev. Lett.* **88**, 023902 (2001).

[14] Hu, X. H., Shen, Y. F., Liu, X. H., Fu, R. T. & Zi, J. Superlensing effect in liquid surface waves. *Phys. Rev. E* **69**, 030201(R) (2004).

[15] Zhang, X. & Liu, Z. Negative refraction of acoustic waves in two-dimensional phononic crystals. *Appl. Phys. Lett.* **85**, 341 – 343 (2004).

[16] Yang, S. *et al*. Focusing of sound in a 3D phononic crystal. *Phys. Rev. Lett.* **93**, 024301 (2004).

[17] Feng, L. et al. Acoustic backward-wave negative refractions in the second band of a sonic crystal. *Phys. Rev. Lett.* **96**, 014301 (2006).

[18] Ke, M. et al. Negative-refraction imaging with two-dimensional phononic crystals. *Phys. Rev. B* **72**, 064306 (2005).

[19] Veselago, V. G. The electrodynamics of substances with simultaneously negative values of ε and μ. *Sov. Phys. Usp.* **10**, 509–514 (1968).

[20] Pendry, J. B., Holden, A. J., Robbins, D. J. & Stewart, W. J. Magnetism from conductors and enhanced nonlinear phenomena. *IEEE Trans. Microwave Theory Tech.* **47**, 2075–2084 (1999).

[21] Smith, D. R., Padilla, W. J., Vier, D. C., Nasser-Nemat, S. C. & Schultz, S. Composite medium with simultaneously negative permeability and permittivity. *Phys. Rev. Lett.* **84**, 4184–4187 (2000).

[22] Shelby, R. A., Smith, D. R. & Schultz, S. Experimental verification of a negative index of refraction. *Science* **292**, 77–79 (2001).

[23] Luo, C., Johnson, S. G., Joannopoulos, J. D. & Pendry, J. B. All-angle negative refraction without negative effective index. *Phys. Rev. B* **65**, 201104(R) (2002).

[24] Cubukcu, E., Aydin, K., Ozbay, E., Foteinopoulou, S. & Soukoulis, C. M. Negative refraction by photonic crystals. *Nature* **423**, 604–605 (2003).

[25] Foteinopoulou, S. & Soukoulis, C. M. Negative refraction and left-handed behavior in two dimensional photonic crystals. *Phys. Rev. B* **67**, 235107 (2003).

[26] Berrier, A. et al. Negative refraction at infrared wavelengths in a two-dimensional photonic crystal. *Phys. Rev. Lett.* **93**, 073902 (2004).

[27] Foteinopoulou, S. & Soukoulis, C. M. Electromagnetic wave propagation in two-dimensional photonic crystals: A study of anomalous refractive effects. *Phys. Rev. B* **72**, 165112 (2005).

[28] Ozbay, E. Plasmonics: Merging photonics and electronics at nanoscale dimensions. *Science* **311**, 189–193 (2006).

[29] Hänsch, T.W. & Walther, H. Laser spectroscopy and quantum optics. *Rev. Mod. Phys.* **71**, S242–S252 (1999).

[30] The work was jointly supported by the National Basic Research Programme of China (973 Programme, grant no. 2007CB613202) and the National Nature Science Foundation of China (grant no. 50632030). We also acknowledge support from the Changjiang Scholars and Innovative Research Team in the university (PCSIRT).

[31] Supplementary Information accompanies this paper on www.nature.com/naturematerials.

[32] Sample fabrication: A two-dimensional triangular-lattice SC was built up. An aluminium plate was designed to be drilled as a triangular array of holes with the lattice constant of 4.5 mm. By inserting 250-mm-long steel cylinders with a radius of 1.0 mm into the periodically drilled plate, a two-dimensional triangular SC was constructed with steel cylinders in air background ($\rho_{steel} = 7,800$ kg·m^{-3} (mass density of steel), $\rho_{air} = 1.21$ kg·m^{-3} (mass density of air), $c_{steel} = 6,100$ m·s^{-1} (longitudinal sound velocity in the steel), $c_{air} = 344.5$ m·s^{-1} (sound velocity in air at 20℃)). Steel rods were arranged in the Γ - K direction with 99 layers in the lateral direction of the plate, and Γ - M with 16 layers in the perpendicular direction.

[33] Experimental set-up and data acquisition: The measurement set-up of the negative refraction and the imaging experiment is illustrated in Supplementary Information, Fig. S1. A fairly good point source is obtained, which was confirmed by the experimental measurement (see Supplementary Information, Fig. S2). This point source was placed 15 mm in front of the SC and the detector was positioned behind

the SC. By scanning the distributions of the intensity and the phase of the pressure field using a 1/8 inch transducer receiver (Brüel & Kjær Company, Denmark), the refraction of beams and focused point images behind the SC were mapped out. The experiments were carried out at three different frequencies: 49 kHz, 60 kHz, located only in the second band, and 73 kHz, located in the second and the third bands simultaneously.

Extraordinary Acoustic Transmission through a 1D Grating with Very Narrow Apertures[*]

Ming-Hui Lu, Xiao-Kang Liu, Liang Feng, Jian Li, Cheng-Ping Huang, Yan-Feng Chen, Yong-Yuan Zhu, Shi-Ning Zhu, and Nai-Ben Ming

National Laboratory of Solid State Microstructures and Department of Materials Science and Engineering, Nanjing University, Nanjing, 210093, People's Republic of China

    Recently, there has been an increased interest in studying extraordinary optical transmission (EOT) through subwavelength aperture arrays perforated in a metallic film. In this Letter, we report that the transmission of an incident acoustic wave through a one-dimensional acoustic grating can also be drastically enhanced. This extraordinary acoustic transmission (EAT) has been investigated both theoretically and experimentally, showing that the coupling between the diffractive wave and the wave-guide mode plays an important role in EAT. This phenomenon can have potential applications in acoustics and also might provide a better understanding of EOT in optical subwavelength systems.

    Recently, the extraordinary transmission of light through artificially micro-structured metal surface has attracted much attention[1—3]. It has been demonstrated that much more light can transmit through one or two dimensional metallic gratings [1—11] than the prediction by the conventional aperture theory[12]. Much effort devoted to the research on such an issue stems from both theoretical and practical interests. In practice, the extraordinary transmission can lead to a wide range of future applications such as subwavelength photolithography[13,14]. In principle, the physical origin of the effect has sparked many discussions[15—17]. Some models have been proposed to explain the underlying physics, such as surface plasmon polariton (SPP)[1—7], waveguide mode[8], cavity resonances[9], and dynamical diffraction[10]. More recently, a two-wave model has been proposed to analyze the properties of the waves scattered by nanoslit apertures. Two types of surface waves, SPP and a free-space surface creep wave with radiative and evanescent character[17], both can contribute to the transmission modulation[18,19]. The detailed physical mechanism for the transmission enhancement still needs further investigation[19,20]. Among these models, the SPP has been widely accepted. SPP is surface electromagnetic waves with collective electron oscillations, due to coupling between light and surface charges, capable of propagating along the metal surface with the amplitudes decaying into both sides. For other classical waves, however, such surface collective

---

[*] Phys.Rev.Lett.,2007,99(17):174301

oscillations do not exist. Therefore, the investigation of enhanced transmission in these classical waves can be helpful for understanding the possible mechanisms, in addition to SPP, in extraordinary optical transmission (EOT).

The transmission through acoustic gratings has been studied for several decades[21]. In the subwavelength regime, the acoustic transmission coefficient of power is tiny and can be expressed as[22]

$$t_P \approx \left(\frac{2\pi a}{\lambda}\right)^2 \qquad (1)$$

where a is the width of the aperture, and λ is the wavelength of the incident acoustic wave, respectively. According to Eq. (1), the acoustic transmission will drop rapidly as λ increases. Rayleigh once pointed out that the acoustic resonant phenomena should appear at some special frequencies when acoustic wave impinges normally on a steel surface perforated by arrays of subwavelength slits with proper depths. The intensity of acoustic waves in these slits can be dramatically enhanced[23], which might result in extraordinary acoustic transmission (EAT). The surmises have been preliminarily studied by multiplescattering numerical simulations[24], but without any experimental evidence and analytically physical explanation. In this Letter, we experimentally demonstrated the EAT phenomena through 1D acoustic gratings with subwavelength apertures. We also exploited a complete analytical model to investigate the physical origin of EAT, which is attributed to the strong coupling between the diffractive surface wave and waveguide modes.

A rectangular acoustic grating with very narrow apertures as shown in Fig. 1(a) has been considered. The grating consisted of 120 4 mm-wide square steel rods with the period of $d=4.5$ mm; thus the aperture size is 0.5 mm and the thickness of the grating is 4 mm. The transmission spectra were measured by an analysis package of the 3560C Brüel & Kjær Pulse Sound and Vibration Analyzers. The zero-order transmission for normal incidence is shown in Fig. 1(a) as a function of λ. The Rayleigh minimum, corresponding to Wood's anomaly[23], is observed at the wavelength ($\lambda=d$), when Bragg diffraction is established. In contrast, the resonance-assisted enhanced transmission is also observed. Within the range of the scanned wavelength, there are two transmission peaks for the grating. For the peak at $\lambda=2.02d$, the experimental transmission efficiency attains 92% (8.28 if normalized by the filling ratio of the aperture). Compared with the result predicted by Eq. (1) for a single aperture (about 0.11), the acoustic transmission has a 75-fold enhancement. Considering a 0.5 mm-wide, 4 mm-thick aperture in a steel plate, the normalized transmission is about 0.6, so it still gets enhanced about 15 times through the grating.

By implementing the finite element simulation with Comsol Multiphysics 3.3, EAT has been numerically reproduced, concerning the spectrum shape and the position of transmission minima and maxima [See the black curve in Fig. 1(a)]. Notice that these two resonant peaks in the experiment are broader than those in the calculation, which may

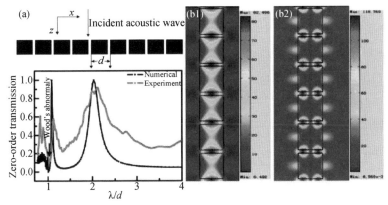

FIG. 1. (color online) (a) Numerical (black curve) and experimental (gray curve, red online) zero-order transmission spectra for a rectangular grating with the size of 4 mm, the period of 4.5 mm. The Wood's anomaly is indicated by an arrow. The upper panel is the sketch of the gratings. (b) The spatial intensity distribution of the pressure field in the grating at the wavelength of $2.02d$ (b1) and $1.09d$ (b2).

result from the dissipative loss, diffraction of finite size of the sample, and other experimental errors. The spatial intensity distributions of the pressure fields in the grating at the wavelength of $2.02d$ and $1.09d$, corresponding to two resonant transmission peaks, have been calculated to show in Figs. 1(b1) and 1(b2), respectively. The intensity of the fields inside the grating is about 80–119 times larger than the original incidence (note that the fields are not zero inside the steel rods; indeed, for the real material, a weak penetration of the wave into rods can be expected). From the analysis of the spatial intensity distribution of the pressure field, the two resonant peaks are believed to originate from the Fabry-Perot resonance inside narrow apertures. These two peaks correspond to the fundamental and the second order resonant modes. From Figs. 1(b1) and 1(b2), it can be clearly seen that the surface diffractive waves are also excited on both sides of the grating.

The incident angular dependence of transmission spectra has been also studied. Figure 2 exhibits the zero-order transmission spectra of acoustic waves with different incident angles $\theta = 0°, 5°, 10°, 15°, 20°$, in which the measurement [Fig. 2(b)] is consistent with theoretical calculation [Fig. 2(a)]. The two resonant peaks show different angular dependence: the peak at the longer wavelength shows insensitive angular dependence, but the peak close to Wood's anomaly is very sensitive to incident angles. As shown in the inset of Fig. 2(b), the increase of the incident angle leads to the redshift of this peak with its amplitude rapidly fading. This angular dependence of zero-order transmission suggests that diffractive waves along the periodic direction must be responsible for the EAT. The wave vector of the transverse diffractive waves can be expressed as $k_{xm} = \dfrac{2\pi}{\lambda} \sin\theta \pm mG_x$ (here G_x

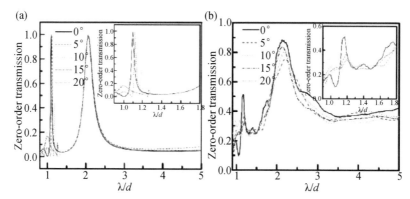

FIG. 2. (color online) (a) Theoretical and (b) experimental angular dependences of zero-order transmission spectra for the rectangular grating. Insets are the zoomed-in figures.

$=\dfrac{2\pi}{d}$ is the grating momentum, and m is the order of the diffraction). The change of the incident angle θ may influence the diffractive surface waves and thus the resonant peak. The different behaviors of two transmission peaks with the incident angle will be discussed in details subsequently.

The zero-order transmission as a function of the thickness of the grating has been shown in Fig. 3. More resonant modes appear as the thickness of the grating increases. In both the calculation [Figs. 3(a) and 3(c) (black curve)] and the measurement [Figs. 3(b) and 3(c) (gray curve, red online)], as the thickness increases from 2 mm to 12 mm, the fundamental frequency of resonance decreases monotonously. However, when the grating is thin, the frequency of the fundamental resonance is close to and affected by the Wood's anomaly, thereby deviating further from the Fabry-Perot resonant mode. If the thickness is large, the fundamental frequency of the resonant mode tends to be saturated, which also deviates from the Fabry-Perot resonant mode. Here, we stress that the acoustic transmission spectrum has a strong dependence on the period but is much less sensitive to the width of the aperture. When the width of aperture becomes larger, in spite of the decrease of the intensity of the transmission, the positions of the transmission maximum and minimum are almost invariable.

To further understand the EAT phenomenon, an original rigorous analysis model has been developed. First, the pressure field inside the grating is assumed as the superposition of two counter propagating waves, forward and backward zero-order acoustic modes, in the air aperture surrounded by steel walls. The monomodal approximation is then used in our analytical model[25]. Second, the composition of the diffractive waves above and below the grating should be considered to match the boundary condition on each side of the grating. So the pressure field p_w in the grating can be expressed as the rectangular waveguide mode:

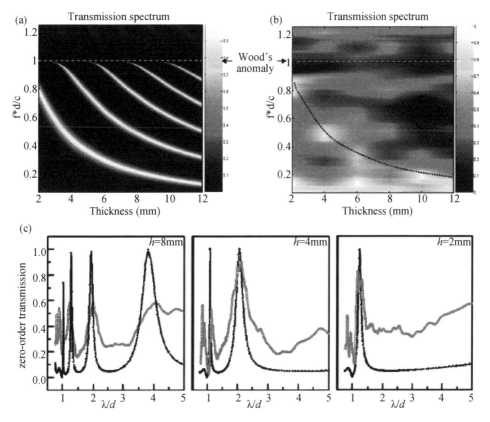

FIG. 3. (color online) (a) Theoretical and (b) experimental zero-order transmission spectra with different grating thickness. The Wood's anomaly is denoted by white dotted line and the black dotted line denotes the track of the fundamental resonant mode. (c) Theoretical (black curve) and experimental (gray curve, red online) zero-order transmission spectra with grating thickness $h = 8$ mm, 4 mm, 2 mm.

$$p_w = \cos(\beta x)[A\exp(-jqz) + B\exp(jqz)], \quad (-a/2 \leqslant x \leqslant a/2), \quad (2)$$

where A and B are the unknown amplitudes of the forward and backward waves; $q = \sqrt{k^2 - \beta^2}$ is the propagation constant of waveguide mode. The eigenwave vector of the aperture waveguide mode β can be determined by the dispersion equation as

$$\beta\tan(\beta a) = jk \frac{\rho_0 c_0}{j\rho_{st}(d-a)\omega + \rho_0 c_0}, \quad (3)$$

where ρ_0, ρ_{st} are the mass densities of air and steel, respectively; c_0 is the acoustic velocity in air, and ω, k are the frequency and wave vector in air of the incident wave. Within the scope of this model, the zero-order transmittance is

$$t_0 = \left| \frac{qfs_0}{k_{z0}} \cdot \frac{4g_0\left[\mathrm{sinc}\left(\frac{\beta a}{2}\right)\right]}{\Gamma_1^2 e^{jqh}} \cdot \frac{1}{1-\gamma^2} \right|^2, \quad (4)$$

where $\Gamma_1 = \sum_m \frac{g_m s_m qf}{k_{zm}} + \left[\mathrm{sinc}\left(\frac{\beta a}{2}\right)\right]$, $\Gamma_2 = \sum_m \frac{g_m s_m qf}{k_{zm}} - \left[\mathrm{sinc}\left(\frac{\beta a}{2}\right)\right]$, and $\gamma = \frac{\Gamma_2}{\Gamma_1} e^{-jqh}$

is defined as the resonant factor. Herein, $g_m = \left[\sin c\left(\dfrac{k_{xm}a}{2}\right)\right]$, $s_m = \dfrac{1}{a} \times \int_{-a/2}^{a/2} \cos(\beta x) e^{jk_{xm}x} dx$, $k_{xm} = \dfrac{2\pi}{\lambda}\sin\theta - m\dfrac{2\pi}{d}$, $k_{zm} = (k^2 - k_{xm}^2)^{1/2}$, and f is the filling ratio of the aperture. More details of this model are elucidated in the Supplementary materials[26].

The zero-order transmission predicted by the model was marked by the black curve in Fig. 4. The resonances of the zero-order transmission should appear when the imaginary part of the resonant factor γ is zero. These zeros occur when the "Fabry-Perot resonance" condition is satisfied:

$$\arg(\gamma) = qh - \arg\left(\dfrac{\Gamma_2}{\Gamma_1}\right) = m\pi, \tag{5}$$

where m is an integer, indicating the mth-order Fabry-Perot resonance. Figure 4(a) shows the principle phase of the resonant factor. The phase becomes zero at the wavelength of the fundamental resonant peak. At the second order resonant peak, π or $-\pi$ presents, meaning the switching from the first Brillouin zone to the second one. The first term in Eq. (5) is from the expression of the waveguide mode and is independent of the incident angle θ. The second term is an additional phase shift due to the composition of the diffractive waves along the grating surfaces. That is, the waveguide resonance has been dressed through the coupling of waveguide mode to the diffractive waves. As shown in Fig. 4(a), when incident angle $\theta = 0°$, the phase shift changes tardily if λ is relatively larger than the period, but rapidly if $\lambda \approx d$. This means the position of the resonant peak strongly depends on the additional phase shift. So the position of the fundamental resonant peak is mainly dependent on the waveguide mode and nearly independent of incident angles. Equation (4) indicates that zero transmission can occur when Γ_1 becomes infinite, which means $k = k_{xm}$, corresponding to Wood's anomaly.

With narrow apertures, the characteristics of the acoustic grating are quite different from the conventional one with larger apertures[23]. The scattering potential for the wide aperture is small so that the first-order Born's approximation is proper and the zero-order transmissivity can be expressed as[27]

$$T \sim \sqrt{t_p}\left[\text{sinc}\left(\dfrac{k_{x0}a}{2}\right)\text{comb}\left(\dfrac{k_{x0}d}{2}\right)\right] = \sqrt{t_p}\left[g_0\text{comb}\left(\dfrac{k_{x0}d}{2}\right)\right]. \tag{6}$$

However, for the narrow apertures with subwavelength dimensions, the multiple scattering and the dynamical diffraction effect have to be considered. So in our grating, many diffractive modes can be simultaneously excited[11,18,19]. Thus, the strongly excited diffractive waves should play a crucial role in the extraordinary transmission. Comparing Eq. (4) with Eq. (6), we can define $F(\lambda) = (\Gamma_1^2 e^{jqh} - \Gamma_2^2 e^{-jqh})^{-1}$ as the enhancement factor of diffractive waves. Figure 4(b) illustrates the wavelength dependence of the enhancement factor $F(\lambda)$. When $F(\lambda)$ is maximal, t_0 also reaches the maximum. The intensity enhancement $|F(\lambda)|^2$ for the fundamental resonant peak is about 25 times, consistent with the simulation and experimental results. Moreover, the normalized intensity of the pressure filed in the aperture was also calculated to show in Fig. 4(b). The maximum of the field intensity is also corresponding to the maximum of the zero-order transmission, and reaches 110 for the second order resonant mode and 84 for the fundamental resonant mode, which also agree well with the finite element simulations [Figs. 1(b1) and 1(b2)].

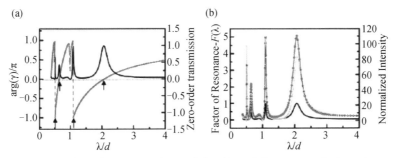

FIG. 4. (color online) (a) The dependence of the phase of the resonant factor γ (gray curve, red online) on the wavelength, and the zero-order transmission spectra (black curve) for the grating. (b) The dependence of the enhancement factor of diffraction $F(\lambda)$ on the wavelength (gray curve, red online), and the zero-order transmission spectra (black curve) for the grating. The light gray curve (green online) is the wavelength dependence of the normalized intensity of the pressure field inside the aperture.

In conclusion, similar EAT phenomena in acoustic subwavelength systems have been investigated. Based on rigorous analysis, the resonant enhancement of the zero-order transmission is believed to be attributed to the coupling between the compositions of diffractive waves excited on the surfaces and the Fabry-Perot resonant modes inside the apertures. This enhancement stems from dynamic diffractions[11] and is also valid for the optical counterpart. However, the discrepancies between EAT and EOT are obvious: (1) there is no SPP excited in acoustic systems; (2) the zero-order propagating waveguide mode exists in acoustic narrow apertures, but in optic subwavelength systems, the cutoff frequency can lead to the truncated evanescent waveguide modes. So in the visible and near infrared regimes, SPP plays a key role in EOT[17−19]. However, the contribution from the dynamic diffractions cannot be neglected. Although more studies are necessary, especially for 2D metallic gratings, we believe that our work provides a profound basis for such extraordinary transmission in subwavelength systems. This remarkable effect will bring a great impact on ultrasonic devices and applications such as thickness selected frequency acoustic filters, acoustic collimators, and compacted acoustic devices with subwavelength, etc.

References and Notes

[1] T. W. Ebbesen, H. J. Lezec, H. F. Ghaemi, T. Thio, and P. A. Wolff, Nature (London) **391**, 667 (1998).

[2] H. F. Ghaemi, T. Thio, D. E. Grupp, T. W. Ebbesen, and H. J. Lezec, Phys. Rev. B **58**, 6779 (1998).

[3] C. Genet and T. W. Ebbesen, Nature (London) **445**, 39 (2007).

[4] F. J. Garcia-Vidal, H. J. Lezec, T. W. Ebbesen, and L. Martin-Moreno, Phys. Rev. Lett. **90**, 213901 (2003).

[5] A. Barbara, P. Quemerais, E. Bustarret, and T. Lopez-Rios, Phys. Rev. B **66**, 161403(R) (2002).

[6] D. C. Skigin and R. A. Depine, Phys. Rev. Lett. **95**, 217402 (2005).

[7] K. G. Lee and Q. H. Park, Phys. Rev. Lett. **95**, 103902 (2005).

[8] Y. Kurokawa and H. T. Miyazaki, Phys. Rev. B **75**, 035411 (2007).

[9] Q. Cao and P. Lalane, Phys. Rev. Lett. **88**, 057403 (2002).

[10] M. M. J. Treacy, Appl. Phys. Lett. **75**, 606 (1999).

[11] W. Srituravanich, N. Fang, C. Sun, Q. Luo, and X. Zhang, Nano Lett. **4**, 1085 (2004).

[12] H. A. Bethe, Phys. Rev. **66**, 163 (1944).

[13] D. B. Shao and S. C. Che, Appl. Phys. Lett. **86**, 253 107 (2005).

[14] J. Bravo-abad, A. Degiron, F. Przybilla, C. Genet, F. J. Garcia-vidal, L. Martin-Moreno, and T. T. Ebbesen, Nature Phys. **2**, 120 (2006).

[15] J. A. Porto, F. J. Garcia-Vidal, and J. B. Pendry, Phys. Rev. Lett. **83**, 2845 (1999).

[16] L. Martin-Moreno, F. J. Garcia-Vidal, H. J. Lezec, K. M. Pellerin, T. Thio, J. B. Pendry, and T. W. Ebbesen, Phys. Rev. Lett. **86**, 1114 (2001).

[17] L. Aigouy, P. Lalane, J. P. Hugonin, G. Julié, V. Mathet, and M. Mortier, Phys. Rev. Lett. **98**, 153902 (2007).

[18] L. Chen, J. T. Robinson, and Michal Lipson, Opt. Express **14**, 12 629 (2006).

[19] P. Lalanne and J. P. Hugonin, Nature Phys. **2**, 551 (2006).

[20] G. Gay, O. Alloschery, B. Viaris De Lesegno, C. O'Dwyer, J. Weiner, and H. J. Lezec, Nature Phys. **2**, 262 (2006).

[21] P. Filippi, D. Habault, J.-P. Lefebvre, and A. Bergassoli, *Acoustics: Basic Physics, Theory and Methods* (Academic, London, 1999).

[22] L. E. Kinsler, A. R. Frey, A. B. Coppens, and J. V. Sanders, *Fundamentals of Acoustics* (John Wiley & Sons, Inc, New York, 1982).

[23] L. Rayleigh, Philos. Mag. **39**, 225 (1920).

[24] X. D. Zhang, Phys. Rev. B **71**, 241102 (2005).

[25] R. C. Mcphedran and D. Maystre, Applied Physics (Berlin) **14**, 1 (1977).

[26] See EPAPS Document No. E-PRLTAO-99-002743 for detailed elucidation of the analytical model and the schematic of the experiment setup. For more information on EPAPS, see http://www.aip.org/pubservs/epaps.html.

[27] See, for example, M. Born and E. Wolf *Principles of Optics* (Pergamon, Oxford, UK, 1975).

[28] Discussions and technical assistance from Professor X. D. Zhang are gratefully acknowledged. The work was jointly supported by the National Basic Research Program of China and the National Nature Science Foundation of China (Grant No. 50632030). We also acknowledge the support from the Changjiang Scholars and Innovative Research Team in University (PCSIRT) and the Nature Science Foundation of Jiangsu Province (BK2007712).

Acoustic Surface Evanescent Wave and its Dominant Contribution to Extraordinary Acoustic Transmission and Collimation of Sound[*]

Yu Zhou, Ming-Hui Lu, Liang Feng, Xu Ni, Yan-Feng Chen,
Yong-Yuan Zhu, Shi-Ning Zhu, and Nai-Ben Ming

National Laboratory of Solid State Microstructures and Department of Materials Science and Engineering,
Nanjing University, Nanjing, 210093, People's Republic of China

We demonstrate both theoretically and experimentally the physical mechanism that underlies extraordinary acoustic transmission and collimation of sound through a one-dimensional decorated plate. A microscopic theory considers the total field as the sum of the scattered waves by every periodically aligned groove on the plate, which divides the total field into far-field radiative cylindrical waves and acoustic surface evanescent waves (ASEWs). Different from the well-known acoustic surface waves like Rayleigh waves and Lamb waves, ASEW is closely analogous to a surface plasmon polariton in the optical case. By mapping the total field, the experiments well confirm the theoretical calculations with ASEWs excited. The establishment of the concept of ASEW provides a new route for the integration of subwavelength acoustic devices with a structured solid surface.

The interactions between matter and waves have been systematically investigated for centuries in natural materials. Recently, the study of matter-wave interactions has been further extended to artificial metamaterials in various realms such as electronics [1], optics [2,3], acoustics [4], and even plasmonics [5]. Among them, the surface plasmon polariton (SPP) has received a lot of attention because of its close relation with a number of intriguing optical phenomena, such as superlens [6], surface enhanced Raman scattering [7], and extraordinary optical transmission (EOT) [8], which have enormous potential applications in the next generation nanoscale optoelectronics. However, recent progress shows that other types of surface excitation mechanisms, such as creep waves [9,10] or cylindrical waves [11,12], also play equally important roles in subwavelength metallic nanostructures.

Because of the similarities between sound and light, the rapid development of optical metamaterials intrigues extensive studies of acoustic counterparts. Acoustic negative refraction [13] and birefraction [14] have been experimentally demonstrated in acoustic

---

[*] Phys.Rev.Lett.,2010,104(16):164301

metamaterials. Extraordinary acoustic transmissions (EAT) through both onedimensional (1D) and two-dimensional (2D) subwavelength acoustic gratings were demonstrated and investigated in both experiment and theory[15-18]. The physical origin of EAT is the coupling of collective surface Bloch modes and Fabry-Perot resonance[15,16] (this coupled mode is also called leaky surface guided mode[17]). Sound collimation was also achieved by engineering the acoustic surface modes[19], which is similar to the light beaming phenomenon supported by SPP[20].

However, the study on EAT so far is not comprehensive since its theory is usually based on a macroscopic viewpoint concerning a mode expansion of pressure field. In this Letter, we will for the first time experimentally and theoretically investigate the rich physics behind EAT and sound collimation by considering microscopic scattering dynamics on a 1D acoustic subwavelength structure. In our microscopic description, we show that the total pressure field involves dynamic scattering processes of two types of waves: One is acoustic surface evanescent waves (ASEWs) bound to the interface of a periodically grooved subwavelength structure; the other is cylindrical waves (CWs) that radiate into far field. Experimental characterization of EAT and sound collimation supported by this structure validates our analytical microscopic theory.

Our sample is a steel plate with periodically imperforated grooves on both sides and a single slit at the center, as depicted in Fig. 1(a). The thickness of the plate and the period of the grooves are $H=4$ mm and $d=5$ mm, respectively. The width of the central slit and the grooves w, and the depth of the grooves h are both 0.5 mm. The transmission spectrum through the central slit of the sample was measured with an acoustic plane wave incidence[21]. EAT has been observed in the measured transmission spectrum as shown in Fig. 1(b). Within the range of the scanned frequencies (from 20 kHz to 90 kHz), there are two resonance peaks located at $\lambda=1.902d$ ($f=36.07$ kHz) and $\lambda=1.035d$ ($f=66.28$ kHz), respectively. Finite element simulations have been implemented to model the corresponding intensity field distributions at two resonance frequencies as shown in Figs. 1(c) and 1(d), respectively. At $f=36.07$ kHz, the acoustic field is strong inside the central slit but weak in the vicinities of the grooves [see Fig. 1(c)]. This indicates that the transmittance is barely affected by the periodicity of the grooves. Fabry-Perot resonance in a single aperture is responsible for EAT at this frequency[15]. However, at $f=66.28$ kHz, the acoustic field is stronger around the grooves due to the excitation of ASEWs, as shown in Figs. 1(d) and 1(e). It is evident that the effect of ASEWs is dominant when the phases are matched (the periodicity of the grooves is close to the wavelength $\lambda=1.035d$. The related frequency is denoted as the "resonance frequency" in the following text). It is worth noting that within the above experiment, two kinds of waves, ASEWs and radiative CWs, exist simultaneously. CWs are results of diffractions and scattering processes, while

ASEWs are supported by the finite periodic subwavelength structure[22]. In order to understand the roles of these two different waves in the EAT and collimation effects, we developed a microscopic analytical method to calculate the transmission spectrum and the diffracted pressure field.

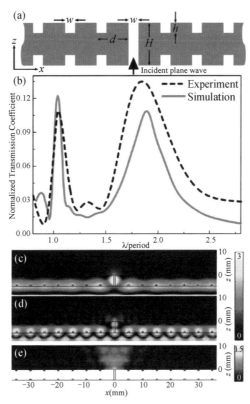

FIG.1.(color online) (a) A schematic of the sample. There are $9 \times 4 = 36$ grooves in total; (b) Experimental (dashed-black line) and simulated (solid-red line) transmission spectra; (c) and (d) FEM simulated near-field intensity pattern in the vicinity of the sample at $\lambda = 1.902d$ and $\lambda = 1.035d$; (e) Experimental near-field intensity pattern in the vicinity of the sample at $\lambda = 1.035d$.

The central slit and the grooves can be treated as independent subwavelength cavities. Acoustic scattering depends only on the depth of the cavities and their geometric shapes. We first considered a more fundamental process in which waves are scattered by a cavity with infinite depth[23]. The transmission and reflection coefficients through this cavity are calculated using a Fourier modal method[24] with defining t_n, r_n, t_g, r_g as normal and grazing incidences from free space to cavity, respectively, and t_c, r_c as incidence from cavity to free space [Fig. 2(a)]. Because of boundary continuities, the expansions of the pressure and velocity fields of the fundamental cavity modes must match the plane wave expansions outside the cavity at the interface. Therefore, these coefficients under the condition of $w \ll \lambda$ can be expressed as $t_n = 2/(1+G)$, $r_n = t_n - 1$, $t_g = F/(1+G)$, $r_g = t_g -$

$1, r_c = (G-1)/(G+1)$, and $t_c = r_c + 1$, where $G = \dfrac{k_0}{\pi w}\displaystyle\int_{-\infty}^{+\infty} dQ \, \dfrac{1-\cos(wQ)}{Q^2 \sqrt{k_0^2 - Q^2}}$, $F = \mathrm{sinc}\left(\dfrac{k_0 w}{2}\right)$, and k_0 is the wave vector. Herein, G means the transmission properties of 1D single slit, originating from the coupling between the fundamental eigenmode and all diffractive waves[15,17]. Scattered pressure field of more complicated geometries, such as the slit and grooves in our experiments, can be calculated using these coefficients. With a normal incidence P_0 scattered by a cavity of depth H (or h), the pressure P_s at the center of the slit (or groove) on the interface can be expressed as $P_s = P_0 (t_n r_c G_{\mathrm{slit}} + r_n)$ for a slit, where $G_{\mathrm{slit}} = t_c \exp(2ik_0 H)/[1 - \exp(2ik_0 H) r_c^2]$, and $P_s = P_0 (t_n G_{\mathrm{gv}} + r_n)$ for a groove, where $G_{\mathrm{gv}} = t_c \exp(2ik_0 h)/[1 - \exp(2ik_0 h) r_c]$ (for details, see [21]).

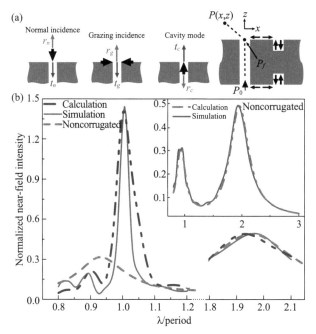

FIG. 2. (color online)(a) The left frames are schematics of the fundamental transmission and reflection coefficients defined in the first part of the calculation. The right frame is a schematic of the calculation process; (b) Calculated (dashed-blue line) and simulated (solid-red line) results of the normalized near-field intensity spectrum of the corrugated structure at $x=0$ and $z=2$ mm. Inset is a comparison of the noncorrugated structure.

The pressure P_s is regarded as a new point source and thus excites both CWs and ASEWs. For a CW that can propagate along the interface, its pressure field can be expressed as

$$P_{\mathrm{CW}}(x) = P_s \cdot A_{\mathrm{CW}}(x) = P_s \alpha_{\mathrm{CW}} \exp(ik_0 x)\left(\dfrac{x}{\lambda}\right)^{-m}, \qquad (1)$$

where α_{CW} is the scattering coefficient of the CW, $m \approx 0.5$ is the attenuation exponent of the CW[12], and k_0 is the propagation constant of sound in air. ASEW can also be excited if this

series of periodic P_s on the surface are in phase and its expression is
$$P_{\text{ASEW}}(x) = P_s \cdot A_{\text{ASEW}}(x) = P_s \alpha_{\text{ASEW}} \exp(ik_{\text{ASEW}} x), \tag{2}$$
where α_{SEW} is the scattering coefficient of the ASEW and its propagation constant is
$$k_{\text{ASEW}} = k_0 \sqrt{1 + \left[\frac{w}{d} \tan(k_0 h)\right]^2}. \tag{3}$$

Since all scattered waves can only be funneled through the central slit, it is necessary to express the field inside the central slit. For simplicity, an assumption has been applied in calculation: All waves are scattered twice at most by the structure (this has turned out to be a good approximation for sound waves; however, multiscattering should be considered when dealing with EM waves). The pressure field of the acoustic wave inside the central slit can be derived by combining Eqs.(1) and (2) with the fundamental transmission and reflection coefficients [see Fig. 2(a)]:
$$P_f = \left[P_0 t_n + 2t_g \sum_{n=1}^{9} P_n A_{\text{tot}}(nd)\right] G_{\text{slit}} \exp(-ik_0 H), \tag{4}$$
where $A_{\text{tot}}(x) = A_{\text{ASEW}}(x) + A_{\text{CW}}(x)$, and P_n considers the contribution from the nth groove on the input side,
$$P_n = P_0(t_n G_{\text{gv}} + r_n) + (t_g G_{\text{gv}} + r_g)\left[P_0(t_n r_c G_{\text{slit}} + r_n) \times A_{\text{tot}}(nd) + \sum_{\substack{m=-9 \\ m \neq 0, n}}^{9} P_0(t_n G_{\text{gv}} + r_n) A_{\text{tot}}(|m-n|)\right]. \tag{5}$$

Transmitted field through the central slit again excites CWs and ASEWs on the output interface, which can be calculated in a similar way. Since ASEWs are near-field waves, the transmitted pressure field mainly attributes to CWs (suppose $z=0$ at the exit of the central slit):
$$P(x,z) = P_f A_{\text{CW}} \sqrt{x^2 + z^2} + \sum_{\substack{n=-9 \\ n \neq 0}}^{9} P'_n A_{\text{CW}} \sqrt{(x-nd)^2 + z^2}, \tag{6}$$
where P'_n is the contribution of the nth groove on the output interface (for details, see [21])
$$P'_n = P_f (t_g G_{\text{gv}} + r_g) A_{\text{tot}}(nd). \tag{7}$$

With our experimental configuration, the transmission spectrum of field intensity at $x=0$ and $z=2$ mm has been calculated and compared to the case in which there is only the central slit in the plate [see Fig. 2(b), P_0 normalized to 1]. In these two different cases, the second peak almost remains the same, indicating a localized Fabry-Perot resonance without contribution from ASEWs, as shown in Fig. 1(c). At the first peak, however, the transmission is drastically enhanced (i.e., EAT is established) since the phase matching condition is satisfied and ASEWs are excited as seen in Figs. 1(d) and 1(e). According to Eqs. (4)–(7), although the near-field ASEWs can not directly contribute to the transmitted far field, they can modify the amplitude of CWs at the interface and affect the diffracted pattern of the far-field radiation.

To determine the relative weights of CWs and ASEWs in the total field, we analyzed the relation between ASEWs and CWs. From our calculation, it is shown that α_{ASEW} is much smaller than α_{CW} ($\alpha_{ASEW} \approx 1/7 \alpha_{CW}$ at the resonance frequency). This is different from the optical case, where SPP is the dominant factor[10,12]. But ASEWs can not be ignored because they barely decay with distance in our current length scale (a damping term $\left(\frac{x}{\lambda}\right)^{-0.1}$ due to scattering from shallow grooves has been properly retrieved from simulations and added to Eq. (2) to precisely describe ASEWs' propagation). To demonstrate this, we calculated the pressure field distribution at $z=0$ and $0<x<40$ mm at the resonance frequency using Eq. (6). At the input surface of the sample, ASEWs indeed directly contribute to the total field, so A_{CW} in Eq. (6) should be replaced by A_{tot}. As shown in Fig. 3, we found that field generated solely by the CWs does not agree with the simulation result; however, the total field including the contribution of ASEWs does. The phase matching condition is established for all the ASEWs generated by the grooves on the interface, resulting in a coherent transmission enhancement at the resonance frequency. Therefore, despite the fact that ASEWs have a relatively low weight in the total field, they are able to collect parts of the pressure from the grooves and funnel them through the central slit by modifying the amplitude of far-field radiative waves, which enhances the transmitted intensity many times larger than the case without grooves.

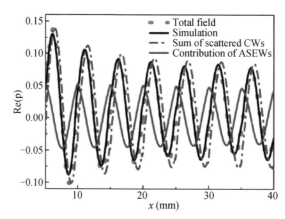

FIG. 3. (color online) Calculated pressure field on the right output side of the central slit ($z=0$) at the resonance frequency. The total field (red dots, grey dots in the white-black line) is composed of the sum of all the CWs (dashed-dotted-blue line, black dashed-dotted blue line) and of the generated ASEWs (solid-green line, grey in the white-black line), and is compared with the simulated result (solid-black line).

At the output side, the grooves act as radiators to effectively couple ASEWs to radiative waves in free space when the phases are matched. These radiators can slightly change the phase of ASEWs according to the position and depth of grooves, and thus modify the far-field diffracted patterns of the acoustic field[25]. Full range intensity field patterns at the resonance frequency have been mapped with an acoustic plane wave normally impinging on the sample, showing sound collimation in Fig. 4(a). The acoustic wave is well collimated after propagating $80\lambda_0$. This agrees well with simulations and calculations using our microscopic theory as shown in Figs. 4(b) and 4(c). We also did calculations at other frequencies. However, such collimation effects are barely observable without excitation of ASEWs, which is consistent with our theoretical explanation.

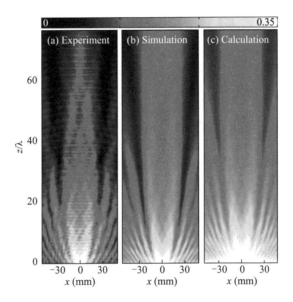

FIG. 4. (color online) (a) Experimental result of full-range intensity field pattern in the xz plane at the resonance frequency, showing the collimation effect; (b) Simulated intensity pattern at the resonance frequency. (c) Calculated intensity pattern at the resonance frequency using our microscopic theory.

In conclusion, we have experimentally observed and theoretically analyzed ASEWs and radiative CWs on the interface of our periodically grooved subwavelength structure as well as their associated EAT and sound collimation. More importantly, we have built a microscopic analytical model to describe the interaction between ASEWs and radiative CWs, which explains how ASEWs modulate and control far-field radiation by modifying the amplitudes of radiative waves. Although ASEW does not have a relatively high weight as SPP does in the optical case, it is still a key factor in both the EAT and sound collimation at the resonance frequency due to its impact on radiative CWs. These two kinds of waves make a contribution together and lead to novel acoustic effects[15—19,26—28] as well as potential acoustic devices and applications.

References and Notes

[1] J. B. Pendry, Science **315**, 1226 (2007).
[2] R.A. Shelby, D.R. Smith, and S. Schultz, Science **292**, 77 (2001).
[3] J. Valentine, S. Zhang, T. Zentgraf, E. Ulin-Avila, D. A. Genov, G. Bartal, and X. Zhang, Nature

(London) **455**, 376 (2008).

[4] Z.Y. Liu, X.X. Zhang, Y.W. Mao, Y.Y. Zhu, Z.Y. Yang, C.T. Chan, and P. Sheng, Science **289**, 1734 (2000).

[5] L. Feng, M. H. Lu, V. Lomakin, and Y. Fainman, Appl. Phys. Lett. 93, 231105 (2008).

[6] X. Zhang and Z.W. Liu, Nature Mater. **7**, 435 (2008).

[7] Y. Fang, N. H. Seong, and D. D. Dlott, Science **321**, 388 (2008).

[8] T.W. Ebbesen, H.J. Lezec, H.F. Ghaemi, T. Thio, and P. A. Wolff, Nature (London) **391**, 667 (1998).

[9] C. Genet and T.W. Ebbesen, Nature (London) **445**, 39 (2007).

[10] P. Lalanne and J.P. Hugonin, Nature Phys. **2**, 551 (2006).

[11] G. Gay, O. Alloschery, B.V.D. Lesegno, C. O'dwyer, J. Weiner, and H. J. Lezec, Nature Phys. **2**, 262 (2006).

[12] X.Y. Yang, H.T. Liu, and P. Lalanne, Phys. Rev. Lett. **102**, 153903 (2009).

[13] S. Zhang, L. L. Yin, and N. Fang, Phys. Rev. Lett. **102**, 194301 (2009).

[14] M.H. Lu, C. Zhang, L. Feng, J. Zhao, Y.F. Chen, Y.W. Mao, J. Zi, Y.Y. Zhu, S.N. Zhu, and N.B. Ming, Nature Mater. **6**, 744 (2007).

[15] M.H. Lu, X.K. Liu, L. Feng, J. Li, C.P. Huang, Y.F. Chen, Y.Y. Zhu, S.N. Zhu, and N.B. Ming, Phys. Rev. Lett. **99**, 174301 (2007).

[16] B. Hou, J. Mei, M.Z. Ke, Z.Y. Liu, J. Shi, and W. J. Wen, J. Appl. Phys. **104**, 014909 (2008).

[17] J. Christensen, L. Martín-Moreno, and F. J. García-Vidal, Phys. Rev. Lett. **101**, 014301 (2008).

[18] J. Mei, B. Hou, M. Z. Ke, S.S. Peng, H. Jia, Z.Y. Liu, J. Shi, W. J. Wen, and P. Sheng, Appl. Phys. Lett. **92**, 124106 (2008).

[19] J. Christensen, A. I. Fernandez-Dominguez, F. de Leon-Perez, L. Martín-Moreno, and F. J. García-Vidal, Nature Phys. **3**, 851 (2007).

[20] N.F. Yu, J. Fan, Q.J. Wang, C. Pflügl, L. Diehl, T. Edamura, M. Yamanishi, H. Kan, and F. Capasso, Nat. Photon. **2**, 564 (2008).

[21] See supplementary material at http://link.aps.org/supplemental/10.1103/PhysRevLett.104.164301 for detailed elucidation of the analytical model and the schematic of the experiment setup.

[22] L. Kelders, J.F. Allard, and W. Lauriks, J. Acoust. Soc. Am. **103**, 2730 (1998).

[23] P. Lalanne, J.P. Hugonin, and J.C. Rodier, Phys. Rev. Lett. **95**, 263902 (2005).

[24] Y. Takakura, Phys. Rev. Lett. **86**, 5601 (2001).

[25] D. R. Matthews, H. D. Summers, K. Njoh, S. Chappell, R. Errington, and P. Smith, Opt. Express **15**, 3478 (2007).

[26] H. Estrada, P. Candelas, A. Uris, F. Belmar, F. J. García de Abajo, and F. Meseguer, Phys. Rev. Lett. **101**, 084302 (2008).

[27] J. Christensen, P.A. Huidobro, L. Martín-Moreno, and F.J. García-Vidal, Appl. Phys. Lett. **93**, 083502 (2008).

[28] M.Z. Ke, Z. J. He, S.S. Peng, Z.Y. Liu, J. Shi, W.J. Wen, and P. Sheng, Phys. Rev. Lett. **99**, 044301 (2007).

[29] Supplementary information accompanies this paper at http://www.nature.com/srep.

[30] The work was jointly supported by the National Basic Research Program of China (Grant No. 2007CB613202) and the National Nature Science Foundation of China (Grant No. 50632030). We also acknowledge the support from the Changjiang Scholars, the Nature Science Foundation of China (Grant No. 10804043), and of Jiangsu Province (Grant No. BK2009007).

Tunable Unidirectional Sound Propagation through a Sonic-Crystal-Based Acoustic Diode[*]

Xue-Feng Li,[1] Xu Ni,[1] Liang Feng,[2] Ming-Hui Lu,[1] Cheng He,[1] and Yan-Feng Chen[1]

[1] *National Laboratory of Solid State Microstructures and Department of Materials Science and Engineering, Nanjing University, Nanjing 210093, People's Republic of China*

[2] *Department of Electrical Engineering, California Institute of Technology, Pasadena, California 91125, USA*

> Nonreciprocal wave propagation typically requires strong nonlinear materials to break time reversal symmetry. Here, we utilized a sonic-crystal-based acoustic diode that had broken spatial inversion symmetry and experimentally realized sound unidirectional transmission in this acoustic diode. These novel phenomena are attributed to different mode transitions as well as their associated different energy conversion efficiencies among different diffraction orders at two sides of the diode. This nonreciprocal sound transmission could be systematically controlled by simply mechanically rotating the square rods of the sonic crystal. Different from nonreciprocity due to the nonlinear acoustic effect and broken time reversal symmetry, this new model leads to a one-way effect with higher efficiency, broader bandwidth, and much less power consumption, showing promising applications in various sound devices.

Electrical diodes, due to their capability of rectification of current flux, have significantly revolutionized fundamental science and advanced technology in various aspects of our routine life. Motivated by this one-way effect of electric currents, considerable effort has been dedicated to the study of the unidirectional nonreciprocal transmission of electromagnetic waves, showing important promise in optical and rf communications[1−6]. The realization of such nonreciprocal and unidirectional propagation requires either a broken time reversal symmetry[1−4] or a broken spatial inversion symmetry[5,6] in the artificial photonic structures (e.g., photonic crystals).

Sonic crystals (SCs), in an analogy with the electronic and photonic band structures of semiconductors and photonic crystals, have shown promising impacts in acoustic devices and applications that can efficiently trap, guide, and manipulate sound[7−15]. In the past two decades, with rapid developments in SCs ranging from engineering of band structure for bulk acoustic waves to design of acoustic grating for surface waves, a series of fascinating acoustic effects are consequently demonstrated, such as acoustic band gaps[8,9], negative refractions[10−14], and extraordinary transmission[15]. It is therefore expected, with a sophisticated SC design, that the exotic properties of SCs can lead to more

---

[*] Phys. Rev. Lett., 2011, 106(8):084301

counterintuitive sound manipulation, for example, the realization of acoustic diodes that can break down the conventional transmission reciprocity[16−19]. Similar to electromagnetic wave, sound usually propagates reciprocally back and forth along a given path. Unidirectional flux transmission requires considering the breaking of parity and time symmetry simultaneously in uniform media[20] that do not typically exist in nature. Therefore, SCs are currently considered good candidates to implement nonreciprocal and unidirectional sound propagation. The previous studies proposed the utilization of acoustic nonlinear effects combined with SCs to implement the broken time reversal symmetry as shown in the upper panel of Fig. 1(a)[16,17]. The nonlinear medium induces the frequency conversion as the solid red line in Fig. 1(b) indicates, and the adjacent SC acts as a frequency filter to block the incidence from the right since only the fundamental frequency locates within the band gap. However, the unidirectional transmission from left to right is quite low due to the inherent low conversion efficiency in acoustic nonlinear activities.

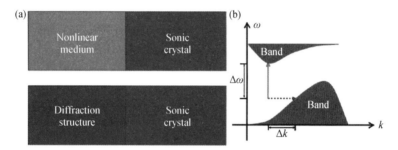

FIG. 1. (color online) (a) Illustration for two diode models, with the top one coupling nonlinear medium to SC and the bottom one using diffraction structure. (b) Schematic for transitions between different modes: the solid red line represents the transition with frequency change and the dashed purple line represents the transition between different spatial modes.

In this Letter, we will break the spatial inversion symmetry in the constructed SCs with a sophisticated design approach, and both experimentally and numerically demonstrate unidirectional transmission in such SCs. The constructed SC-based acoustic diodes can be further controlled by simply mechanically manipulating the unit cell of SCs[14] to support either reciprocal or nonreciprocal sound transmission. Our SC-based acoustic diode is a completely linear system without any acoustic nonlinearity. The acoustic diode is designed with an asymmetric periodic corrugated SC as shown in the lower panel of Fig. 1(a) that consists of a diffraction structure and a regular SC. The diffraction structure causes the transition between two spatial modes with different spatial frequencies as indicated by the dashed purple line in Fig. 1(b). The adjacent SC thus behaves as a spatial filter due to its intrinsic anisotropy of acoustic band structures, such that different spatial modes, especially high-order diffraction modes with different parallel wave vectors, can be either transferred or prohibited in the designed SC.

The concrete design of the acoustic diode, as shown in Fig. 2(a), consisted of a two-dimensional (2D) SC arranged in a square mesh with a lattice constant of $a=7$ mm and a corrugated diffraction structure with a modulation period of $L=6a$ in the y direction to the left of the SC. Both the SC and the diffraction structure were composed of square steel rods with the width of $d=4$ mm in an air background. Finite element simulation was implemented to evaluate the unidirectional transmission property of our acoustic diode, as shown in Fig. 2(d), where the transmission spectra clearly demonstrates nonreciprocal transmission efficiencies for the left incidence (LI) and the right incidence (RI) with a normally incident plane wave. The unidirectional frequency band is indicated by the blue shaded region in Fig. 2(d). In the experiment, the manufactured acoustic diode had 7 corrugation periods in the y direction, and acoustic field scanning measurement was carried out in two ranges of frequencies from 15.0 kHz to 25.0 kHz and from 35.0 kHz to 50.0 kHz, due to the limited response range of transducers. In spite of limitations of the finite size in the transverse direction and slight imperfection of the plane wave source in the experiment, the measurement results shown in Fig. 2(e) still agreed with the simulation. Especially within 17.5 – 19.5 kHz and 38.7 – 47.5 kHz, LI was associated with high transmission efficiency, but transmission was not allowed for RI, showing a relatively broadband unidirectional transmission only for LI as indicated by the blue shaded region in Fig. 2(e). In order to illustrate this unidirectional transmission phenomenon more clearly, spatial intensity distributions of the acoustic pressure field are mapped out both numerically and experimentally as shown in Figs. 2(b) and 2(c) at two frequencies of 18.0 kHz and 47.0 kHz, respectively. Because of the SC's directional band gap in the x direction(Γ-X), RI was almost completely reflected without any transmission. In the case of LI, however, strong acoustic field can be observed in the output area; for example, the transmission at 18.0 kHz was about 69%. But since the outgoing waves are not parallel to the incident waves, it is evident that energy in the normal incidence was converted, through high-order diffractions, to other spatial modes that have different spatial frequencies to overcome the barrier imposed by the Γ-X directional band gap. The unidirectional transmission was therefore established and the output field was actually the interference of outgoing beams from different diffraction orders.

To clearly quantify the performance of the acoustic diode, we defined the contrast ratio (R_c) as

$$R_c = \frac{T_L - T_R}{T_L + T_R}, \quad (1)$$

where T_L and T_R are the transmissions for LI and RI, respectively. The absolute value of R_c represents the relative transmission weight between these two incident cases. R_c was also evaluated as a function of frequency as shown in Fig. 2(f). Within the specific frequency ranges where unidirectional transmission was clearly observed, R_c reaches 1 with a good agreement with experiments and simulations, evidently manifesting great performance of our acoustic diode.

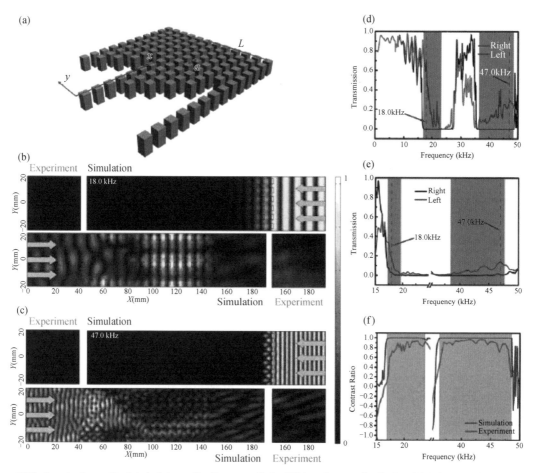

FIG. 2. (color online) (a) Schematic diagram of the SC-based acoustic diode which is periodic in y direction. (b), (c) Simulated and experimental field distribution mappings with the incident wave frequency at 18.0 kHz and 47.0 kHz, respectively. Green arrows indicate the propagation directions. (d), (e) Numerically calculated and experimentally measured transmission spectra of LI and RI, respectively. Green arrows indicate the frequency at which field distribution is mapped. (f) Contrast transmission ratio of the acoustic diode.

In addition to the broken spatial symmetry resulting from the diffraction structure, the building block of our SC, the steel square rod also introduces a broken rotational symmetry to the unit cell itself[14]. The introduced broken rotational symmetry can cause the change of the SC's band dispersion by simply mechanically rotating all the square rods with the same lattice configuration, as different rotations affect the effective scattering section for acoustic waves significantly. It is therefore expected to effectively control the sound rectification (i.e., efficiently tune the relative transmission weight between LI and RI) with our acoustic diode by rotating the rods. In the experiment, we rotated all the square rods 45° with keeping other parameters the same in the diode configuration, as shown in Fig. 3(a). Intrinsic characteristics of multiple scatterings in the SC were consequently dramatically changed, leading to a broader first band gap of SC as shown in

Fig. 3(d) due to the increase of the effective scattering section. The unidirectional transmission could still be observed from 16.2 kHz to 22.3 kHz. But from 40.0 kHz to 50.0 kHz, transmission was allowed for both LI and RI and the one-way phenomenon was deactivated. Therefore, it is evident that the sound rectification effect could be effectively turned on or off in our acoustic diode by rotating the square rods of the SC. This tunable acoustic diode was further confirmed with mappings of spatial intensity distribution of the acoustic pressure field at 17.25 kHz and 47.0 kHz, as shown in Figs. 3(b) and 3(c), respectively. Consistent with transmission spectra in Figs. 3(d) and 3(e), pronounced unidirectional transmission can be seen at 17.25 kHz, but equivalent transmission was obtained in both directions at 47.0 kHz. The contrast ratio in Fig. 3(f) also confirmed that the relative transmission weight is 1 from 16.5 kHz to 22.5 kHz, and the unidirectional

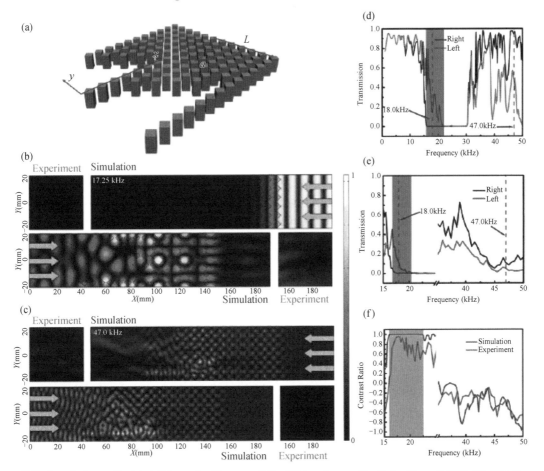

FIG. 3. (color online)(a) Schematic diagram of the SC-based acoustic diode after rotating the square rods. (b),(c) Simulated and experimental field distribution mappings with the incident wave frequency at 17.25 kHz and 47.0 kHz, respectively. Green arrows indicate the propagation directions. (d),(e) Numerically calculated and experimentally measured transmission spectra of LI and RI, respectively. Green arrows indicate the frequency at which field distribution is mapped. (f) Contrast transmission ratio of the acoustic diode.

character thus remains the same (compared to Fig. 2), while the weight was around 0 from 35.0 kHz to 50.0 kHz, showing the change from unidirectional to bidirectional transmission. In addition, this tunable unidirectional transmission effect could also be influenced by different rotation angles and filling fractions (see Fig. S4 in [21]). In principle, with a more systematic design of the square-rod-based SC, this tunable acoustic diode could be constructed either in the first or the second band (as demonstrated above), or both. And the frequency range of unidirectional transmission might be anticipated to be effectively controlled with different rotational angles of these square rods.

To gain deeper insight, we analytically investigated the mechanism underlying this unidirectional transmission phenomenon. With a plane wave incidence the acoustic pressure field in the input and output half-spaces can be Fourier expanded as

$$p(x,y) = \exp(jk_0 x) + \sum_{n=-\infty}^{\infty} \rho_n \exp(j\alpha_n y + j\beta_n x), \qquad (2)$$

$$p(x,y) = \sum_{n=-\infty}^{\infty} \tau_n \exp(j\alpha_n y + j\beta_n x), \qquad (3)$$

respectively, where α_n and β_n are wave numbers in the x and y directions, respectively, and $\beta_n = \sqrt{k_0^2 - \alpha_n^2}$, $\alpha_n = 2n\pi/L$ (n is the diffraction order), k_0 is the free-space wave number, ρ_n and τ_n are amplitudes of the nth-order diffraction beams in reflection and transmission, respectively. The mechanism of our acoustic diode is closely relevant to these two diffraction terms. These diffraction terms represent the energy conversion among different diffraction orders. The difference of structure geometries on two sides resulted in different diffraction orders, and only the orders locating within the propagating band of the SC could transmit the energy from one side to the other. Notice that, however, the diode in our study cannot be completely analytically described by Eqs. (2) and (3), due to its complex diffraction construction. Therefore, simplified diode geometry was studied and compared [21]. As shown in Figs. S2(b) and S2(d) of [21], tunable unidirectional transmission effect is also realized with a much weaker diode behavior. The optimized complex geometry in this letter has advantages over this simplified diode in two aspects. (1) With the adiabatic change in the complex asymmetric corrugation, the reflection of the acoustic wave incident from the left side could be efficiently reduced. (2) Compared to diffractions through the simpler corrugation, it is more efficient to generate higher order diffraction modes through the complex asymmetric corrugation due to the large slope angle.

In order to illustrate nonreciprocal sound propagation in the acoustic diode, we further analyze the equifrequency surfaces (EFSs) at different frequencies of the SC as shown in Fig. 4. For the case of RI in Figs. 4(a) and 4(b), high diffraction orders completely fell in the evanescent regime (ρ_n and τ_n are 0) in the interested frequency ranges, and thus all the energy was still conserved in the zero order, i.e., normal incidence to the SC, which was located within the directional band gap of the SC. Therefore, sound transmission for RI was prohibited by the SC. However, for the case of LI, the existence of the diffraction structure before the SC converted the normal incidence mode to high-order diffraction modes. This can be more clearly visualized in Fig.S3 of [21] where an incident beam with a finite width was considered. The transmitted beams propagated along different directions, corresponding to different positive and negative diffraction orders due to the complex diffraction structure. Consistent with EFSs in Figs. 4(a) and 4(b), ±1 and ±2 modes significantly contributed to transmission through the SC and resulted in unidirectional sound transmission. After the rotation of the square rods, EFS from 17 kHz to 19 kHz remained almost the same as shown in Fig. 4(c), demonstrating a nonreciprocal feature as all the diffracted beams easily transmit through the SC for LI as shown in Fig. S3(f) of [21]. But from 45 kHz to 49 kHz, EFS was drastically changed as shown in Fig. 4(d), and the original directional band gap in Fig. 4(b) disappeared. Therefore, sound transmission was also allowed for RI, and the nonreciprocal sound propagation is thus expected to be simply mechanically controlled by rotating the rods of our acoustic diode.

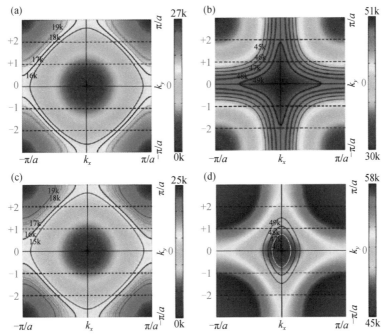

FIG. 4.(color online) (a),(b) EFSs of the SC in first Brillouin zone from 16 kHz to 19 kHz and from 45 kHz to 49 kHz, respectively. The numbers (−2, −1, 0, +1, +2) on left and right sides correspond to the diffraction orders. (c),(d) EFSs of the SC after the rotation of rods in two frequency ranges.

To summarize, we have theoretically proposed and experimentally constructed a sonic-crystal-based acoustic diode with a broken spatial inversion symmetry in which nonreciprocal propagation and unidirectional transmission of sound were clearly observed. Furthermore, the introduced broken rotational symmetry of the unit cell of the sonic crystal results in a sophisticated mechanical manipulation of the observed nonreciprocal sound transmission. Compared to previously proposed acoustic diodes using nonlinearities, our system is a completely linear system, showing advantages such as broadband operation, high conversion efficiency, and much less power consumption. The same concept is also expected to implement an on-chip isolator for various types of acoustic waves, such as surface acoustic waves[22] and Lamb waves[23].

References and Notes

[1] Z. Wang et al., Phys. Rev. Lett. **100**, 013905 (2008).
[2] F. D. M. Haldane and S. Raghu, Phys. Rev. Lett. **100**, 013904 (2008).
[3] C. He et al., Appl. Phys. Lett. **96**, 111111 (2010).
[4] Z. Yu and S. Fan, Nat. Photon. **3**, 91 (2009).
[5] A. E. Serebryannikov, Phys. Rev. B **80**, 155117 (2009).
[6] A. E. Serebryannikov and E. Ozbay, Opt. Express **17**, 13335 (2009).
[7] M.H. Lu, L. Feng, and Y.F. Chen, Mater.Today **12**, 34 (2009)
[8] M. S. Kushwaha et al., Phys. Rev. Lett. **71**, 2022 (1993).
[9] R. Martinez-Sala et al., Nature (London) **378**, 241 (1995).
[10] M. H. Lu et al., Nature Mater. **6**, 744 (2007).
[11] L. Feng et al., Phys. Rev. Lett. **96**, 014301 (2006).
[12] L. Feng et al., Phys. Rev. B **72**, 033108 (2005).
[13] X. D. Zhang and Z.Y. Liu, Appl. Phys. Lett. **85**, 341 (2004).
[14] L. Feng et al., Phys. Rev. B **73**, 193101 (2006).
[15] Y. Zhou et al., Phys. Rev. Lett. **104**, 164301 (2010).
[16] B. Liang, B. Yuan, and J. C. Cheng, Phys. Rev. Lett. **103**, 104301 (2009).
[17] B. Liang et al., Nature Mater. **9**, 989 (2010).
[18] B.W. Li, L. Wang, and G. Casati, Phys. Rev. Lett. **93**, 184301 (2004).
[19] V.F. Nesterenko et al., Phys. Rev. Lett. **95**, 158702 (2005).
[20] C. E. Ruter et al., Nature Phys. **6**, 192 (2010).
[21] See supplemental material at http://link.aps.org/ supplemental/10.1103/PhysRevLett.106.084301 for detailed elucidation of the contribution of different diffraction modes to the one-way effect and influence of rotation angles and filling fraction on the tunability.
[22] J.H. Sun and T.T. Wu, Phys. Rev. B **74**, 174305 (2006).
[23] C.Y. Huang, J.H. Sun, and T.T.Wu, Appl. Phys. Lett. **97**, 031913 (2010).
[24] The work was jointly supported by the National Basic Research Program of China (Grant No. 2007CB613202). We also acknowledge the support from the Nature Science Foundation of China (Grant No. 10874080) and the Nature Science Foundation of Jiangsu Province (Grant No. BK2007712).

Acoustic Asymmetric Transmission Based on Time-dependent Dynamical Scattering[*]

Qing Wang[1], Yang Yang[1], Xu Ni[1], Ye-Long Xu[1], Xiao-Chen Sun[1], Ze-Guo Chen[1],
Liang Feng[2], Xiao-ping Liu[1], Ming-Hui Lu[1] and Yan-Feng Chen[1]

[1] *National Laboratory of Solid State Microstructures & Department of Materials Science and Engineering,
Nanjing University, Nanjing 210093, China*

[2] *Department of Electrical Engineering, The State University of New York at Buffalo,
Buffalo, NY 14260, USA*

An acoustic asymmetric transmission device exhibiting unidirectional transmission property for acoustic waves is extremely desirable in many practical scenarios. Such a unique property may be realized in various configurations utilizing acoustic Zeeman effects in moving media as well as frequency-conversion in passive nonlinear acoustic systems and in active acoustic systems. Here we demonstrate a new acoustic frequency conversion process in a time-varying system, consisting of a rotating blade and the surrounding air. The scattered acoustic waves from this time-varying system experience frequency shifts, which are linearly dependent on the blade's rotating frequency. Such scattering mechanism can be well described theoretically by an acoustic linear time-varying perturbation theory. Combining such time-varying scattering effects with highly efficient acoustic filtering, we successfully develop a tunable acoustic unidirectional device with 20 dB power transmission contrast ratio between two counter propagation directions at audible frequencies.

The phenomenon of unidirectional motion for matters or particles in a gradient potential has inspired physicists for many centuries and fruitful discoveries by these physicists have led to multiple extraordinary inventions including hydroelectric power and electric diodes, which have undoubtedly revolutionized our society. Recently, the research of such important phenomenon has been extended to include even more particles and quasi-particles, e.g., photons represented by optical waves[1–3], phonons represented by acoustic or elastic waves[4] and thermal phonons represented by thermal waves[5,6]. For photons, several physical phenomena can be exploited to break the reciprocity to allow for unidirectional transmission and even optical isolation, e.g., non-reciprocal Faraday effect in a linear system[7,8] and direction-dependent frequency conversion in a time-varying modulation system[3,9]. Similarly, for phonons or acoustic waves, acoustic diodes[10] exhibiting unidirectional acoustic wave transmission[11–19] may be realized in several configurations. For example, acoustic diodes or circulators have been realized in linear

---

[*] Sci.Rep.,2015,5:10880

systems with magneto-acoustic coupling effects[20] or with acoustic analogue of Zeeman effects in moving media[4]. Besides, acoustic rectifiers and diodes have also been demonstrated using frequency conversion processes in passive acoustic nonlinear systems[21,22] and a hybrid acoustic/electric active system with nonlinear functionality implemented in the electric domain[23,24]. However, practical implementations of these systems have several drawbacks including the need of a high input energy, or limited tunability and operation bandwidth. In this letter, we present a new acoustic frequency conversion mechanism through the acoustic wave's passive interaction with a rotating object, i.e., an elliptical blade with its surrounding air. Starting with an acoustic linear time-varying perturbation theory, we develop a method to solve the problem of acoustic wave scattering through a time-varying medium, by which we can fully describe the underlying mechanism responsible for the frequency conversion. Our theoretical results show an excellent agreement with simulation results obtained from a time-varying finite element method. By combining such time-varying medium with a deliberately designed acoustic filtering structure in an acoustic guiding system, we develop an acoustic asymmetric device with unidirectional propagation exhibiting 20 dB contrast ratio between two propagation directions at audible frequencies.

Results

Time-varying medium scattering effect. The interaction of waves and static matter has been widely explored since the early days of physics. In recent years, the interaction of waves with non-stationary medium[25], whose mass density and wave velocity are assumed to be time-varying, has raised considerable interests. In this letter, we propose a new kind of time-varying medium, composed of a rotating elliptical blade with its surrounding air, of which the schematic is shown in Fig. 1(a) and the photograph is shown in Fig. 1(b). According to the full wave simulation, the effective mass density of the rotating blade are dependent on the rotating angle, which is defined as an angle between the propagation direction of plane waves and the majoraxis of elliptical blade (shown in Fig. S.1. in Supplementary Information). Because the time-varying angle is a linear function of the blade's rotating frequency, the material elastic parameters are time-varying and modulated correspondingly. Such periodically varying material properties of the scatters placed in an acoustic waveguide will change the propagation behavior of the input acoustic waves and result in a harmonic scattering phenomenon. Consequently, this would lead to the energy transfer from the fundamental frequency to the harmonic waves, which acquire a lower and higher frequency as indicated in Fig. 1(c). Since the acoustic field of the effective time-varying region might be inhomogeneous, measuring the field in the time-varying region at several discrete locations is insufficient in figuring out the parameters of the whole time-varying region. And the limits of present measurement techniques also prevent us from directly measuring the whole sound field in the time-varying region. Furthermore, because of the uncertainty principle in the time-domain measurement, it is of huge difficulties using the current transient measurement method to deduce the time-varying parameters from the recorded transmission and reflection waves with only finite recording length of signals. Taking all these factors into consideration, we resort to a quasi-static approximation method to deduce the effective parameters of modulation region from reflected and transmitted waves. (see Theory in Supplementary Information for details).

Herein, the shape of the rotating blade is chosen to be an elliptic instead of other common shapes, because the air turbulence[26] caused by the rotational motion is smaller for the elliptical blade than, for example, a rectangular blade. Slow varying approximation can thus still be preserved in the theoretical analysis. In this paper, we develop a method by utilizing an acoustic linear time-varying perturbation theory to study the scattering process of acoustic waves transmitting through a time-varying medium. Using such method we can fully describe the frequency conversion effect of scattering waves as follows.

Replacing velocity field v by velocity potential $v = \nabla \phi$, the acoustic equations can be described as

$$(H_0 + H_1(t)) | \psi \rangle = \omega | \psi \rangle \qquad (1)$$

where ω is the eigen frequency, with H_0 and $H_1(t)$ expressed as

$$H_0 = \begin{bmatrix} 0 & c_0^2 \rho_0 \partial_x & c_0^2 \rho_0 \partial_y \\ H_{210} & 0 & 0 \\ H_{310} & 0 & 0 \end{bmatrix}$$

$$H_1(t) = \begin{bmatrix} 0 & c_0^2 \rho_1(t) \partial_x & c_0^2 \rho_1(t) \partial_y \\ H_{21}(t) & 0 & 0 \\ H_{31}(t) & 0 & 0 \end{bmatrix}$$

$$| \psi \rangle = \begin{pmatrix} p \\ iv_x \\ iv_y \end{pmatrix} \qquad (2)$$

In the above analysis, we have set

$$\rho = \begin{bmatrix} \rho_{xx} & \rho_{xy} \\ \rho_{yx} & \rho_{yy} \end{bmatrix} = \begin{bmatrix} \rho_{xx0} + \rho_{xx}(t) & \rho_{xy0} + \rho_{xy}(t) \\ \rho_{yx0} + \rho_{yx}(t) & \rho_{yy0} + \rho_{yy}(t) \end{bmatrix} = \rho_0 + \rho_1(t), \rho_0 \text{ and } \rho_1(t) \text{ are}$$

tensors.

In the presence of rotation and considering the two-folded symmetry of the elliptical blade, the effective Hamiltonian $H_1(t)$ can be written as $H_1(t) = V\cos(2\pi\omega_r t)$ (refer to Theory in Supplementary Information for more theoretical details), where ω_r is the rotation angular frequency of the blade. The new eigenvectors can be written as a linear superposition of $|\psi_n\rangle$ as

$$(H_0 + V\cos(2\omega_r t)) | \psi \rangle = \omega | \psi \rangle. \qquad (3)$$

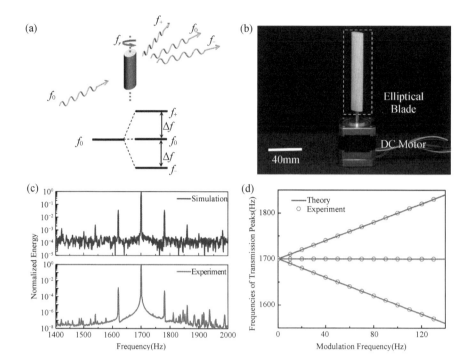

FIG.1. Time-varying medium scattering effect. (a) Illustration of time-varying acoustic scattering from an acoustic time-varying medium, represented by an elliptical-shaped blade rotating around its center axis in the air. Acoustic time-varying scattering interaction produces extra harmonic frequency components (f_- and f_+). (b) Photography of the time-varying medium which is represented by a nylon elliptical-shaped blade driven by a precisely controlled DC motor. (c) The top plot shows time-varying full-wave acoustic transmission simulation results (blue solid line) for an input acoustic wave with a frequency $f_0 = 1700$ Hz transmitting through a rotating elliptical blade (8 mm long semi-major axis and 4 mm long semi-minor axis) centered in an acoustic waveguide with a width of 22.5 mm with its rotating frequency $f_r = 40$ Hz. Note that the rotating motion of the blade is modeled as a sinusoidal time-varying density parameter in the simulation (see Fig. S.1 in Supplementary Information for details). The 1620 Hz and 1780 Hz frequency components observed in the transmission spectrum correspond to the scattering products. There are also secondary cascaded scattering products peaked at 1540 Hz and 1860 Hz. The bottom plot shows the experimental transmission spectrum (red solid line) obtained using a fabricated time-varying medium with the same geometric parameters as in the simulation. Without taking into account the higher order cascaded scattering terms, the experimental spectrum agrees well with the numerical simulated result. Note the harmonic scattering components on the high frequency side of the spectrum, which might originate from the nonlinear effect of the forced motion of the blade's surrounding air. (d) The frequency of the primary high order scattering components as a function of blade's rotation frequency f_r. Circles correspond to experimental data while lines correspond to the results obtained from the theory outlined in Supplementary Information.

By solving the above equation (refer to Theory in Supplementary Information for more theoretical details), we find that the initial system energy level will be split to form two additional sub-energy levels as shown in Fig. 1(a). The corresponding angular frequencies for these two sub-energy levels are determined according to the following expression:

$$\omega_{\pm}=\omega_0 \pm 2\omega_r. \tag{4}$$

Shown in Fig.1(c) (blue solid line) is the transmission spectrum of the acoustic waves propagating through a waveguide obtained from full-wave simulations conducted in COMSOL Multiphysics (see Methods). Clearly when the time-varying medium in the waveguide is brought to a time-varying state corresponding to the rotation motion of the blade, the obtained acoustic transmission spectrum shows several important features. There is still a large unconverted spectral component at the input frequency of 1700 Hz, indicating relatively low conversion efficiency for this single-pass time-varying scattering process. Nevertheless, the transmission spectrum contains newly generated scattering frequency components with two major peaks at 1620 Hz and 1780 Hz, which are converted from the input acoustic wave. Notice that the first order frequency shift is twice the rotating frequency of the blade, which is expected from the two-folded geometric symmetry of the elliptical blade. In addition, due to a secondary order scattering effect, there are also much weaker peaks at 1540 Hz and 1860 Hz.

Figure 1(b) shows a picture of the fabricated time-varying medium consisting of an elliptical blade made of nylon driven by a precisely controlled brushless DC motor. The acoustic transmission property through this time-varying modulated medium was experimentally investigated (see Methods) with results shown in Fig. 1(c) (red solid line). The frequency shifts are clearly visible in the experiment results, which coincides with the simulated data. Both experiments and simulations indicate that time-varying modulation can in fact lead to a dramatically altered transmission properties of the fundamental frequency waves. Note that due to a noisy background in our experiments, the spectral position of the secondary cascaded scattering products cannot be unambiguously distinguished. The frequency conversion efficiency observed in the experiments is somehow lower than that in the simulations. This is primarily due to the non-ideal experimental conditions, e.g., the nonlinear effects caused by the moving air around the blade and energy loss in the waveguide. One of the advantages of using time-varying modulation lies in the fact that the frequency of scattered waves can be directly manipulated by controlling the blade's rotation frequency (f_r) as shown in Fig.1(d). More specifically, the frequency of scattered waves shows a linear relationship with f_r, which is also consistent with the theoretical result shown in Eq. 4. Exploiting such dependence enables an extra design freedom for acoustic devices exhibiting unidirectional transmission characteristics.

Unidirectional acoustic system design. Shown in Fig. 2 is the schematic of our proposed unidirectional acoustic transmission system consisting of two major components: a time-varying medium region as described in the above section and an acoustic band-stop filter constructed with an array of cascaded Helmholtz resonators. Usually Helmholtz resonators can have high quality factors and their resonance frequencies can be accurately tailored by changing their geometry. Such unique property of Helmholtz resonators allows one to design a filter with high extinction ratio and with desired bandwidth and center frequency.

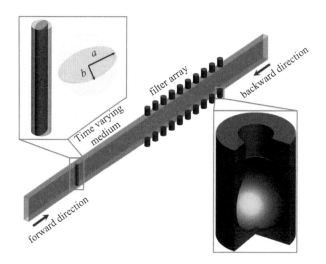

FIG.2. Schematic illustration of a unidirectional transmission system. It is a two-component device consisting of a time-varying medium region and an acoustic filter region, both of which are enclosed in a rectangular acoustic waveguide. A rotating elliptical-shaped blade represents the time-varying medium. The acoustic filter is a band-stop filter constructed with an array of Helmholtz resonators. Propagation directions in this paper are defined as follows: forward propagation defines the direction along which the input acoustic wave first encounters the time-varying medium and then the acoustic filter, and the opposite propagation direction is defined as backward propagation. In a forward propagation, part of the input acoustic wave is scattered into high order scattering components, and if the frequency of these scattering products are out of the filter's stop-band, they can pass through the filter leading to certain power transmission. But in a backward propagation, if the frequency of input acoustic wave falls into the filter's stop-band, it will be filtered out and attenuated, leading to nearly no power transmission regardless of the scattering process occurred at the blade. The power transmission contrast in these two propagation directions characterizes the unidirectional behavior of this system.

Therefore the transmitted acoustic power through this two-component system for the forward (from time-varying medium to band-stop filter) and backward propagation (from band-stop filter to time-varying medium) can be expressed as follows:

$$P_{\text{forward}} = \left(1 - \sum_{n=\pm 1}^{n=\pm\infty} \eta_n\right) P_0 T(f_0) + P_0 \sum_{n=\pm 1}^{n=\pm\infty} \eta_n T(f_n), \tag{5}$$

$$P_{\text{backward}} = \left(1 - \sum_{n=\pm 1}^{n=\pm\infty} \eta_n\right) P_0 T(f_0) + P_0 T(f_0) \sum_{n=\pm 1}^{n=\pm\infty} \eta_n T(f_n). \tag{6}$$

Here, P_0 and f_0 are power and frequency of the input acoustic wave respectively, $T(f)$ is the frequency-dependent normalized power transmission function of the filter, η_n and f_n are the conversion efficiency (defined as the ratio between converted power and input power: P_n/P_0) and frequency of the nth order acoustic scattering product. In the above equations, the first and second term describes the power transmission for the input acoustic wave and for all converted multiple-order scattering components respectively.

Clearly the power transmitted for the input acoustic wave is exactly the same for both propagation directions. However, for the frequency band of interest, i.e., the filter's stop-band, the transmitted multiple-order scattering components' power is different due to the fact that the incident power for the time-varying acoustic scattering process at the elliptical blade highly depends on the propagation direction. In a forward propagation case, the input acoustic wave encounters the time-varying medium first, indicating that the incident power for the time-varying scattering process is approximately the same as the input power, while in a backward propagation case, the input acoustic wave is first filtered and attenuated heavily by the Helmholtz filter before it encounters the time-varying medium, in which case the incident power is much lower. The difference in the total transmitted power ΔP for the two directions can be expressed as

$$\Delta P = P_{\text{forward}} - P_{\text{backward}} = P_0 (1 - T(f_0)) \sum_{n=\pm 1}^{n=\pm\infty} \eta_n T(f_n). \tag{7}$$

Depending on the property of the band-stop filter $T(f)$, especially its bandwidth and center-frequency, and the high order scattering frequency f_n governed by the rotating frequency of the blade f_r, as well as the input frequency f_0, the value of ΔP can either be zero or positive, corresponding to symmetric transmission or unidirectional transmission.

Our proposed unidirectional system is implemented and numerically studied in COMSOL Multiphysics (See Methods). When the elliptical blade is at rest meaning a zero rotation frequency $f_r = 0$, as shown in Fig. 3(a) this system functions merely as a regular filter with numerical-error limited symmetric transmission property, and the obtained power transmission spectrum reflects exactly the filter's frequency response. This result is consistent with our above analysis and is expected from Eq. (7) because there is no scattering term and the system only has a filter response. In the case for a rotating blade, the resulted scattering process greatly modifies the transmission spectrum as illustrated in Fig. 3(b)-(d) corresponding to blade rotation frequencies f_r of 15 Hz, 40 Hz and 65 Hz, respectively. First of all, when the input frequency falls out of the stop-band of the filter (e.g. frequency > 1800 Hz), this two-component system acts as if the filter did not even exist and there is no direction-dependent power transmission, suggesting symmetric transmission spectrum for such frequencies. However, for input frequencies falling within the filter's stop-band, the transmitted power increases with the increasing f_r in a forward propagation configuration, but remains almost unchanged in a backward configuration. This observation can be well explained by our theoretical framework outlined in the previous section. As indicated by Fig. 1(d), the change of f_r causes a linear shift for the scattering frequency components. According to our analysis and Eq. (7), increasing the rotating frequency f_r pushes the generated scattering frequency components to the edge of or even out of the filter's stop-band, leading to increased transmitted power difference for the two propagation directions.

Measurement results. A full experimental test system is built based upon the simulated

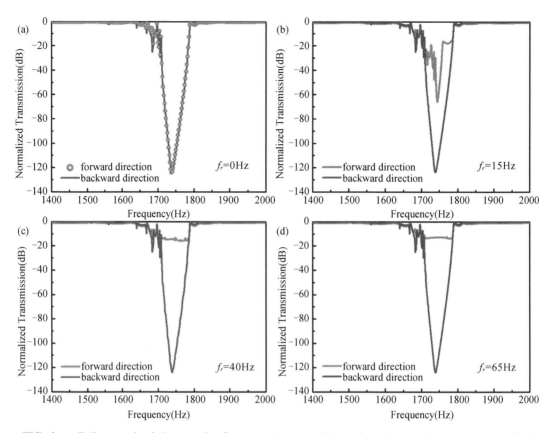

FIG. 3. Full-wave simulation results for acoustic transmission through our designed system. (a) Normalized transmission spectra when blade is at rest with rotating frequency $f_r = 0$. The spectra for forward and backward propagation directions are nearly identical within the numerical error in the simulation, and they correspond to the filter's intrinsic frequency response. In this case, the system transmission property is symmetric. (b-d) Normalized transmission spectra corresponding to three different excited states/rotating frequencies of the blade: (b) $f_r = 15$ Hz, (c) $f_r = 40$ Hz, (d) $f_r = 65$ Hz. With the increasing rotating frequency of the blade, the system starts to develop asymmetric response with increasing power transmission for the forward propagation direction but almost unchanged power transmission power for the backward propagation, which leads to an increased power transmission contrast ratio for these two propagation directions. At $f_r = 15$ Hz asymmetric transmission is observed primarily for input frequency near the center of the filter's stop-band, while at higher frequencies $f_r = 40$ Hz and $f_r = 65$ Hz asymmetric frequency response occurs for most of input frequencies falling into the filter's stop-band. At the highest rotation frequency of $f_r = 65$ Hz used in our experiment, the power transmission contrast ratio is close to 100 dB.

structure. Photographs of the blade unit (time-varying medium) and the whole device are presented in Fig.4(a), (b). Multiple Helmholtz resonators are made of plastic and cascaded together to form an array with a resonance frequency around 1700 Hz. More details of this system are provided in Supplementary Information. Without the physical presence of the elliptical blade in the waveguide system, as shown in Fig. 4(c) nearly symmetric normalized power for the forward and backward propagation direction transmission (corresponding to the filter response) is obtained. Notice that the bandwidth of this

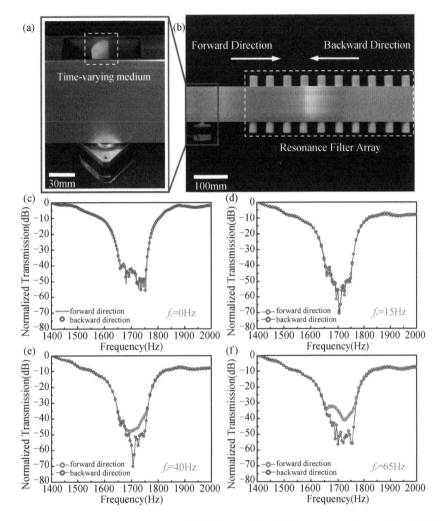

FIG. 4. Experimental system and results. (a) Photography of the time-varying medium installed in an acoustic waveguide with an opening at the top. The time-varying medium is represented by a nylon elliptical-shaped blade driven by a precisely controlled DC motor. (b) Photography of the two-component acoustic waveguiding system, consisting of a time-varying medium region and a Helmholtz-resonator-array-based acoustic band-stop filter. The definition of forward and backward propagation direction in the experiment is the same as that in the simulation shown in Fig. 2(c)-(f) Measured normalized power transmission spectra of this system for four different configurations: (c) without the presence of the blade in the system, (d), (e) and (f) rotation frequency $f_r = 15$ Hz, $f_r = 40$ Hz, and $f_r = 65$ Hz respectively. Similar to our numerical investigations, this experimental system experiences a transition from symmetric transmission behavior into asymmetric behavior and consequently an increase in contrast ratio. However, compared with our numerical simulation results, this transition occurs at a higher rotating frequency, because the fabricated filter has a broader bandwidth than the numerically modeled one and a larger rotating frequency is necessary to induce a larger frequency shift so that the high order scattering frequency components can acquire frequencies out of the filter's stop-band. At the maximum rotation frequency (65 Hz), this device shows a unidirectional transmission property with a contrast ratio up to 20 dB.

fabricated filter array's stop-band is broader than that of the numerically modeled one shown in Fig.3(a) (3 dB bandwidth: 100 Hz vs. 20 Hz). The discrepancy can be ascribed to a lower resonant Q factor for the experimentally fabricated Helmholtz cavity due to acoustic propagation loss in air (as shown in Supplementary Information Fig. S.8.) and the non-ideal rigid wall condition. When the time-varying medium is introduced into the system but is at rest ($f_r=0$), the forward and backward acoustic power transmission spectra are slightly different particularly around the resonant frequency of the filter as shown in Supplementary Information Fig. S.9. Such difference is originated from the non-ideal acoustic absorption condition at the two ends of our wave-guiding system and the resulted stand-wave fluctuations recorded at the acoustic probe/detector.

As discussed theoretically above, our proposed system can experience a transition from symmetric transmission into unidirectional transmission with the blade's increasing rotation frequency. When the blade is rotating at a frequency of 15 Hz (Fig. 4(d)), acoustic power transmission spectra for the two directions is nearly identical. Such experimental behavior is different from what is observed in the simulation shown in Fig. 3(b), which is resulted from the broader bandwidth of the Helmholtz filter used in the experiment as discussed above. When the rotation frequency is increased to $f_r=40$ Hz and $f_r=65$ Hz, the generated acoustic first order scattering frequency components start to acquire acoustic frequency falling out of the filter's stop-band, which results in distinct asymmetric transmitted power for the two propagation directions as shown in Fig. 4(c,d). Compared with the ~100 dB power transmission contrast ratio in the simulation at the filter's center frequency as shown in Fig. 3(d), the contrast ratio for the two propagation directions obtained in the experiment is much smaller than the theoretical predictions due to the relative low scattering conversion efficiency and large propagation loss. Nevertheless, at the blade's highest rotation frequency of 65 Hz in our experiment, it still can be as large as 20 dB, which is adequate for many practical applications.

Discussion

We demonstrate an acoustic scattering effect from a time-varying medium (a rotating elliptical blade). Both our theoretical study and experimental investigation suggest that the state of the blade plays a critical role in achieving novel acoustic transmission behavior. When this blade is at rest without any motion, the system is linear and time-independent, which only exhibits symmetric power transmission for two propagation directions as indicated in Figs. 3(a), 4(c). However, as soon as the blade starts to rotate, acoustic scattering takes place and part of the input acoustic wave is scattered into harmonic frequency components as shown in Fig. 1, which subsequently transmit through the Helmholtz filter in a forward propagation configuration. However, in a backward propagation, for the input acoustic wave with frequency falling into the filter's stop-band, the filter greatly attenuates the input power leading to almost no power transmission. This asymmetric direction-dependent power transmission phenomenon manifests the acoustic unidirectionality in our system.

In our system design, the use of the time-varying medium is very advantageous, because it allows for great flexibility in manipulating acoustic scattering process and in optimizing the performance of our unidirectional system. The scattering frequency shift is determined by the rotating frequency of the blade, which can be intentionally modified to a certain extent, the frequency of scattering frequency components can also be tuned accordingly. This unique feature in our system implies that the contrast ratio can also be deliberately designed (see Supplementary Information Fig. S.7.). In addition, such kind of time-varying medium can be exploited to construct complicated acoustic structures, e.g., a phase-conjugated time-varying medium array or an acoustic resonator consisting of acoustic time-varying medium, to solve the low conversion efficiency issue in the time-varying acoustic scattering process in our current configuration and consequently to further improve the contrast ratio for the unidirectional transmission.

Since different rotating angles of the elliptical blade correspond to different effective densities (see Supplementary Information Fig. S.1.) and thus different effective phases, the acoustic time-varying medium concept can also be extended to investigate many dynamic phase modulation related physical phenomena. These could include quantum geometric phase caused by non-adiabatic evolution, and gauge magnetic potential[27,28] with Aharonov-Bohm effects in electronic or photonic time-varying systems. Moreover, by utilizing phase conjugated time-varying modulation, nonreciprocal transmission devices such as acoustic isolators[3,27] also might be expected in the future.

In conclusion, we demonstrate a frequency shift effect in a linear time-varying system based on the interaction of an acoustic wave with a rotating elliptical blade. In such a system, an incident acoustic wave is scattered by a rotating blade, and the resulted scattering wave experiences harmonic frequency shift. We show that such frequency conversion process can be described by a method that we developed based on the linear time-varying scattering theory, although there are some discrepancies between theory and experiment, which needs further investigation. Furthermore, by combining such frequency shift effect with efficient acoustic filtering, we realize acoustic unidirectional propagation with 20 dB contrast ratio between two counter propagation directions at audible frequencies, which can be tuned by varying the blade's rotation frequency. It is worth noting that our concept of using an acoustic time-varying medium can be equally applied to frequency ranges other than audible frequencies. We believe our concept of linear time-varying acoustic scattering and the design of such unidirectional transmission device may find applications in relevant fields such as non-destructive testing of turbulent flow pipe, acoustic imaging, audible signal processing, and etc. Moreover, the underlying mechanism of this time-varying acoustic scattering is linked to acoustic dynamic phase modulation process. In other words, by taking advantage of dynamic phase modulation, novel physical phenomena such as gauge magnetic potential[3,9] in condensed matter might be simulated by a carefully designed macroscopic acoustic time-varying system.

References and Notes

[1] Wang, Z., Chong, Y., Joannopoulos, J. D. & Soljacic, M. Observation of unidirectional backscattering-immune topological electromagnetic states. *Nature* **461**, 772–775 (2009).

[2] Wang, Z. & Fan, S. Magneto-optical defects in two-dimensional photonic crystals. *Appl. Phys. B* **81**, 369–375 (2005).

[3] Yu, Z. F. & Fan, S. H. Complete optical isolation created by indirect interband photonic transitions. *Nat. Photon.* **3**, 91–94 (2009).

[4] Fleury, R., Sounas, D. L., Sieck, C. F., Haberman, M. R. & Alù, A. Sound Isolation and Giant Linear Nonreciprocity in a Compact Acoustic Circulator. *Science* **343**, 516 (2014).

[5] Li, B. W., Wang, L. & Casati, G. Thermal diode: Rectification of heat flux. *Phys. Rev. Lett.* **93**, 184301 (2004).

[6] Li, B. W., Lan, J. H & Wang, L. Interface thermal resistance between dissimilar an harmonic lattices. *Phys. Rev. Lett.* **95**, 104302 (2005).

[7] Saleh, B. E. A. & Teich, M. C. [Chapter 6: Polarization and Crystal Optics] *Fundamentals of Photonics* (Wiley-Interscience, 2007).

[8] Kodera, T., Sounas, D. L. & Caloz, C. Artificial Faraday rotation using a ring metamaterial structure without static magnetic field. *Appl. Phys. Lett.* **99**, 031114 (2011).

[9] Fang, K. J., Yu, Z. F. & Fan, S. H. Realizing effective magnetic field for photons by controlling the phase of dynamic modulation. *Nat. Photon.* **6**, 782–787 (2012).

[10] Liang, B., Yuan, B. & Cheng, J. C. Acoustic diode: Rectification of acoustic energy flux in one-dimensional systems. *Phys. Rev. Lett.* **103**, 104301 (2009).

[11] Li, X.-F. *et al*. Tunable unidirectional sound propagation through a sonic crystal-based acoustic diode. *Phys. Rev. Lett.* **106**, 084301 (2011).

[12] Zhu, X., Zou, X., Liang, B. & Cheng, J. C. One-way mode transmission in one-dimensional phononic crystal plates. *J. Appl. Phys.* **108**, 124909 (2010).

[13] He, Z. *et al*. Asymmetric acoustic gratings. *Appl. Phys. Lett.* **98**, 083505 (2011).

[14] Sun, H., Zhang, S. & Shui, X. A tunable acoustic diode made by a metal plate with periodical structure. *Appl. Phys. Lett.* **100**, 103507 (2012).

[15] Cicek, A., Kaya, O. A. & Ulug, B. Refraction-type sonic crystal junction diode. *Appl. Phys. Lett.* **100**, 111905 (2012).

[16] Oh, J. H., Kim, H. W., Ma, P. S., Seung, H. M. & Kim, Y. Y. Inverted biprism phononic crystals for one-sided elastic wave transmission applications. *Appl. Phys. Lett.* **100**, 213503 (2012).

[17] Yuan, B., Liang, B., Tao, J., Zou, X. & Cheng, J. C. Broadband directional acoustic waveguide with high efficiency. *Appl. Phys. Lett.* **101**, 043503 (2012).

[18] Xu, S. J., Qiu, C. Y. & Liu, Z. Y. Acoustic transmission through asymmetric grating structures made of cylinders. *J. Appl. Phys.* **111**, 094509 (2012).

[19] Jia, H., Ke, M. Z., Li, C. H., Qiu, C. Y. & Liu, Z. Y. Unidirectional transmission of acoustic waves based on asymmetric excitation of lamb waves. *Appl. Phys. Lett.* **102**, 153508 (2013).

[20] Kittel, C. Interaction of spin waves and ultrasonic waves in ferromagnetic crystals. *Phys. Rev.* **110**, 836–841 (1958).

[21] Liang, B., Guo, X. S., Guo, X. S., Zhang, D. & Cheng, J. C. An acoustic rectifier. *Nat. Mater.* **9**, 989–992 (2010).

[22] Boechler, N., Theocharis, G. & Daraio, C. Bifurcation-based acoustic switching and rectification. *Nat. Mater.* **10**, 665–668 (2011).

[23] Popa, B.-I., Zigoneanu, L. & Cummer, S. A. Tunable active acoustic metamaterials. *Phys. Rev. B* **88**, 024303 (2013).

[24] Popa, B.-I. & Cummer, S. A. Non-reciprocal and highly nonlinear active acoustic metamaterials. *Nat. Commun.* **5**, 3398 (2014).

[25] Hayrapetyan, A. G., Grigoryan, K. K., Petrosyan, R. G. & Fritzsche, S. Propagation of sound waves through a spatially homogeneous but soomthly time-varying medium. *Ann. Phys.* **333**, 47–65 (2013).

[26] Schaffarczyk, A. P. [Chapter 3: Basic Fluid Mechanics] *Introduction to Wind Turbine Aerodynamic*, (Springer, Berlin, 2014).

[27] Fang, K. J., Yu, Z. F. & Fan, S. H. Photonic Aharonov-Bohm Effect Based on Dynamic Modulation. *Phys. Rev. Lett.* **108**, 153901 (2012).

[28] Zanjani, M. B., Davoyan, A. R., Mahmoud, A. M., Engheta, N. & Lukes, J. R. One-way phonon isolation in acoustic waveguides. *Appl. Phys. Lett.* **104**, 081905 (2014).

[29] The work was jointly supported by the National Basic Research Program of China (Grant No. 2012CB921503, No. 2013CB632904 and No. 2013CB632702) and the National Nature Science Foundation of China (Grant No. 11134006). We also acknowledge the support from Natural Science Foundation of Jiangsu Province (BK20140019), Academic Program Development of Jiangsu Higher Education (PAPD) and China Postdoctoral Science Foundation (Grant No. 2012M511249 and No. 2013T60521).

[30] Q.W., X.N., and M.-H.L. conceived the idea. X.-Y.L., Y.Y., X.-C.S. and Z.-G.C performed the numerical calculation. Q.W. carried out the experiments. Y.-L.X., X.-C.S., Z.-G.C and M.-H.L. carried out the theoretical derivation and analysis. Y.-F.C., X.-P.L. and L.F. assisted in analyzing the results. All the authors contributed to discussion of the project. M.-H.L. and Y.-F.C. supervised all of the work and guided this project. Q. W., X.-P.L. and M.-H.L. wrote the manuscript with revisions from other authors.

[31] Supplementary information accompanies this paper at http://www.nature.com/srep

[32] Methods: Full-wave simulations. The full-wave numerical simulations of the acoustic wave transmission were performed using the aeroacoustics module of the commercially available finite-element software COMSOL Multiphysics. The rotating blade was modeled as a mathematically imposed and time-varying area inside the waveguide. Rigid wall boundary conditions were used for the waveguide except for the two ends, which were terminated with plane-wave scattering boundary conditions. For acoustic excitation, a plane-wave pressure field was used at the left end boundary of the waveguide. The time varying acoustic scattering process of the blade was studied in time domain with a transient solver in COMSOL using a time step 10^{-6} s. Resonance frequency of the Helmholtz resonator array was determined with a steady state solver. Transmission spectra were obtained by performing a set of frequency domain simulations with a frequency increment of 1 Hz.

Acoustic measurement methods. Our experimental investigation was carried out in an acoustic waveguide. In order to reduce the reflection at the end of the waveguide caused by impedance mismatch, the end was sealed with sound absorption sponges. The measured absorption efficiency of the absorption layer is provided in Supplementary Information. We used an acoustic pressure probe (B & K - 4939 - 2670 microphone) and a signal analyzer (B & K - 3560 - C) for acoustic signal acquisition and frequency analysis. More details for measurements can be found in Supplementary Information.

Acoustic Cloaking by a Near-zero-index Phononic Crystal[*]

Li-Yang Zheng,[1] Ying Wu,[2] Xu Ni,[1] Ze-Guo Chen,[1] Ming-Hui Lu,[1] and Yan-Feng Chen[1]

[1] *National Laboratory of Solid State Microstructures and Department of Materials Science and Engineering, Nanjing University, Nanjing 210093, China*

[2] *Computer, Electrical and Mathematical Sciences and Engineering, King Abdullah University of Science and Technology (KAUST), Thuwal 23955-6900, Saudi Arabia*

Zero-refractive-index materials may lead to promising applications in various fields. Here, we design and fabricate a near Zero-Refractive-Index (ZRI) material using a phononic crystal (PC) composed of a square array of densely packed square iron rods in air. The dispersion relation exhibits a nearly flat band across the Brillouin zone at the reduced frequency $f = 0.5443c/a$, which is due to Fabry-Perot (FP) resonance. By using a retrieval method, we find that both the effective mass density and the reciprocal of the effective bulk modulus are close to zero at frequencies near the flat band. We also propose an equivalent tube network model to explain the mechanisms of the near ZRI effect. This FP-resonance-induced near ZRI material offers intriguing wave manipulation properties. We demonstrate both numerically and experimentally its ability to shield a scattering obstacle and guide acoustic waves through a bent structure.

Zero-refractive-index (ZRI) materials are unconventional materials that exhibit zero refractive indices.[1-14] In a ZRI material, waves do not experience spatial phase changes, because the phase velocity, inversely proportional to the refractive index, approaches infinity and the wavelength becomes very long even at high frequencies.[1] This special characteristic enables unprecedented wave properties, such as tunneling of electromagnetic energy through sub-wavelength channels and bend,[2] tailoring of the radiation phase pattern of electromagnetic sources,[3] and super-reflection or cloaking with different defect loadings.[4-7] In electromagnetic wave propagation, according to the relationship that the refractive index is directly related to the permittivity, ε, and permeability, μ, via $n = \sqrt{\varepsilon\mu}$, two kinds of ZRI materials are categorized: (Ⅰ) single-zero-index materials, in which only one parameter is zero. One such material, known as the epsilon-near-zero (ENZ) material, can be achieved by utilizing metamaterials,[7,8] plasma,[9] and metal-clad waveguides.[1] However, this strategy depends heavily on a few specific plasmonic materials or complex metamaterials. (Ⅱ) double-zero-index (DZI) materials, with two parameters being zero simultaneously. There are two ways to achieve a DZI material in

---

[*] Appl.Phys.Lett.,2014,104(16):161904

electromagnetics. One is to achieve matched zero-index material by embedding resonant non-magnetic inclusions into an ENZ host medium.[10] The other is to utilize a photonic crystal possessing a Dirac-like cone inducted by accidental degeneracy, at which the effective permittivity and permeability are zero simultaneously.[11]

The acoustic analog of ZRI materials has also been explored extensively. For instance, by installing thin tensioned circular membranes to the rigid plane perforated with subwavelength holes, one can make the mass of the air in the holes effectively vanish because the restoring force from the membrane adds a negative term to the effective mass.[12] The design suits better for extraordinary acoustic transmission, but applications on cloaking have seldom been reported. Another example is the acoustic analogy of the electromagnetic DZI material, in which the effective mass density and the reciprocal of the effective bulk modulus vanish simultaneously at the Dirac-like point.[13,14] Although the problem for acoustic waves can be mapped from its electromagnetic counterpart mathematically, the requirement on low wave velocity in the inclusions greatly limits the application of zero-index materials in airborne sound. This limitation may be overcome due to the rapid progress in metamaterials.[15—24]

In this work, we propose a simple method to achieve near zero quantities in both the effective mass density and the reciprocal of the effective bulk modulus for airborne sound in a 2D phononic crystal (PC). Different from the previous approaches,[7—11] no materials with extreme material parameters are required, and the near-zero effective medium parameters come from the zeroth order Fabry-Perot (FP) resonance which exhibits an infinite phase velocity. We numerically and experimentally demonstrated that the phononic crystal can hide an obstacle. This FP resonance induced cloaking effect is fundamentally different from transformation acoustics[25,26] or scattering cancelation.[27,28] We also experimentally observe the bending effect that sound wave can pass through a bend waveguide with our PC embedded in it.

The 2D PC considered in this study is composed of a square array of square steel rods in air. A schematic of the structure, with four units, is shown in the inset of Fig. 1(a). The dimension of the steel rod is infinite along Z direction. While in X and Y directions, the steel rod has the same size of $l = 0.89a$, where a is the lattice constant. The mass densities of steel and air are $\rho = 7870$ kg/m^3 and $\rho_0 = 1.25$ kg/m^3, respectively, and their corresponding sound velocities are $c = 5960$ m/s and $c_0 = 343$ m/s. Fig. 1(a) shows the band structure of the PC calculated by COMSOL Multiphysics, a finite-element-based commercial software capable of full-wave simulations. Apparent in Fig. 1(a) is a flat branch, near the frequency of $0.5c/a$ and across the whole Brillouin zone, that does not intersect with any other branches.

Field distribution of the eigenmode on the flat branch at the Γ point is plotted in Fig. 1(c). Due to the large impedance mismatch between steel and air, the acoustic wave field is mainly concentrated in the narrow air channels. To better understand the underlying physics, the iso-frequency contour (IFC) of the flat band is analyzed. As shown in Fig. 1 (b), the IFC is nearly circular in the vicinity of the Γ point, while it becomes more rectangular when the frequency decreases. But at the working frequency $0.5443c/a$, the IFC is a small circle (blue line). It indicates that the phononic crystal can be regarded as an isotropic effective medium at that frequency. Therefore, to better understand the interaction of sound waves in the interfaces, we can use its effective medium parameters to describe the wave propagation.

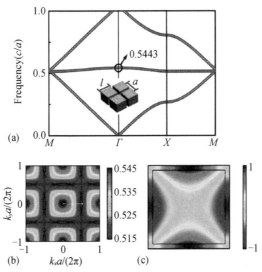

FIG. 1. (a) The band structure of the PC. A flat band is noticed around $f = 0.544c/a$. The inset is a schematic of the PC structure. A square array of square steel rods of the same size, l, are placed in air with the lattice constant a. (b) The IFCs of the flat band. At the working frequency, IFC is a small circle noted as blue line. The red circle represents the IFC of air. (c) The pressure field pattern of one unit cell at $f = 0.5443c/a$.

Near the center of the Brillouin zone, the system may be characterized by an effective medium theory (EMT) because k is almost zero.[29] Here, the effective mass density, ρ_{eff} and bulk modulus, B_{eff}, are calculated from the transmission and reflection coefficients,[30] denoted as T and R, respectively, of a plane wave normally incident on a finite-sized slab of the PC evaluated by COMSOL. The results of ρ_{eff} and $1/B_{eff}$ are plotted in Fig. 2(b) as blue and red solid curves, respectively. This figure shows that both the effective mass density and reciprocal of bulk modulus undergo a jump near the frequency of $0.5437c/a$. Though the reciprocal of the effective bulk modulus is negative and very close to zero in this frequency regime, the effective mass density varies drastically from positive value to a negative one. It then gradually increases and becomes positive again. When the frequency reaches $f = 0.5443c/a$, $\rho_{eff} = 0$. Both of the effective parameters are negative between frequencies $0.5437c/a$ and $0.5443c/a$, which implies that there is a negative band within this narrow frequency regime.

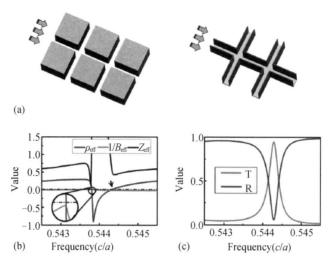

FIG. 2. (a) An equivalent tube network model of the PC. (b) Effective medium parameters of the PC. The effective mass density and reciprocal of bulk modulus undergo a drastic change near the frequency of $0.5437c/a$. Then, the effective mass density gradually increases as the frequency increases. At $f = 0.5443c/a$, it equals zero. (c) The transmittance (reflectance) changes as the frequency increases. The transmittance comes to a peak value (roughly 0.8) at $0.5443c/a$.

To explain the mechanisms of near zero mass density and reciprocal of bulk modulus, we propose an equivalent tube network model, shown in Fig. 2(a), of the PC. The PC in this work is constructed by square steel rods embedded in air background with a high filling ratio of 0.79. It leads to a sub-wavelength air channel between two adjacent steel unit cells. Considering the rigidness of steel cells, which can be treated as hard boundary, the sub-wavelength air channel is equivalent to a sub-wavelength short tube. As the number of the steel rod layers increases along the propagating direction, it is similar to the case when we put many identical short tubes in series. From this perspective, the whole PC can be viewed as a sub-wavelength tube network. For each short horizontal tube along the propagating direction, the vibration velocity will be generated as the air in the internal hollow FP cavity vibrates back and forth. At resonant frequency, large amount of energy is stored in the internal cavity, which causes the acceleration of the air medium in the opposite direction of the excited sound pressure and induces the negative effective mass density. On the other hand, the tube network also has vertical sub-wavelength tubes which are perpendicular to the propagation direction. These vertical tubes have the same size of the horizontal ones and can lead to negative effective bulk modulus at resonant frequencies because the air in the horizontal channels can flow in and out of the vertical tubes when pressure is imposed. Fundamentally, the principle is similar to the previously investigated acoustic systems consisting of arrays of side-attached tubes, Helmholtz resonators, in which closed cavities are connected to the channel.[21,31] Although the vertical tubes in this model are opened cavities, they still serve similar purpose as the Helmholtz resonators did because this two kinds of cavities in the side-attached cases act as the role to store wave energy and cause vibrations when sound wave is applied

From this tube network model, there are two kinds of tubes, the horizontal ones, which provide negative effective mass density, and the vertical ones, which can change effective bulk modulus, interacting together at resonant frequency $0.5437c/a$ to open a negative band for propagation of sound wave. Due to the identical sizes of all the tubes, resonance occur at the same frequency of $0.5437c/a$, leading to both effective mass density and bulk modulus undergo a sharp drop simultaneously (leap for the reciprocal of the effective bulk modulus as shown in Fig. 2(b)). When working frequency increases, effective mass density varies gradually and comes to zero at $0.5443c/a$. Near this frequency, propagation of sound wave is still permitted, while both effective mass density and reciprocal of bulk modulus are very close to zero, leading to the fantastic phenomenon of ZRI.

A ZRI material can give rise to various intriguing phenomena such as cloaking and beam-shaping.[1,11] Here, we demonstrate in both simulations and experiments the cloaking of an object by the PC. The experimental set-up is shown in Fig. 3(a), which exhibits an obstacle made of a steel block of size $4a \times 2a$ (red rectangle) and is embedded in the PC containing 10×8 steel rods (green) with the lattice constant $a = 4.5$ mm. The waveguide was constructed by sponges (yellow) located at four sides of the PC that can minimize reflections. We used a Function Generator (Agilent 33120A 15 MHz Function Waveform Generator) as sound source and an ultrasonic transducer, whose central frequency is 40 kHz, attached closely to the waveguide port to generate the incident wave (working frequency in the experiment is 39.1 kHz). We placed the transducer 100 mm away from the left side of the PC to mimic a good plane wave source in the sponge waveguide. A Brüel & Kjær free field 1/8 Inch microphone (Brüel & Kjær Type 2670) is placed 45 mm away from the right side of the PC to detect the sound wave. The phase and amplitude of sound waves are recorded by the oscilloscope (Zolix DPO 2012) and the lock-in amplifier (EG&G MODEL 5302 Lock-in Amplifier). With the help of computer, experimental data then can be postprocessed. In this work, sound waves were incoming normally to the sample, thus, we did not have to consider multiple reflections in the waveguide.

The normalized pressure field and the phase pattern of a 32×40 mm^2 area in the detection area are shown in the right panel of Fig. 3(b). In the experiment, acoustic waves with frequencies of $0.513c/a$ (39.1 kHz) was incoming from the left and impinging on the PC containing the obstacle. The simulation is also shown in the left panel of Fig. 3(b), where the operating frequency $0.545c/a$ (41.54 kHz) is slightly higher than $0.5443c/a$ (41.47 kHz) due to the small change in the impedance when obstacle is embedded in the PC. Despite the imperfection from the plane wave source and boundary reflections, it is obvious that the wave front preserves its original pattern after it passes through the PC as if the steel block was not there. The measured results agree well with the simulation. For

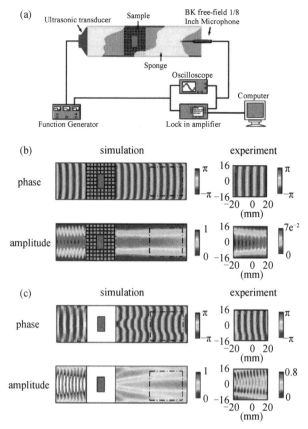

FIG. 3. (a) Schematics of the experimental setup used to map the sound field behind the sample. The phase and normalized amplitude patterns with (b) and without (c) the PC in the waveguide. Left: simulated result. Right: the experimental result for the region corresponding to the dashed box shown in left. Note that there exist some white areas in the amplitude patterns in simulations. It means that the amplitude is larger than 1.

comparison, the simulated and measured results for the case when the PC is removed and the steel block remains are plotted in Fig. 3(c). A clear shadow cast by the steel block and a distorted wavefront can be observed in both figures, suggesting that the cloaking effect shown in Fig. 3(c) does not exist. The PC therefore plays a crucial role in hiding the object inside a waveguide channel.

The transmittance of the PC-based ZRI material in this work is high for a lossless system (shown in Fig. 2(c)). But in the experiment, we observed the cloaking effect at $0.513c/a$ and the transmittance was only about 2% (shown in Fig. 4(b)), which implies that the attenuation is big, and the transmittance is small. Theoretically, a perfect flat band leads to large attenuation.[32,33] In particular, loss becomes much more complicated at working frequency when strong FP resonance occurs. Therefore, although sound waves still can travel through the PC, the loss leads to a significant decrease in the transmittance and a shift of the working frequency for cloaking effect. This phenomenon can be easily verified in both theory and experiment as shown in Fig. 4. For simplicity, only the viscous force of air in the narrow channel was considered in calculation and the calculated transmittance of attenuation coefficient $\alpha = 8.16$ is shown in Fig. 4(a), which demonstrate a shift in the transmission peaks and a drop in the amplitudes. Comparing the experiment result (Fig. 4(b)) with the calculation, we can see that number of peaks and the profile of transmittance are consistent with each other. It implies that losses would reduce the transmission amplitudes and lead to a shift of the working frequency. However, it would not change the physical phenomenon. Thus, we can extend this idea to small loss system such as water sound. Since loss in water is much smaller than that in air, transmittance will increase largely and the cloaking effect will be much more perfect.

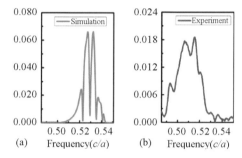

FIG. 4. The transmission of sound waves when loss was taken into account. (a) Theoretical (red curve) and (b) experimental (blue curve) pressure intensity transmission spectra with PC and the obstacle in the waveguide.

Another fascinating property of our PC-based ZRI material is shown in Fig.5, which demonstrates that waves can pass through a bent waveguide with an object embedded in the PC. Fig. 5(a) shows a cutaway view of the experimental set up, where the same PC and the same steel block as shown in Fig. 3(a) are used and the walls of the bent waveguide are again made of sponge. The plane waves at the frequency of $0.513c/a$ is incident on the PC from the lower left channel; they turn through the 90° end in the PC channel; they then exit from the upper right channel. The simulated and measured phase and amplitude patterns are shown in Fig. 5(b). Although there are slight distortions, the main features of the incident plane wave are preserved. Thus, the bending of acoustic waves is achieved with the PC and the cloaking effect is also realized. The steel object inside the PC does not distort the plane wave front after sound waves pass through the PC, creating the illusion that the object is not there. The phase and amplitude patterns for the case without the PC are shown in Fig. 5(c). It can be seen that plane wave front does not exist at the upper right channel.

In conclusion, different from the previously studied ZRI materials, we have theoretically designed and experimentally fabricated a phononic-crystal-based ZRI material. The PC consists of closely packed identical square steel rods. It exhibits the peculiar dispersion characteristic of a flat branch across the Brillouin zone near $f = 0.5443c/a$. The

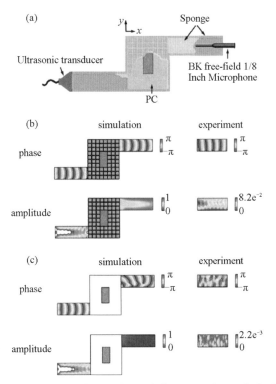

FIG. 5. Numerical and experimental demonstrations of the bending effect. (a) A sketch of the same configurations as shown in Fig. 3(a). The incident wave comes from the lower left channel and the phase pattern at the upper right channel is measured by the scanning detector. Simulated and experimental phase and normalized amplitude patterns with (b) and without (c) the PC in the waveguide.

physical origin of this flat band is the zeroth order FP resonance, which gives rise to the infinite phase velocity in the air channels. It ensures no phase change when sound waves pass through the PC. Consequently, the PC exhibits an effective ZRI. This FP-induced ZRI material also offers unprecedented wave properties. We explicitly illustrate, through numerical simulation and experiments, its abilities to shield an object and guide waves through a bent structure. There are many possible applications for this easily fabricated PC in manipulating acoustic waves through ZRI materials.

References and Notes

[1] N. Engheta, Science **340**, 286 (2013).
[2] M. G. Silveirinha and N. Engheta, Phys. Rev. B **76**, 245109 (2007).
[3] A. Alù, M. G. Silveirinha, A. Salandrino, and N. Engheta, Phys. Rev. B **75**, 155410 (2007).
[4] V. C. Nguyen, L. Chen, and K. Halterman, Phys. Rev. Lett. **105**, 233908 (2010).
[5] J. Hao, W. Yan, and M. Qiu, Appl. Phys. Lett. **96**, 101109 (2010).
[6] Y. Xu and H. Chen, Appl. Phys. Lett. **98**, 113501 (2011).
[7] Y. Wu and J. Li, Appl. Phys. Lett. **102**, 183105 (2013).

[8] R. Liu, Q. Cheng, T. Hand, J. J. Mock, T. J. Cui, S. A. Cummer, and D. R. Smith, Phys. Rev. Lett. **100**, 023903 (2008).
[9] E. J. R. Vesseur, T. Coenen, H. Caglayan, N. Engheta, and A. Polman, Phys. Rev. Lett. **110**, 013902 (2013).
[10] M. Silveirinha and N. Engheta, Phys. Rev. B **75**, 075119 (2007).
[11] X. Huang, Y. Lai, Z. H. Hang, H. Zheng, and C. T. Chan, Nature Mater. **10**, 582 (2011).
[12] J. J. Park, K. J. B. Lee, O. B. Wright, M. K. Jung, and S. H. Lee, Phys. Rev. Lett. **110**, 244302 (2013).
[13] F. Liu, X. Huang, and C. T. Chan, Appl. Phys. Lett. **100**, 071911 (2012).
[14] F. Liu, Y. Lai, X. Huang, and C. T. Chan, Phys. Rev. B **84**, 224113 (2011).
[15] Z. Liang and J. Li, Phys. Rev. Lett. **108**, 114301 (2012).
[16] H. Estrada, P. Candelas, F. Belmar, A. Uris, J. Garcia de Abajo, and F. Meseguer, Phys. Rev. B **85**, 174301 (2012).
[17] J. Christensen and F. Javier Garcia de Abajo, Phys. Rev. Lett. **108**, 124301 (2012).
[18] Z. Liu, C. T. Chan, and P. Sheng, Phys. Rev. B **62**, 2446–2457 (2000).
[19] Y. Cui, K. H. Fung, J. Xu, H. Ma, J. Yi, S. He, and N. X. Fang, Nano Lett. **12**, 1443 (2012).
[20] J. Zhu, Y. Chen, X. Zhu, F. J. Garcia-Vidal, X. Yin, W. Zhang, and X. Zhang, Sci. Rep. **3**, 1728 (2013).
[21] K. J. B. Lee, M. K. Jung, and S. H. Lee, Phys. Rev. B **86**, 184302 (2012).
[22] S. Yang, J. H. Page, Z. Liu, M. L. Cowan, C. T. Chan, and P. Sheng, Phys. Rev. Lett. **93**, 024301 (2004).
[23] C. Goffaux and J. P. Vigneron, Phys. Rev. B **64**, 075118 (2001).
[24] Z. Liu, X. Zhang, Y. Mao, Y. Y. Zhu, Z. Yang, C. T. Chan, and P. Sheng, Science **289**, 1734–1736 (2000).
[25] S. Zhang, C. Xia, and N. Fang, Phys. Rev. Lett. **106**, 024301 (2011).
[26] N. Stenger, M. Wilhelm, and M. Wegener, Phys. Rev. Lett. **108**, 014301(2012).
[27] S. A. Cummer, B. I. Popa, D. Schurig, D. R. Smith, J. Pendry, M. Rahm, and A. Starr, Phys. Rev. Lett. **100**, 024301 (2008).
[28] L. Sanchis, V. M. García-Chocano, R. Llopis-Pontíveros, A. Climente, J. Martínez-Pastor, F. Cervera, and J. Sánchez-Dehesa, Phys. Rev. Lett. **110**(12), 124301 (2013).
[29] Y. Wu, J. Li, Z. Q. Zhang, and C. T. Chan, Phys. Rev. B **74**, 085111 (2006).
[30] V. Fokin, M. Ambati, C. Sun, and X. Zhang, Phys. Rev. B **76**, 144302 (2007).
[31] N. Fang, D. J. Xi, J. Y. Xu, M. Ambati, W. Srituravanich, C. Sun, and X. Zhang, Nature Mater. **5**, 452 (2006).
[32] E. N. Economou and M. Sigala, J. Acoust. Soc. Am. **95**(4), 001734 (1994).
[33] J. V. Sánchez-Pérez, D. Caballero, R. Mártinez-Sala, C. Rubio, J. Sánchez-Dehesa, F. Meseguer, J. Llinares, and F. Gálvez, Phys. Rev. Lett. **80**, 5325–5328 (1998).
[34] This work was jointly supported by the National Basic Research Program of China (Grant Nos. 2012CB921503 and 2013CB632904) and the National Natural Science Foundation of China (Grant No. 1134006). We also acknowledge the Academic Development Program of Jiangsu Higher Education (PAPD) and KAUST Baseline Research Funds.

Acoustic Phase-reconstruction near the Dirac Point of a Triangular Phononic Crystal[*]

Si-Yuan Yu, Qing Wang, Li-Yang Zheng, Cheng He, Xiao-Ping Liu, Ming-Hui Lu, and Yan-Feng Chen

College of Engineering and Applied Sciences and National Laboratory of Solid State Microstructures, Nanjing University, Nanjing 210093, China

In this work, acoustic phase-reconstruction is studied and experimentally demonstrated in a triangular lattice two-dimensional phononic crystal (PnC) composed of steel rods in air. Owning to the fact that two bands of this triangular lattice PnC touch at the K/K' point and thus give rise to a conical Dirac cone, acoustic waves transmitting through this PnC can exhibit a pseudo-diffusion transportation feature, producing a reconstructed planar wavefront in the far field away from the interface of the PnC. Such phase reconstruction effect can be utilized in many applications, and here we demonstrate experimentally two important applications: an acoustic collimator and an acoustic cloak operating at a Dirac frequency of 41.3 kHz.

To control precisely the acoustic wave has always been an enduring pursuit. In the past two decades, with the advent of phononic crystal (PnCs),[1—3] i.e., periodic structures composed of scattering inclusions in a homogeneous host, scientists have been able to control the acoustic wave in surprising ways, which include acoustic collimating,[4—6] focusing,[7,8] negative refraction,[9,10] waveguiding,[11,12] super directivity,[13,14] unidirectional transmission,[15,16] seismic prevention,[17] etc. Recently, the extraordinary transportation properties near the Dirac point have caught world-wide attention in science community. In 2010, Huang and Lai first demonstrated spatial phase-reconstruction using degenerate Bloch modes at the Dirac point in a photonic crystal.[18] They observed that the transmitting wave experiences no spatial phase change through the crystal as a result of the excitation of an accidental threefold degenerate Bloch mode at the high symmetric Γ point. Later, this same phenomenon but with an acoustic wave in the PnC was theoretical studied[19] and numerically demonstrated[20] by Liu using the similar Dirac cone at the Γ point. Recently, Zheng demonstrated experimentally an acoustic cloaking phenomenon based on the same mechanism but with a nearly flat dispersion around the Γ point of the PnC.[21]

However, these previous attempts rely on either accidental mode degeneracy, flat dispersion, or both of them. As a result, they introduce fabrication challenges for practical realizations. Besides, the flat dispersion greatly reduces the operation bandwidth and

---

[*] Appl.Phys.Lett.,2015,106(15):151906

energy transmission. Moreover, the Bloch modes at the Γ point can only be effectively converted in pure two-dimensional (2D) systems and in quasi-2D systems with a finite extension in the third dimension only modes under the light cone[22]/sound line[23] (the slowest bulk dispersion indicating the boundary of the bulk and surface modes) can be effectively excited. This limitation simply means that the concept of phase-reconstruction phenomena based on Γ point degeneracy cannot be extended to the well-confined surface acoustic or surface plasmonic wave. It is thus highly desired to explore options beyond the Dirac point at Γ point. For example, Zhang has reported the pseudo-diffusion behavior for an acoustic wave transmitting through an underwater PnC with conical dispersions inherently degenerate at the high symmetric K/K' point and demonstrated the acoustic beating effect at this extremal point.[24] However, it still remains a question whether the spatial phase-reconstruction occurs at this particular symmetric point.

In this work, we study acoustic transportation behavior at this extremal point using a two-dimensional triangular lattice PnC composed of rigid steel rods in air and show phase-reconstruction phenomenon in both finite element method (FEM) simulations and experiments. At the high symmetric K/K' point of such triangular PnC, there are two inherently degenerate dispersion bands, giving rise to a linear Dirac cone. At the Dirac frequency, the pseudo-diffusion phenomenon occurs when a plane wave is injected into the PnC along the Γ-K or Γ-K' direction, and the phase-reconstruction phenomenon can be observed for the transmitted field through the PnC, which is characterized by a planar wavefront in the far field. By exploiting this phenomenon, we demonstrate an acoustic collimator and an acoustic cloak at the Dirac frequency of 43.1 kHz.

The PnC used in our experiments consists of identical stainless steel (ST 304) rods arranged in a triangular lattice placed in air as shown in Fig. 1(a). The lattice constant and the radius of the steel rods are 4.5 mm and 1.1 mm, respectively. Acoustic band structure of this PnC shown in Fig. 1(b) is calculated using commercial FEM software COMSOL Multiphysics. Material parameters used in the calculations include a density of 7800 kg/m³ and a longitudinal velocity of 6010 m/s for the stainless steel, and a density of 1.20 kg/m³ and a sound velocity of 343 m/s for air. Our calculated band structure clearly indicates that two acoustic dispersion bands merge at the high symmetric K point at a frequency of 43.1 kHz. For a better illustration, an equi-frequency contour (EFC) in the reciprocal space for the low frequency band is also calculated and shown in Fig. 1(c). The degeneracy points appear at the equivalent Brillouin zone corners, K and K' points. The band structure near these points exhibits concentric circle shaped EFCs, which result in Dirac cones. Note that the occurrence of Dirac cones here is inherent and entirely due to the threefold rotational symmetry of the triangular lattice.

Detailed band structures near the Dirac frequency along the Γ-K (x) direction are shown in Fig. 2(a), along with the calculated plane-wave transmission spectrum for a PnC ribbon sample with a thickness of 67.5 mm (15 layers) and a width of 155.8 mm (40

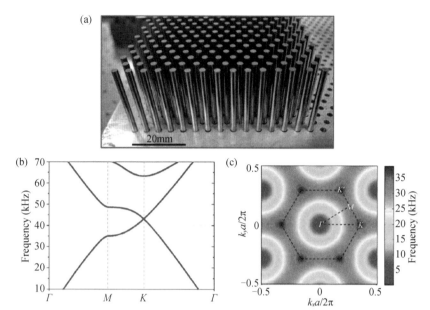

FIG. 1. (a) Two-dimensional triangular lattice PnC made of identical steel rods in air. The lattice constant $a = 4.5$ mm, and the radius of rods $r = 1.1$ mm. (c) Calculated band structure of the PnC. A Dirac point appears at K point with a Dirac frequency of 43.1 kHz. (b) EFC of the lower band of the PnC.

layers). Curve I and curve II in the figure refer to the two distinct dispersion bands, giving rise to the Dirac point. Upon examining the acoustic field distribution of the Bloch modes shown in the insets of Fig. 2(a), it can be verified that curve-I describes a symmetric mode, because the acoustic field in one unit cell along the propagating (x) direction possesses mirror symmetry (with the same vibration phases), while curve-II describes an anti-symmetric mode characterized by the relative motion inside one unit cell as a result of an opposite phase. This mode classification is also consistent with the simulated transmission spectrum. In the frequency range of interest from 25 kHz to 60 kHz, the transmission initially maintains a relatively high value but gradually declines when the frequency approaches 40 Hz. The transmission becomes vanishingly small when the frequency is above 49 kHz due to the fact that the incident acoustic wave falls into the forbidden band or deaf band,[25] since the symmetric plane wave cannot efficiently excite the anti-symmetric Bloch mode represented by curve-II. Noticeably, there exists a transmission dip centered at the Dirac frequency determined by the crossing point of the calculated Dirac cone in Fig. 2(a). This transmission dip is a direct consequence of the pseudo-diffusion transportation of energy fluxes at the Dirac point, which was first discovered in electron transportation through graphene.[26,27] An intuitive explanation for the underlying mechanism responsible for such phenomenon is summarized as follows. Inside the PnC, the direction of energy flux coincides with the direction of the group velocity determined by $\partial\omega/\partial k$, i.e., the slope of the band. At the non-differentiable conical

Dirac point, the energy flux direction is undefined and can thus exist in every direction in the two dimensional space. Consequently, the Bloch modes converted from the directional incident wave along the Γ-K direction (or the Γ-K' direction) can diffuse into everywhere inside the PnC.

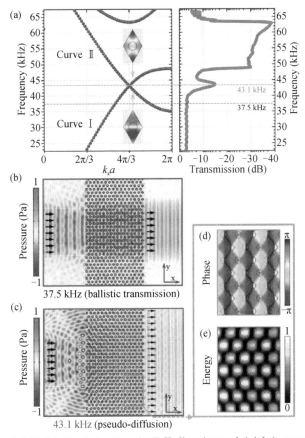

FIG. 2. (a)(Left) Band structure in Γ-K direction and (right) corresponding transmission spectrum around the Dirac frequency. Pressure field distribution with a plane wave incidence at frequency of (b) 37.5 kHz (c) and 43.1 kHz. (d) and (e) Zoom-in view of the pressure field's phase and energy distribution for an off-center interface region of the PnC.

Figs. 2(b) and 2(c) show the normalized pressure field distribution for planar acoustic incident waves of two different frequencies (37.5 kHz vs. 43.1 kHz) transmitting through a 15 layers PnC ribbon sample arranged along the Γ-K direction. Obviously, at the frequency of 37.5 kHz, far from the Dirac frequency of 43.1 kHz, the propagation of the incident beam through the PnC exhibits a ballistic feature. The beam width of the transmitted beam remains more or less the same because of the directional band gap in Γ-M direction at this frequency. However, at the Dirac frequency (43.1 kHz), the propagation of the directional incident beam exhibits a prominent pseudo-diffusion pattern inside the PnC. As a result, the beam width of the transmitted beam is greatly broadened. This phenomenon, occurring even in the strictly periodic systems, exhibits similarity to the diffusion behavior of waves propagating through a disordered medium. It has been observed in various systems including graphene, photonic crystal,[28—30] and underwater phononic crystal.[24]

However, the phase distribution inside these periodic structures under the pseudo-diffusion condition is seldom discussed in literatures. Here, we take a look at this issue by examining the result shown in Fig. 2(c). In the far field, the transmitted acoustic field along the y direction is in phase forming a plane wave. According to Huygens-Fresnel principle, the "diffusing" acoustic waves along y direction near the right interface should be all in phase. Figs. 2(d) and 2(e) show the zoomed phase and energy distribution for the pressure field for an off-center interface region in Fig. 2(c). The phase exhibits a perfect spatial harmonic along x direction but more importantly an almost uniform distribution along y direction, while the energy is mostly confined between two nearest rods in the x direction. It is worth noting that in regions near the left interface the phase distribution is complicated, and the phase uniformity along y direction may not appear until the acoustic waves propagate several wavelengths inside the PnC. If a defect were deliberately placed inside the PnC (not shown in the figure), the phase of the acoustic wave should be disturbed around the defect, but the phase uniformity would quickly emerge as the acoustic wave propagates away from the defect for several wavelengths. The mechanism responsible for the phase-reconstruction is explained as follows. At the Dirac frequency, the corresponding Bloch mode can only support by three wave vectors (ΓK, $\Gamma K'$, and $\Gamma K - \Gamma K'$). Consequently, the scattering process taking place at an interface or a defect site cannot generate any propagating Bloch mode with its wave vector other than these three. In other words, the spatial field distribution inside the PnC away from the scattering site has to be the coherent summation of the same Bloch mode propagating along three directions, which ultimately produces uniform phase distribution.

This intriguing phenomenon shows great potentials in practical applications, e.g., an acoustic collimator shown in Fig. 3. A ribbon PnC sample is used in our demonstration. Its thickness L (in x direction) and width W (in y direction) equal to 45 mm and 116.9 mm, both of which are much larger than the lattice constant of 4.5 mm. An ultrasonic transducer driven by a function waveform generator is centered along y direction and placed next to the left interface of the PnC. An excitation frequency of 43.1 kHz corresponds to an acoustic wavelength of 7.9 mm, which is comparable to emitting diameter of 6 mm for the ultrasonic transducer. Under such condition, this acoustic source can be considered as a pseudo point source. A detector, pressure field 1/8 in. microphone probe (Brüel & Kjær Type 4138), is mounted on a motorized 2-axis translation stage placed to the right of the PnC. The phase and amplitude of the ultrasonic wave are recorded independently by an oscilloscope and a lock-in amplifier. The translation stage is scanned to map out the transmitted pressure filed distribution. The whole measurement rig is enclosed in a chamber made of highly absorbing sponge to prevent any undesired acoustic reflections as well as ambient noise.

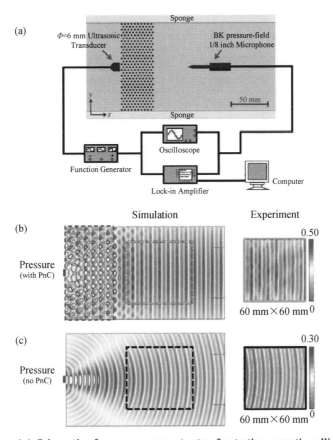

FIG. 3. (a) Schematic of our measurement setup for testing acoustic collimation. (b) and (c) FEM simulation (left) and experimental measurement (right) of pressure field distribution corresponding to the case with and without the presence of the PnC under a pseudo-point source excitation at a frequency of 41.3 kHz. Yellow boxes indicate the spatial location of the incident beams, and black arrows indicate the direction of the transmitted acoustic beams.

The measured pressure field distribution in a 60 mm × 60 mm area is shown in the right panel of Fig. 3(b). For comparison, the FEM simulation result for the same area is shown in the left panel. Clearly, the experimental result matches well with the simulation despite slight imperfection of the wavefront possibly due to boundary reflections in experiments. Both results show that the transmitted acoustic wave resembles a plane wave with a planar wavefront, indicating a collimating process occurred when the acoustic wave transmits through the PnC. To confirm this, the PnC sample is removed and the pressure field distribution is recorded and plotted in Fig. 3(c). The transmitted acoustic wave in this case has a curved wavefront, consistent with the point source excitation condition discussed above. Therefore, the presence of the PnC plays a crucial role in reconstructing the phase to form a collimated output plane wave.

Another fascinating application related to the phase-reconstruction phenomenon at the

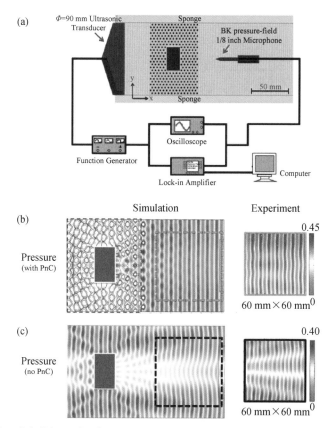

FIG. 4. (a) Schematic of our measurement setup for testing acoustic cloak of a rectangular steel block. (b) and (c) FEM simulation (left) and experimental measurement (right) of pressure field distribution corresponding to the case with and without the presence of the PnC under a plane wave excitation at a frequency of 41.3 kHz. Yellow boxes indicate the spatial location of the incident beams, and black arrows indicate the direction of the transmitted acoustic beams.

Dirac frequency is the acoustic cloaking, which has drawn extensive attention over recent years.[21,31—33] As shown in Fig. 4, a 18 mm×31.2 mm rectangular steel block, as an object to be cloaked, is embedded in the middle of a PnC ribbon with its thickness of 67.5 mm (in x direction) and width of 116.9 mm (in y direction), respectively. A similar measurement rig is used except for the ultrasonic transducer, which in this case has a 90 mm emitting diameter and is placed 40 mm away from the left interface of the PnC ribbon to mimic a plane wave incidence instead of a point excitation. The measurement and simulation results at Dirac frequency of 43.1 kHz are shown in Fig. 4(b). Again, they agree remarkably well with each other. The acoustic wave could transmit through the PnC as if the block were not there, since the presence of this large block inside the PnC does not distort the planar wavefront of the transmitted acoustic wave. In other words, the steel block is cloaked in this case. However, when the PnC is removed, the planar wavefront of the incident acoustic wave is severely distorted by the steel block as illustrated in Fig. 4(c), which indicates that the cloaking effect no longer exists without the PnC.

In summary, we investigate and demonstrate the phase-reconstruction phenomenon in a two dimensional PnC made of rigid steel rods in a triangular lattice embedded in air. By taking advantage of inherent band degeneracy at the K and K' points and the resulted conical dispersion, pseudo-diffusion acoustic transportation behavior can be realized. We show both experimentally and numerically that such pseudo-diffusion behavior can be exploited in applications such acoustic collimating and cloaking. Our demonstration shows that phase-reconstruction takes place inside the PnC due to acoustic pseudo-diffusion and the transmitted acoustic wave can thus acquire a planar wavefront.

References and Notes

[1] M. M. Sigalas and E. N. Economou, J. Sound Vib. **158**, 377 (1992).

[2] M. S. Kushwaha, P. Halevi, L. Dobrzynski, and B. Djafari-Rouhani, Phys. Rev. Lett. **71**, 2022 (1993).

[3] F. R. Montero de Espinoza, E. Jimenez, and M. Torres, Phys. Rev. Lett. **80**, 1208 (1998).

[4] L. S. Chen, C. H. Kuo, and Z. Ye, Appl. Phys. Lett. **85**, 1072 (2004).

[5] J. Shi, S. S. Lin, and T. J. Huang, Appl. Phys. Lett. **92**, 111901 (2008).

[6] J. Bucay, E. Roussel, J. O. Vasseur, P. A. Deymier, A-C. Hladky-Hennion, Y. Pennec, K. Muralidharan, B. Djafari-Rouhani, and B. Dubus, Phys. Rev. B **79**, 214305 (2009).

[7] S. Yang, J. H. Page, Z. Liu, M. L. Cowan, C. T. Chan, and P. Sheng, Phys. Rev. Lett. **93**, 024301 (2004).

[8] T. T. Wu, Y. T. Chen, J. H. Sun, S. S. Lin, and T. J. Huang, Appl. Phys. Lett. **98**, 171911 (2011).

[9] X. Zhang and Z. Liu, Appl. Phys. Lett. **85**, 341 (2004).

[10] M. H. Lu, C. Zhang, L. Feng, J. Zhao, Y. F. Chen, Y. W. Mao, J. Zi, Y. Y. Zhu, S. N. Zhu, and N. B. Ming, Nat. Mater. **6**, 744 (2007).

[11] A. Khelif, A. Choujaa, S. Benchabane, B. Djafari-Rouhani, and V. Laude, Appl. Phys. Lett. **84**, 4400 (2004).

[12] P. H. Otsuka, K. Nanri, O. Matsuda, M. Tomoda, D. M. Profunser, I. A. Veres, S. Danworaphong, A. Khelif, S. Benchabane, V. Laude, and O. B. Wright, Sci. Rep. **3**, 3351 (2013).

[13] C. Qiu, Z. Liu, J. Shi, and C. T. Chan, Appl. Phys. Lett. **86**, 224105 (2005).

[14] A. Hakansson, D. Torrent, F. Cervera, and J. Sanchez-Dehesa, Appl. Phys. Lett. **90**, 224107 (2007).

[15] B. Liang, X. S. Guo, J. Tu, D. Zhang, and J. C. Cheng, Nat. Mater. **9**, 989 (2010).

[16] X. F. Li, X. Ni, L. Feng, M. H. Lu, C. He, and Y. F. Chen, Phys. Rev. Lett. **106**, 084301 (2011).

[17] S. Brûlé, E. H. Javelaud, S. Enoch, and S. Guenneau, Phys. Rev. Lett. **112**, 133901 (2014).

[18] X. Huang, Y. Lai, Z. H. Hang, H. Zheng, and C. T. Chan, Nat. Mater. **10**, 582 (2011).

[19] F. Liu, Y. Lai, X. Huang, and C. T. Chan, Phys. Rev. B **84**, 224113 (2011).

[20] F. Liu, X. Huang, and C. T. Chan, Appl. Phys. Lett. **100**, 071911 (2012).

[21] L. Y. Zheng, Y. Wu, X. Ni, Z. G. Chen, M. H. Lu, and Y. F. Chen, Appl. Phys. Lett. **104**, 161904 (2014).

[22] S. Fan, P. R. Villeneuve, and J. D. Joannopoulos, Phys. Rev. Lett. **78**, 3294 (1997).

[23] S. Benchabane, A. Khelif, J.-Y. Rauch, L. Robert, and V. Laude, Phys. Rev. E **73**, 065601(R) (2006).

[24] X. Zhang and Z. Liu, Phys. Rev. Lett. **101**, 264303 (2008).

[25] J. V. Sánchez-Pérez, D. Caballero, R. Mártinez-Sala, C. Rubio, J. Sánchez-Dehesa, F. Meseguer, J. Llinares, and F. Gálvez, Phys. Rev. Lett. **80**, 5325 (1998).

[26] M. I. Katsnelson, Eur. Phys. J. B **51**, 157 (2006).

[27] J. Tworzydlo, B. Trauzettel, M. Titov, A. Rycerz, and C. W. J. Beenakker, Phys. Rev. Lett. **96**, 246802 (2006).

[28] R. A. Sepkhanov, Ya. B. Bazaliy, and C. W. J. Beenakker, Phys. Rev. A **75**, 063813 (2007).

[29] S. R. Zandbergen and M. J. A. de Dood, Phys. Rev. Lett. **104**, 043903 (2010).

[30] S. Bittner, B. Dietz, M. Miski-Oglu, and A. Richter, Phys. Rev. B **85**, 064301 (2012).

[31] S. Zhang, C. Xia, and N. Fang, Phys. Rev. Lett. **106**, 024301 (2011).

[32] N. Stenger, M. Wilhelm, and M. Wegener, Phys. Rev. Lett. **108**, 014301 (2012).

[33] L. Sanchis, V. M. García-Chocano, R. Llopis-Pontiveros, A. Climente, J. Martínez-Pastor, F. Cervera, and J. Sánchez-Dehesa, Phys. Rev. Lett. **110**, 124301 (2013).

[34] The work was jointly supported by the National Basic Research Program of China (Grant Nos. 2012CB921503 and 2013CB632702) and the National Natural Science Foundation of China (1134006 and 11474158). We also acknowledge the support of Natural Science Foundation of Jiangsu Province (BK20140019) and the support from Academic Program Development of Jiangsu Higher Education (PAPD).

Topologically Protected One-way Edge Mode in Networks of Acoustic Resonators with Circulating Air Flow*

Xu Ni[1], Cheng He[1], Xiao-Chen Sun[1], Xiao-ping Liu[1], Ming-Hui Lu[1], Liang Feng[2] and Yan-Feng Chen[1]

[1] *National Laboratory of Solid State Microstructures & Department of Materials Science and Engineering, Nanjing University, Nanjing 210093, People's Republic of China*

[2] *Department of Electrical Engineering, University at Buffalo, The State University of New York, Buffalo, NY 14260, USA*

Recent explorations of topology in physical systems have led to a new paradigm of condensed matters characterized by topologically protected states and phase transition, for example, topologically protected photonic crystals enabled by magneto-optical effects. However, in other wave systems such as acoustics, topological states cannot be simply reproduced due to the absence of similar magnetics-related sound-matter interactions in naturally available materials. Here, we propose an acoustic topological structure by creating an effective gauge magnetic field for sound using circularly flowing air in the designed acoustic ring resonators. The created gauge magnetic field breaks the time-reversal symmetry, and therefore topological properties can be designed to be nontrivial with non-zero Chern numbers and thus to enable a topological sonic crystal, in which the topologically protected acoustic edge-state transport is observed, featuring robust one-way propagation characteristics against a variety of topological defects and impurities. Our results open a new venue to non-magnetic topological structures and promise a unique approach to effective manipulation of acoustic interfacial transport at will.

1. Introduction

While topology is a pure mathematic concept, it has now become a powerful freedom in designing physical systems and their intrinsic symmetries. For example, the conventional Landau symmetry breaking theory alone failed to describe recent developments in condensed matter physics such as the quantum Hall effect[1-6], where topological descriptions become crucial and the corresponding topological phases can be characterized by topological invariants such as Chern numbers. Remarkably, the local perturbation to the edge cannot affect bulk properties, due to the presence of topologically protected edge states at the interface between two media with different topological phases,

---

* New J.Phys.,2015,17(5):053016

which leads to the robust one-way edge state transport in recent emerging fields of Chern insulators[7—9] (systems with nonzero Chern numbers and associated with integer quantum Hall effects) and topological insulators[10, 11] (associated with quantum spin Hall effects). A typical condition to observe these unique topological phenomena is the broken time-reversal symmetry, e.g., by means of applied magnetic fields in the two-dimensional electron gas for the quantum Hall effect. Similarly, nontrivial photonic topological phases have also been demonstrated in a variety of photonic crystals based on magneto-optical materials in which time-reversal symmetry is broken with external magnetic fields[12—17]. Similar to the electronic system, the bands of photonic crystals can also be characterized by Chern numbers[13, 18], and gap Chern numbers[19], defined as the sum of the Chern numbers of all the bands below the photonic bandgap, can be used to describe the topological property of the corresponding photonic bandgaps. When such gap Chern number changes across an interface separating two media, there will exist gapless surface states with robust one-way light propagation against impurities and perturbations, characterizing the non-trivial topological properties of photonic crystals due to the breaking of the time reversal symmetry.

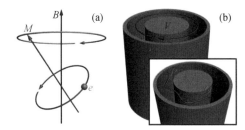

FIG. 1. Time-reversal symmetry breaking due to air-flow circulation. (a) Illustration of the mechanism of gyromagnetic materials showing the precession of the magnetic moment M of the electron e under the magnetic field B. (b) Illustration of the air-flow-circulated ring resonator containing clockwise flowing air of speed V with broken time-reversal symmetry similar to gyromagnetic materials. The inset shows an arbitrary propagation path of acoustic waves inside the ring.

However, due to the absence of similar magnetics-introduced breaking of time reversal symmetry in acoustics[20, 21], it is therefore necessary to develop an effective strategy by which a gauge magnetic field can be created for acoustics to break the time-reversal symmetry of sound propagation, and acoustic metamaterials[22—29] or sonic crystals[30] may be exploited to solve this issue. Recently Alu *et al* proposed a good solution to solve the problem with a compact design[31]. They designed a circular structure containing flowing air, attached with three channels for signal input/output. When the air inside the ring is flowing circularly, the signal in any two channels of the device exhibits non-reciprocal propagation due to the broken time-reversal symmetry induced by an effective gauge magnetic bias, similar to the Zeeman effect observed in magneto-optical medium. In this paper, we propose to utilize the air-flow circulation[31] to break the time-reversal symmetry from the 'meta-atom' level and thus to control the fundamental symmetries in order to address this grand challenge. Specifically, we apply circulating air flow to a designed 'meta-atom'—an acoustic ring resonator—to spatially modulate the effective sound velocity in air. Similar to how circulating electrons produce magnetic field, the circulating air flow creates an effective gauge magnetic field that breaks the time-reversal symmetry of sound propagation. We further demonstrate a topological sonic crystal based on the ring resonators with circulating air inside, in which the gapless acoustic edge-state transport occurs in the band gap between two adjacent bands with nontrivial Chern invariants.

2. Time-reversal Symmetry Breaking Due to Air-flow Circulation

In both topological electronic and photonic crystals and their associated quantum Hall effects, the external magnetic field is required to break the time-reversal symmetry. For example, the Lorentz force is induced by the magnetic field, such that electrons preferentially circulate only in one direction, as shown in figure 1(a). Similarly in an acoustic case, the clockwise air circulation applied to the acoustic ring cavity as shown in figure 1(b) creates an equivalent time-reversal symmetry breaking for sound. Intuitively, if the velocity of the clockwise air flow is V and the sound speed in air is c, the speeds of sound transiting through circulating air flow become different: $c+V$ for the clockwise and $c-V$ for the counter-clockwise directions, respectively. It is therefore obvious that sound transports with reverse phase shift in clockwise and counter-clockwise directions, leading to the time-reversal symmetry breaking. More rigorously, the wave equation for sound in such a circulating air flow can be expressed using an aeroacoustic[32] model,

$$-\frac{\rho}{c^2}i\omega(i\omega\phi + \boldsymbol{V}\cdot\nabla\phi) + \nabla\cdot\left[\rho\nabla\phi - \frac{\rho}{c^2}(i\omega\phi + \boldsymbol{V}\cdot\nabla\phi)\boldsymbol{V}\right] = 0, \quad (1)$$

where ϕ is the velocity potential, ω is the angular frequency of sound, and ρ is the density of air. Since the air flow is confined in the clockwise direction, the corresponding flow velocity only contains an azimuthal component, $\boldsymbol{V} = v\boldsymbol{e}_\theta$, where v is the amplitude and $\boldsymbol{e}_\theta$ is the azimuthal unit vector. As a result, $\nabla\cdot\boldsymbol{V}=0$ and equation (1) can thus be simplified as

$$[(\nabla - i\boldsymbol{A}_{\text{eff}})^2 + \omega^2/c^2 + (\nabla\rho/2\rho)^2 - \nabla^2\rho/(2\rho)]\psi = 0, \quad (2)$$

where the term $(\nabla\rho/2\rho)^2 - \nabla^2\rho/(2\rho)$ represents the scalar potential, while $\boldsymbol{A}_{\text{eff}} = -\omega|\boldsymbol{V}|\boldsymbol{e}_\theta/c^2$ denotes[33] the vector potential, showing an effective magnetic field in such a non-magnetic structure due to the introduced circulating air flow. As a result of the existence of such non-zero vector potential, sound circulating inside the resonator experiences a phase shift if taking different paths, similar to the well-known Aharanov-Bohm effect. For example, between two points Q and P in the resonator in the inset of figure 1(b), the additional phase due to the vector potential is $\xi_{QP} = \int_Q^P \boldsymbol{A}_{\text{eff}} \cdot \mathrm{d}\boldsymbol{l}$ when sound travels from Q to P, whereas it becomes $\xi_{PQ} = \int_P^Q \boldsymbol{A}_{\text{eff}} \cdot \mathrm{d}\boldsymbol{l}'$ from P to Q. Because the selected paths are $\mathrm{d}\boldsymbol{l} = -\mathrm{d}\boldsymbol{l}'$, the additional phases become $\xi_{QP} = -\xi_{PQ}$. This vector potential-enabled phase difference clearly shows the breaking[2, 15] of time-reversal symmetry if simply reversing the sound propagating direction, i.e. the time-reversal symmetry and reciprocity are simultaneously broken for sound transport in such air-flow-circulated acoustic ring resonator.

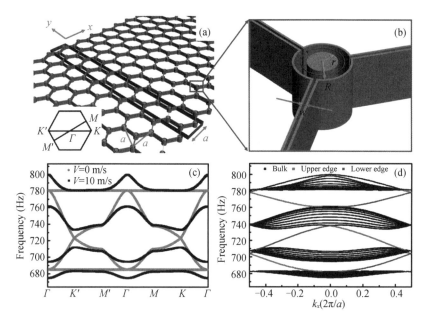

FIG.2. Topological sonic crystal made of air-flow-circulated units. (a) Illustration of the sonic crystal consisting of the air-flow-circulated ring resonators. The supercell (highlighted by the red rectangle) is selected to calculate the projected frequency band along the zigzag edge and the inset shows the first Brillouin zone. (b) The zoomed-in illustration of the detailed structural parameters. (c) The bulk frequency bands of the sonic crystal without air flow ($V=0$, denoted by red circles) and with clockwise air flow ($V=10$ m·s^{-1}, denoted by black squares). (d) The projected frequency bands of the sonic crystal with clockwise air flow ($V=10$ m·s^{-1}) along x direction (zigzag edge). See appendix A for the case of smaller V ($V=3$ m·s^{-1}).

3. Topological Sonic Crystal Made of Air-flow-circulated Units

The topological sonic crystal we propose here is constructed with a graphene-like or honeycomb lattice[34] as depicted in figure 2(a) using the acoustic 'meta-atoms' demonstrated above. The graphene-like structure is to introduce the deterministic Dirac-like degeneracy at the boundaries of the Brillouin zone due to the intrinsic three-fold rotation symmetry of honeycomb lattice. In the designed topological sonic crystal, the neighboring rings are connected by non-circulated subwavelength waveguides ($w = 9.5$ mm) in which only the acoustic plane wave (i.e. the fundamental waveguide mode) is supported in the studied frequency range [figure 2(b)]. The lattice constant of the honeycomb-lattice sonic crystal is chosen to be $a = 0.6\sqrt{3}$ m for potential applications related to audible sound. The inner and outer radii of the ring resonators are $r = 51.0$ mm and $R = 92.0$ mm, respectively. The corresponding band structures with and without the circulating air flow, i.e. $\mathbf{V} = 0$ and $\mathbf{V} \neq 0$, respectively, are shown in figure 2(c). It is clear that the degeneracies of a Dirac point at K or K' and two quadratic degenerate points at Γ are lifted due to the time-reversal symmetry breaking caused by the circulating air flow. Such degeneracy lift leads to Chern number exchange of adjacent acoustic bands and their associated special topological states.

4. Tight-binding Model to Calculate the Chern Numbers

Here, we employed the tight-binding approximation theory to analyze the Chern numbers of two bands around the Dirac point, e.g., at 723 Hz, by recasting equation (2) into the Dirac equation. In the tight-binding model, each ring resonator directly couples to the nearest neighbor through the subwavelength waveguide. As a result, such coupling is not affected by the introduced air flow and its resulted vector potential, leading to the first-neighbor coupling coefficient of t_1 without any additional phase shift. The outer rings can only be coupled through the adjacent neighbor rings in which the air flow induces an additional phase shift due to the vector potential, leading to the second-neighbor coupling coefficient of $t_2 e^{i\varphi}$, where the additional phase term is expressed as[2]

$$\varphi = 2m\pi \frac{F_h}{F_0}, \quad (3)$$

where $F_0 = |\oint \mathbf{A}_{\text{eff}} \cdot d\mathbf{l}|$ is the total effective flux through an acoustic ring, $F_h = \frac{1}{3}\oint \mathbf{A}_{\text{eff}} \cdot d\mathbf{l}$ is the effective flux occupied by the second-neighbor hopping path (only one third of an acoustic ring due to the three-fold rotation symmetry), and $2m\pi$ is the propagation phase of the mth-order azimuthal mode[31] in the ring resonator. Therefore, the Hamiltonian based on location wave functions of the two atoms in each unit cell of the graphene-like structure in the vicinity of the K and K' point can be expressed as[2] $H_K = \begin{bmatrix} t_2 \sin \varphi & t_1 \Delta k' \\ t_1 \Delta k & -t_2 \sin \varphi \end{bmatrix}$ and $H_{K'} = \begin{bmatrix} -t_2 \sin \varphi & t_1 \Delta k \\ t_1 \Delta k' & t_2 \sin \varphi \end{bmatrix}$ (see appendix B for the derivation of the Hamiltonian without air flow), where $\Delta k = \frac{\sqrt{3}}{2}|\mathbf{k} - \mathbf{K}|e^{i\theta}$ and $\Delta k' = \frac{\sqrt{3}}{2}|\mathbf{k} - \mathbf{K}|e^{-i\theta}$

are small quantities associated with the wave vectors k expanded near the Dirac point at K (θ is the angle between the vectors $k-K$ and k_x). Based on the eigen vector $|u(k)\rangle$ of the Hamiltonians, the resulted Berry phase near the Dirac point can be derived with its definition $\Phi = \oint d\mathbf{k} \cdot i\langle u(k)|\nabla_k|u(k)\rangle$. The total Berry phase of the lower frequency band including the contributions from phase changes of both K and K', denoted as Φ_K and $\Phi_{K'}$, becomes $\Phi = \Phi_K + \Phi_{K'} = 2\pi i \dfrac{t_2 \sin\varphi}{\sqrt{t_1^2 \Delta k \Delta k' + (t_2 \sin\varphi)^2}}$, which can be further simplified as $\Phi = 2\pi i \dfrac{\sin\varphi}{|\sin\varphi|}$ due to the very small Δk and $\Delta k'$. Hence, the topological invariant of the gap Chern number[13, 19] associated with the bandgap induced by the time-reversal symmetry breaking is

$$C_{\text{gap}} = \frac{\Phi}{2\pi i} = \frac{\sin\varphi}{|\sin\varphi|}. \tag{4}$$

In the discussed frequency range around 723 Hz, the dipole mode of the ring resonator [see figure C1(a) in appendix C] is excited corresponding to $m = 1$, and according to equation (3) (defining $\oint \mathbf{A}_{\text{eff}} \cdot d\mathbf{l} < 0$ in the case of clockwise air flow) $\varphi = -\dfrac{2\pi}{3}$ is obtained consequently leading to $C_{\text{gap}} = -1$ associated with the bandgap around the Dirac degeneracy at K or K'. Similarly the gap Chern number of -1 can also be obtained for the other two bandgaps around the quadratic degeneracies at Γ, analogous to the photonic counterpart[35]. Therefore, in our sonic crystal, the Chern numbers associated with the 4 bulk frequency bands can be easily calculated: $-1, 0, 0, 1$, respectively [from lower to upper bands in figure 2(c)]. The obtained nonzero Chern invariants clearly indicate the nontrivial topological properties of the designed sonic crystal due to the degeneracy lift around the Dirac and quadratic degeneracy points.

5. Robust One-way Acoustic Edge Modes

A signature of topological states is the presence of topologically protected gapless edge states inside the bulk frequency band gap at the interface between two media with different topological phases. In our design, such edge states exist by placing the air-flow-circulated graphene-like sonic crystal next to a rigid boundary [figure 2(a)]. Since they possess different gap Chern numbers (for example, -1 for the sonic crystal and 0 for the rigid boundary) and the local topological perturbations due to the rigid boundary cannot affect the bulk properties of the sonic crystal, the topological phase transition occurs at the interface, leading to the continuous spectra of the gapless edge states with the neutralized Chern numbers inside the bulk frequency band gap of the sonic crystal, as shown in figure

2(d) where the projected frequency dispersion of the zigzag edge state is calculated. It is worth noting that the obtained topology-protected property is attributed to the bulk topological invariants across the interface and the number of the supported edge-modes is consistent with the corresponding difference in gap Chern numbers between adjacent materials. Since gap Chern number is the acoustic/photonic analogue[13] of Hall conductance whose sign denotes the transport direction of edge currents in integer quantum Hall effect, similarly the sign of the gap Chern number difference can also indicate the propagation direction of the supported edge mode. In our case the topological sonic crystal with $C_{gap}=-1$ is surrounded by a rigid wall with $C_{gap}=0$ leading to a negative sign of gap Chern number difference, which indicates a counterclockwise propagation of edge modes along the interface, corresponding to a rightward ($+x$) propagation of the lower edge mode [blue lines in figure 2(d)] and a leftward ($-x$) propagation of the upper edge mode (red lines) respectively.

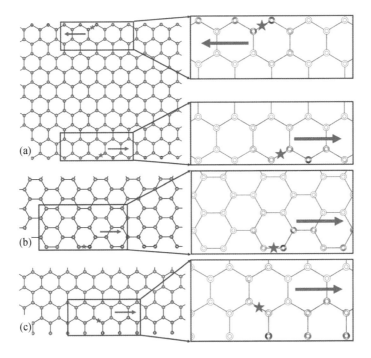

FIG.3. One-way acoustic edge modes. (a) The velocity-potential field distribution of the upper and lower unidirectional zigzag edge mode at 770 Hz with clockwise air flow ($V=10$ m · s^{-1}). The red star shows the position of the acoustic source and the arrows show the propagation direction. The color from blue to red represents the value from minimum value to maximum value. (b) And (c) the velocity-potential field distribution of the propagating edge mode along the lower (b) armchair and (c) bearded edge at 770 Hz with clockwise air flow ($V=10$ m · s^{-1}). The insets show the zoomed-in field distributions.

We performed finite element simulations to examine the properties of the edge mode supported by the designed topological sonic crystal. Figure 3 shows the sound transport of the edge mode with the edge of the sonic crystal terminated at different locations. If the graphene structure is well preserved at the edge (i.e. the zigzag edge) [figure 3(a)], the expected unidirectional counterclockwise sound transport is clearly observed: sound propagates from right to left at the upper edge, but travels from left to right at the lower edge, well consistent with the theoretical predictions in figure 2(d). Although how we geometrically determine the edge of the sonic crystal leads to different geometrical forms and may slightly modify the acoustic dispersion of the edge[36], the global topological property of the sonic crystal is still maintained. The edge mode intrinsically supported by different topological phases between two different media therefore remains, for example, with the armchair edge [figure 3(b)] and the bearded edge [figure 3(c)] (see appendix D for associated projected frequency bands). The fact that the supported edge modes possess the same sound transport direction (from left to right at the lower edge of the sonic crystal) regardless of the specific geometric forms of interfaces evidently demonstrates the robustness of the edge-mode sound transport as a result of inherent topological protection.

Such sound transport robustness of the edge mode of the topological sonic crystal remains even if a variety of interface defects are introduced. This is because the system is immune to any backscattering, which is topologically forbidden with nonzero gap Chern numbers as shown in figure 2(d). Here, we introduced four types of interface defects and investigated their effects on the edge mode of the topological sonic crystal: a vacuum void formed by removing one ring resonator [figure 4(a)], a strong local dislocation in the lattice [figure 4(b)], an exaggeratedly enlarged ring resonator [figure 4(c)], and a random-shaped object hidden inside an enlarged circle [figure 4(d)]. Remarkably, sound transport is immune to backscattering by self-detouring around the disordered region or reorganizing the field distribution in the local defects. It is therefore evident that the supported edge-mode is robust against a large number of defects, which is a universal characteristic caused by the intrinsic topological feature of the topological sonic crystal and its enabled topological protection. It should be noticed that in the simulation above the air flow is well-isolated within each ring, and does not leak to neighbors through the coupling waveguides. The inter-cell flows may slightly modify the dispersions of the frequency bands but won't change the topological properties of the associated bands, and our further simulation (see appendix E) has verified that the topologically protected one-way edge mode is highly robust to both cases of the inter-cell flows in the waveguides and the inhomogeneous flow in the rings, making this design of topological sonic crystal actually feasible in experiments, which still need further investigation considering the complex air-flow in real systems.

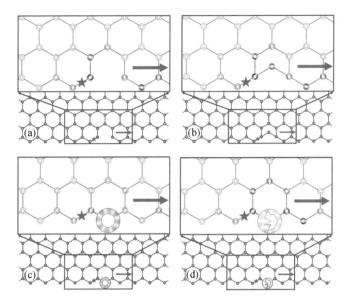

FIG. 4. Robust one-way acoustic edge modes. The velocity-potential field distribution of the acoustic edge mode along the lower zigzag edge robust to interfacial defects caused by (a) removing one ring, (b) moving one ring's position, (c) enlarging one ring's size, and (d) hiding one arbitrary-shape object inside the enlarged circle. The excitation frequency is chosen to be 770 Hz with clockwise air flow ($V = 10$ m·s^{-1}). The red star shows the position of the acoustic source and the arrows show the propagation direction. The color from blue to red represents the value from minimum value to maximum value.

6. Discussion

In summary, we have designed a topologically protected sonic crystal by using air-flow-circulated unit cells. Such topological acoustic phenomenon originates from the broken time-reversal symmetry caused by effective gauge magnetic field of circulating air flow and its related nontrivial topological properties with non-zero Chern numbers. Interestingly, the corresponding topological Chern number is relevant to the azimuthal order of the resonant mode in the ring resonator. For example, the topological Chern properties in the study above are within the frequency range around 770 Hz and with the azimuthal order of 1 for the resonant mode in the ring. If the frequency range is doubled (see appendix F), the corresponding azimuthal order of the resonant mode [see figure C1(b) in appendix C] becomes 2, which can flip the sign of the gap Chern number of the sonic crystal and thus reverse the direction of the edge-mode sound transport (from counterclockwise in figures 3 and 4 to clockwise). Under the topological protection, the one-way edge propagation is robust against various kinds of structural defects or disorders. Our findings may inspire novel designs of multi-channel and controllable one-way acoustic devices, and may pave the way for the investigation of acoustic topological phenomena, such as topological insulators.

References and Notes

[1] Klitzing K V, Dorda G and Pepper M 1980 *Phys. Rev. Lett.* **45** 494

[2] Haldane F D 1988 *Phys. Rev. Lett.* **61** 2015

[3] Zhang Y, Tan Y, Stormer H L and Kim P 2005 *Nature* **438** 201

[4] Novoselov K S, Geim A K, Morozov S V, Jiang D, Katsnelson M I, Grigorieva I V, Dubonos S V and Firsov A A 2005 *Nature* **438** 197

[5] Chang C Z et al 2013 *Science* **340** 167

[6] Young A F, Sanchez-Yamagishi J D, Hunt B, Choi S H, Watanabe K, Taniguchi T, Ashoori R C and Jarillo-Herrero P 2014 *Nature* **505** 528

[7] Thouless D J, Kohmoto M, Nightingale M P and den Nijs M 1982 *Phys. Rev. Lett.* **49** 405

[8] Berry M V 1984 *Proc. R. Soc.* A **392** 45

[9] Hatsugai Y 1993 *Phys. Rev. Lett.* **71** 3697

[10] Hasan M Z and Kane C L 2010 *Rev. Mod. Phys.* **82** 3045

[11] Qi X L and Zhang S C 2011 *Rev. Mod. Phys.* **83** 1057

[12] Wang Z, Chong Y D, Joannopoulos J D and Soljacic M 2009 *Nature* **461** 772

[13] Raghu S and Haldane F D M 2008 *Phys. Rev.* A **78** 033834

[14] Haldane F D M and Raghu S 2008 *Phys. Rev. Lett.* **100** 013904

[15] Wang Z, Chong Y D, Joannopoulos J D and Soljacic M 2008 *Phys. Rev. Lett.* **100** 013905

[16] He C, Chen X L, Lu M H, Li X F, Wan W W, Qian X S, Yin R C and Chen Y F 2010 *Appl. Phys. Lett.* **96** 111111

[17] Poo Y, Wu R X, Lin Z, Yang Y and Chan C T 2011 *Phys. Rev. Lett.* **106** 093903

[18] Lu L, Joannopoulos J D and Soljacic M 2014 *Nat. Photonics* **8** 821

[19] Skirlo S A, Lu L and Soljacic M 2014 *Phys. Rev. Lett.* **113** 113904

[20] Kittel C 1958 *Phys. Rev.* **110** 836

[21] Brekhovskikh L M and Lysanov I P 2003 *Fundamentals of Ocean Acoustics* (Berlin: Springer)

[22] Zhu J, Christensen J, Jung J, Martin-Moreno L, Yin X, Fok L, Zhang X and Garcia-Vidal F J 2011 *Nat. Phys.* **7** 52

[23] Lee S H, Park C M, Seo Y M, Wang Z G and Kim C K 2010 *Phys. Rev. Lett.* **104** 054301

[24] Li J, Fok L, Yin X, Bartal G and Zhang X 2009 *Nat. Mater.* **8** 931

[25] Jacob X, Aleshin V, Tournat V, Leclaire P, Lauriks W and Gusev V E 2008 *Phys. Rev. Lett.* **100** 158003

[26] Fang N, Xi D, Xu J, Ambati M, Srituravanich W, Sun C and Zhang X 2006 *Nat. Mater.* **5** 452

[27] Zhang S, Zhou J, Park Y S, Rho J, Singh R, Nam S, Azad A K, Chen H T, Yin X, Taylor A J and Zhang X 2012 *Nat. Commun.* **3** 942

[28] Wu Y, Lai Y and Zhang Z Q 2011 *Phys. Rev. Lett.* **107** 105506

[29] Ma G, Yang M, Xiao S, Yang Z and Sheng P 2014 *Nat. Mater.* **13** 873

[30] Liu Z, Zhang X, Mao Y, Zhu Y Y, Yang Z, Chan C T and Sheng P 2000 *Science* **289** 1734

[31] Fleury R, Sounas D L, Sieck C F, Haberman M R and Alù A 2014 *Science* **343** 516

[32] Goldstein M E 1976 *Aeroacoustics* (New York: McGraw-Hill)

[33] Yang Z J, Gao F, Shi X H, Lin X, Gao X, Chong Y D and Zhang B L 2015 *Phys. Rev. Lett.* **114** 114301

[34] Polini M, Guinea F, Lewenstein M, Manoharan H C and Pellegrini V 2013 *Nat. Nanotechnology* **8** 625

[35] Chong Y D, Wen X G and Soljacic M 2008 *Phys. Rev.* B **77** 235125

[36] He C, Lu M, Wan W, Li X and Chen Y 2010 *Solid State Commun.* **150** 1976

[37] The work was jointly supported by the National Basic Research Program of China (Grant No. 2012CB921503, and No. 2013CB632702) and the National Nature Science Foundation of China (Grant No. 11134006, No. 11474158, and No. 11404164). We also acknowledge the support of Natural Science Foundation of Jiangsu Province (BK20140019) and the support from Academic Program Development of Jiangsu Higher Education (PAPD).

[38] Appendix A. The bulk and projected frequency bands with smaller V ($V=3$ m·s^{-1})

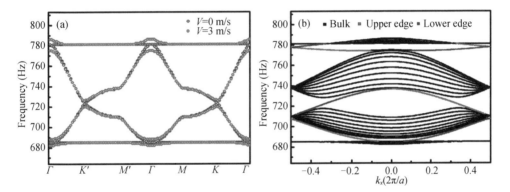

FIG.A1. (a) The bulk frequency band of the sonic crystal without air flow ($V=0$, denoted by green circles) and with clockwise air flow ($V=3$ m·s^{-1}, denoted by magenta circles). (b) The projected frequency band of sonic crystal with clockwise air flow ($V=3$ m·s^{-1}) along the x direction (zigzag edge). The blue and red lines correspond to the lower and upper edge states, respectively.

Appendix B. The derivation of the Hamiltonian

$$\hat{H}_{0K}(\boldsymbol{k}) = \begin{pmatrix} 0 & \frac{\sqrt{3}}{2}t_1 |\boldsymbol{k}-\boldsymbol{K}| e^{-i\theta} \\ \frac{\sqrt{3}}{2}t_1 |\boldsymbol{k}-\boldsymbol{K}| e^{i\theta} & 0 \end{pmatrix} \text{ without air flow}$$

In the honeycomb-lattice sonic crystal, rings are located at positions $\boldsymbol{R}_{n\alpha}=\boldsymbol{R}_n+\boldsymbol{R}_\alpha$, where n runs for all the lattice vectors ($\boldsymbol{r}_1=\frac{a}{2}\boldsymbol{e}_x+\frac{\sqrt{3}a}{2}\boldsymbol{e}_y$ and $\boldsymbol{r}_2=\frac{a}{2}\boldsymbol{e}_x-\frac{\sqrt{3}a}{2}\boldsymbol{e}_y$ are basic lattice vectors, where a is the lattice constant, $\boldsymbol{e}_x$ and $\boldsymbol{e}_y$ are unit vectors in x and y direction) and α ($\alpha=1,2$) runs for the two rings within the unit cell. Because of the axial symmetry, the Hamiltonian based on location wave functions of the two rings in each unit cell can be written as $\hat{H}_{0K} = \begin{pmatrix} \varepsilon & t_a+t_b+t_c \\ (t_a+t_b+t_c)^* & \varepsilon \end{pmatrix}$,

in which

$\varepsilon = E_0 + \int [\sum_n \sum_\alpha \hat{q}_{n\alpha}(\boldsymbol{r})\delta(\boldsymbol{r}-\boldsymbol{R}_{n\alpha}) - \hat{q}_{n\alpha}(\boldsymbol{r})\delta(\boldsymbol{r}-\boldsymbol{R}_{01})]\psi^*(\boldsymbol{r}-\boldsymbol{R}_{01})\psi(\boldsymbol{r}-\boldsymbol{R}_{01})d\boldsymbol{r}$ indicates the onsite energy of the ring at $\mathbf{R}_{01}$ (E_0 is the eigen value of a single unit cell), while the overlap integrals $t_a = t_1 = e^{-i\mathbf{k}\cdot(\mathbf{r}_{01}-\mathbf{r}_{02})} \int [\sum_{R_n} \sum_a \hat{q}_{na}(\mathbf{r})\delta(\mathbf{r}-\mathbf{R}_{na}) - \hat{q}_{na}(\mathbf{r})\delta(\mathbf{r}-\mathbf{R}_{01})]\psi^*(\mathbf{r}-\mathbf{R}_{02})\psi(\mathbf{r}-\mathbf{R}_{01})d\mathbf{r}$ ($\mathbf{k}$ is the reciprocal lattice vector), $t_b = e^{-i\mathbf{k}\cdot\mathbf{r}_1}t_1$ and $t_c = e^{-i\mathbf{k}\cdot\mathbf{r}_2}t_1$ indicate the hopping terms between the ring at $\mathbf{R}_{01}$ and three nearest ones at $\mathbf{R}_{02}$, $\mathbf{R}_{02}-\mathbf{r}_1$ and $\mathbf{R}_{02}-\mathbf{r}_2$, where $\hat{q}_{na}(\mathbf{r}) = \nabla_{na}^2 + (\nabla_{na}\rho/2\rho)^2 - \nabla_{na}^2\rho/(2\rho)$ represents the operator of local potential. Consequently, we obtain

$$\hat{H}_{0K}(\mathbf{k}) = \begin{pmatrix} \varepsilon & t_1(1+e^{-i\mathbf{k}\cdot\mathbf{r}_1}+e^{-i\mathbf{k}\cdot\mathbf{r}_2}) \\ t_1(1+e^{i\mathbf{k}\cdot\mathbf{r}_1}+e^{i\mathbf{k}\cdot\mathbf{r}_2}) & \varepsilon \end{pmatrix}.$$

Further expanding $\hat{H}_{0K}(\mathbf{k})$ near the K point and removing the energy origin point, we finally come to the Dirac equation: $\hat{H}_{0K}(\mathbf{k})\psi = \nu(\hbar\kappa_x\sigma_x + \hbar\kappa_y\sigma_y)\psi = \begin{pmatrix} 0 & \hbar\nu\kappa e^{-i\theta} \\ \hbar\nu\kappa e^{i\theta} & 0 \end{pmatrix}\begin{pmatrix} \psi_1 \\ \psi_2 \end{pmatrix} = E(\mathbf{k})\begin{pmatrix} \psi_1 \\ \psi_2 \end{pmatrix}$, in which $\kappa_x = \kappa\cos(\theta)$ and $\kappa_y = \kappa\sin(\theta)$ ($\kappa = |\mathbf{k}-\mathbf{K}|$) are x and y components of the vector $\mathbf{k}-\mathbf{K}$ (θ is the angle between the vectors $\mathbf{k}-\mathbf{K}$ and $\mathbf{k}_x$), $\hbar\nu = \frac{\sqrt{3}}{2}t_1$ is proportional to the nearest coupling term, σ_x and σ_y are Pauli matrices, and $\psi = \begin{pmatrix} \psi_1 \\ \psi_2 \end{pmatrix}$ represents the amplitudes of two degenerate Bloch states at one of the corners of the hexagonal first Brillouin zone.

Appendix C. Different azimuthal modes of the ring resonator

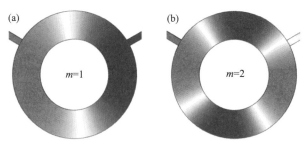

FIG. C1. The velocity-potential distribution of the (a) dipole and (b) quadrupole mode of the ring resonator corresponding to the azimuthal order $m=1$ and $m=2$, respectively. The color from blue to red represents the value from minimum value to maximum value.

Appendix D. The projected frequency bands of the armchair and bearded edges

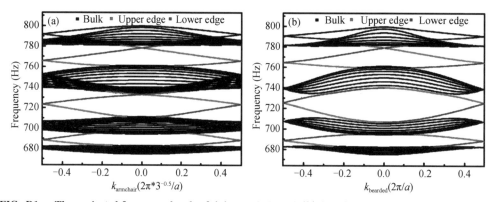

FIG. D1. The projected frequency bands of (a) armchair and (b) bearded edges with clockwise air flow ($V=10\ \text{m}\cdot\text{s}^{-1}$). The blue and red lines correspond to the lower and upper edge states respectively.

Appendix E. Simulation consideration of the inter-cell flows in the waveguides and the inhomogeneous flow in the rings

As to the details of simulation, we utilize the aeroacoustics module of COMSOL Multiphysics, which is a finite-element-method based commercial simulation software. This aeroacoustics module is based on solving the equation below

$$-\frac{\rho}{c^2}i\omega(i\omega\phi + \mathbf{V}\cdot\nabla\phi) + \nabla\cdot\left[\rho\nabla\phi - \frac{\rho}{c^2}(i\omega\phi + \mathbf{V}\cdot\nabla\phi)\mathbf{V}\right] = 0,$$

where ϕ is the velocity potential, ω is the angular frequency of sound, ρ is the density of air, c is the sound speed of air, and $\mathbf{V}=(V_x, V_y)$ is the speed of air flow. In the simulation [figure E1(a)], we set the air-flow speed inside the Nth ring to be $\mathbf{V}(x,y) = \frac{\alpha R + |x - x_N|(1-\alpha)}{R}\left(\frac{(y-y_N)\nu}{\sqrt{(x-x_N)^2+(y-y_N)^2}}, \frac{-(x-x_N)\nu}{\sqrt{(x-x_N)^2+(y-y_N)^2}}\right)$, where $\nu = 10$ m·s^{-1} is the maximum amplitude, $\alpha\nu = 5$ m·s^{-1} is the minimum amplitude, $R = 92$ mm is the ring's outer radius, and (x_N, y_N) is the location of the center of the Nth ring. The speed of air flow inside the waveguides are $\mathbf{V}_1 = \left(\frac{\sqrt{3}}{2}\nu_1, \frac{1}{2}\nu_1\right)$, $\mathbf{V}_2 = (0, \nu_2)$, $\mathbf{V}_3 = \left(\frac{\sqrt{3}}{2}\nu_3, -\frac{1}{2}\nu_3\right)$ with the amplitudes $\nu_1 = \nu_2 = \nu_3 = 2$ m·s^{-1}. The resulted field distribution [figure E1(b)] still shows a robust one-way edge mode propagating to the right side with the acoustic source (770 Hz) located at the blue-star position. This simulation result suggests that the one-way edge mode is highly robust to both the inhomogeneous air flow inside the rings and the inter-cell flows in the waveguides.

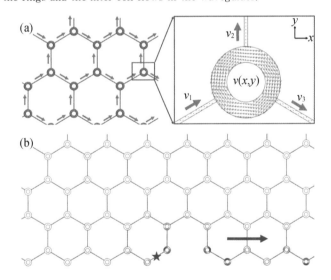

FIG. E1. The robust one-way edge modes tolerant of the inter-cell flows in the waveguides and the inhomogeneous flow in the rings. (a) The air-flow distribution of the topological sonic crystal system. The speed of air flow inside the rings is inhomogeneous, while that inside the waveguides is homogeneous. The red arrows represent the amplitude and direction of the air flow. (b) The velocitypotential distribution of a one-way edge mode robust to a void defect. The blue star shows the location of the acoustic source with frequency of 770 Hz, and the blue arrow gives the propagating direction of the edge mode. Color from red to blue represents the value from maximum to minimum.

Appendix F. The higher-frequency bands corresponding to the ring's azimuthal mode of $m=2$

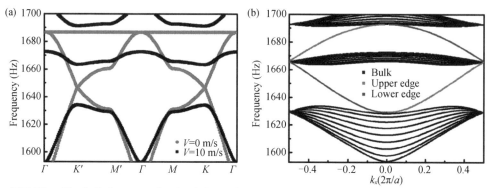

FIG.F1. The bulk frequency bands of the sonic crystal without air flow ($V=0$, denoted by red circles) and with clockwise air flow ($V=10$ m·s^{-1}, denoted by black squares) in a higher frequency range corresponding to the ring's azimuthal mode of $m=2$. (b) The projected frequency bands of the sonic crystal with clockwise air flow ($V=10$ m·s^{-1}) along x direction (zigzag edge) in a higher frequency range.

图书在版编目（CIP）数据

介电体超晶格. 上 / 朱永元等编著. —南京：南京大学出版社，2017.3
ISBN 978-7-305-17840-5

Ⅰ.①介… Ⅱ.①朱… Ⅲ.①超晶格半导体-研究 Ⅳ.①TN304.9

中国版本图书馆 CIP 数据核字（2016）第 262155 号

出版发行	南京大学出版社
社　　址	南京市汉口路 22 号　　邮　编 210093
出 版 人	金鑫荣
书　　名	**介电体超晶格·上**
编　著	朱永元　王振林　陈延峰　陆延青　祝世宁
策划编辑	吴　汀
责任编辑	王南雁　　　　　　　　编辑热线　025-83593962
照　　排	南京紫藤制版印务中心
印　　刷	南京爱德印刷有限公司
开　　本	787×1092　1/16　印张 29.75　字数 876 千
版　　次	2017 年 3 月第 1 版　　2017 年 3 月第 1 次印刷
ISBN	978-7-305-17840-5
定　　价	198.00 元
网　　址	http://www.njupco.com
官方微博	http://weibo.com/njupco
官方微信	njupress
销售热线	025-83594756

* 版权所有，侵权必究
* 凡购买南大版图书，如有印装质量问题，请与所购图书销售部门联系调换

定价:198.00元